The Plant Viruses

Volume 2

THE ROD-SHAPED PLANT VIRUSES

THE VIRUSES

THE VIRUSES: Catalogue, Characterization, and Classification
Heinz Fraenkel-Conrat

THE ADENOVIRUSES
Edited by Harold S. Ginsberg

THE HERPESVIRUSES,
Volumes 1–3 • Edited by Bernard Roizman
Volume 4 • Edited by Bernard Roizman and Carlos Lopez

THE PAPOVAVIRIDAE
Volume 1 • Edited by Norman P. Salzman

THE PARVOVIRUSES
Edited by Kenneth I. Berns

THE PLANT VIRUSES
Volume 1 • Edited by R. I. B. Francki
Volume 2 • Edited by M. H. V. Van Regenmortel and Heinz Fraenkel-Conrat

THE REOVIRIDAE
Edited by Wolfgang K. Joklik

THE TOGAVIRIDAE AND FLAVIVIRIDAE
Edited by Sondra Schlesinger and Milton J. Schlesinger

The Plant Viruses

Volume 2

THE ROD-SHAPED PLANT VIRUSES

Edited by

M. H. V. VAN REGENMORTEL

Institute of Molecular and Cellular Biology
Strasbourg, France

and

HEINZ FRAENKEL-CONRAT

Department of Molecular Biology and Virus Laboratory
University of California
Berkeley, California

PLENUM PRESS • NEW YORK AND LONDON

Library of Congress Cataloging in Publication Data

The Plant viruses.

(Viruses)
Contents: v. 2. The rod-shaped plant viruses/edited by M. H. V. Van Regenmortel and Heinz Fraenkel-Conrat.
Bibliography: v. 2, p.
Includes index.
1. Plant viruses. 2. Tobacco mosaic virus. I. Van Regenmortel, M. H. V. II. Fraenkel-Conrat, Heinz, 1910- . III. Series.
QR402.P57 1986 576'.6483 86-9517
ISBN 0-306-42258-1 (v. 2)

A Division of Plenum Publishing Corporation
233 Spring Street, New York, N.Y. 10013

Printed in the United States of America

Contributors

J. G. Atabekov, Department of Virology, Moscow State University, Moscow 119899, USSR

A. C. Bloomer, MRC Laboratory of Molecular Biology, Cambridge CB2 2QH, England

Alan A. Brunt, Glasshouse Crops Research Institute, Littlehampton, West Sussex BN17 6LP, England

P. J. G. Butler, MRC Laboratory of Molecular Biology, Cambridge CB2 2QH, England

T. W. Carroll, Department of Plant Pathology, Montana State University, Bozeman, Montana 59717

R. G. Christie, Plant Virus Laboratory, Agronomy Department, University of Florida, Gainesville, Florida 32611

V. V. Dolja, Department of Virology, Moscow State University, Moscow 119899, USSR

J. R. Edwardson, Plant Virus Laboratory, Agronomy Department, University of Florida, Gainesville, Florida 32611

H. Fraenkel-Conrat, Virus Laboratory and Department of Molecular Biology, University of California, Berkeley, California 94720

Adrian Gibbs, Research School of Biological Sciences, Australian National University, Canberra, Australian Capital Territory, 2601, Australia

G. V. Gooding, Jr., Department of Plant Pathology, North Carolina State University, Raleigh, North Carolina 27695

B. D. Harrison, Scottish Crop Research Institute, Invergowrie, Dundee DD2 5DA, Scotland

Yoshimi Okada, Department of Biophysics and Biochemistry, Faculty of Science, University of Tokyo, Hongo Bunkyo-ku, Tokyo, 113, Japan

Peter Palukaitis, Department of Plant Pathology, Cornell University, Ithaca, New York 14853

D. J. Robinson, Scottish Crop Research Institute, Invergowrie, Dundee DD2 5DA, Scotland

Satyabrata Sarkar, Institute of Plant Medicine, University of Hohenheim, 7000 Stuttgart 70, West Germany

Eishiro Shikata, Department of Botany, Faculty of Agriculture, Hokkaido University, Sapporo, Japan

M. H. V. Van Regenmortel, Institut de Biologie Moléculaire et Cellulaire du CNRS, 67084 Strasbourg Cédex, France

Anupam Varma, Division of Mycology and Plant Pathology, Indian Agricultural Research Institute, New Delhi 110012, India

Carl Wetter, Department of Botany, University of Saarland, D-6600 Saarbrücken, West Germany

Milton Zaitlin, Department of Plant Pathology, Cornell University, Ithaca, New York 14853

F. W. Zettler, Plant Pathology Department, University of Florida, Gainesville, Florida 32611

Preface

This volume of the series The Plant Viruses is devoted to viruses with rod-shaped particles belonging to the following four groups: the tobamoviruses (named after tobacco mosaic virus), the tobraviruses (after tobacco rattle), the hordeiviruses (after the latin *hordeum* in honor of the type member barley stripe mosaic virus), and the not yet officially recognized furoviruses (fungus-transmitted rod-shaped viruses, Shirako and Brakke, 1984).

At present these clusters of plant viruses are called groups instead of genera or families as is customary in other areas of virology. This peculiarity of plant viral taxonomy (Matthews, 1982) is due to the fact that the current Plant Virus Subcommittee of the International Committee of Taxonomy of Viruses is deeply split on what to call the categories or ranks used in virus classification.

Some plant virologists believe that the species concept cannot be applied to viruses because this concept, according to them, necessarily involves sexual reproduction and genetic isolation (Milne, 1984; Murant, 1985). This belief no doubt stems from the fact that these authors restrict the use of the term *species* to biological species. According to them, a collection of similar viral isolates and strains does constitute an individual virus, i.e., it is a taxonomy entity separate from other individual viruses. However, instead of calling these elementary units of classification: different virus species, these authors call them viruses, refusing to acknowledge the fact that there is a need for a word to signify that one is not referring to a viral object but to a taxonomic construct (an abstract concept as opposed to a collection of material objects).

Some of the protagonists of this viewpoint (Milne, 1984) go as far as to deny that there is a difference between the existence of a concept and the existence of material objects, and they reject the view that classifications are conceptual constructions. It is not clear why the word *species* should be used only in the sense of biological species defined by gene pools and breeding barriers, and why a phenetic or morphological definition of species, as practiced by numerical taxonomists, should not prove acceptable for classifying viruses.

The material objects, i.e., the viruses discussed in the present volume, have been grouped according to the divisions embodied in the CMI/AAB descriptions of plant viruses. In the past, most tobamoviruses such as ribgrass mosaic virus or cucumber virus 4 were considered strains of tobacco mosaic virus (TMV).

As new CMI/AAB descriptions of tobamoviruses are published and the corresponding taxonomic entity is given a name, it has become standard practice to consider that these names now stand for individual viruses and not for individual strains of TMV. According to standard taxonomic parlance, these individual viruses have thus become *de facto* species (Matthews, 1985). In addition to the species TMV, six other species of tobamoviruses are discussed at length in the present volume (Chapters 9–14). Following standard taxonomic categories and terminology, the tobamovirus group can be considered a genus made up of a serologically homogeneous cluster of related species. Similarly the furo-, tobra-, and hordeivirus groups can be taken to represent three separate plant virus genera.

About a third of this volume is devoted to TMV, a deliberate choice in view of the importance of this virus for the development of virology. Although the common TMV strain is perhaps the best studied object in the whole of virology, the reader will discover that the boundaries of our knowledge concerning this virus are still being extended and that the end is not yet in sight. The open-ended nature of scientific knowledge is never more apparent than in the description of a subject that has been very extensively researched.

M. H. V. Van Regenmortel
H. Fraenkel-Conrat

Strasbourg
Berkeley

REFERENCES

Milne, R. G., 1984, The species problem in plant virology, *Microbiol. Sci.* **1**:113–117.
Matthews, R. E. F., 1982, Classification and nomenclature of viruses, fourth report of the international committee on taxonomy of viruses, *Intervirology* **17**:1–199.
Matthews, R. E. F., 1985, Viral taxonomy, *Microbiol. Sci.* **2**:74–76.
Murant, A. F., 1985, Taxonomy and nomenclature of viruses, *Microbiol. Sci.* **2**:218–220.
Shirako, Y., and Brakke, M. K., 1984, Two purified RNAs of soil-borne wheat mosaic virus are needed for infection, *J. Gen. Virol.* **65**:119–127.

Contents

I. Tobamoviruses

Chapter 6

Tobacco Mosaic Virus: Epidemiology and Control

G. V. Gooding, Jr.

Chapter 7

Tobacco Mosaic Virus: Cytopathological Effects

J. R. Edwardson and R. G. Christie

Chapter 8

Tobamovirus Classification

Adrian Gibbs

Chapter 9

Tomato Mosaic Virus

Alan A. Brunt

Chapter 10

Tobacco Mild Green Mosaic Virus

Carl Wetter

Chapter 13

Sunn-Hemp Mosaic Virus

Anupam Varma

Chapter 14

Cucumber Green Mottle Mosaic Virus

Yoshimi Okada

Chapter 15

Miscellaneous Tobamoviruses

Alan A. Brunt

II. Fungus-Transmitted and Similar Labile Rod-Shaped Viruses

Chapter 16

Fungus-Transmitted and Similar Labile Rod-Shaped Viruses

Alan A. Brunt and Eishiro Shikata

III. Tobraviruses

Chapter 17

Tobraviruses

B. D. Harrison and D. J. Robinson

Chapter 19

Hordeiviruses: Structure and Replication

J. G. Atabekov and V. V. Dolja

The Plant Viruses

Volume 2

PART I

TOBAMOVIRUSES

CHAPTER 1

TOBACCO MOSAIC VIRUS

The History of Tobacco Mosaic Virus and the Evolution of Molecular Biology

H. FRAENKEL-CONRAT

During the second half of the 19th century, bacteria were recognized as the causative agents of many infectious diseases of plants and animals, including humans. The assumption came to be that bacteria or other microorganisms would be found to be responsible for all transmissible diseases. The main technique used to separate such infectious agents from extracts or exudates was that of filtration. It thus came as a great surprise or shock to Ivanowski, working on the tobacco mosaic disease in the 1890s in the Crimea (see 1903), that his filtrates of infected plant extracts retained their infectivity. The same observations were made by Beijerinck (1898) in Holland. But only Beijerinck believed the results of his own experiments, and concluded that he had discovered a new type of infectious agent, "Contagium vivum fluidum." This term was abandoned when it became apparent that the contagium was neither alive nor fluid. But the concept of the existence of filtrable infectivity, soon called filtrable virus, became quickly accepted, since similar results were obtained by Loeffler and Frosch (1898) working in Germany on the foot-and-mouth disease of cattle, and later by others working on other diseases.

H. FRAENKEL-CONRAT • Virus Laboratory and Department of Molecular Biology, University of California, Berkeley, California 94720.

This new type of infectious agent, soon termed simply virus, was characterized as being too small to be retained by bacterial filtration methods, too small to be seen using the microscope, and not cultivatable *in vitro* by any of the methods used by bacteriologists to "grow" microorganisms. Although small, compared to microorganisms, they were nevertheless soon found to be large, compared to even the largest known chemical molecules. This recognition was put on a quantitative basis by the use of filters of varying porosity (Elford, 1932). It was particularly the development of the ultracentrifuge and of the electron microscope that confirmed this fact, and gave us definitive information about the size and shape of viruses.

The general rodlike shape of tobacco mosaic virus (TMV) had been derived from various physical data (Takahashi and Rawlins, 1932; Bernal and Fankuchen, 1941), but these did not necessarily convince the biologists and biochemists. However, any doubts were removed by the electron microscopic images of cigarette-shaped virus particles (Kausche and Ruska, 1939; Stanley and Anderson, 1941), the dimensions of which agreed remarkably well with those postulated on the basis of ultracentrifugal data (Lauffer, 1938). There remained some question about the particle's mass, although a value of about 40×10^6 daltons became generally accepted.

Aside from size and shape, what about the other difference between viruses such as TMV and microorganisms, namely that viruses cannot be cultivated *in vitro*? Since some viruses are almost as large as the smallest microorganism and *can* be seen using the microscope, and all viruses can be retained by appropriate filters, the property of noncultivatability is the only one that clearly separates viruses from organisms, and makes them a new type of infectious agent, incontrovertibly qualitatively different from living organisms. We now know that this is due to the viruses lacking the metabolic machinery to produce the energy and the physical matter needed for their replication. They are obligatory parasites, requiring living cells for their replication.

Leaving this brief review of the history of virology in general, let us return to TMV. The first virus to be identified, it has many more firsts in its chronicle. In the 1920s and 1930s, methods then being developed for protein purification were used in several laboratories to purify TMV. The trouble with many of the experiments was the lack of a good infectivity assay system. When the local lesion assay method was developed by Holmes in 1929, the evaluation of purification procedures became much more reliable. Crystalline preparations of TMV were reported in that year by Vinson and Petre. These, however, appeared to contain much ash, i.e., salts, were variously infective, and were not obtained reproducibly. In general, there was much skepticism and resistance to the concept that a biological function as complex as infectivity would be purifiable and crystallizable. Efforts to crystallize TMV were not abandoned, however. The head of the Boyce–Thompson Institute, where Vinson was working, became the director of a newly formed plant pathology department at the Rockefeller Institute for Medical Research at Princeton and he hired a young organic chemist, Wendell M. Stanley, and suggested that he work on the purification of TMV. This was next door to Northrop's department, who with his famous group of collaborators (Kunitz, Herriott, Anson, and others) was developing new gentle methods of purifying and then crystallizing proteins, such as the enzymes pepsin, trypsin, trypsinogen, ribonuclease, and so on.

With this new methodology, and a reliable bioassay at hand, Stanley was able to bring the Vinson and Petre pilot experiments to successful fruition. In 1935 he reported the isolation of pure crystalline, and repeatedly recrystallizable TMV, and he had the courage and the personality to convince others of the validity of his achievements. Thus, within a few years the resistance was overcome, and TMV, and soon other viruses were accepted as being chemical substances, particulate in nature, and upon isolation sufficiently homogeneous to be crystallized like smaller chemicals, even though they were known from sedimentation and other data to be 2 to 3 orders of magnitude larger than any of the proteins being crystallized at that time.

As far as the TMV crystals are concerned, it was soon realized that they were not truly three-dimensional crystals, as formed by small molecules, by several viruses, or even by TMV intracellularly. Instead, they are termed paracrystalline, the rod-shaped particles forming needles by aggregating side by side. But TMV remained the first pure virus to be isolated and characterized.

Concerning the chemical nature of TMV, Stanley thought it to be proteinaceous, and that the presence of a small amount of phosphorus (0.5%) had no particular significance. But much more phosphorus was found in other viruses and it was soon clearly shown by Bawden and Pirie (1937) that the phosphorus in TMV and other plant viruses studied was due to the presence of RNA. We now know that all viruses are complexes of protein and RNA or DNA.

Another historically interesting aspect of the notion that viruses were exclusively proteinaceous was that it gave rise to the hypothesis that viruses might replicate by an autocatalytic process, similar to the conversion of trypsinogen, in the presence of a trace of trypsin, to trypsin. Subsequently, it became clear that virus replication, induced by fractions of micrograms and yielding many grams of virus in the plant (see Fig. 1), had nothing in common with conversion of a precursor to an active compound.

During the two decades after the crystallization of TMV, several X-ray studies demonstrated that the virus rod consisted of many orderly, helically spaced, protein molecules (Bernal and Fankuchen, 1941; Watson, 1954). The location of its RNA was established by the X-ray crystallographic studies of Franklin, as well as Caspar, in 1956. Since the genetic information in its RNA was quite insufficient to code for a protein

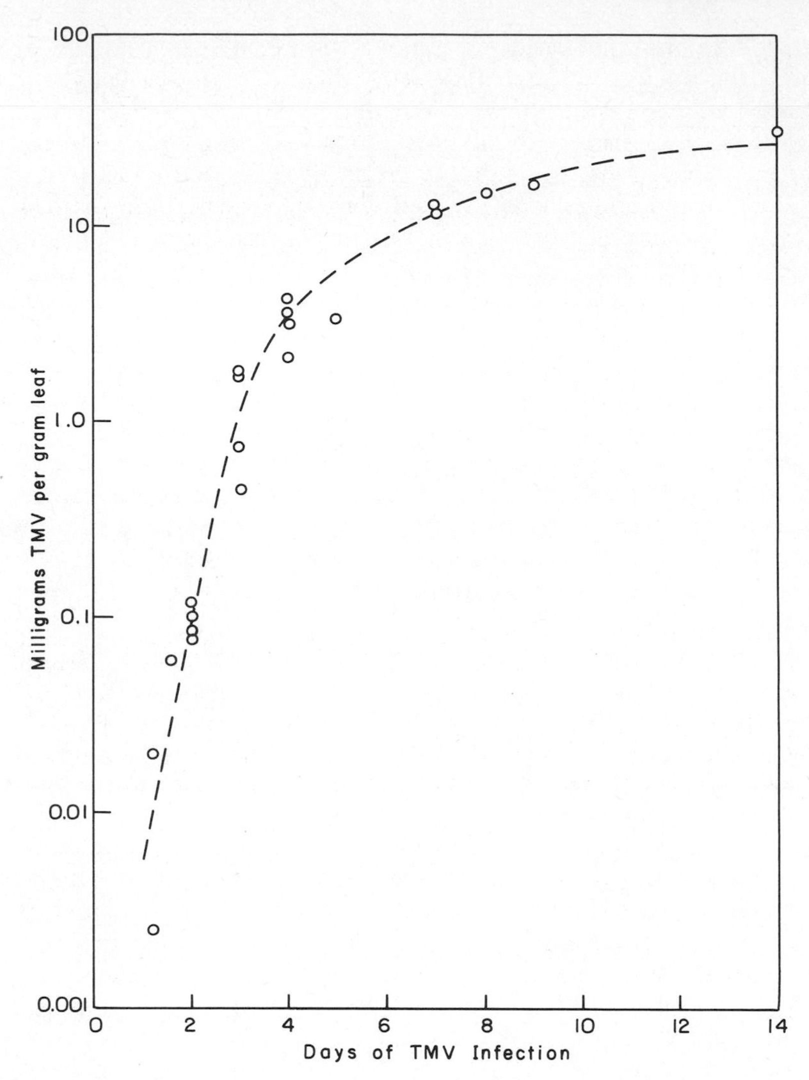

FIGURE 1. Virus replication, induced by fractions of micrograms and yielding many grams of virus in the plant.

of 40×10^6 daltons, it was deduced that TMV had to be built up of one or several species of identical protein molecules. The same had to be true for other viruses.

However, chemical evidence concerning the presence of protein subunits in TMV had been obtained several years earlier. Thus, in 1951, protein chemists, concerned about carboxypeptidase's specificity of attacking only the ends of peptide chains, utilized as a presumably negative control TMV with a presumed 40×10^6-dalton protein, the end group of which would be below the threshold of detection. To the great surprise of Harris and Knight (1952), they found about 2000 threonine molecules (and nothing else) released from the virus by this enzyme. That this was really one terminal threonine per subunit peptide chain was shown by hydrazinolysis, and by finding a terminal hexapeptide after enzymatic digestion of the protein (Niu and Fraenkel-Conrat, 1955a,b). More difficult was the elucidation of the N-terminus of the viral coat protein, because this was acetylated, a feature not previously observed in protein chemistry (Narita, 1958). Here also the terminal amino acid sequence revealed itself together with the end group. Two years later, the methods for protein sequencing had been sufficiently well developed (thanks to F. Sanger, and S. Moore and W. Stein) so that the entire TMV coat protein sequence of 158 amino acids could be established simultaneously in Berkeley and in Tübingen (Tsugita *et al.*, 1960; Anderer *et al.*, 1960). This was the third protein after insulin (51 amino acids) and ribonuclease (124 amino acids) to be sequenced, and only very minor corrections were needed to bring the Berkeley and Tübingen sequences to agreement; in contrast, the ribonuclease sequence needed extensive later corrections. Wittmann's group in Tübingen sequenced the coat proteins of several other TMV-related viruses in subsequent years, and thus supplied one of the first series of comparative data for related proteins (Wittmann-Liebold and Wittmann, 1963, 1965, 1967).

Another important question in protein chemistry at that time was the mechanism that gave proteins their specifically folded conformation or tertiary structure. It had been shown with pancreatic ribonuclease that a protein rich in disulfide bonds could be completely denatured with concomitant reduction of these to –SH groups, and that this so-called "random coil" would refold itself properly *in vitro* upon gentle reoxidation of the –SH groups to –S–S bonds. Although TMV protein has only one cysteine, and thus no disulfide-bonded secondary structure, Anderer (1959) could show that this protein after complete denaturation was able to renature itself under appropriate conditions to the proper conformation, and again become able to aggregate to rods. The realization that the primary structure, i.e., the amino acid sequence of a protein, alone suffices to give proteins their functional conformation was a most important milestone in the development of molecular biology.

In following the history of TMV coat protein structural elucidation, we have chronologically gone past the most important discoveries of TMVology, if not virology in general. In the early 1950s, Schramm and co-workers dissociated TMV with alkali, isolated the protein, and observed that under appropriate conditions it was able to aggregate to form rods that, except for very variable lengths, looked exactly like TMV rods (Schramm *et al.*, 1955; Schramm and Zillig, 1955). They were disappointed by the finding that this material was quite noninfectious. DNA, but not RNA, was recognized at that time as the genetic material, and the mode of infectivity of RNA viruses was a complete mystery. When this author in 1953 proposed to produce TMV mutants by adding amino acids to the coat protein chain, none of his colleagues regarded this as a ridiculous line of research. However, 2 years later we focused our attention on the viral RNA. Schramm's alkali treatment surely degraded RNA, and so we began experiments to isolate undegraded RNA. Detergents such as sodium dodecyl sulfate (SDS) dissociated the virus at neutrality and allowed the isolation of intact RNA. The protein–SDS complex did not form rods, but by isolating the protein by alkali, or preferably by 67% acetic acid treatment (Fraenkel-Conrat, 1957), native (or renatured) protein could be recovered, and the interaction of this protein with SDS-isolated RNA gave infective TMV rods, a reaction we termed virus reconstitution (Fraenkel-Conrat and Williams, 1955; Fraenkel-Conrat, 1956a; Fraenkel-Conrat and Singer, 1957). This caused a considerable stir in the scientific community.

To test our working hypothesis that particle formation was a prerequisite for viral infectivity, we had to show that neither of the two components was by itself infectious. This was indubitably true for the protein, but the RNA often showed low-level infectivity (Fraenkel-Conrat, 1956b; Fraenkel-Conrat *et al.*, 1957). By the time we learned how to safeguard the RNA against all traces of nucleases, this infectivity was regularly obtained. Parallel work by Gierer and Schramm (1956), who isolated their TMV RNA by phenol extraction, gave similar results. Without doubt, TMV RNA was by itself infectious, and, as we soon showed, carried all the genetic information of the virus, even if reconstituted with the needed 20-fold higher amount of the protein of other related viruses such as the Holmes ribgrass virus (Fraenkel-Conrat and Singer, 1957). Thus, RNA was established to be an alternate genetic material. Several physical methods soon showed that the TMV rod contained 1 molecule of RNA of 2.1×10^6 daltons and that it was the RNA that actually determined and stabilized the length of the rod. As stated, the viral protein can aggregate to any length, but these protein rods are stable only over a small range of pH, whereas TMV, natural or reconstituted, is stable over the range of pH 3–9.

Reconstitution protects the RNA, and it is for this reason that the reconstituted product shows 500- to 1000-fold higher infectivity than obtained upon inoculation with the free RNA. We have therefore continued to use reconstitution as a tool in our subsequent nucleic acid studies.

Reconstitution proceeds more reproducibly in high yield in pyrophosphate buffer (at pH 7.1) than in other buffers. This discovery represents a typical case of serendipity. When Severo Ochoa spent a summer month in Berkeley, we were wondering whether his newly discovered enzyme from *E. coli*, polynucleotide phosphorylase, might be able to replicate TMV RNA. The mixture of "template" viral RNA, the four nucleoside diphosphates, enzyme, and any synthesized copy was then treated with TMV protein to reconstitute virus prior to the infectivity assay, and increased infectivity was actually obtained. Our (naive) excitement diminished, however, when we found the same to be true upon omitting one, then two, or three of the diphosphates. Finally, diphosphate alone, i.e., pyrophosphate, was found to be active, and it became evident that the favorable chelating effect of the diphosphate linkage had caused higher reconstitution yields, rather than that RNA replication had been achieved by this enzyme. A less effective alternative to reconstitution in increasing the infectivity of viral RNA is the addition of nuclease-binding clays, bentonite being most effective (Singer and Fraenkel-Conrat, 1961).

Many years after the first reconstitution of TMV, active work on the mechanism of this reaction was taken up in several laboratories in England, France, and Japan and yielded unexpected results, which will be described or mentioned in subsequent chapters.

When the biological activity of RNA was discovered, much of the research focused on its chemistry. As stated, the infectivity depended on the integrity of the RNA molecule, a single break, caused by a nuclease, alkali, or other means, in the 6390-nucleotide chain causing loss of infectivity (Gierer, 1957; Schuster and Schramm, 1958). Many subsequent studies of the action of various nondegrading chemicals showed that all of them caused more or less loss of infectivity. However, certain of these modification reactions caused the appearance of mutants among the surviving molecules. This was first noted by Gierer and Mundry (1958) as a consequence of nitrous acid treatment (Schuster, 1960), which besides other effects caused obvious mutagenic events such as C → U and A → I transitions. The mutagenicity of hydroxylamine and methoxyamine is accounted for by the change in the predominant tautomer of the resultant 4-hydroxyamino C to resemble U (Fraenkel-Conrat and Singer, 1972). Although most alkylating agents are poor mutagens (Fraenkel-Conrat, 1961; Singer and Fraenkel-Conrat, 1967), it was (and remains) surprising that nitrosoguanidine was very highly mutagenic when acting on the virus, but not when acting on the RNA (Singer and Fraenkel-Conrat, 1967, 1969a,b,c).

Hundreds of TMV mutants were subsequently obtained in Berkeley, Tübingen, and elsewhere. The coat proteins of these mutants, the only easily accessible gene product of plant viruses, were then analyzed for amino acid composition, and some for partial sequences (Tsugita and Fraenkel-Conrat, 1960, 1962; Funatsu and Fraenkel-Conrat, 1964; Funatsu *et al.*, 1964). As expected, many mutants showed no changes in the coat protein, since it represents only about 8% of the total potential gene products. When mutants did show one to three amino acid replacements, these were considered in terms of the codon dictionary that was being deciphered at the same time in Nierenberg's and Ochoa's laboratories. Most of the exchanged amino acids could be attributed to the C → U or A → I (≅G) nucleotide exchanges that would be expected from chemical modification of the RNA (Singer and Fraenkel-Conrat, 1974). Thus, prolines were often replaced by leucines and serines, and these by phenylalanines, all well accounted for by the C → U exchanges that mutagens such as nitrous acid would cause (see Chapter 3). Of special interest was the fact that a particular mutation "hot" spot changed the proline near the C-terminus to leucine (Tsugita and Fraenkel-Conrat, 1960). It is the presence of that proline that prevented carboxypeptidase from degrading TMV protein beyond the C-terminal threonine (see pp. 8–9). Harris and Knight (1952) would have obtained a confusing pattern of amino acids, rather than the single threonine per chain, if they had used such a mutant instead of wild-type TMV. This location of a proline near the C-terminus, as well as the acetyl group on the N-terminus, probably represent favorable evolutionary developments in producing a virus coat resistant to exopeptidases. The tight folding of the peptide chain actually makes the virus surprisingly resistant also to endopeptidases and proteases.

Studies of the primary structure of the RNA began in 1960 when Sugiyama developed the methodology to detect and separate one nucleoside among 6390 nucleotides (Sugiyama and Fraenkel-Conrat, 1961, 1963). He thus was able to show that the 3′-end of the RNA of TMV and several related viruses carried unphosphorylated adenosine. The next six residues of TMV were determined by stepwise degradation (Steinschneider and Fraenkel-Conrat, 1966). The absence of the expected 5′-terminal phosphates was shown by Fraenkel-Conrat and Singer in 1962. Later, the presence of a typical 5′-terminal cap structure was discovered in most mRNAs ($m^7G^{5'}ppp^{5'}Gp$–) and this type of chain end was shown by Keith and Fraenkel-Conrat (1975) to also exist in TMV, although lacking additional methylated terminal bases.

TMV RNA presented an additional problem when its *in vitro* translation failed to yield evidence for the formation of the only known product, the coat protein. This mystery was solved 15 years later when it was recognized that the long RNA of TMV was processed *in vivo* to yield the relatively short coat protein mRNA (Hunter *et al.*, 1976), which only then was available for translation. That certain TMV-related viruses (e.g., the cowpea or SM virus; see Chapter 7) separately encapsidate this RNA (Higgins *et al.*, 1976) led to the discovery of many other similar instances. Another oddity of the translation of TMV RNA and other large plant viral RNAs was the appearance of two long translation products, amounting to more than the genome length of the virus. This was shown by Pelham (1978) to be due to the existence in TMV RNA of a leaky termination codon. All these topics will be dealt with in detail in other chapters of this book. I mention them only as illustrations of the multitude of molecular biological phenomena that have been, often first, studied with TMV. Among these should also be mentioned the presence in all plants of RNA-dependent RNA polymerases, greatly increased in amount by virus infection. Our work on the TMV-stimulated production of this tobacco enzyme (Ikegami and Fraenkel-Conrat, 1979; Fraenkel-Conrat,

TABLE I. Features of Our Understanding of Molecular Biology Gained through the Use of TMV

TMV	Feature	References[a]
Particle	Architecture	1–9
	Infectivity	10, 11
	Dissociation	12, 13
	Reconstitution, assembly	14–17
Coat protein	End groups, terminal sequences	18–21
	Complete sequences	22–26
	Reversibility of denaturation	27
	Conformation	28
	Mutant sequences, evolution	29–31
RNA	Infectivity	32–34
	Molecular weight	35, 36
	End groups, nucleotide sequences	37–48
	Chemical modification, mutagenesis	44–53
	Silent genes, subgenomic RNAs, readthrough	54, 55
	RNA-dependent RNA polymerases of plants	56, 57

[a] 1, Takahashi and Rawlins (1932); 2, Bernal and Fankuchen (1941); 3, Watson (1954); 4, Crick and Watson (1956); 5, Franklin (1956); 6, Caspar (1956); 7, Lauffer (1938); 8, Kausche and Ruska (1939); 9, Stanley and Anderson (1941); 10, Vinson and Petre (1929); 11, Stanley (1935); 12, Schramm *et al.* (1955); 13, Schramm and Zillig (1955); 14, Fraenkel-Conrat and Williams (1955); 15, Fraenkel-Conrat (1956a); 16, Fraenkel-Conrat and Singer (1957); 17, Lebeurier *et al.* (1977); 18, Harris and Knight (1952); 19, Niu and Fraenkel-Conrat (1955a); 20, Niu and Fraenkel-Conrat (1955b); 21, Narita (1958); 22, Tsugita *et al.* (1960); 23, Anderer *et al.* (1960); 24, Wittmann-Liebold and Wittmann (1963); 25, Wittmann-Liebold and Wittmann (1965); 26, Wittmann-Liebold and Wittmann (1967); 27, Anderer (1959); 28, Bloomer *et al.* (1978); 29, Tsugita and Fraenkel-Conrat (1962); 30, Funatsu and Fraenkel-Conrat (1964); 31, Funatsu *et al.* (1964); 32, Fraenkel-Conrat (1956b); 33, Gierer and Schramm (1956); 34, Fraenkel-Conrat *et al.* (1957); 35, Gierer (1957); 36, Schuster and Schramm (1958); 37, Sugiyama and Fraenkel-Conrat (1961); 38, Fraenkel-Conrat and Singer (1962); 39, Sugiyama and Fraenkel-Conrat (1963); 40, Steinschneider and Fraenkel-Conrat (1966); 41, Keith and Fraenkel-Conrat (1975); 42, Richards *et al.* (1978); 43, Goelet *et al.* (1982); 44, Schuster (1960); 45, Gierer and Mundry (1958); 46, Fraenkel-Conrat (1961); 47, Singer and Fraenkel-Conrat (1967); 48, Singer and Fraenkel-Conrat (1969a); 49, Singer and Fraenkel-Conrat (1969b); 50, Singer and Fraenkel-Conrat (1969c); 51, Singer *et al.* (1968); 52, Fraenkel-Conrat and Singer (1972); 53, Singer and Fraenkel-Conrat (1974); 54, Hunter *et al.* (1976); 55, Pelham (1978); 56, Ikegami and Fraenkel-Conrat (1979); 57, Fraenkel-Conrat (1983).

1983) has led to intensive studies of these plant enzymes in our and many other laboratories.

CONCLUSIONS

The study of TMV has led to a great number of findings that have had general importance in the development not only of virology but also of our molecular understanding of biological processes in general. These are presented in Table I (incompletely referenced) to illustrate what the TMV particle, its protein, and its RNA have taught us over the years. The main topics are:

1. Self-assembly of proteins to particulate structures
2. Renaturability of proteins
3. Genetic capability and infectivity of RNA
4. Self-assembly of protein and RNA to a specific structure
5. Acetylation of the N-terminus of proteins
6. Blocking of exopeptidase action by N-terminal acetyl group, and by proline near the C-terminus
7. Support, from the study of protein changes in chemically produced mutants, in the gradual elucidation of the genetic code
8. Mechanisms of chemical mutagenesis
9. Some of the complexities of the translation process, such as readthrough of termination codons, and the need for processing of RNA to activate silent genes
10. Presence of RNA-dependent RNA polymerases in plants

REFERENCES

Anderer, F. A., 1959, Reversible Denaturierung des Proteins aus Tabakmosaikvirus, *Z. Naturforsch.* **14b**:642.

Anderer, F. A., Uhlig, H., Weber, E., and Schramm, G., 1960, Primary structure of the protein of tobacco mosaic virus, *Nature (London)* **186**:922.

Bawden, F. C., and Pirie, N. W., 1937, The isolation and some properties of liquid crystalline substances from solanaceous plants infected with three strains of tobacco mosaic virus, *Proc. R. Soc. London* **123**:274.

Beijerinck, M. W., 1898, Über ein Contagium vivum fluidum als Ursache der Fleckenkrankheit der Tabaksblätter, *Zentralbl. Bakteriol. Parasitenkd. Infektionskr. Hyg. Abt. 2* **5**:27.

Bernal, J. D., and Fankuchen, I., 1941, X-ray and crystallographic studies of plant virus preparations, *J. Gen. Physiol.* **25**:147.

Bloomer, A. C., Champness, H. N., Bricogne, G., Staden, R., and Klug, A., 1978, Protein disk of tobacco mosaic virus at 2.8 Å resolution showing the interactions within and between subunits, *Nature (London)* **276**:362.

Caspar, D. L. D., 1956, The radial density distribution in the tobacco mosaic virus particle, *Nature (London)* **177**:928.

Crick, F. H. C., and Watson, J. D., 1956, Structure of small viruses, *Nature (London)* **177**:473.

Elford, W. J., 1932, The principles of ultrafiltration as applied in biological studies, *Proc. R. Soc. London Ser. B* **112**:384.

Fraenkel-Conrat, H., 1953, The reaction of proteins with ^{14}C-labelled N-carboxyleucine anhydride, *Biochim. Biophys. Acta* **10**:180.

Fraenkel-Conrat, H., 1956a, Rebuilding a virus, *Sci. Am.* **194**:42.

Fraenkel-Conrat, H., 1956b, The role of the nucleic acid in the reconstitution of active tobacco mosaic virus, *J. Am. Chem. Soc.* **78**:882.

Fraenkel-Conrat, H., 1957, Degradation of tobacco mosaic virus with acetic acid, *Virology* **4**:1.

Fraenkel-Conrat, H., 1961, Chemical modification of viral ribonucleic acid. I. Alkylating agents, *Biochim. Biophys. Acta* **49**:169.
Fraenkel-Conrat, H., 1983, RNA-dependent RNA polymerases of plants, *Proc. Natl. Acad. Sci. USA* **80**:422.
Fraenkel-Conrat, H., and Singer, B., 1957, Virus reconstitution. II. Combination of protein and nucleic acid from different strains, *Biochim. Biophys. Acta* **24**:540.
Fraenkel-Conrat, H., and Singer, B., 1962, The absence of phosphorylated chain ends in tobacco mosaic virus ribonucleic acid, *Biochemistry* **1**:120.
Fraenkel-Conrat, H., and Singer, B., 1972, The chemical basis for the mutagenicity of hydroxylamine and methoxyamine, *Biochim. Biophys. Acta* **262**:264.
Fraenkel-Conrat, H., and Williams, R. C., 1955, Reconstitution of active tobacco mosaic virus from its inactive protein and nucleic acid components, *Proc. Natl. Acad. Sci. USA* **41**:690.
Fraenkel-Conrat, H., Singer, B., and Williams, R. C., 1957, Infectivity of viral nucleic acid, *Biochim. Biophys. Acta* **25**:87.
Franklin, R. E., 1956, Location of the RNA in the TMV particle, *Nature (London)* **177**:928.
Funatsu, G., and Fraenkel-Conrat, H., 1964, Location of amino acid exchanges in chemically evoked mutants of tobacco mosaic virus, *Biochemistry* **3**:1356.
Funatsu, G., Tsugita, A., and Fraenkel-Conrat, H., 1964, Studies on the amino acid sequence of tobacco mosaic virus protein. V. Amino acid sequences of two peptides from tryptic digests and location of amide group, *Arch. Biochem. Biophys.* **105**:25.
Gierer, A., 1957, Structure and biological function of ribonucleic acid from tobacco mosaic virus, *Nature (London)* **179**:1297.
Gierer, A., and Mundry, K. W., 1958, Production of mutants of TMV by chemical alteration of its ribonucleic acid *in vitro*, *Nature (London)* **182**:1457.
Gierer, A., and Schramm, G., 1956, Die Infektiosität der Ribonukleinsäure des Tabakmosaikvirus, *Z. Naturforsch.* **11b**:138.
Goelet, P., Lomonossoff, G. P., Butler, P. J. G., Akam, M. E., Gait, M. J., and Karn, J., 1982, Nucleotide sequence of tobacco mosaic virus RNA, *Proc. Natl. Acad. Sci. USA* **79**:5818.
Harris J. I., and Knight C. A., 1952, Action of carboxypeptidase on TMV, *Nature (London)* **170**:613.
Higgins, T. J. V., Goodwin, P. B., and Whitfeld, P. R., 1976, Occurrence of short particles in beans infected with cowpea strain of TMV. II. Evidence that short particles contain the cistron for coat protein, *Virology* **71**:286.
Holmes, F. O., 1929, Local lesions in tobacco mosaic, *Bot. Gaz.* **87**:39.
Hunter, T. R., Hung, T., Knowland, J., and Zimmern, D., 1976, Messenger RNA for the coat protein of TMV, *Nature (London)* **260**:759.
Ikegami, M., and Fraenkel-Conrat, H., 1979, Characterization of the RNA-dependent RNA polymerase of tobacco leaves, *J. Biol. Chem.* **254**:149.
Ivanowski, D. I., 1903, Über die Mosaik-Krankheit der Tabakpflanze, *Zbl. Bakteriol. 2E* **5**:250.
Kausche, G. A., and Ruska, H., 1939, Die Struktur der "kristallinen Aggregate" des Tabakmosaikvirus-Proteins, *Biochem. Z.* **303**:211.
Keith, J., and Fraenkel-Conrat, H., 1975, Tobacco mosaic virus RNA carries 5'-terminal triphosphorylated guanosine blocked by 5'-linked 7-methylguanosine, *FEBS Lett.* **57**:31.
Lauffer, M. A., 1938, The molecular weight and shape of tobacco mosaic virus (protein), *Science* **87**:469.
Lebeurier, G., Nicolaieff, A., and Richards, K. E., 1977, Inside-out model for self-assembly of tobacco mosaic virus, *Proc. Natl. Acad. Sci. USA* **74**:149.
Loeffler, F., and Frosch, P., 1898, Berichte der Kommission zur Erforschung der Maul und Klauenseuche bei dem Institut für Infektionskrankheiten, *Zentralbl. Bakter. Abstract*, **23**:371.
Narita, K., 1958, Isolation of acetylpeptide from enzymic digests of TMV protein, *Biochim. Biophys. Acta* **28**:184.

Niu, C.-I., and Fraenkel-Conrat, H., 1955a, C-terminal amino-acid sequence of tobacco mosaic virus protein, *Biochim. Biophys. Acta* **16**:597.

Niu, C.-I., and Fraenkel-Conrat, H., 1955b, C-terminal amino acid sequences of four strains of tobacco mosaic virus, *Arch. Biochem. Biophys.* **59**:538.

Pelham, H. R. B., 1978, Leaky UAG termination codon in TMV RNA, *Nature (London)* **272**:469.

Richards, K., Guilley, H., Jonard, G., and Hirth, L., 1978, Nucleotide sequence at the 5′ extremity of TMV RNA. 1. The noncoding region (nucleotides 1–68), *Eur. J. Biochem.* **84**:513.

Schramm, G., and Zillig, W., 1955, Über die Struktur des Tabakmosaikvirus. IV. Die Reaggregation des nucleinsaurefreien Proteins, *Z. Naturforsch.* **10b**:493.

Schramm, G., Schumacher, G., and Zillig, W., 1955, Über die Struktur des Tabakmosaikvirus. III. Der Zerfall in alkalischer Losung, *Z. Naturforsch.* **10b**:481.

Schuster, H., 1960, Die Reaktionsweise der Desoxribonucleinsäure mit salpetriger Säure, *Z. Naturforsch.* **15b**:298.

Schuster, H., and Schramm, G., 1958, Bestimmung der biologisch wirksamen Einheit in der Ribosenucleinsäure des Tabakmosaikvirus auf chemischen Wege, *Z. Naturforsch.* **13b**:697.

Singer, B., and Fraenkel-Conrat, H., 1961, Effects of bentonite on infectivity and stability of TMV-RNA, *Virology* **14**:59.

Singer, B., and Fraenkel-Conrat, H., 1967, Chemical modification of viral RNA. VI. The action of N-methyl-N′-nitro-N-nitrosoguanidine, *Proc. Natl. Acad. Sci. USA* **58**:234.

Singer, B., and Fraenkel-Conrat, H., 1969a, Chemical modification of viral ribonucleic acid. VII. The action of methylating agents and nitrosoguanidine on polynucleotides including tobacco mosaic virus ribonucleic acid, *Biochemistry* **8**:3260.

Singer, B., and Fraenkel-Conrat, H., 1969b, Chemical modification of viral ribonucleic acid. VIII. The chemical and biological effects of methylating agents and nitrosoguanidine on tobacco mosaic virus, *Biochemistry* **8**:3266.

Singer, B., and Fraenkel-Conrat, H., 1969c, Mutagenicity of alkyl and nitroso-alkyl compounds acting on tobacco mosaic virus and its RNA, *Virology* **39**:395.

Singer, B., and Fraenkel-Conrat, H., 1974, Correlation between amino acid exchanges in coat protein of TMV mutants and the nature of the mutagens, *Virology* **60**:485.

Singer, B., Fraenkel-Conrat, H., Greenberg, J., and Michelson, A. M., 1968, Reaction of nitrosoguanidine (N-methyl-N′-nitro-N-nitrosoguanidine) with tobacco mosaic virus and its RNA, *Science* **160**:1235.

Stanley, W. M., 1935, Isolation of a crystalline protein possessing the properties of the tobacco mosaic virus, *Science* **81**:644.

Stanley, W. M., and Anderson, H. F., 1941, A study of purified viruses with the electron microscope, *J. Biol. Chem.* **139**:325.

Steinschneider, A., and Fraenkel-Conrat, H., 1966, Studies of nucleotide sequences in tobacco mosaic virus ribonucleic acid. IV. Use of aniline in stepwise degradation, *Biochemistry* **5**:2735.

Sugiyama, T., and Fraenkel-Conrat, H., 1961, Identification of 5′ linked adenosine as end group of TMV-RNA, *Proc. Natl. Acad. Sci. USA* **47**:1393.

Sugiyama, T., and Fraenkel-Conrat, H., 1963, The end-groups of tobacco mosaic virus RNA. II. Nature of the 3′-linked chain end in TMV and of both ends in four strains, *Biochemistry* **2**:332.

Takahashi, W. N., and Rawlins, T. E., 1932, Method for determining shape of colloidal particles: Application in study of TMV, *Proc. Soc. Exp. Biol. Med.* **30**:155.

Tsugita, A., and Fraenkel-Conrat, H., 1960, The amino acid composition and C-terminal sequence of a chemically evoked mutant of TMV, *Proc. Natl. Acad. Sci. USA* **46**:636.

Tsugita, A., and Fraenkel-Conrat, H., 1962, The composition of proteins of chemically evoked mutants of TMV-RNA, *Colloq. Int. CNRS* **106**:241.

Tsugita, A., Gish, D. T., Young, J., Fraenkel-Conrat, H., Knight, C. A., and Stanley, W. M.,

1960, The complete amino acid sequence of the protein of tobacco mosaic virus, *Proc. Natl. Acad. Sci. USA* **46:**1463.
Vinson, C. G., and Petre, A. W., 1929, Mosaic disease of tobacco, *Bot. Gaz.* **87:**14.
Watson, J. D., 1954, The structure of TMV. I. X-ray evidence of a helical arrangement of subunits around the longitudinal axis, *Biochim. Biophys. Acta* **13:**10.
Wittmann-Liebold, B., and Wittmann, H. G., 1963, Die primäre Proteinstruktur von Stämmen des Tabakmosaikvirus. Aminosäuresequenzen des Proteins des Tabakmosaikvirus stammes Dahlemense. Teil III. Diskussion der Ergebnisse, *Z. Vererbungsl.* **94:**427.
Wittmann-Liebold, B., and Wittmann, H. G., 1965, Lokalisierung von Aminosäureaustauschen bei Nitritmutanten des Tabakmosaikvirus, *Z. Vererbungsl.* **97:**305.
Wittmann-Liebold, B., and Wittmann, H. G., 1967, Coat proteins of strains of two RNA viruses: Comparison of their amino acid sequences, *Mol. Gen. Genet.* **100:**358.

CHAPTER 2

TOBACCO MOSAIC VIRUS

Structure and Self-Assembly

A. C. BLOOMER AND P. J. G. BUTLER

I. INTRODUCTION

Tobacco mosaic virus (TMV) has become a classical object for studies on the structure and assembly of viruses. Shortly after the first purification of virus by Stanley (1935), structural studies were begun using the methods of biochemistry (Bawden and Pirie, 1937) and X-ray diffraction (Bernal and Fankuchen, 1941). The difficulties encountered in elucidating a structure of this size led to many developments in both techniques and instrumentation which enabled the structure to be determined at steadily increasing resolution. Since the virus has a helical structure (Watson, 1954) and forms highly ordered gels rather than single crystals, the three-dimensional structure must be determined by deconvolution from two-dimensional diffraction patterns of the virus gels which are azimuthally disordered. Notwithstanding the difficulties imposed by this, the virus structure has been solved to a resolution approaching 0.4 nm (Stubbs *et al.*, 1977). In the absence of RNA, the protein will form true crystals of one of its aggregates (the disk; see Section II.B.1) whose structure has been solved to a resolution beyond 3 nm which allowed an atomic model to be built (Bloomer *et al.*, 1978).

The reassembly of TMV was studied in parallel with its structure. Fraenkel-Conrat and Williams (1955) showed that the isolated protein and RNA could reassemble to give infective virus particles, the first time

A. C. BLOOMER AND P. J. G. BUTLER • MRC Laboratory of Molecular Biology, Cambridge CB2 2QH, England.

FIGURE 1. Schematic drawing of a TMV particle, showing about one-seventh of its total length. Adapted from Caspar (1963).

that such assembly had been demonstrated *in vitro*. This led to the concept of the "self-assembly" of viruses, with the nucleic acid and protein containing all the genetic information necessary to assemble an infectious virion. The concept has been confirmed and extended for TMV although it is now clear that the assembly of some other viruses requires additional factors (for a review, see Butler, 1979).

The vast literature on TMV, and even on the restricted topics of structure and assembly, precludes a full coverage of the earlier literature. The limited coverage attempted here is restricted to the seminal papers, as fuller references to the early literature can be found in several other reviews (e.g., Caspar, 1963; Lauffer, 1975; Butler and Durham, 1977; Hirth and Richards, 1981) which also include fuller consideration than can be given here of the structural studies of the polymorphic aggregation of TMV protein and the thermodynamics of its interactions.

II. STRUCTURE

TMV is a rod-shaped virus (Fig. 1) of length 300 nm and radius 9 nm, with a central hole of radius about 2 nm (Caspar, 1963). The subunits are arranged in a single helix (Watson, 1954), which is right-handed (Finch, 1972) with $16\frac{1}{3}$ protein subunits per turn (Franklin and Holmes, 1958). The single-stranded RNA binds at a radius of about 4 nm (Franklin, 1956) with three nucleotides per protein subunit. The RNA conformation is thus fairly extended as it intercalates between adjacent turns of the protein helix. The length of the virus rod is determined by the length of the RNA, which is fully coated. There is probably an extra turn of protein subunits at each end since the RNA is fully protected from nuclease attack in the virion. The RNA from the bulk preparation used for sequencing exhibited slight polymorphism, especially near the 5′-terminus, and the lengths were either 6395 or 6398 nucleotides (Goelet *et al.*, 1982) (discussed in Section II.B.3) corresponding to approximately 2140 protein subunits and 130 turns of the viral helix.

A. Polymorphic Aggregates of Protein

The aggregation of TMV is highly polymorphic and many different forms have been described. Hydrophobic interactions provide most of the energy driving the aggregation, giving larger aggregates at higher temperatures (for review, see Lauffer, 1975). At any given temperature, the dominant variable controlling the mode of aggregation is the pH, with increasing ionic strength causing higher degrees of polymerization but not affecting the basic mode (Durham *et al.*, 1971; Durham, 1972; Durham and Klug, 1972). The regions of occurrence of the main aggregates are shown in Fig. 2. Helical polymers are found at low pH; at neutral or alkaline pH the dominant form is the "two-layer" aggregate together with some smaller aggregates. The switch between these two forms is the partial protonation of two groups with abnormal pK values (around 7.1) in the virus and helical forms but not in the two-layered forms (Caspar, 1963; Butler *et al.*, 1972).

Helical aggregates of TMV protein are very similar to the virus, but without the RNA. The main difference is that the protein helix may have either $16\frac{1}{3}$ or $17\frac{1}{3}$ subunits per turn (Mandelkow *et al.*, 1976, 1981) whereas only $16\frac{1}{3}$ have ever been found in the virus. The main form of the protein at high pH and low ionic strength is a mixture of monomer and small aggregates, known collectively as "A-protein." Theoretical considerations led Caspar (1963) to predict that this mixture would contain both two- and three-layered aggregates. However, analysis of the observed equilibrium distribution of sizes in this mixture indicated that the two-layered form is strongly dominant (Durham and Klug, 1972). More recently, spectroscopic and circular dichroism measurements (carried out at higher temperatures), in conjunction with analysis of sedimentation velocity and equilibrium, suggest that two families of aggregates may be present: two-layered ones starting from 4 S material and a three-layered series from 8 S material (Vogel *et al.*, 1979). The three-layered aggregates have not been characterized in any detail and no larger closed aggregates involving three layers have been observed.

The two-layered disk of TMV protein is central among the aggregates, occurring at temperatures above about 15°C at neutral pH and most ionic strengths, and also at higher pH and ionic strengths where disks stack on each other. This stacking is of two types: either polar to give "limited stacks" of disks or with twofold symmetry as seen in the crystals of TMV

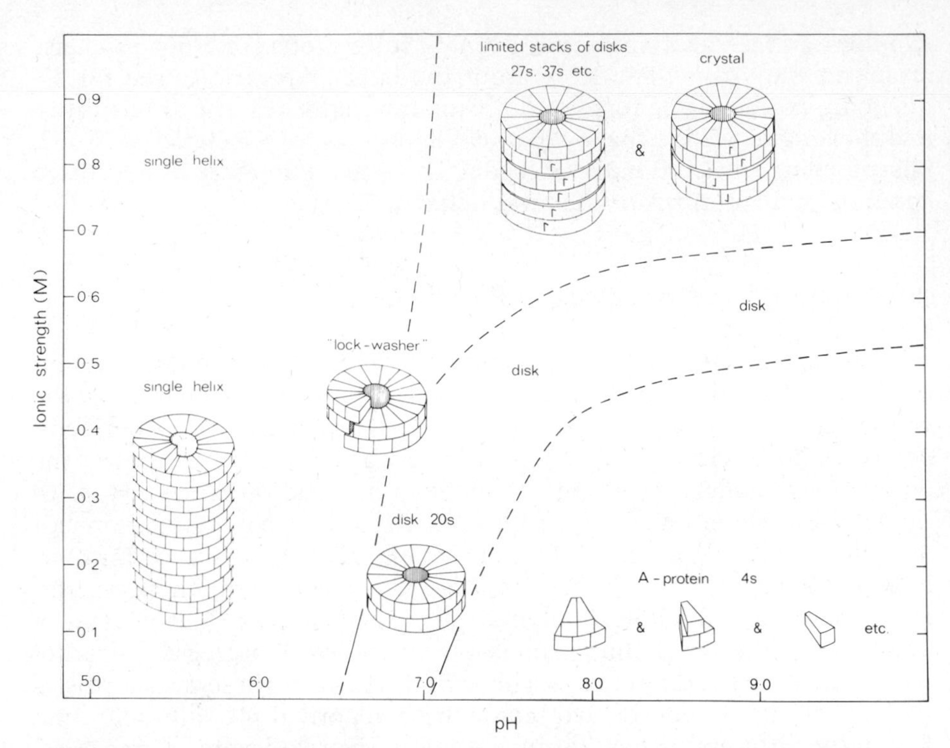

FIGURE 2. Diagram showing the regions of dominant occurrence of the polymorphic aggregates of TMV protein. Multiple species exist under most conditions and only the major aggregates are shown. This is not a conventional phase diagram: a boundary is drawn where a larger species becomes detectable and does not imply that the smaller species disappears sharply. Boundaries are approximately correct for a protein concentration of 5 mg/ml and 20°C but not all species which have been observed are shown. The "lock washer" indicated on the boundary between the 20 S disk and helix is not well defined and represents a metastable transitory state observed when not enough time is allowed for the transition (see Fig. 3). Adapted from Durham *et al.* (1971).

protein which proved suitable for a high-resolution crystallographic analysis (Bloomer *et al.*, 1978). This led to the atomic model for the protein subunit, as described in the next section. Disks are also essential for the nucleation of assembly of TMV (Butler and Klug, 1971) as discussed in Section III.A.1.

Because the protein disk is a reversibly equilibrating aggregate, disks do not occur without an equilibrium concentration of smaller aggregates being present. The proportions of these two forms in a "disk preparation" have been measured for a range of concentrations (Durham, 1972) and the aggregation shown to be a *quasi*-crystallization with a critical concentration of A-protein above which further protein assembles into disks.

The mechanism which controls aggregation of A-protein into larger aggregates remains unclear. The pH dependence suggests that proton binding is involved, but titration data indicated that two-layered aggregates bound very few protons as the pH decreased from 8 to 6 (Butler *et al.*, 1972). Notwithstanding earlier doubts as to whether the short protohelix (or "lock washer") would exist free in solution, the observed change in sedimentation coefficient from 19 S at pH 7 to 20 S or higher at lower pH values (Durham, 1972) has now been shown to correlate with small but measurable proton binding (Schuster *et al.*, 1980) so that the increase from 19 S to larger values can be interpreted as a change in aggregation state from disks to protohelices with a similar number of subunits. Very recently, further measurements have been reported on the average particle weights of the 20 S:4 S mixture (Correia *et al.*, 1985) and on the circular dichroism (Raghavendra *et al.*, 1985). From these it was concluded that the 20 S aggregate is not in fact a "disk" but rather a protohelix containing 39 ± 2 subunits. This conclusion is possibly rather simplistic, since the data were fitted by models in which only a single 20 S species was allowed, i.e., a 34-subunit disk or a variable-sized protohelix, and models containing a 34-subunit disk in equilibrium with protohelices do not appear to have been tested, even though they were indicated as likely in describing the aggregation from earlier data (Schuster *et al.*, 1980). Moreover, as described below, "pH drop" experiments (Durham *et al.*, 1971; Durham and Finch, 1972) also appear to show that the "disk preparations" contain aggregates of about this size which are distinct from protohelices, and which have previously been thought to be disks. Further work will therefore be required to resolve the question of exactly what aggregates are present in the 20 S boundary and, assuming that it in fact contains a mixture, which play the particular roles found for 20 S material in the virus assembly. Because of the very recent publication of this new interpretation and its uncertainty, we have continued to treat the 20 S species involved in assembly as the disk. While the detailed mechanisms would be altered, and probably more complex, if a protohelix interacts rather than a disk (because the RNA-binding site is less accessible in the helical aggregate than in the disk; see discussion of structures below), the overall picture will be unchanged and the important point remains of the involvement of the 20 S aggregates in these processes.

The formation of disks from A-protein is a slow process taking several hours at 20°C whether at pH 7.0 and ionic strength 0.1 M (Durham and Klug, 1971) or at higher pH and ionic strength (Durham, 1972) where disks are the dominant species present at equilibrium, although it can occur rapidly at very high ionic strengths, as used for the crystallization. The rate of disk breakdown is also slow under these conditions (Butler, 1976), although it is much faster at high pH or lower temperature. However, there is a rapid microequilibrium of protein subunits between the two aggregates, as evidenced by the rapid equilibration of radioactively labeled subunits between disk and A-protein (Richards and Williams, 1972). Subunits, probably individually, dissociate from and are recaptured by disks without complete loss of structure of the larger aggregate, though

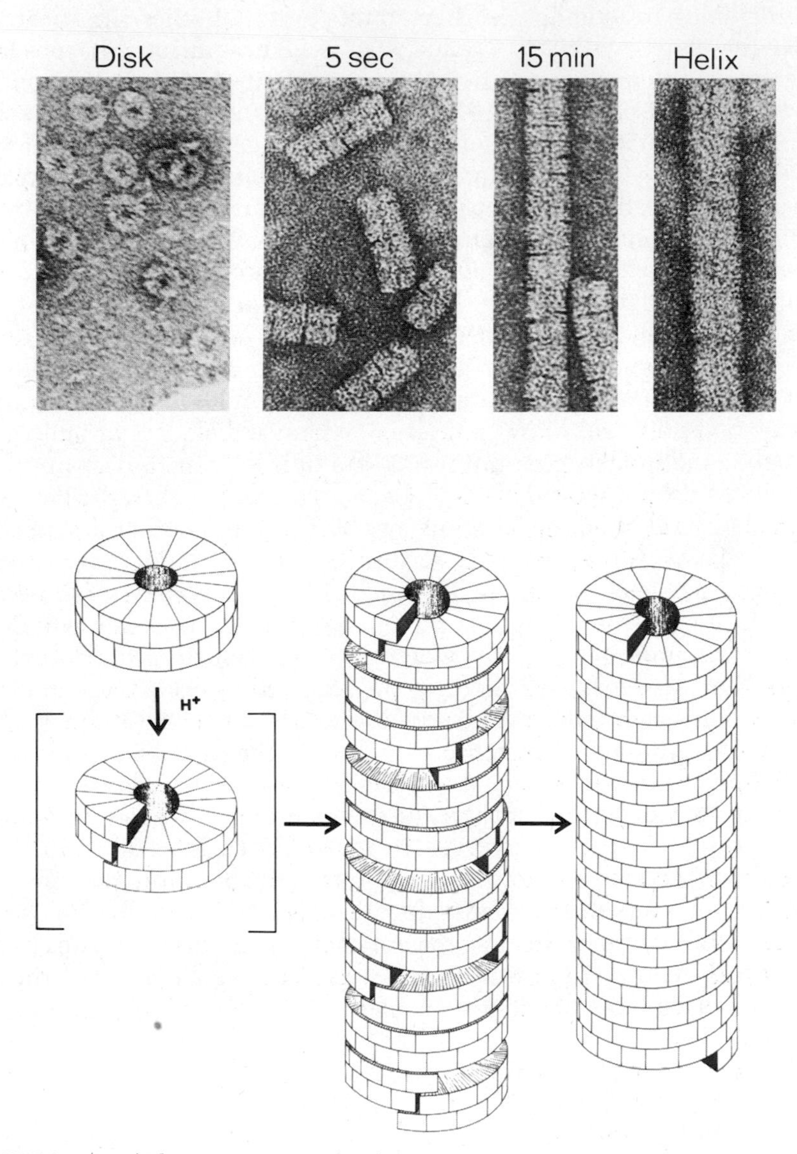

FIGURE 3. (Top) Electron micrographs of TMV protein disk preparation on pH drop. Initial disk preparation was diluted from pH 7 into pH 5 buffer (at 20°C) and specimens prepared at the times indicated. (Bottom) Schematic interpretation of the structures. The disks are "dislocated" into short helices, of just over two turns, without dissociation into subunits, and these stack together in random azimuthal orientation to give longer, imperfect or "nicked" helices. The imperfections anneal over a further period of hours to give continuous helical rods. Samples were negatively stained with uranyl acetate. Electron micrographs courtesy of Dr. J. T. Finch, adapted from Durham *et al.* (1971).

it remains unclear as to whether this involves a rapid equilibration between disk and protohelix. Protohelices are not closed structures and so can readily gain or lose subunits from their ends and probably do not maintain a fixed number of subunits.

Below neutral pH a rapid breakdown of disks to helices is observed. The helices are long and apparently perfect if the pH change occurs over several minutes. However, if the pH is dropped rapidly, then short, imperfect ("nicked") helices are observed (Fig. 3) with the nicks annealing out over a period of several hours (Durham *et al.*, 1971; Durham and Finch, 1972). Electron micrographs indicate that the initial disks become protohelices, which rapidly stack in a poorly ordered manner. These short rods stack further but still in imperfect register, giving a nicked appearance. The separation of the nicks supports the view that disks are converted directly to protohelices, without breakdown to subunits and reaggregation.

A similar picture has emerged for the nucleation of helical aggregates when these are assembled from A-protein. With a slow rise in temperature (at constant pH) to drive the aggregation, the resultant helices have a length distribution similar to that at equilibrium, whereas if the temperature is increased rapidly, then the lengths of the helices overshoot the equilibrium size. Analysis of the kinetics of cycling the temperature up and down shows the critical factor to be a limiting supply of nuclei for the helices, with very rapid elongation on all available nuclei (Schuster *et al.*, 1979). The nuclei, which are relatively stable even at low temperature and sediment at around 20 S, have been identified as disks or protohelices (Shire *et al.*, 1979a).

All of the preceding discussion of the rates of conversion of the several aggregates of TMV protein relates to reversibly equilibrating protein, even though some of the reactions may show hysteresis when the rate of conversion is slower than the rate of change of the controlling variable. However, disaggregated protein is readily proteolysed and the damaged molecules then will aggregate, according to the initial conditions, into a variety of forms which may no longer be readily disaggregated and hence the aggregation is no longer an equilibrium reaction (Durham and Finch, 1972).

B. Detailed Molecular Structures

1. Protein

The structure of the protein disk of TMV has been determined in several steps at successively higher resolution (Fig. 4). Electron micrographs of single disks viewed perpendicular to the axis revealed the presence of 17-fold rotational symmetry which became clear on applying suitable filtering to the image (Crowther and Amos, 1971). This confirmed

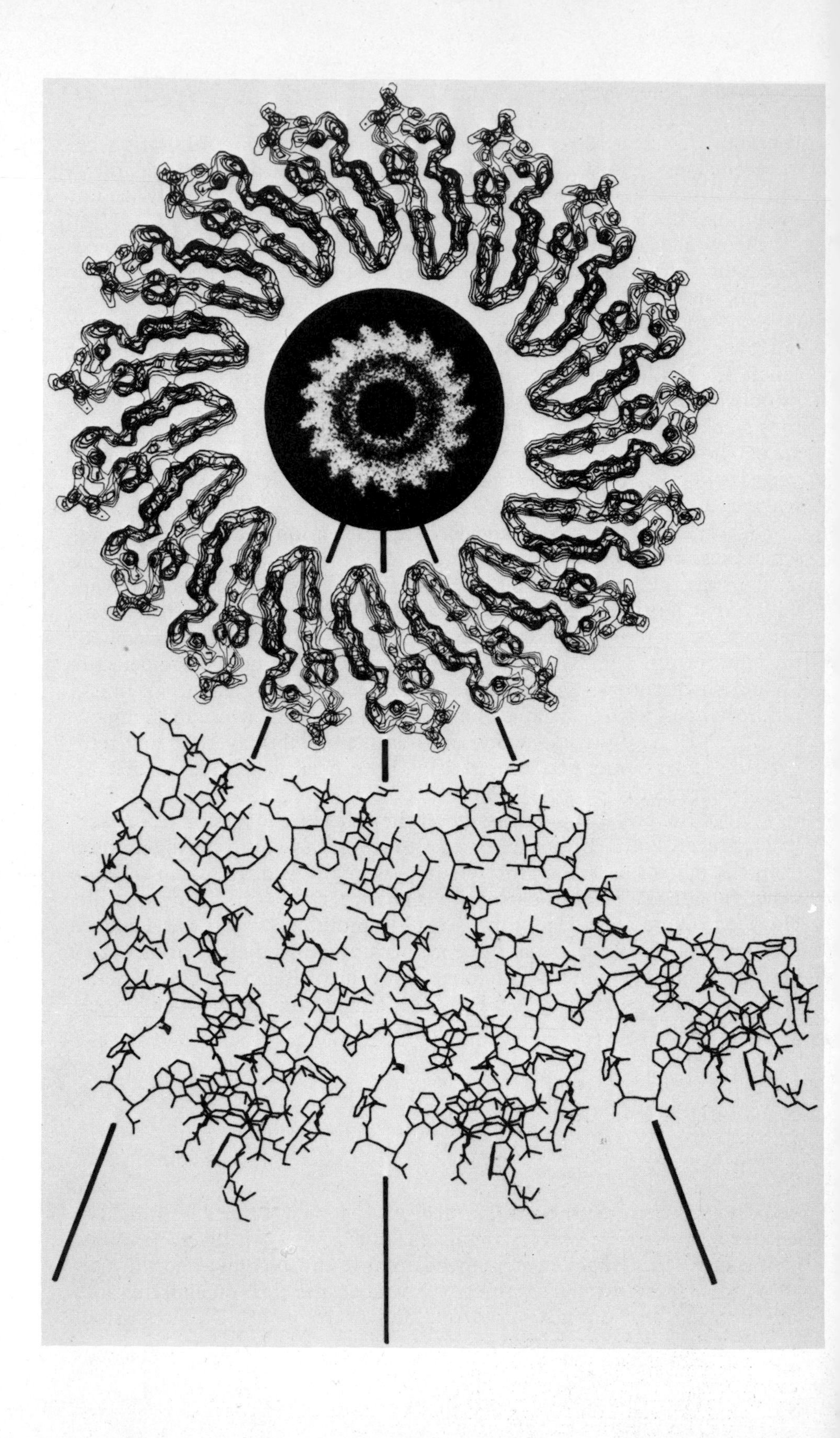

the results of preliminary X-ray studies (Finch *et al.*, 1966). Crystals of the disk have since been extensively analyzed by X-ray diffraction. At 1.5-nm resolution the subunit was seen as a curved, elongated molecule (Gilbert and Klug, 1974) which is slewed azimuthally relative to the radius of the disk. Later, at 0.5-nm resolution, individual α-helices were clearly seen within the subunit (Champness *et al.*, 1976). Then, at 0.28-nm resolution, a detailed atomic model was built of the protein subunit within the disk (Bloomer *et al.*, 1978). The atomic coordinates from this initial model have been refined crystallographically (Mondragon, 1984) to give the most accurate model so far available.

The α-carbon backbone of the subunits, represented as a ribbon, is shown for four subunits within one ring of the disk viewed from above (Fig. 5a) and for a vertical section through a disk (Fig. 5b). The central part of the subunit between 4 nm and 7 nm comprises a bundle of four α-helices. At their outer ends, these are joined by a short β-sheet around the circumference, beyond which lies an extended loop between the second and third α-helices and the amino and carboxy termini of the polypeptide, both of which lie on the outside of the subunit. At the inner end of the α-helical bundle, the first two helices are linked by a short loop (hairpin bend) whereas the two lower helices are joined by 24 residues for which there is no clear electron density visible in the crystallographic map. This lack of density for residues 90–113 could be due to either dynamic disorder (movement of the loop) or static disorder (differences in folding between the 17 subunits of a ring). However, NMR studies of mutant protein (Thr-107 replaced by Met) showed that the region around residue 107 was in rapid motion in A-protein or disks but not in the protein helix or the virus (Jardetsky *et al.*, 1978). This supports the notion of dynamic disorder within the central loop, which thus has no well-defined conformation within the disk. This is in contrast to the virus where this part of the molecule appears to be as well ordered as the rest. However, a preferred conformation for residues 90–92 and 107–113 has been determined during the crystallographic refinement (Mondragon, 1984).

The packing of the subunits within the disk and virus is compared in Fig. 6 from which it is seen that the lateral contacts within one ring of the disk or one turn of the helix are essentially the same in the two aggregates whereas the axial contacts between rings or turns are quite different. The lateral contacts are shown schematically in Fig. 5a as alternating patches of polar and hydrophobic interactions. The polar contacts include both hydrogen bonds and charged electrostatic interactions.

←

FIGURE 4. The protein disk aggregate of TMV viewed from above at successive stages of resolution. From the center outward are (1) a rotationally filtered electron microscope image at about 2.5-nm resolution (Crowther and Amos, 1971); (2) a slice through the 0.5-nm electron density map of the disk obtained by X-ray crystallographic analysis (Champness *et al.*, 1976); and (3) part of the atomic model built from the 0.28-nm map (Bloomer *et al.*, 1978).

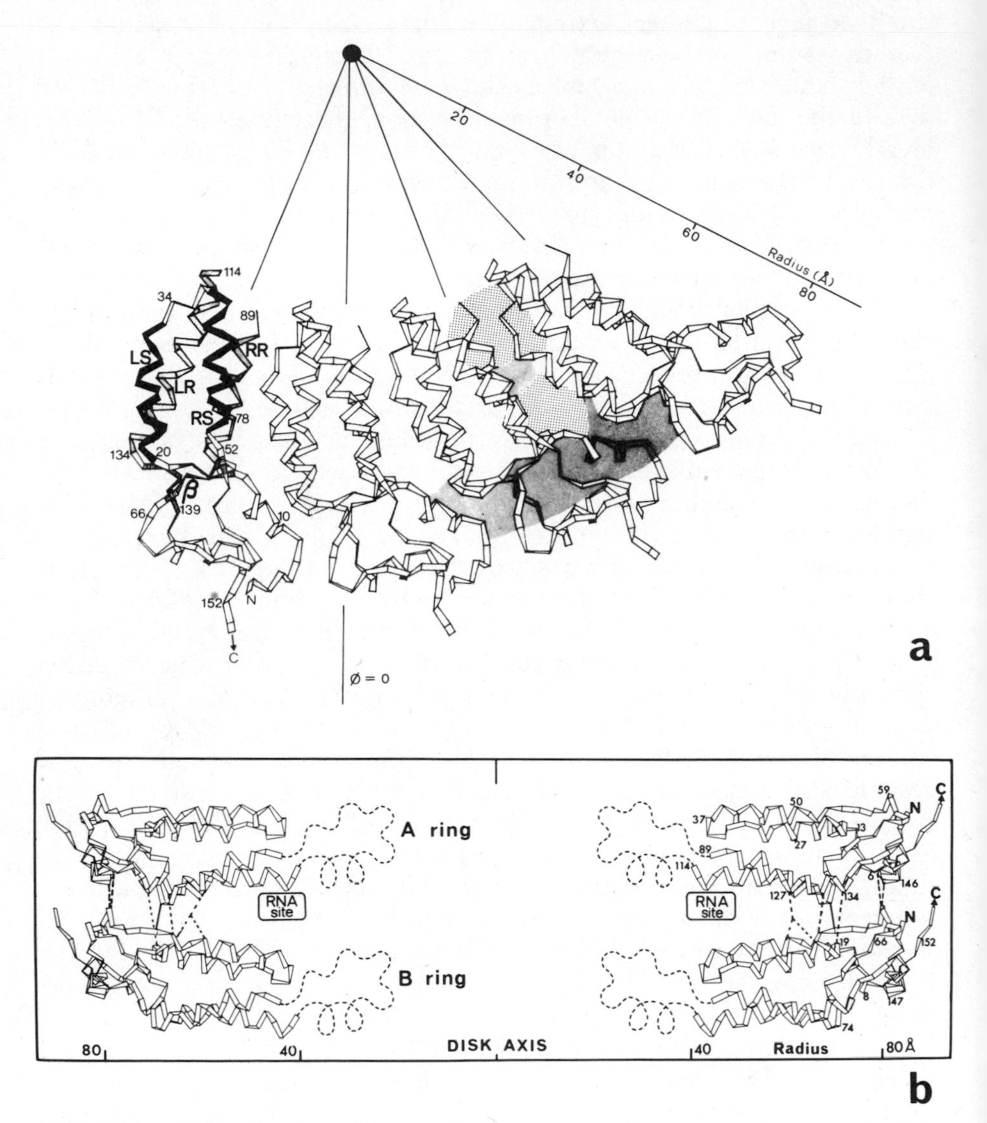

FIGURE 5. Ribbon drawing of the α-carbon positions in the protein subunits in the disk. Numbers refer to amino acid position in the primary sequence. (a) Four adjacent subunits in one ring viewed from above. Regions of α-helix (labeled LS and RS for left and right slewed and LR and RR for left and right radial, respectively) and β-sheet are indicated on the leftmost subunit, and the alternating patches of polar (light shading) and hydrophobic (heavy shading) contact are indicated in the interface. (b) Side view of subunits in an axial section through the disk showing vertical contacts between subunits in the two rings (labeled A and B). The flexible loop at the inner radius is shown dotted and the RNA-binding site is indicated. Adapted from Bloomer *et al.* (1978).

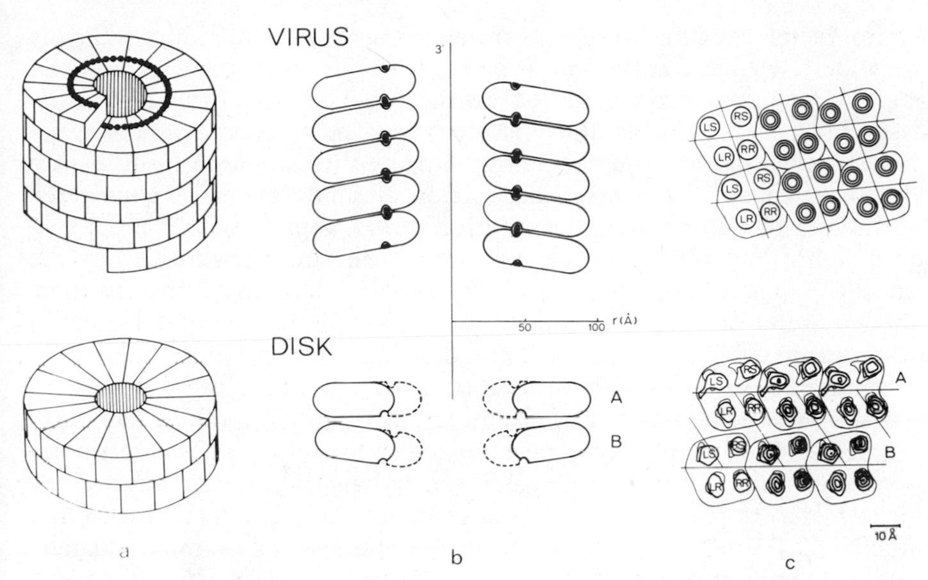

FIGURE 6. Packing of subunits in the TMV protein disk (below) and in the virus (above). (a) The disk contains 34 subunits in two rings of 17, and the virus, $16\frac{1}{3}$ subunits per turn of the helix. RNA is present in the virus particle, sandwiched between the helically arranged subunits as shown. (b) Subunits in the A (top) and B (bottom) layers of the disk have different tilts with respect to the axis, and subunits in the virus have a still different tilt. The innermost part of the subunit appears to be disordered in the disk (dashed outline), perhaps to facilitate incorporation of RNA in the transition from disk to helix that occurs during assembly, when this part of the protein becomes ordered around the RNA. (c) Cylindrical sections through subunits in the disk and in the virus, viewed end-on. The disk sections are from the 0.5 nm electron density map; the virus section is schematic. The major part of each subunit consists of four α-helical rods, running in a roughly radial direction, and denoted LS, RS, RR, LR; the sections have been taken through these rods. The figure shows that in the transition from a 17-fold disk to a $16\frac{1}{3}$-fold helix, subunits slide over each other by about nm. Because of the change in subunit tilt, lateral contacts will be bent somewhat, but not completely altered like the axial bonds.

Hydrophobic interactions are found not only in the lateral interface but also within each subunit. The core of the α-helical bundle is hydrophobic as is the region on the outside of the β-sheet. In this high-radius region, the hydrophobic contacts are continuous across both the width of a subunit and the interfaces with its lateral neighbors giving a "hydrophobic girdle" of interactions around the circumference of the disk.

The vertical contacts between the two rings of a disk are all at high radius (Fig. 5b) whereas there is a more extended axial interface in the virus (Fig. 6b). Here, molecules in successive turns of the helix are essentially parallel and the axial contacts can be repeated indefinitely (the length of the viral helix being determined solely by the length of RNA to be coated). Contrastingly, the two rings of subunits in a disk show quite different orientations with respect to the axis and this polar structure is thus limited to two layers, as the same pattern of axial contacts cannot be repeated for a further complete ring of subunits. The two layers are widely separated at low radius facing into the central hole from which there is easy access to the RNA-binding site. The role of the disk in the assembly of TMV, when RNA is entrapped between the two rings of protein subunits, is discussed below. During the transition from disk to virus, the changes in relative disposition of subunits can be considered as the sum of a circumferential rotation of 11°, approximately half a subunit width (Fig. 6c), and a rotation about the tangent which alters the tilt of the long axis of the molecule by 14° for the A ring of the disk (Fig. 6b). Any rotations about the other two molecular axes are much smaller (Mondragon, 1984) as are any internal deformations within the molecule.

Subunits from different strains of TMV, or from closely related viruses,* will frequently not coassemble (e.g., Sarkar, 1960) or they give aggregates with anomalous properties (Novikov *et al.*, 1974). On the assumption that the basic folding of the polypeptide is the same in all viruses closely related to TMV, these effects are seen to be the consequence of altered interactions between subunits. For example, changes in the lateral interface move the boundaries between the alternating patches of polar and hydrophobic contacts (Bloomer *et al.*, 1981). The hydrophobic contacts between subunits probably provide the driving energy for protein aggregation but the specificity (both spatially within an aggregate and between species) derives from the hydrogen bonding and charge neutralization networks, i.e., the hydrophilic interactions.

Structural analysis of the helical aggregates of TMV protein has shown that the flexible loop seen in the center of the disk is more ordered in the helix, but is still more mobile than in the intact virus (Mandelkow *et al.*, 1981). The other main difference is the freedom to form helices of either $16\frac{1}{3}$ or $17\frac{1}{3}$ subunits per turn. The control of this phenomenon is not understood, although it appears that in any single gel of the protein helix, all of the particles have the same number of subunits per turn (Mandelkow *et al.*, 1976).

2. Virus

Notwithstanding the technical difficulties involved in the analysis of X-ray diffraction data from ordered gels, the structure of the virus has also been solved in stages of successively increasing resolution covering first one (Barrett *et al.*, 1971) and then two overlapping Bessel function orders of diffraction (Holmes *et al.*, 1975). Most recently, three such functions were separated, leading to a map which included all data to 0.4 nm (Stubbs *et al.*, 1977). The quality of this map was much reduced toward

* Conforming to the nomenclature now recommended for the tobamovirus group, only very close relatives of TMV *Vulgare* are here referred to as strains of TMV. All of the others, which are here denoted as separate viruses, are commonly described as strains of TMV in both past and present literature.

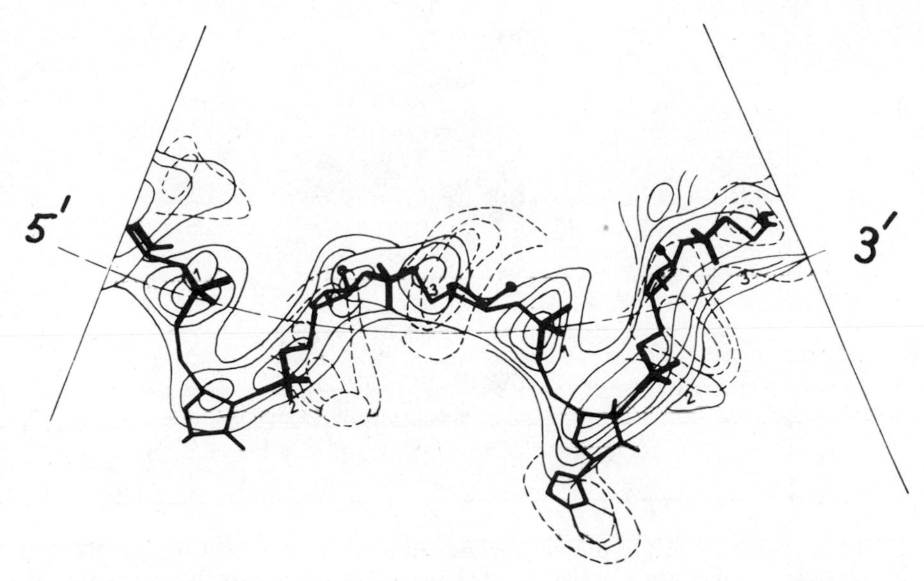

FIGURE 7. Configuration of the RNA in TMV. Two full repeats (six nucleotides) are shown with the 5′-end on the left, looking down the axis of the virus. The nucleotides are numbered 1, 2, and 3 at the phosphate groups (shown by heavy lines). Bases 1 and 3 are vertically above the plane of the drawing and are omitted for clarity. The electron density for the bases is shown by dotted contour lines. From Holmes (1979), adapted from Mandelkow *et al.* (1981).

the outside of the virus because of technical factors and so the atomic model fitted to the map was subsequently adjusted slightly (Holmes, 1979) in the light of the higher-resolution protein structure then available from the disk (Bloomer *et al.*, 1978). Considerable improvements to the analysis of the diffraction pattern have now been made (Namba and Stubbs, 1985) which should lead to a refined model in the near future.

The main differences between the protein subunits as now modeled in the disk and the virus are at low radius (below about 4 nm). The flexible loop of the disk becomes ordered with all subunits adopting the conformation identified as that preferred in the disk: residues 107–113 form the lower part of a vertical column and two additional turns of the α-helix which continues from residue 114 to residue 134 whereas residues 90–92 are more extended (Mondragon, 1984). The exact conformation of residues 93–106 remains ambiguous but the polypeptide chain must follow a vertical path for much of this region (shown schematically in Fig. 5b). The initial virus atomic model in which this vertical feature was an α-helix forming part of a cluster of carboxylate groups (Stubbs *et al.*, 1977) requires modification in the light of these more recent results.

The conformation of the RNA chain was identified first by the higher electron density of the phosphate groups (Stubbs *et al.*, 1977) and more recently by difference maps between the virus and protein helix (Mandelkow *et al.*, 1981). These allow the total density corresponding to the

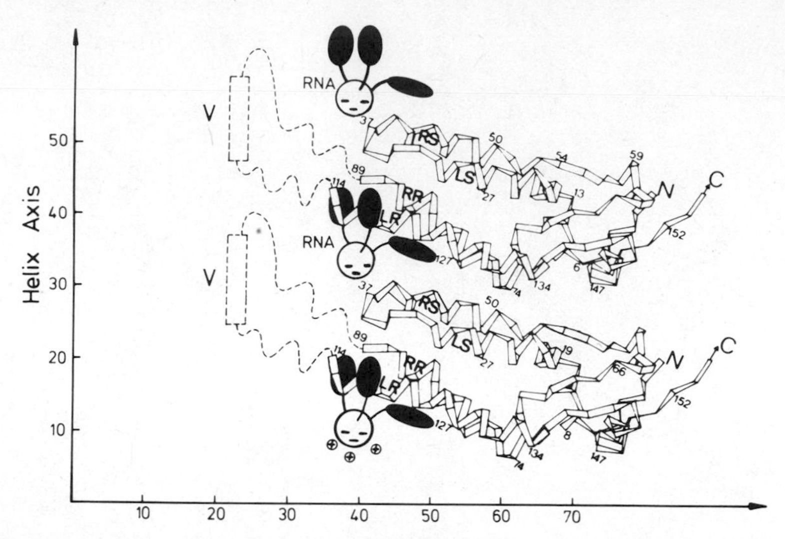

FIGURE 8. Side view of a section through two turns of the virus showing the three nucleotide bases wrapping about the left radial helix LR. The structure of the protein subunit is taken from Bloomer *et al.* (1978). Its position in the virus and the clawlike structure of the RNA are from Stubbs *et al.* (1977) and Holmes (1979). Adapted from Holmes (1983).

RNA to be identified and an atomic model has been built for the RNA, with an "average" nucleotide in each of the three positions (Stubbs and Stauffacher, 1981). This shows a relatively extended conformation of the ribose-phosphate backbone (Fig. 7) and the disposition of the three nucleotide bases around the left radial α-helix of the protein (Fig. 8). Interactions between protein and RNA can be deduced by combining the separate models. Arginine residues 90, 92, and 113, which are invariant in all known sequences of TMV and closely related viruses, probably interact with the three phosphate groups, neutralizing their charge. The three bases lie against the surface of the α-helix between residues 114 and 123 where the interactions are hydrophobic. Specific hydrogen bonds have been suggested, e.g., from the 2′-hydroxyl groups of the riboses in positions 1 and 3 to the invariant pair of aspartate residues 115 and 116 (Holmes, 1979). However, recent analysis (Mondragon, 1984; Mondragon *et al.*, 1985) suggests that such bonds are not unequivocal and the precise interactions remain to be determined.

3. RNA

The conformation of the RNA present in the virus (Mandelkow *et al.*, 1981; Stubbs and Stauffacher, 1981) has just been described (Figs. 7, 8). However, since the coat protein must be able to encapsidate all of the RNA while there are only three distinct binding sites on the protein, it is necessary for the interaction with most of the nucleotides to occur in a manner independent of the particular base sequence and yet the virus

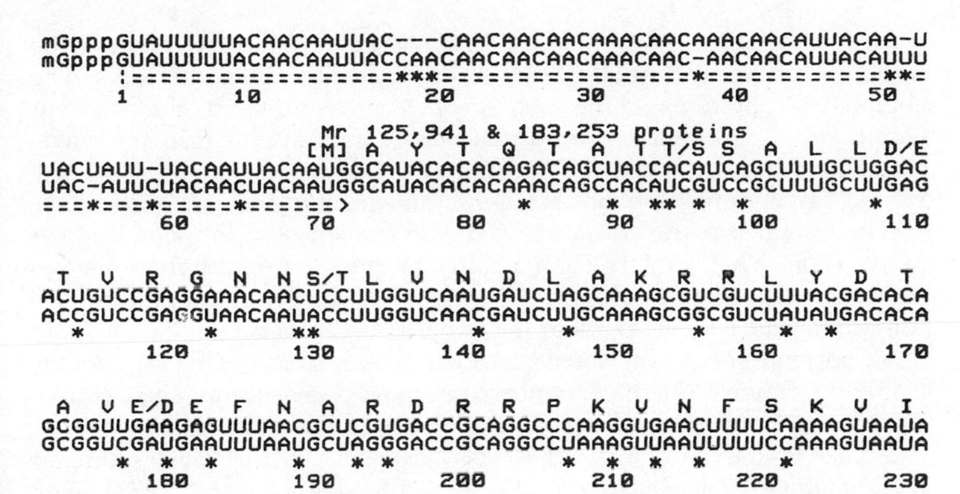

FIGURE 9. Nucleotide sequence of the 5′-end of TMV RNA. The sequences of approximately 290 nucleotides are shown for each of the two paradigm RNAs, as found by Goelet *et al.* (1982) in their bulk preparations of RNA, aligned to show the insertions or deletions and with asterisks under nucleotide differences. The 5′-nontranslated region is underlined and the codon translation is given above the RNA sequence in the single-letter code. All insertions or deletions are outside the coding region, within which most substitutions do not affect the coding capacity of the RNA because of redundancy in the genetic code. The four amino acid differences are all conservative mutations. From Goelet *et al.* (1982).

does have a distinct nucleotide sequence. Determination of the complete sequence (Goelet *et al.*, 1982) followed the partial sequences of the origin of assembly (Zimmern, 1977; Jonard *et al.*, 1977), the 5′-end (Richards *et al.*, 1978; Jonard *et al.*, 1978), and the 3′-end (Guilley *et al.*, 1979), and even earlier work to identify the six nucleotides at the 3′-end (Steinschneider and Fraenkel-Conrat, 1966). These sequences were all determined for the *Vulgare* strain of TMV. The sequence is also known for the related tomato mosaic L virus (Ohno *et al.*, 1984) and partial sequences are available for other strains and related viruses (Meshi *et al.*, 1982, 1983a).

One question which arises from the complete sequence is the extent of polymorphism present within the RNA population. Ohno *et al.* (1984) observed no sequence heterogeneity within the tomato mosaic L virus except in the length of the adenine cluster near the 3′-end (which is not present in TMV). However, from 400 clones, each independently derived from their bulk preparations of *Vulgare* TMV RNA, Goelet *et al.* (1982) found some sequence diversity. There were many changes in the "wobble" position of codons throughout the genome, but these have no effect upon the coding for proteins. The variation was especially concentrated toward the 5′-end with two distinct families of sequence (Fig. 9). These were identified by Meshi *et al.* (1983b) as being homologous with the Japanese common strain of TMV (OM) and tomato mosaic L virus. The coat and 30K proteins for these two viruses are significantly different but Goelet *et al.* (1982) detected no major polymorphism within the corresponding RNA sequence. The sequencing technique which they used, relying largely upon random sequence nonspecific primers, would have detected such sequence variation if it were present within the bulk RNA preparation. They explain their results by postulating a hybrid species rather than, as suggested by Meshi *et al.* (1983b), a contamination with tomato mosaic L virus. While it may appear a surprising coincidence that most polymorphism was detected to the 5′-side of a specific sequencing primer common to both sequences, one would expect any such recombination to have occurred at a region of maximum homology.

The 5′-end of TMV RNA is "capped" having the sequence $m^7G^{5'}ppp^{5'}Gp$ (Zimmern, 1975; Keith and Fraenkel-Conrat, 1975). Such capping is typical of a eukaryotic mRNA. This is an essential early function of TMV RNA and is necessary for infectivity, as removal of the cap leads to loss of infectivity (Ohno *et al.*, 1976). The removal did not, however, have a major effect on the reassembly of RNA with TMV coat protein, although clearly the packaging of RNA at the extreme 5′-end of the particles may have been different.

III. SELF-ASSEMBLY*

Understanding of the assembly process for TMV has grown considerably since the pioneering studies of Fraenkel-Conrat and Williams (1955) which demonstrated that self-assembly occurred *in vitro* from the separate protein and RNA constituents. Detailed aspects of the process are considered here for the nucleation and elongation stages separately. The definition of these two stages differs among those working in the field and so, while their descriptions of the stage under investigation are adopted here without change, attention is drawn to the instances where confusion may have arisen within the literature. Considerable debate has focused around several details of the nucleation and elongation stages and, especially in the latter case, this remains unresolved.

A. Nucleation

1. Requirement for Disks

The classic demonstration of the assembly of TMV from its protein and RNA components (Fraenkel-Conrat and Williams, 1955) incubated

* This topic is also discussed in Chapter 14.

the reassembly mixture "at 3° for at least 24 hours" under conditions of undefined ionic strength and pH about 6 to 7. In later experiments, the conditions were varied so as to define the optimum (Fraenkel-Conrat and Singer, 1959); 0.1 M pyrophosphate buffer was superior to 0.1 M phosphate and the optimum pH was 7.3. Under these conditions, reassembly was essentially complete in about 6 hr at 25°C and 3 hr at 30°C starting with A-protein and RNA. This rate is very slow for the assembly of a virus but is similar to the rate of aggregation of TMV A-protein into disks.

The nucleation step is a major problem in the formation of any of the larger aggregates of TMV. For the virus, this requires at least 17 subunits to be aligned next to each other along the RNA strand before there is any possibility of axial interactions between subunits in adjacent turns of the viral helix. These interactions would supplement the lateral contacts which, until the first turn of the virus has been completed, are the only possible bonds between the protein subunits. The second and subsequent turns of the helix would thus be expected to form more rapidly than the first turn, the completion of which could well be kinetically very slow. The protein disk of TMV does, however, provide an ideal candidate for the role of former or "jig" upon whose surface the protein and RNA could aggregate to overcome the kinetic block. Butler and Klug (1971) tested this by adding a disk preparation to the reassembly mixture.

The effect of adding disks to the mixture of RNA and A-protein was to increase the rate of assembly by more than an order of magnitude. Electron micrographs taken after 1 min showed many growing rods but no full-length ones, suggesting that nucleation was occurring quite rapidly and that the remainder of time before completion of assembly was primarily for elongation of these partially assembled rodlets rather than for further nucleation to give an increasing number of particles.

The surprising feature of the reassembly using disk preparations was that disks had to be added at similar concentrations to those of the A-protein and also that omitting the extra A-protein entirely had no effect on the rate of assembly. A-protein was of course still present in the reassembly mixture as it forms about 20% by weight of an equilibrium disk preparation. The reaction had proceeded to such an extent that significant amounts of disks must have been involved and measurements of the residual free protein in the analytical ultracentrifuge showed depletion of the 20 S boundary but little change in the amount of 4 S material.

From these results, Butler and Klug (1971) concluded that disks were not only required for the nucleation reaction but were actually incorporated during this step rather than simply acting as a former. It also appeared that significant elongation could occur using disks as the main source of protein, though it was not then clear whether or not the disks were acting simply as a "delivery package." The requirement of disks for nucleation was confirmed by other groups (Okada and Ohno, 1972; Richards and Williams, 1972) though their role in any subsequent elongation continues to be a topic for discussion (see Section III.B.2). The much slower rates of assembly found in the earliest reassembly experiments are thus probably due to the protein being supplied purely as A-protein, with consequent delay before sufficient disks would form for the nucleation.

2. Specificity of Interaction with RNA

The assembly of TMV protein into nucleoprotein rods was found to be selective for the RNA species supplied (Fraenkel-Conrat and Singer, 1959, 1964). The most efficient reassembly was with homologous RNA and also the homopolymers poly(A) and poly(I), whereas heterologous RNAs from other viruses closely related to TMV assembled less efficiently if at all, and other homopolymers and yeast RNA failed to assemble. With a disk preparation as the source of protein for assembly, Butler and Klug (1971) showed that the homologous RNA nucleated assembly more than an order of magnitude faster than either poly(A) or poly(I). The homopolymer reaction is different from that with homologous TMV RNA as nucleation on poly(A) was found to show second-order dependence on protein concentration, suggesting two disks interacting per RNA molecule (Butler, 1972), whereas nucleation on TMV RNA shows first-order kinetics, requiring only a single disk in the rate-limiting step (Butler, 1974a).

Selectivity for homologous RNA is a biological requirement of the assembly process. The nucleation step is the most efficient point for this to occur because it prevents protein being used up abortively coating RNA molecules nonspecifically. Even within the homologous RNA, selectivity is essential to avoid the production of defective particles due to the packing fault which would arise from nucleation at two sites on the same RNA molecule.

The site for nucleation of assembly on TMV (*Vulgare*) RNA was identified by Zimmern and Butler (1977), who reassembled very limited amounts of a disk preparation with TMV RNA and then removed the uncoated, and unprotected, RNA tails by nuclease digestion before reisolating the protected RNA fragments. These consisted of a family of overlapping fragments, all with a common core sequence but extended by different amounts at the two ends. These fragments would rebind to disks rapidly, quantitatively and with high affinity, so becoming more nuclease-resistant. This rebinding would not occur with A-protein, thus suggesting that the fragments all contained the origin of assembly which must be within the common core.

The nucleotide sequence of the fragments was determined (Zimmern, 1977) and the minimum protected sequence found to be about 55 to 65 nucleotides in length, consisting of a sequence capable of forming a weakly base-paired stem with a loop of sequence AGAAGAAGUUG-UUGAUGA (Fig. 10). The occurrence of G in every third position looks significant in view of the known binding of three nucleotides per protein

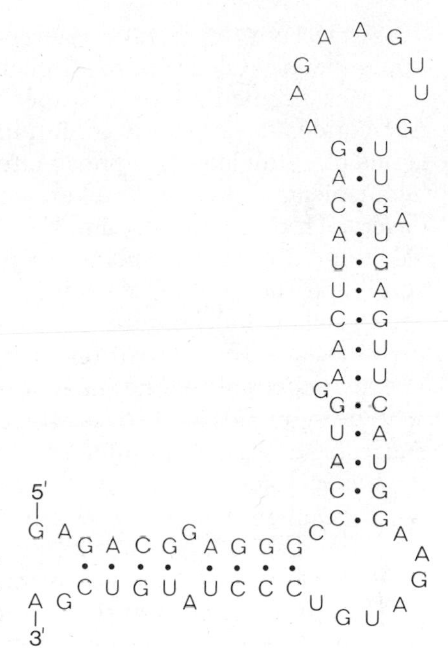

FIGURE 10. Nucleotide sequence and possible secondary structure of the origin of assembly on TMV RNA. Adapted from Zimmern (1977).

subunit. The nucleation region was located between 900 and 1300 nucleotides from the 3′-end of the RNA by Zimmern and Wilson (1976) who looked for the presence of characteristic nucleotides in 3′-fragments of TMV RNA. The nucleotide sequencing of TMV has since confirmed that the origin of assembly is just over 900 nucleotides from the 3′-end (Guilley *et al.*, 1979; Goelet *et al.*, 1982).

Notwithstanding the very high specificity of nucleation on intact TMV RNA, this is partially lost when the RNA is degraded into shorter fragments. After digestion of TMV RNA by ribonuclease A into fragments of about 3 S, many of these will bind to disks (Tyulkina *et al.*, 1975), though with only half the normal ratio of RNA to protein. A similar low ratio is also found for some of the fragments generated with ribonuclease T1 (Jonard *et al.*, 1975). Sequencing of these fragments, which do show some specificity in binding, revealed them to be derived from within the coat protein cistron (Guilley *et al.*, 1975a,b). When the true nucleation region was sequenced, these fragments were found to have some homology with it.

The existence of a region within the coat protein cistron with both favorable binding to the disk and some homology to the nucleation region on TMV *Vulgare* RNA led Zimmern (1977) to suggest that this might act as an alternative origin of assembly in some viruses, especially cowpea mosaic virus. This has been confirmed for both sun-hemp virus (Fukuda *et al.*, 1980) and cucumber green mottle mosaic virus (Fukuda *et al.*, 1981). In both cases nucleation occurs at about 300 nucleotides from the 3′-end.

This selectivity probably depends on the coat protein as well as on the nucleotide sequences of possible assembly origins.

Measurements have also been made of the binding of short oligonucleotides to TMV protein. Binding constants have been measured for a series of trinucleotide diphosphates to the protein at neutral or acid pH (Schuster *et al.*, 1980; Steckert and Schuster, 1982). Weak binding was observed for AAG, CAG, and UAG at pH 7 although far stronger binding was found in the range pH 5 to 6 (where the protein will be in the helical state) and the binding constants were calculated for this lower pH. All the trimers which bound contained a G residue and binding was strongest if this was in the 3′-position of the trimer. The binding constants were not related simply to the nucleotide composition of the trimer but to the specific sequence within the trimer. Although an attempt was made to suggest the "phase" for binding of protein to the nucleation region (Steckert and Schuster, 1982), the conclusions are logically suspect and, anyway, are derived from binding to helical protein whereas nucleation occurs with the disk and it is not known if the same relationships would hold in both the helical and disk forms of the protein.

The binding constants of short oligonucleotides to protein disks of TMV have been determined in our laboratory. When disk preparations were used as the only source of protein, at pH 7 and ionic strength 0.1 M, the binding of the hexanucleotide AAGAAG was some orders of magnitude stronger than any binding observed at the lower pH by Schuster. The binding constant for this nucleotide to the disk preparation was determined as 1.5 μM and the saturation level was equivalent to between eight and nine hexanucleotides bound per disk (Turner *et al.*, 1986). Preliminary studies with other nucleotides indicate that the nonamer AAGAAGUUG binds slightly more strongly but with a lower saturation level. Binding of the trimer AAG is much weaker than that of the hexamer and also than that observed by Schuster at lower pH. When the equivalent deoxynucleotide was used with the disk preparation, the binding of d(AAGAAG) was scarcely detectable. In similar experiments but with the protein supplied in the form of crystals of the disk, the binding of the ribonucleotide AAGAAG was very greatly reduced in comparison to that of the same nucleotide to the disk preparation in solution.

It is known from crystallographic studies of the disk (Bloomer *et al.*, 1978) that there should be unhindered access, even in the crystal, of nucleotides to the binding site. Thus, the crystal lattice forces appear to be stopping all but the initial binding of hexanucleotide to the crystalline disks, perhaps by preventing a change in molecular conformation or packing which would otherwise occur in isolated disks.

3. Mechanism

The structure of the protein disk, with two rings of subunits spaced widely apart toward the central hole of the disk (Fig. 5b), together with

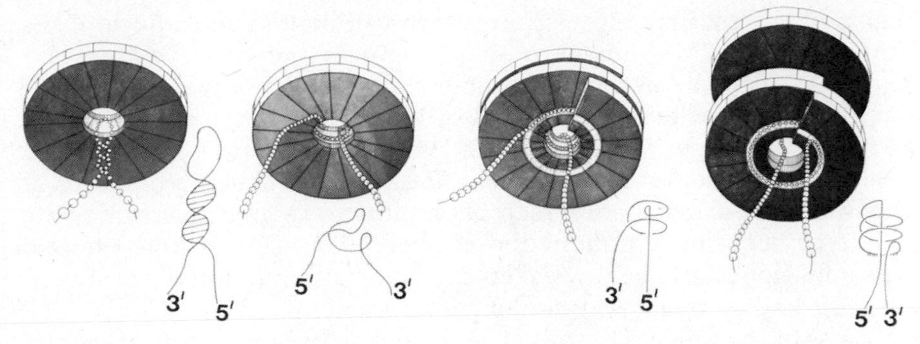

FIGURE 11. Possible mechanism for nucleation of assembly of TMV. The protein and RNA are drawn diagrammatically and the expected configuration of the RNA backbone at each stage is shown. Details are discussed in the text. From Butler (1984).

the possible secondary structure of the origin of assembly on TMV RNA (Fig. 10), suggested that nucleation might occur by insertion of this RNA hairpin loop into the central hole of the nucleating disk (Butler *et al.*, 1976). Figure 11 shows the stages in this proposed mode of nucleation. Following insertion of the nucleation loop into the hole of the disk, the single-stranded RNA at the tip of the loop binds between the rings of the disk, at an RNA-binding site which forms around it from the subunits both above and below. The weakly base-paired loop of RNA then melts to permit binding of a complete turn of RNA, about 50 nucleotides. During this interaction, some feature of it could cause dislocation of the disk into a protohelix, with the accompanying closing together of the two layers of subunits. There is a major rearrangement of the protein molecules during which the RNA is entrapped between the adjacent layers of molecules.

This proposed mechanism predicts that both tails of the RNA should be at the same end of the growing rod, notwithstanding nucleation occurring internally on the RNA. This was indeed found to be the case and, moreover, it was shown that the longer 5′-tail was the one looped back down the length of the rods (Butler *et al.*, 1977; Lebeurier *et al.*, 1977).

During nucleation the rods tend to grow rapidly until about 300 nucleotides are encapsidated. If only a limited amount of protein is supplied, these rods tend to dominate even at the expense of leaving some RNA molecules completely unnucleated (Zimmern and Butler, 1977; Zimmern, 1977). This suggests that there may be other features of the RNA in this region, in addition to the loop of the actual origin of assembly, which facilitate the interaction with further disks after the rate-limiting step involving the first disk (see above). Zimmern (1983) observed that, in the nucleation region of TMV and all strains and related viruses for which the sequence is known, two further weakly base-paired stems and loops occur to the 5′-side of the origin of assembly, spaced by 75 to 80 nucleotides. A potentially similar structure can also be predicted in the region of the coat protein cistron which can bind specifically to disks. Furthermore, a TMV mutant, which is temperature-sensitive for assembly (Ni 2519), has been shown to nucleate at nonpermissive temperatures simultaneously at both the normal origin and the extra location in the coat protein cistron (Taliansky *et al.*, 1982; Kaplan *et al.*, 1982). Sequence analysis of the nucleation region of Ni 2519 RNA shows a mutation in the third base-paired stem which back mutates to give reversion to the wild-type behavior (Zimmern and Hunter, 1983). One obvious effect of this mutation would be to weaken the base-pairing in this third stem of Ni 2519 until it might well no longer be stable.

Loss of the third stem thus appears to decrease the selectivity for the nucleation region until other favorable regions can compete equally with it, with lethal consequences for assembly of the virus. One proposal is that these equally spaced loops may serve to guide one or two further disks into interaction with that already bound to RNA in the origin of assembly (Zimmern, 1983), though the mechanism for this is not clear. However, it cannot be by the interaction of the single-stranded loop with the protein subunits in a strictly analogous manner to the nucleation, as the packing would become topologically impossible without later dissociation of this binding. One possible mechanism would involve the loops binding loosely to disks, holding them in a stack which might even be responsible for facilitating the dislocation of the first and nucleating disk. If this were a kinetic effect, it could explain the role of the loops in providing specificity without contradicting the observed rebinding of RNA fragments lacking one or both of these loops (Zimmern and Butler, 1977).

B. Elongation

1. Growth is Bidirectional

A consequence of the internal location of the nucleation region (Zimmern and Wilson, 1976) is that elongation must occur in both directions. There are about 900 nucleotides to the 3′-side of the nucleation region and 5500 to the 5′-side. There remains some debate as to whether the coating of these two tails is separated in time (i.e., "biphasic") or occurs concurrently. The resolution of this is complicated by the use of different criteria for studying elongation. Thus, Okada and his colleagues have used nuclease-resistant infectivity and protection of the RNA termini from nuclease digestion—both of which will actually measure the *finishing* of the particle—while Butler and his colleagues have measured the physical elongation and the progress toward longer nucleoprotein helices containing a greater length of RNA.

Butler and his colleagues have shown that the 5′-tail is substantially coated within about 6 min, such that about 15 nucleotides/sec must be encapsidated in the major direction under optimal conditions. With a substantial excess of disk preparation present during reassembly, Butler and Finch (1973) observed rods in the size class 260 to 300 nm after about 6 min. In more recent experiments reported from Okada's laboratory (Fukuda *et al.*, 1978), full-length rods were not seen for 30 to 50 min whereas rods of up to 260 nm were seen after 5 to 7 min. The difference between these two results may be explained by the relatively low resolution of the earlier histogram of rod lengths, which could fail to show such an apparent stop at 260 nm to rapid elongation, although there are other differences between the conditions for the two experiments. In support of the earlier observation, the RNA protected after 6 min of reassembly did contain a fraction which was indistinguishable in gel electrophoresis from full-length viral RNA (Butler and Lomonossoff, 1978). Despite the different interpretations, it is clear that protection of *most* of the RNA towards the 5′-end from the nucleation region occurs within only a few minutes.

In the earliest measurements where two tails could be distinguished (Lebeurier *et al.*, 1977), no shortening of the shorter (3′) tail was observed while the longer tail was considerably reduced. It was thus suggested that elongation was occurring in only one direction in the first phase of assembly. However, the measurement of lengths of single-stranded RNA in the electron microscope is very difficult and the attainable precision must render this conclusion open to question (Butler *et al.*, 1977).

The conclusion that no elongation toward the 3′-end occurs until the 5′-tail is completely coated, was also reached by others (Otsuki *et al.*, 1977; Fukuda *et al.*, 1978). However, the use by the Tokyo group of both TMV strain OM (a common strain in Japan) and high ionic strength (about 0.25 M) makes these results not directly comparable to those obtained by other groups from TMV *Vulgare* under other conditions (ionic strength 0.1 M). It is also doubtful whether the experiments described in either of these papers would have detected a relatively slow elongation toward the 3′-end. In the earlier work the elongation was observed by "electron microscopic serology" which the authors state is imprecise for quantitative measurements (Otsuki *et al.*, 1977). Fukuda *et al.* (1978) used various methods but most experiments measured nuclease-resistant infectivity or protection of the extreme termini from nuclease. These results will thus be susceptible to any final steps or "finishing events" which could be slowing encapsidation of the extreme ends of the RNA; e.g., incorporation of the final turn at each end of the virus helix where protein interacts with only one turn of RNA and in particular of the distinctive structure of the cap at the 5′-end. Lomonossoff and Butler (1979) pointed out logical inconsistencies in these results which lead to questions on their precision. The other experiments involved length measurements of rods in the electron microscope in which rods reached a length of only 260 nm within 5 to 7 min and then elongation appeared to stop for about 15 min. This was interpreted as slow elongation toward the 3′-tail occurring only after completion to the 5′-end. However, a rod length of 260 nm corresponds to uncoated RNA about 850 nucleotides long and, if all of this were located at the 3′-end, the protection in this direction would have reached less than 50 nucleotides beyond the origin of assembly. This interpretation is inconsistent with the results of Lomonossoff and Butler (1979) who measured the coating of the 3′-tail directly.

Lomonossoff and Butler measured the protection against nuclease of characteristic nucleotides, distributed on each side of the nucleation region, during a time course of elongation. This revealed that, while elongation in the 5′-direction was substantially faster, nucleotides on the 3′-side were being protected before those on the 5′-side were totally coated and that several hundred nucleotides of the 3′-tail were protected within 5 min (Lomonossoff and Butler, 1979).

Recently, experiments were performed employing RNA which had been hybridized and cross-linked at specific sites to cloned cDNA (Fairall *et al.*, 1986). Such reaction blocks the reassembly at the region of hybridization and thus allows the effect of a 5′-tail, which cannot be fully encapsidated, to be determined. The results showed that even though this tail was still looped back through the central hole of the rod, elongation had proceeded along most, if not all, of the 3′-tail.

These experiments therefore show that elongation can occur in both directions simultaneously, but is faster in the 5′-direction. This is compatible with the relative lengths of the RNA tails and, for maximum efficiency of RNA packaging, a fivefold difference in rates would be expected. Moreover, it does mean that any measurement of the overall rate of elongation will be dominated by that in the 5′-direction such that measurements in the 3′-direction are only possible with specially designed techniques.

2. 5′-Direction

Unlike the wide agreement on the requirement for disks in nucleation, their possible role in elongation remains subject to debate. When Butler and Klug (1971) first showed the involvement of disks in assembly and observed that these were used up stoichiometrically, they concluded that disks must be active in the elongation process. This unexpected point was emphasized in their discussion of elongation to the exclusion of any comment about the possible role of A-protein, though it was always considered likely that A-protein could add directly under appropriate conditions (e.g., Butler, 1972). The fact that it can do so does not preclude addition from disks, if these are present, either as a protein source or as a delivery package which breaks down at the site of, or even during, the actual incorporation of protein into the growing helix. The initial omission of any discussion of A-protein was taken by others as implying that Butler and Klug thought that A-protein could not participate in the elongation (Richards and Williams, 1972; Okada and Ohno, 1972; Ohno *et al.*, 1972) and these groups demonstrated that elongation could occur with A-protein as source, concluding that it was the sole source of protein. However, Richards and Williams (1973) have since noted that the ability of A-protein to add does not preclude the involvement of disks, but others have shown again that A-protein can add and concluded that disks are therefore not involved (Ohno *et al.*, 1977; Fukuda *et al.*, 1978; Fukuda and Okada, 1982). In contrast, it has repeatedly been shown that elongation can occur more quickly in the presence of disks than of A-protein alone (Butler and Finch, 1973; Butler, 1974a,b) and that the resulting rods are both more infective and more nuclease-resistant (Lebeurier and Hirth, 1973), hence supporting the hypothesis that disks do have a role in elongation.

Part of the disagreement may well arise from the different starting materials and conditions used. Most groups have standardized conditions using TMV *Vulgare* and working at 20°C, pH 7.0, and ionic strength 0.1 M. However, Okada's group works under conditions that differ in several respects. Firstly, they use TMV OM strain and they also perform heterologous reactions with the protein from cucumber green mottle mosaic virus (CGMMV). The choice of a TMV strain could be significant since even a single mutation is known to render the protein nonfunctional and alter dramatically its aggregation properties: for example, PM6 protein discussed by Hubert *et al.* (1976), and the mutant Ni 2519 (discussed above for the nucleation) which even shows altered interaction between *Vulgare* protein and a mutant of the RNA. Secondly, and probably far more significantly, the Tokyo group frequently works at 25°C with ionic strength in the range 0.25 to 0.5 M, although it is known that both temperature and ionic strength markedly affect the aggregation of TMV protein: for example, it has been shown that raising the temperature or ionic strength causes the 20 S disks to stack into 28 S aggregates (Durham and Finch, 1972) and that these interconvert only with difficulty into protein helix. The likely effect of these aggregates on the rate of assembly and probably also its pathway, was pointed out by Butler and Klug (1973) who found that assembly was much slower than under the standardized conditions. Recent experiments by Schön and Mundry (1984) report very different behavior and rates of assembly under "high salt" conditions (pyrophosphate, pH 7.25, ionic strength 0.5 M) and under the standardized conditions.

A further complication in assessing the relationship between the results and conclusions from different laboratories is that Okada's group has often followed the assembly by measurements, after nuclease treatment, of infectivity or of protection of the RNA termini. As discussed above, these assays will be strongly influenced by any "finishing events" which might occur in packaging the RNA and thus may well not be measuring reliably the main process of physical elongation, though they are indeed excellent assays for the completion of assembly. For example, measurement of the protection of full-length RNA by electrophoresis in polyacrylamide gels is complicated by the observation of an anomalous behavior with increasing concentrations of RNA (Asselin and Zaitlin, 1978).

Measurement of the kinetics of assembly (Butler, 1972, 1974a) showed that the rate of reaction saturates with increasing concentration of disk preparation, just as in an enzyme–substrate reaction. The protein concentration for half-maximal rate is about 1 mg/ml disk preparation. Following the demonstration that the elongating particles contain a complex RNA configuration (Butler *et al.*, 1977; Lebeurier *et al.*, 1977), it became clear that the major RNA tail (i.e., the 5′-tail) might have a special (and possibly labile) structure. On the plausible assumption that the *in vivo* system will follow the most advantageous pathway, it thus seems appropriate to rely on only those kinetic measurements which were made under conditions where rapid growth was occurring, so as not to allow time for loss of any special structure. A compilation of such measurements (Butler and Lomonossoff, 1980) showed average rates of elongation of 6.5 subunits/sec from disk preparations and 2.1 subunits/sec from A-protein. These figures are derived from several different techniques, at different times, and the standard deviations of both are less than 15%, suggesting that they are good estimates of the actual rates. The measurements from A-protein utilized prenucleated rods. The results thus indicate that, under optimal *in vitro* conditions, elongation from the disk preparation is about three times faster than from A-protein alone. Only under other conditions, with slower elongation, do the rates of growth from the two different sources become more nearly equal.

Two different approaches have been followed in an attempt to avoid the difficulties inherent in arguing from kinetics. In one the RNA was added to A-protein–disk preparation mixtures and the residual free protein species were measured, after allowing reaction (Shire *et al.*, 1979b, 1981; Schuster *et al.*, 1980). In all of the experiments the protein was simply described as "4 S" or "20 S" and there was no distinction, even where this was possible, between the different species in the same sedimentation band: e.g., disks and protohelices both occur at around 20 S (Schuster *et al.*, 1980), though it is, in our opinion, unlikely that protohelices could participate directly in any assembly reaction, if the subunits are already in their "final" positions just as in an infective virus particle (see Fig. 6c). It is unlikely that disks would exist under the conditions of the first set of experiments (pH 6.5, ionic strength 0.1 M, 6.5°C), where any 20 S material is probably present as protohelices (Durham, 1972), so it is perhaps not surprising that preferential addition of A-protein was found (Shire *et al.*, 1979b). However, in the second set of experiments (pH 7.0, ionic strength 0.1 M, 20°C), where disks are a significant component of the 20 S material, almost all of the protein could come from this source. Measurements were made of the fraction of protein depleted from each size class at different starting ratios of the classes. The quantification is complex because with the known mixture of species sedimenting at around 20 S it is unclear what the actual amount is of any potentially active species present, but this will be less than the total 20 S. This is important since only a 20% excess of protein was present, so that at lower 20 S/4 S ratios there may be an insufficiency of any of the active species in the 20 S material to supply much protein for the elongation.

Despite the problem of quantification, two clear observations do emerge: first, the protein source for over 90% of the reassembly *can* come from the 20 S peak; and second, at least 25% of the protein *must* come from the 20 S material, corresponding to a coating of over 1500 nucleotides for which the delivery package must be the disk breaking down, if at all, only during the actual interaction with the growing rod. Although Schuster and his colleagues chose to interpret all of this as nucleation (Schuster *et al.*, 1980; Shire *et al.*, 1981), this seems to be a semantic issue, owing much to a residual view that disks are involved only in nucleation. A different definition of nucleation is the formation of the product of the rate-limiting step or else the first stable ribonucleoprotein complexes. From the first-order dependence of the nucleation rate on the disk concentration (Butler, 1974a), only a single disk is required for the rate-limiting step, while the stable complex appears to require about three disks (Zimmern and Butler, 1977). Taking this definition of nucleation, the results of Schuster and his colleagues contradict the view that disks are not involved in elongation.

The other alternative to rate measurements has been investigation of the lengths of RNA protected from nuclease attack during reassembly (Butler and Lomonossoff, 1978). In these experiments, rapid growth was ensured by allowing RNA to react with a large excess of the disk preparation at high concentration. Samples were taken at various times, the reaction was stopped, and the RNA tails were removed by nuclease digestion. Following extraction and electrophoresis in polyacrylamide gels, the RNA appeared as a series of bands (Fig. 12) indicative of some quantization of the elongation and protection. By comparison with marker bands, the protected RNA bands appeared to be equally spaced, although certain bands in the series were either very weak or not visible on the gel. In separate experiments, the spacing was about 54 nucleotides over the range from 450 to 1250 nucleotides and 99 nucleotides for the longer fragments within the range from 1260 to 2750 nucleotides. These results compare well with the 51 or 102 nucleotides expected to be protected on the incorporation of one or both rings of 17 subunits each within a disk. Such precision, though, is clearly fortuitous because of both errors in length measurements and the few subunits expected to be lost or gained during any such reaction. Furthermore, no account is being taken of possible protection toward the 3′-tail arising from growth in that direction (see Section III.B.3).

There are two reasonable objections to the interpretation that this

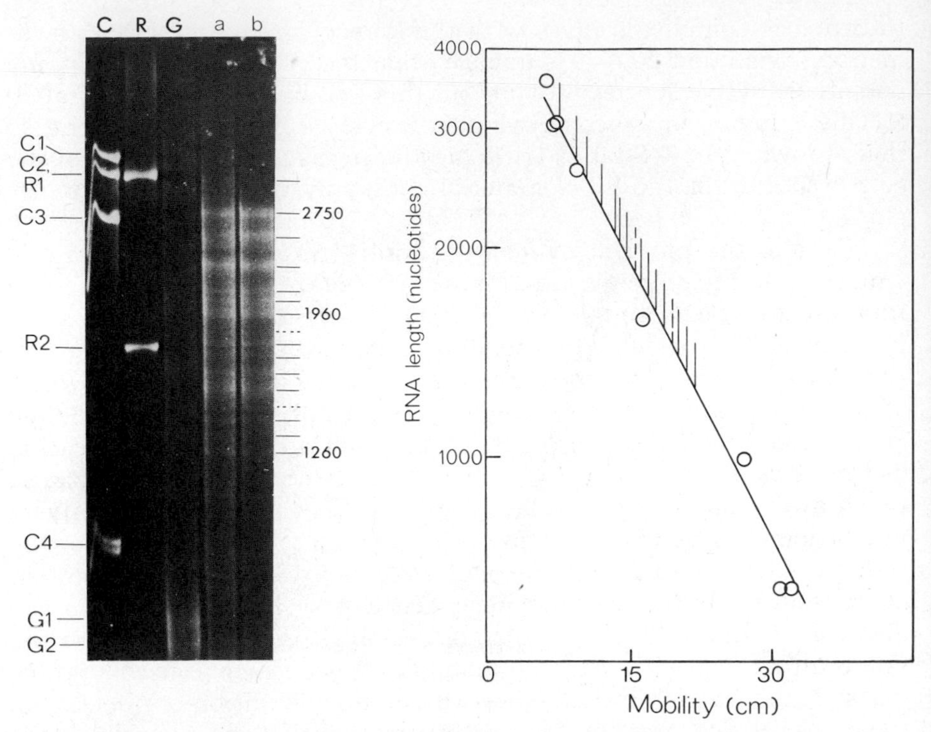

FIGURE 12. Protection of RNA during TMV elongation with the disk preparation as protein source. The left-hand panel shows RNA fractionated in an agarose/acrylamide gel after reassembly for 1 min and digestion with micrococcal nuclease for (a) 5 min and (b) 60 min. Marker RNAs are cowpea chlorotic mottle virus RNA (C), *E. coli* RNA (R), and globin mRNA (G). The right-hand panel shows the plot of RNA length against mobility in the gel. Markers are shown as circles while the positions of bands in the pattern from protected TMV RNA are shown as vertical lines. From Butler and Lomonossoff (1978).

quantization arises from the direct addition of protein from disks. The first is in the analysis of the step size by counting the number of bands, since this requires complicating assumptions about the weak or absent bands. The second objection, granting the existence of the regular steps, is one of interpretation, namely that there could be some feature present in the RNA sequence which repeats with the observed spacing and leads to the apparent quantization. However, both of these points can be objectively analyzed by the use of Fourier series to search for any repeating feature in first the gel pattern and second the complete RNA sequence (Butler, 1984).

In obtaining an objective measurement of any regular spacing of bands, photographic negatives from gels of protected RNA fragments were scanned with a densitometer and those parts of the traces corresponding to regions of the gel where banding was visible were analyzed by a one-dimensional Fourier transform. In calculating the transform shown in Fig. 13a for the shorter material, the range of data was limited to 800

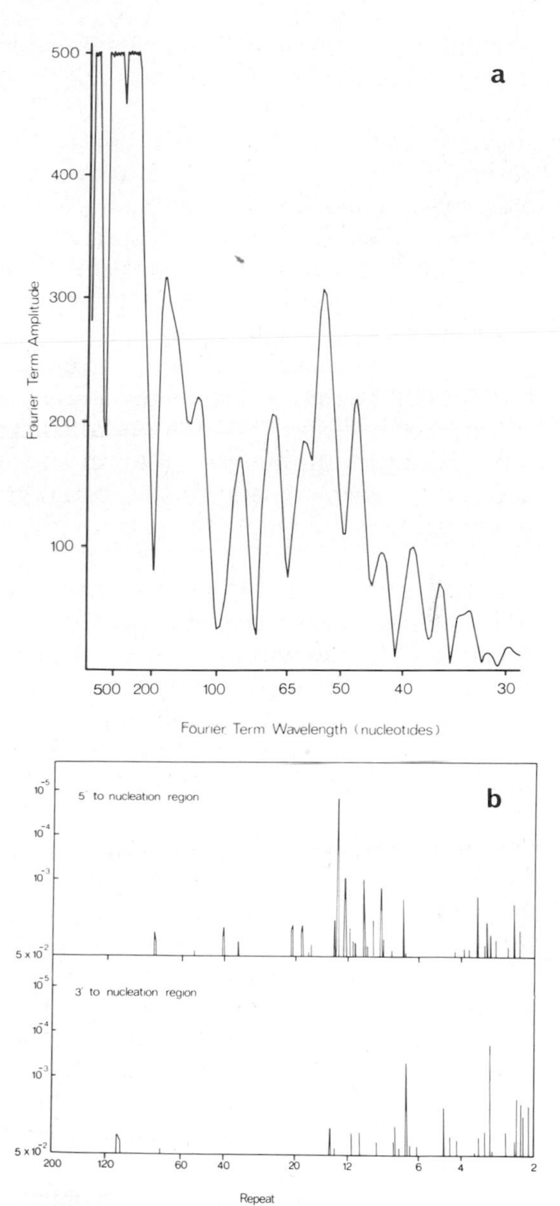

FIGURE 13. Fourier analysis of TMV RNA banding pattern. (a) Transform of densitometer tracing of negative of gel electrophoresis of protected RNA between 450 and 1250 nucleotides long. Vertical scale (in arbitrary units, as it depends upon photographic exposure and densitometer sensitivity) shows the amplitudes of all Fourier components within the observed banding pattern. The most significant peak represents a repeat every 53 nucleotides. The third harmonic of this term also shows strongly at about 161 nucleotides, above the sloping background. Horizontal scale is calibrated from marker RNAs with a maximum estimated error of ±10%. (b) Transform of purine repeats (and, by complement, pyrimidines) in RNA sequences over 2000 nucleotides to 5′-side of nucleation region and 900 nucleotides from nucleation region to 3′-terminus. Vertical scale represents the probability of the extent of repeat observed within the TMV sequence analyzed occurring in random sequence RNA. From Butler (1984).

nucleotides (from 450 to 1250 in length) and the number of sampling points was restricted to 81 (corresponding to intervals of 10 nucleotides), and the transform thus inevitably shows strong low-order terms so that the peaks corresponding to repeats of about 150 nucleotides or longer are not significant. However, the striking feature is the very strong repeat around 53 nucleotides which agrees well with the earlier estimate of 54. The analysis of longer protected fragments (1260 to 2750 nucleotides) does not give such a clear result but the most significant peak in this transform relates to a repeat of about 90 nucleotides. These two results thus confirm the conclusion that the protected RNA fragments form a series of bands with spacings of around 50 or 100 nucleotides, as first suggested on a qualitative basis (Butler and Lomonossoff, 1978). As these authors pointed out, there is no requirement that the full protein complement of a disk be added at each step and, of course, small quantities of A-protein could also be adding to blur a simple regular effect.

A similar analysis for elongation from A-protein alone is not technically feasible because, in order to achieve almost complete nucleation, such an excess of the disk preparation must be used and allowed to react to completion, that the banding pattern already extends to beyond 1000 nucleotides in the nucleated rod preparation. During elongation, after subsequent addition of A-protein, all these bands tend to elongate similarly. The pattern thus moves up in size while still being dominated by the pattern characteristic of the initial growth from disks, and so it is not possible to determine the pattern given by elongation with A-protein alone.

The second point can now be investigated since, with the knowledge of the complete RNA sequence (Goelet *et al.*, 1982), it is possible to look for any regularities within the RNA which differ from a random sequence. The results of an analysis for purines (and, by complement, for pyrimidines) are shown in Fig. 13b, and similar analyses looking for triplets with 3′-terminal G and any combination of A or U in the other places were equally negative (Butler, 1984). In no case is there any significant repeating pattern near a spacing of 50 or 100 nucleotides along either tail of the RNA. While any repetition of a sequence which might be bound favorably to TMV protein is eliminated by these results, the possibility remained of regularly spaced hairpin loops which might obstruct elongation. Potentially stable loops were identified by searching for all loops with stems containing at least six A/U or G/C base pairs (and also allowing G/U pairing) together with single-stranded loops of between 2 and 20 nucleotides. This distribution of loops again showed no significant regularity when tested by Fourier analysis (Butler, 1984). Thus, the RNA sequence cannot be creating the banding, but it may of course be modulating the intensity of individual bands to give the overall pattern observed.

The interpretation that the banding pattern is due to the protein aggregate adding and not to the RNA, has recently been questioned by Fukuda and Okada (1985), who performed experiments that were similar to, but under conditions different from, those of Butler and Lomonossoff (1978). They used both RNA and protein from two different viruses, TMV (OM) and CGMMV, and reported that the pattern depends solely on the RNA used and not on the state of aggregation of the protein. Interpretation of their results is not straightforward and their relationship to those obtained from TMV *Vulgare* is complicated by the different conditions used (Butler and Lomonossoff worked at ionic strength 0.1 M while Fukuda and Okada used 0.1 M or 0.25 M phosphate buffers, giving ionic strengths of about 0.25 to 0.7 M; in addition, the temperatures were not always the same). These new experiments therefore ignore the 20-fold difference in rates found by Schön and Mundry (1984) between reassembly under these different conditions, even when employing a constant species of RNA and protein. Significantly, the higher elongation rate is found at the lower ionic strength, suggesting that it is here that the most efficient process is occurring.

Independent confirmation that the banding pattern does not depend on the RNA sequence comes from the experiments of Schön and Mundry (1984) in which TMV RNA was replaced by poly(A), with defined lengths covering the range from about 50 to 5000 nucleotides, and the poly(A) allowed to interact with TMV *Vulgare* protein in both "low-salt" and "high-salt" conditions under which assembly has been studied. In the low-salt system where disks predominate, Schön and Mundry (1984) observed that for poly(A) lengths from 100 to over 1000 the increments in protected length of poly(A) were 188($\pm$11), whereas for longer polymers in the range 1550 to 4850 nucleotides the increment was 102.5 nucleotides. They concluded that four or two turns of the helix were formed at a time with poly(A) as compared with the two or one turn described for assembly with TMV RNA.

To ensure that what was being measured was incorporation of poly(A) into the normal TMV nucleoprotein helix, Schön and Mundry characterized the products assembled with poly(A) of length around 5000 nucleotides. These showed the same apparent buoyant density, the same ratio of nucleotide to protein, and the same nuclease stability as do both native and reconstituted TMV. Their rate of formation is very similar to that found using TMV RNA when the polymer has a length of around 5000 nucleotides but it does become a function of the chain length for shorter adenine polymers and especially for lengths of poly(A) of about 185 nucleotides or less. They further observed that particles of less than four helical turns of protein subunits are unstable while particles with about 150 nucleotides or less differ in buoyant density and optical properties from mature virus and longer assembly products, thus indicating a changed structure in such short particles. This structure alteration is distinct from the transition of a disk (with two circular layers of subunits) into a protohelix (with two helical turns), which was shown by ultraviolet absorption derivative difference spectroscopy to occur in poly(A)-mediated assembly just as in the normal assembly with TMV RNA. Both varieties of nucleic acid reacted with disks (20 S aggregates) much more rapidly than with the larger 27 S aggregates (probably two-disk stacks or three-turn helices) found in the high-salt conditions.

These results agree with the interpretation that the band spacing seen with TMV RNA is not due to any sequence features within the RNA (whether these are favorable for binding or inhibiting for assembly). The most reasonable suggestion remains that the banding arises, under the

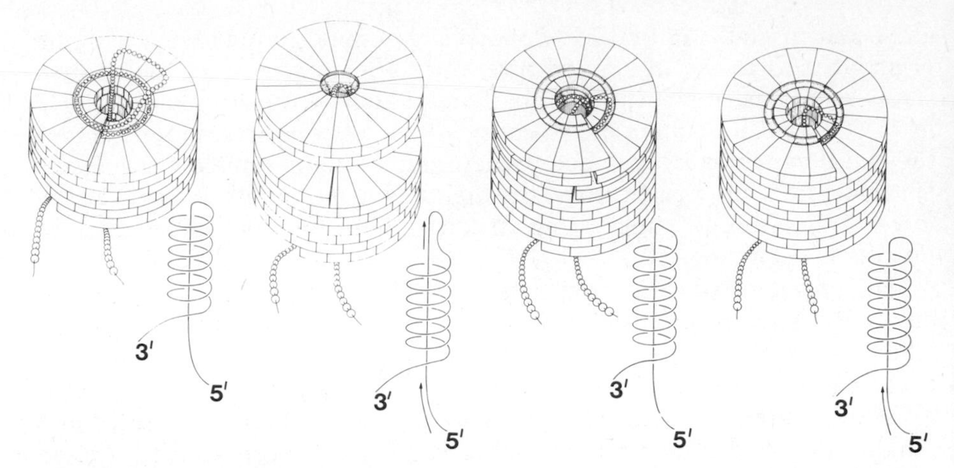

FIGURE 14. Possible mechanism of elongation along 5′-tail from protein disk. The protein and RNA are drawn diagrammatically and the expected configuration of the RNA and its movement up through the central hole of the rod are shown. Details are discussed in the text. From Butler (1984).

conditions for most rapid assembly, from the protein being prepackaged into units which add one or two turns at a time. The most obvious such package is the disk.

One hypothesis for such a mode of elongation along the major 5′-tail of the RNA is illustrated in Fig. 14. This shows the 5′-tail, which is folded back down the central hole of the growing rod, forming a "traveling loop" after binding to the upper surface of the growing rod. This loop can then interact with an incoming disk in analogous fashion to that of the nucleation loop, with the traveling loop inserting into the central hole of the disk and binding round between the two turns of protein, dislocation of the disk being caused by its interaction with both the RNA and the end of the existing nucleoprotein helix. More RNA might then move up through the central hole, binding to the upper surface and regenerating the traveling loop. The size of this loop may well be critical for its interaction with an incoming disk. If it were too large, it would be unable to insert into the disk which might no longer be able to interact. The size of the loop is expected to grow with time as a direct result of random movement of the RNA along the central hole of the rod which, since the RNA will not be pulled away from its interaction with the protein, is essentially linear diffusion with a boundary. Thus, the special structure required for elongation with disks may well be lost either under conditions of slow elongation or else on sitting around or during any adverse handling, any of which may result in the traveling loop becoming too big. This could well explain some of the apparent differences in experimental results, and hence conclusions, depending upon how the elongation is carried out and how rapidly it occurs.

3. 3′-Direction

Few experiments have been carried out on elongation specifically toward the 3′-end and, because of the faster elongation toward the other end, little is known about it. Lomonossoff and Butler (1980) studied such elongation with short RNA from which most of the sequence to the 5′-side of the nucleation region had been removed. This was prepared using partially stripped TMV rods where the 5′-tails were exposed by alkaline conditions (Perham and Wilson, 1976). These tails were removed by nuclease digestion, the RNA isolated and fractionated, and RNA fragments selected which contained the nucleation region and only a short 5′-tail but a complete 3′-tail. Following nucleation with small amounts of the disk preparation, elongation in the 5′-direction would rapidly give rodlets coated to their 5′-ends but with residual uncoated 3′-tails upon which further elongation could be observed.

In contrast to elongation upon intact RNA, these rodlets grew more rapidly with A-protein than with a disk preparation, irrespective of whether this was measured by the turbidity increase, the length of the longest protected RNA, or the protection of specific oligonucleotides. The kinetics showed saturation with increasing protein concentration but the saturating concentration was four times higher when the protein was supplied as the disk preparation than when it was present as A-protein. Thus, it is almost certainly A-protein which is the source of protein for elongation toward the 3′-tail. This view is further supported by the absence of any banding pattern in the RNA protected from nuclease digestion: the RNA migrates as a single broad band when run in gels.

IV. CONCLUSIONS

The assembly of TMV initiates in a unique region on the viral RNA, located in *Vulgare* about 900 nucleotides from the 3′-end. The actual origin of assembly is probably a specific hairpin loop which inserts into the central hole of the disk aggregate of TMV protein, binding between the two rings of subunits which are spaced widely apart toward the center of the disk. Entrapment of the RNA occurs by rearrangement of the protein subunit packing. The two layers of subunits close together around the RNA and the disk is transformed into a protohelix. This initial stage of assembly is stabilized by the interaction of two further hairpin loops on the 5′-side of the origin of assembly. In TMV *Vulgare*, at pH 7, ionic strength 0.1 M, and 20°C, elongation then occurs simultaneously in both directions: the shorter 3′-tail is protected only by subunits from A-protein; the longer 5′-tail interacts, for the most rapid assembly, via a traveling loop which inserts into further disks, in a manner analogous to the nucleation step. However, A-protein aggregates can add in this 5′-direction, at a lower rate, and they do so alone if the structure of the traveling loop is lost for any reason. Nothing is known about finishing events in which the final turns of the RNA are encapsidated and so the elongation processes may change as these ends are approached.

The nucleation reaction shows a high specificity for the most favorable RNA sequence. Depending upon whether the coat proteins are able to coaggregate, phenotypically mixed particles may be formed during double infections (Taliansky *et al.*, 1977; Otsuki and Takebe, 1978), but in general homologous RNA and protein assemble most readily. Evidence for some lowering of specificity also comes from the presence, at about one copy per virion, of H-protein which has the amino- and carboxy-termini of coat protein, but also some additional material, possibly as a branched fusion product (Collmer *et al.*, 1983). While small amounts of H-protein will reconstitute with the coat protein and RNA, particles without it show no differences in stability or infectivity to native virus (Collmer and Zaitlin, 1983) and its function remains unknown.

The polymorphic forms shown by aggregates of TMV protein thus closely reflect its role in virus assembly. The most stable is the protein helix, which is essentially the protein component of the virion, and it is probably too stable to participate directly in any assembly reaction. Central for the assembly is the protein disk, structurally related to the helix, essential for nucleation of assembly, and participating in the most rapid elongation in the major direction. The disk satisfies both the physical requirement for correct helix formation and the biological requirement for selectivity of RNA during the initial stages of assembly. The smaller protein aggregates—A-protein—are also required, though few details are known about their structures. During assembly, both disks and A-protein must be converted into nucleoprotein helix and thus the fine balance between the three polymorphic forms is probably essential for the most efficient growth of TMV. The extent to which these three forms participate in the elongation process remains the subject of vigorous debate.

REFERENCES

Asselin, A., and Zaitlin, M., 1978, An anomalous form of tobacco mosaic virus RNA observed upon polyacrylamide gel electrophoresis, *Virology* **88:**191–193.

Barrett, A. N., Barrington Leigh, J., Holmes, K. C., Leberman, R., von Sengbusch, P., and Klug, A., 1971, An electron-density map of tobacco mosaic virus at 10 Å resolution, *Cold Spring Harbor Symp. Quant. Biol.* **36:**433–448.

Bawden, F. C., and Pirie, N. W., 1937, The isolation and some properties of liquid crystalline substances from solanaceous plants infected with three strains of tobacco mosaic virus, *Proc. R. Soc. London Ser. B* **123:**274–320.

Bernal, J. D., and Fankuchen, I., 1941, X-ray and crystallographic studies of plant virus preparations, *J. Gen. Physiol.* **25:**147–165.

Bloomer, A. C., Champness, J. N., Bricogne, G., Staden, R., and Klug, A., 1978, Protein disk of tobacco mosaic virus at 2.8 Å resolution showing the interactions within and between subunits, *Nature* (*London*) **276:**362–368.

Bloomer, A. C., Graham, J., Hovmoller, S., Butler, P. J. G., and Klug, A., 1981, Tobacco mosaic virus: interaction of the protein disk with nucleotides and its implications for virus structure and assembly, in: *Structural Aspects of Recognition in Biological Macromolecules* (M. Balaban, J. Sussman, W. Traub, and A. A. Yonath, eds.), pp. 851–864, Balaban ISS, Rehovot.

Butler, P. J. G., 1972, Structures and roles of the polymorphic forms of tobacco mosaic virus protein. VI. Assembly of the nucleoprotein rods of tobacco mosaic virus from the protein disks and RNA, *J. Mol. Biol.* **72**:25–35.

Butler, P. J. G., 1974a, Structures and roles of the polymorphic forms of tobacco mosaic virus protein. IX. Initial stages of assembly of nucleoprotein rods from virus RNA and the protein disks, *J. Mol. Biol.* **82**:343–353.

Butler, P. J. G., 1974b, Structures and roles of the polymorphic forms of tobacco mosaic virus protein. VIII. Elongation of nucleoprotein rods of the virus RNA and protein, *J. Mol. Biol.* **82**:333–341.

Butler, P. J. G., 1976, Assembly of tobacco mosaic virus, *Philos. Trans. R. Soc. London Ser. B* **276**:151–163.

Butler, P. J. G., 1979, Assembly of regular viruses, in: *International Review of Biochemistry*, Volume 25/IIB (R. E. Offord, ed.), pp. 205–237, University Park Press, Baltimore.

Butler, P. J. G, 1984, The current picture of the structure and assembly of tobacco mosaic virus, *J. Gen. Virol.* **65**:253–279.

Butler, P. J. G., and Durham, A. C. H., 1977, Tobacco mosaic virus protein aggregation and the virus assembly, *Adv. Protein Chem.* **31**:187–251.

Butler, P. J. G., and Finch, J. T., 1973, Structures and roles of the polymorphic forms of tobacco mosaic virus protein. VII. Lengths of the growing rods during assembly into nucleoprotein with the viral RNA, *J. Mol. Biol.* **78**:637–649.

Butler, P. J. G., and Klug, A., 1971, Assembly of the particle of tobacco mosaic virus from RNA and disks of protein, *Nature New Biol.* **229**:47–50.

Butler, P. J. G., and Klug, A., 1973, Effect of state of polymerisation of the protein component on the assembly of tobacco mosaic virus, *Mol. Gen. Genet.* **120**:91–93.

Butler, P. J. G., and Lomonossoff, G. P., 1978, Quantized incorporation of RNA during assembly of tobacco mosaic virus from protein disks, *J. Mol. Biol.* **126**:877–882.

Butler, P. J. G., and Lomonossoff, G. P., 1980, RNA–protein interactions in the assembly of tobacco mosaic virus, *Biophys. J.* **32**:295–312.

Butler, P. J. G., Durham, A. C. H., and Klug, A., 1972, Structures and roles of the polymorphic forms of tobacco mosaic virus protein. IV. Control of mode of aggregation of tobacco mosaic virus protein by proton binding, *J. Mol. Biol.* **72**:1–18.

Butler, P. J. G., Bloomer, A. C., Bricogne, G., Champness, J. N., Graham, J., Guilley, H., Klug, A., and Zimmern, D., 1976, Tobacco mosaic virus assembly—Specificity and the transition in protein structure during RNA packaging, *Proc. 3rd John Innes Symp.* pp. 101–110.

Butler, P. J. G., Finch, J. T., and Zimmern, D., 1977, Configuration of tobacco mosaic virus RNA during virus assembly, *Nature (London)* **265**:217–219.

Caspar, D. L. D., 1963, Assembly and stability of the tobacco mosaic virus particle, *Adv. Protein Chem.* **18**:37–121.

Champness, J. N., Bloomer, A. C., Bricogne, G., Butler, P. J. G., and Klug, A., 1976, The structure of the protein disk of tobacco mosaic virus to 5 Å resolution, *Nature (London)* **259**:20–24.

Collmer, C. W., and Zaitlin, M., 1983, The H protein isolated from tobacco mosaic virus reassociates with virions reconstructed *in vitro*, *Virology* **126**:449–458.

Collmer, C. W., Vogt, V. M., and Zaitlin, M., 1983, H protein, a minor protein of TMV virions, contains sequences of the viral coat protein, *Virology* **126**:429–448.

Correia, J. J., Shire, S., Yphantis, D. A., and Schuster, T. M., 1985, Sedimentation equilibrium measurements of the intermediate-size tobacco mosaic virus protein polymers, *Biochemistry* **24**:3292–3297.

Crowther, R. A., and Amos, L. A., 1971, Harmonic analysis of electron microscope images with rotational symmetry, *J. Mol. Biol.* **60**:123–130.

Durham, A. C. H., 1972, Structures and roles of the polymorphic forms of tobacco mosaic virus protein. I. Sedimentation studies, *J. Mol. Biol.* **67**:289–305.

Durham, A. C. H., and Finch, J. T., 1972, Structures and roles of the polymorphic forms of tobacco mosaic virus protein. II. Electron microscope observations of the larger polymers, *J. Mol. Biol.* **67**:307–314.

Durham, A. C. H., and Klug, A., 1971, Polymerization of tobacco mosaic virus protein and its control, *Nature New Biol.* **229**:42–46.

Durham, A. C. H., and Klug, A., 1972, Structures and roles of the polymorphic forms of tobacco mosaic virus protein. III. A model for the association of A-protein into disks, *J. Mol. Biol.* **67**:315–332.

Durham, A. C. H., Finch, J. T., and Klug, A., 1971, States of aggregation of tobacco mosaic virus protein, *Nature New Biol.* **229**:37–42.

Fairall, L., Finch, J. T., Hui, C-F., Cantor, C. R., and Butler, P. J. G., 1986, Studies of tobacco mosaic virus reassembly with an RNA tail blocked by a hybridised and cross-linked probe, *Eur. J. Biochem.* (in press).

Finch, J. T., 1972, The hand of the helix of tobacco mosaic virus, *J. Mol. Biol.* **66**:291–294.

Finch, J. T., Leberman, R., Chang, Y.-S., and Klug, A., 1966, Rotational symmetry of the two turn disk aggregate of tobacco mosaic virus protein, *Nature (London)* **212**:349–350.

Fraenkel-Conrat, H., and Singer, B., 1959, Reconstitution of tobacco mosaic virus. III. Improved methods and use of mixed nucleic acids, *Biochim. Biophys. Acta* **33**:359–370.

Fraenkel-Conrat, H., and Singer, B., 1964, Reconstitution of tobacco mosaic virus. IV. Inhibition by enzymes and other proteins, and use of polynucleotides, *Virology* **23**:354–362.

Fraenkel-Conrat, H., and Williams, R. C., 1955, Reconstitution of active tobacco mosaic virus from its inactive protein and nucleic acid components, *Proc. Natl. Acad. Sci. USA* **41**:690–698.

Franklin, R. E., 1956, Structure of tobacco mosaic virus, *Nature (London)* **175**:379–383.

Franklin, R. E., and Holmes, K. C., 1958, Tobacco mosaic virus: Application of the method of isomorphous replacement to the determination of the helical parameters and radial density distribution, *Acta Crystallogr.* **11**:213–220.

Fukuda, M., and Okada, Y., 1982, Mechanism of tobacco mosaic virus assembly: Role of subunit and larger aggregate protein, *Proc. Natl. Acad. Sci. USA* **79**:5833–5836.

Fukuda, M., and Okada, Y., 1985, Elongation in the major direction of tobacco mosaic virus assembly, *Proc. Natl. Acad. Sci. USA* **82**:3631–3634.

Fukuda, M., Ohno, T., Okada, Y., Otsuki, Y., and Takebe, I., 1978, Kinetics of biphasic reconstitution of tobacco mosaic virus *in vitro*, *Proc. Natl. Acad. Sci. USA* **75**:1727–1730.

Fukuda, M., Okada, Y., Otsuki, Y., and Takebe, I., 1980, The site of initiation of rod assembly on the RNA of a tomato and a cowpea strain of tobacco mosaic virus, *Virology* **101**:493–502.

Fukuda, M., Meshi, T., Okada, Y., Otsuki, Y., and Takebe, I., 1981, Correlation between particle multiplication and location on virion RNA of the assembly initiation site for viruses of the tobacco mosaic virus group, *Proc. Natl. Acad. Sci. USA* **78**:4231–4235.

Gilbert, P. F. C., and Klug, A., 1974, X-ray analysis of the disk of tobacco mosaic virus protein. III. A low resolution electron density map, *J. Mol. Biol.* **86**:193–207.

Goelet, P., Lomonossoff, G. P., Butler, P. J. G., Akam, M. E., Gait, M. J., and Karn, J., 1982, Nucleotide sequence of tobacco mosaic virus RNA, *Proc. Natl. Acad. Sci. USA* **79**:5818–5822.

Guilley, H., Jonard, G., Richards, K. E., and Hirth, L., 1975a, Sequence of a specifically encapsidated RNA fragment originating from the tobacco-mosaic-virus-coat protein cistron, *Eur. J. Biochem.* **54**:135–144.

Guilley, H., Jonard, G., Richards, K. E., and Hirth, L., 1975b, Observations concerning the sequence of two additional specifically encapsidated RNA fragments originating from the tobacco-mosaic-virus coat-protein cistron, *Eur. J. Biochem.* **54**:145–153.

Guilley, H., Jonard, G., Kukla, B., and Richards, K. E., 1979, Sequence of 1000 nucleotides at the 3′ end of tobacco mosaic virus RNA, *Nucleic Acids Res.* **6:**1287–1308.

Hirth, L., and Richards, K. E., 1981, Tobacco mosaic virus: Model for structure and function of a simple virus, *Adv. Virus Res.* **26:**145–199.

Holmes, K. C., 1979, Protein–RNA interactions during TMV assembly, *J. Supramol. Struct.* **12:**305–320.

Holmes, K. C., 1983, Flexibility of tobacco mosaic virus, *Ciba Found. Symp.* **93:**116–138.

Holmes, K. C., Stubbs, G. J., Mandelkow, E., and Gallwitz, U., 1975, Structure of tobacco mosaic virus at 6.7 Å resolution, *Nature (London)* **254:**192–196.

Hubert, J. J., Bourque, D. P., and Zaitlin, M., 1976, A tobacco mosaic virus mutant with non-functional coat protein and its revertant: Relationship to the virus assembly process. *J. Mol. Biol.* **108:**789–798.

Jardetsky, O., Akasaka, K., Vogel, D., Morris, S., and Holmes, K. C., 1978, Unusual segmental flexibility in a region of tobacco mosaic virus coat protein, *Nature (London)* **273:**564–566.

Jonard, G., Guilley, H., and Hirth, L., 1975, Specific encapsidation of TMV RNA fragments by 25S TMV protein: Isolation and some properties of the nucleoprotein complexes formed, *Virology* **64:**1–9.

Jonard, G., Richards, K. E., Guilley, H., and Hirth, L., 1977, Sequence for assembly nucleation region of TMV RNA, *Cell* **11:**483–493.

Jonard, G., Richards, K., Mohier, E., and Gerlinger, P., 1978, Nucleotide sequence at the 5′ extremity of tobacco-mosaic-virus RNA. 2. The coding region (nucleotides 69–236), *Eur. J. Biochem.* **84:**521–531.

Kaplan, I. B., Kozlov, Y. V., Pshennikova, E. S., Taliansky, M. E., and Atabekov, J. G., 1982, A study of TMV *ts* mutant Ni 2519. III. Location of the reconstitution initiation sites on Ni 2519 RNA, *Virology* **118:**317–323.

Keith, J., and Fraenkel-Conrat, H., 1975, Tobacco mosaic virus RNA carries 5′-terminal triphosphorylated guanosine blocked by 5′-linked 7-methylguanosine, *FEBS Lett.* **57:**31–33.

Lauffer, M. A., 1975, *Entropy Driven Processes in Biology*, Springer-Verlag, Berlin.

Lebeurier, G., and Hirth, L., 1973, Tobacco mosaic virus reconstitution at low ionic strength, *FEBS Lett.* **34:**19–23.

Lebeurier, G., Nicolaieff, A., and Richards, K. E., 1977, Inside-out model for self-assembly of tobacco mosaic virus, *Proc. Natl. Acad. Sci. USA* **74:**149–153.

Lomonossoff, G. P., and Butler, P. J. G., 1979, Location and encapsidation of the coat protein cistron of tobacco mosaic virus: A bidirectional elongation of the nucleoprotein rod, *Eur. J. Biochem.* **93:**157–164.

Lomonossoff, G. P., and Butler, P. J. G., 1980, Assembly of tobacco mosaic virus: Elongation towards the 3′-hydroxyl terminus of the RNA, *FEBS Lett.* **113:**271–274.

Mandelkow, E., Holmes, K. C., and Gallwitz, U., 1976, A new helical aggregate of tobacco mosaic virus protein, *J. Mol. Biol.* **102:**265–285.

Mandelkow, E., Stubbs, G., and Warren, S., 1981, Structures of the helical aggregates of tobacco mosaic virus protein, *J. Mol. Biol.* **152:**375–386.

Meshi, T., Ohno, T., and Okada, Y., 1982, Nucleotide sequence and its character of cistron coding for the 30K protein of tobacco mosaic virus (OM strain), *J. Biochem. (Tokyo)* **91:**1441–1444.

Meshi, T., Kiyama, R., Ohno, T., and Okada, Y., 1983a, Nucleotide sequence of the coat protein cistron and the 3′ noncoding region of cucumber green mottle mosaic virus (watermelon strain) RNA, *Virology* **127:**54–64.

Meshi, T., Ishikawa, M., Takamatsu, N., Ohno, T., and Okada, Y., 1983b, The 5′-terminal sequence of TMV RNA, *FEBS Lett.* **162:**282–285.

Mondragon, A., 1984, X-ray crystallographic refinement of the disk form of tobacco mosaic virus, Ph.D. thesis, University of Cambridge.

Mondragon, A., Bloomer, A. C., Gebhard, W., and Holmes, K. C., 1985, Preliminary atomic model for tobacco mosaic virus, in: *Proc. 16th FEBS Congress* (Y. A. Ovchinnikov, ed.), Part A, pp. 362–365, VNU Science Press, Utrecht.

Namba, K., and Stubbs, G. J., 1985, Solving the phase problem in fiber diffraction: Application to tobacco mosaic virus at 3.6 Å resolution, *Acta Crystallogr.* **A41**:252–261.

Novikov, V. K., Sarukhan-Bek, K. K., and Atabekov, J. G., 1974, Anomalous stable aggregates in mixture of TMV and cucumber virus 3 proteins, *Virology* **62**:134–144.

Ohno, T., Inoue, H., and Okada, Y., 1972, Assembly of rod-shaped virus *in vitro*: Reconstitution with cucumber green mottle mosaic virus protein and tobacco mosaic virus RNA, *Proc. Natl. Acad. Sci. USA* **69**:3680–3683.

Ohno, T., Okada, Y., Shimotohno, K., Miura, K., Shinshi, H., Miura, M., and Sugimura, T., 1976, Enzymatic removal of the 5′-terminal methylated blocked structure of tobacco mosaic virus RNA and its effects on infectivity and reconstitution with coat protein, *FEBS Lett.* **67**:209–213.

Ohno, T., Takahashi, M., and Okada, Y., 1977, Assembly of tobacco mosaic virus *in vitro*: Elongation of partially reconstituted RNA, *Proc. Natl. Acad. Sci. USA* **74**:552–555.

Ohno, T., Aoyagi, M., Yamanashi, Y., Saito, H., Ikawa, S., Meshi, T.,and Okada, Y., 1984, Nucleotide sequence of the tobacco mosaic virus (tomato strain) genome and comparison with common strain genome, *J. Biochem.* (*Tokyo*) **96**:1915–1923.

Okada, Y., and Ohno, T., 1972, Assembly mechanism of tobacco mosaic virus particle from its ribonucleic acid and protein, *Mol. Gen. Genet.* **114**:205–213.

Otsuki, Y., and Takebe, I., 1978, Production of mixed coated particles in tobacco mesophyll protoplasts doubly infected by strains of tobacco mosaic virus, *Virology* **84**:162–171.

Otsuki, Y., Takebe, I., Ohno, T., Fukuda, M., and Okada, Y., 1977, Reconstitution of tobacco mosaic virus rods occurs bidirectionally from an internal initiation region: Demonstration by electron microscopic serology, *Proc. Natl. Acad. Sci. USA* **74**:1913–1917.

Perham, R. N., and Wilson, T. M. A., 1976, The polarity of the stripping of coat protein subunits from the RNA in tobacco mosaic virus under alkaline conditions, *FEBS Lett.* **62**:11–15.

Raghavendra, K., Adams, M. L., and Schuster, T. M., 1985, Tobacco mosaic virus protein aggregates in solution: Structural comparison of 20S aggregates with those near conditions for disk crystallization, *Biochemistry* **24**:3298–3304.

Richards, K. E., and Williams, R. C., 1972, Assembly of tobacco mosaic virus *in vitro*: Effect of state of polymerization of the protein component, *Proc. Natl. Acad. Sci. USA* **69**:1121–1124.

Richards, K. E., and Williams, R. C., 1973, Assembly of tobacco mosaic virus *in vitro*: Elongation of partially assembled rods, *Biochemistry* **12**:4574–4581.

Richards, K., Guilley, H., Jonard, G., and Hirth, L., 1978, Nucleotide sequence at the 5′ extremity of tobacco-mosaic-virus RNA. 1. The noncoding region (nucleotides 1–68), *Eur. J. Biochem.* **84**:513–519.

Sarkar, S., 1960, Interaction and mixed aggregation of proteins from tobacco mosaic virus strains, *Z. Naturforsch.* **15b**:778–786.

Schön, A., and Mundry, K.-W., 1984, Coordinated two-disk nucleation, growth and properties, of virus-like particles assembled from tobacco-mosaic-virus capsid protein with poly(A) or oligo(A) of different length, *Eur. J. Biochem.* **140**:119–127.

Schuster, T. M., Scheele, R. B., and Khairallah, L. H., 1979, Mechanism of self-assembly of tobacco mosaic virus protein. I. Nucleation-controlled kinetics of polymerization, *J. Mol. Biol.* **127**:461–485.

Schuster, T. M., Scheele, R. B., Adams, M. L., Shire, S. J., Steckert, J. J., and Potschka, M., 1980, Studies on the mechanism of assembly of tobacco mosaic virus, *Biophys. J.* **32**:313–329.

Shire, S. J., Steckert, J. J., and Schuster, T. M., 1979a, Mechanism of self-assembly of tobacco mosaic virus protein. II. Characterization of the metastable polymerization nucleus and the initial stages of helix formation, *J. Mol. Biol.* **127**:487–506.

Shire, S. J., Steckert, J. J., Adams, M. L., and Schuster, T. M., 1979b, Kinetics and mechanism of tobacco mosaic virus assembly: Direct measurement of relative rates of incorporation of 4S and 20S protein, *Proc. Natl. Acad. Sci. USA* **76**:2745–2749.

Shire, S. J., Steckert, J. J., and Schuster, T. M., 1981, Mechanism of tobacco mosaic virus assembly: Incorporation of 4S and 20S protein at pH 7.0 and 20°C, *Proc. Natl. Acad. Sci. USA* **78**:256–260.

Stanley, W. M., 1935, Isolation of a crystalline protein possessing the properties of tobacco-mosaic virus, *Science* **81**:644–645.

Steckert, J. J., and Schuster, T. M., 1982, Sequence specificity of trinucleoside diphosphate binding to polymerized tobacco mosaic virus protein, *Nature (London)* **299**:32–36.

Steinschneider, A., and Fraenkel-Conrat, H., 1966, Studies of nucleotide sequences in tobacco mosaic virus ribonucleic acid. IV. Use of aniline in stepwise degradation, *Biochemistry* **5**:2735–2743.

Stubbs, G., and Stauffacher, C., 1981, Structure of the RNA in tobacco mosaic virus, *J. Mol. Biol.* **152**:387–396.

Stubbs, G., Warren, S., and Holmes, K., 1977, Structure of RNA and RNA binding site in tobacco mosaic virus from 4-Å map calculated from X-ray fibre diagrams, *Nature (London)* **267**:216–221.

Taliansky, M. E., Atabekova, T. I., and Atabekov, J. G., 1977, The formation of phenotypically mixed particles upon mixed assembly of some tobacco mosaic virus (TMV) strains, *Virology* **76**:701–708.

Taliansky, M. E., Kaplan, I. B., Yarvekulg, L. V., Atabekova, T. I., Agranovsky, A. A., and Atabekov, J. G., 1982, A study of TMV *ts* mutant Ni 2519. II. Temperature sensitive behaviour of Ni 2519 RNA upon reassembly, *Virology* **118**:309–316.

Turner, D. R., Mondragon, A., Fairall, L., Bloomer, A. C., Finch, J. T., van Boom, J. H., and Butler, P. J. G., 1986, Oligonucleotide binding to the coat protein disk of tobacco mosaic virus: Possible steps in the mechanism of assembly, *Eur. J. Biochem.* (in press).

Tyulkina, L. G., Nazarova, G. N., Kaftanova, A. S., Ledneva, R. K., Bogdanov, A. A., and Atabekov, J. G., 1975, Reassembly of TMV 20-S protein disks with 3-S RNA fragments, *Virology* **63**:15–29.

Vogel, D., de Marcillac, G. D., Hirth, L., Gregori, E., and Jaenicke, R., 1979, Size distribution in the higher stages of polymerization of the A-protein of tobacco mosaic virus (*vulgare*), *Z. Naturforsch.* **34c**:782–792.

Watson, J. D., 1954, The structure of tobacco mosaic virus. I. X-ray evidence of a helical arrangement of subunits around the longitudinal axis, *Biochim. Biophys. Acta* **13**:10–19.

Zimmern, D., 1975, The 5′ end group of tobacco mosaic virus RNA is $m^7G^{5'}ppp^{5'}Gp$, *Nucleic Acids Res.* **2**:1189–1201.

Zimmern, D., 1977, The nucleotide sequence at the origin for assembly on tobacco mosaic virus RNA, *Cell* **11**:463–482.

Zimmern, D., 1983, An extended secondary structure model for the TMV assembly origin, and its correlation with protection studies and an assembly defective mutant, *EMBO J.* **2**:1901–1907.

Zimmern, D., and Butler, P. J. G., 1977, The isolation of tobacco mosaic virus RNA fragments containing the origin for viral assembly, *Cell* **11**:455–462.

Zimmern, D., and Hunter, T., 1983, Point mutation in the 30K open reading frame of TMV implicated in temperature sensitive assembly and local lesion spreading of mutant Ni 2519, *EMBO J.* **2**:1893–1900.

Zimmern, D., and Wilson, T. M. A., 1976, Location of the origin for viral reassembly on tobacco mosaic virus RNA and its relation to stable fragment, *FEBS Lett.* **71**:294–298.

CHAPTER 3

TOBACCO MOSAIC VIRUS

Mutants and Strains

SATYABRATA SARKAR

I. INTRODUCTION

The properties of a virus as well as of all living organisms are subject to variation giving rise to altered forms, even though the change may not be very conspicuous. Once recognized, a variant is called a *mutant* if it was found under experimental conditions, indicating a probable direct origin from the apparently homogeneous starting population. In fact, preparations of viruses are rarely free of at least a few variants (Bawden, 1950; Garciá-Arenal *et al.*, 1984). The word *strain,* on the other hand, is usually chosen for variants that are found in the field, sometimes in regions widely apart, so that the time and mode of origin of the variant remain unknown (Hennig and Wittmann, 1972). The extent to which a particular variant differs from the original strain can only be known if the detailed structure of both are determined at the molecular level. The results show that the difference between a strain and a mutant is one of degree only.

Among the mutants themselves, some may have a spontaneous origin, whereas others are induced. A large number of mutants of tobacco mosaic virus (TMV) isolated in different laboratories are induced ones. However, it cannot be excluded that certain mutants could have already been present as a spontaneous variant or that they originated independently of the action of the mutagenic agent employed.

As will be evident from the above, the distinction between mutants and strains is made mainly for practical reasons. Both types of variants

SATYABRATA SARKAR • Institute of Plant Medicine, University of Hohenheim, 7000 Stuttgart 70, West Germany.

of TMV have played a significant role in the elucidation of many basic topics in biology. The present chapter will describe some of the major types of mutants and strains of TMV and will stress the role they have played in the development of virology and molecular biology.

Limitation of space necessitated an abridged treatment of the subject matter. The following publications are recommended for additional information: Wittmann (1962), Knight (1963, 1975), Hennig and Wittmann (1972), Gibbs and Harrison (1976), Matthews (1981).

II. ISOLATION OF VARIANTS

The isolation of variants of TMV was facilitated by the introduction of the local lesion hosts, *Nicotiana glutinosa* L. and later on *N. tabacum* L. cv. Xanthi-nc (Holmes, 1929; Takahashi, 1956). Both hosts carry the dominant gene N for hypersensitivity. Two other species, namely *N. sylvestris* L. and *N. tabacum* L. var. Java, in which some TMV strains spread systemically and some produce only localized lesions, have been very useful in the recognition and isolation of variants. In fact, the conclusion that nitrous acid increases the mutation rate, could hardly have been drawn and confirmed without the help of the java-type tobacco (Gierer and Mundry, 1958; Mundry, 1959; see also Section II.B.1). In more recent years, a spontaneous mutant of Samsun (Samsun EN) has been found (Melchers *et al.*, 1966) that seems to contain the N′-type gene in java tobacco. Samsun EN has been successfully used in experiments on cross-protection between some defective mutants of TMV (Sarkar and Smitamana, 1981b). The local lesions produced on tobacco varieties carrying the N′ gene vary considerably in size depending on the TMV strain or mutant. The lesion size was found to be correlated with the thermal stability of the isolated coat proteins of the strains and not with the intrinsic efficiency of the strains to multiply in a systemic host (Fraser, 1983).

With the help of differential hosts, variants of TMV have been isolated (Table I) before the virus had been purified and crystallized (Stanley, 1935). The source and properties of a few TMV strains isolated before the era of chemical mutagenesis are described here, since they laid the foundation of our present knowledge about mutation in viruses.

A. Source and Properties of a Few Strains and Mutants of TMV

The wild strain vulgare (= U1) is the one that is most commonly found in field-grown tobacco. It is very probably the same strain that was reported by Ivanowski in 1899. McKinney (1929) succeeded in obtaining more virulent variants by choosing different parts of the infected plant as inocula. The masked strain M, isolated by Holmes (1934) under high-

TABLE I. Amino Acid Content of TMV Protein[a,b]

	Strain or mutant								
	U1 or M	J14 D1	YA	GA	F	D	Y-TAMV	G-TAMV	HR
Ala	14	—	—	—	15	11	11	18	18
Arg	11	—	12	12	—	9	9	8	10
Asp	18	17	19	19	17	17	—	22	17
Cys	1	—	—	—	—	—	—	—	—
Glu	16	15	—	—	—	19	19	—	21
Gly	6	—	—	5	—	—	—	4	5
His	0	—	—	—	—	—	—	—	1
Ile	9	—	8	8	—	8	7	8	8
Leu	12	—	—	—	—	—	13	11	11
Lys	2	3	—	—	—	—	—	1	—
Met	0	—	—	—	—	1	1	2	3
Phe	8	—	—	—	—	—	—	—	6
Pro	8	—	—	—	—	—	—	10	—
Ser	16	17	14	15	—	—	15	10	13
Thr	16	—	17	17	—	17	17	19	13
Trp	3	—	—	—	—	—	—	2	2
Tyr	4	—	—	—	—	5	5	6	7
Val	14	—	—	—	—	15	15	12	10

[a] Values are number of amino acid residues per subunit.
[b] The data for flavum (F) are from Wittmann (1962), those for Holme's ribgrass strain (HR) are from Jauregui-Adell *et al.* (1969) and Wittmann *et al.* (1969), and most of the others are from Knight (1963). Only those values that differ from those of U1 are shown; dashes indicate no change compared with U1. The masked strain M (Holmes, 1934) has the same total amino acid composition as the wild strain U1, indicating that the symptom produced on a host plant is not necessarily correlated with a change in the viral coat protein. The mutant Ni 54 is another such example, which is also symptomless on Samsun tobacco (Wittmann, 1959).

temperature conditions, is symptomless on Turkish (= Samsun) tobacco, and its coat protein has the same overall amino acid composition as that of U1. The strain J14 D1 is a variant (= derivative No. 1) of J14 isolated by Jensen (1937). Strain J14 produces lesions on Turkish tobacco and does not spread systemically. Isolate J14 D1, which is most probably a mutant of J14, is quite distinct since it not only produces local lesions on the initially inoculated leaf of Turkish tobacco, but also spreads systemically, giving rise to yellow symptoms on the young upper leaves. The J14 D1 mutant produces local necrotic lesions on java tobacco and it causes a severe systemic necrosis on young Turkish tobacco plants that can often be lethal. The yellow aucuba strain (YA) is also necrotic on java but spreads systemically in Turkish tobacco, inducing a strong chlorosis (Bewley, 1924; Kunkel, 1934). The green aucuba mosaic strain (GA) was isolated by Kunkel (1934) from YA-infected *N. sylvestris* plants that had been kept for 3 days at 35°C. On Turkish tobacco it produces practically the same type of symptom as the wild strain U1, but the amino acid composition of its coat protein differs significantly from that of U1 (Table I). The amino acid sequence of U1 is shown in Fig. 1 (Anderer *et al.*, 1960; Tsugita *et al.*, 1960; Wittmann-Liebold and Wittmann, 1967).

5 10 15
Acetyl.Ser - Tyr - Ser - Ile - Thr - Thr - Pro - Ser - Gln - Phe - Val - Phe - Leu - Ser - Ser -

20 25 30
Ala - Trp - Ala - Asp - Pro - Ile - Glu - Leu - Ile - Asn - Leu - Cys - Thr - Asn - Ala -

35 40 45
Leu - Gly - Asn - Gln - Phe - Gln - Thr - Gln - Gln - Ala - Arg - Thr - Val - Gln - Val -

50 55 60
Arg - Gln - Phe - Ser - Gln - Val - Trp - Lys - Pro - Ser - Pro - Gln - Val - Thr - Val -

65 70 75
Arg - Phe - Pro - Asp - Ser - Asp - Phe - Lys - Val - Tyr - Arg - Tyr - Asn - Ala - Val -

80 85 90
Leu - Asp - Pro - Leu - Val - Thr - Ala - Leu - Leu - Gly - Ala - Phe - Asp - Thr - Arg -

95 100 105
Asn - Arg - Ile - Ile - Glu - Val - Glu - Asn - Gln - Ala - Asn - Pro - Thr - Thr - Ala -

110 115 120
Glu - Thr - Leu - Asp - Ala - Thr - Arg - Arg - Val - Asp - Asp - Ala - Thr - Val - Ala -

125 130 135
Ile - Arg - Ser - Ala - Asp - Ile - Asn - Leu - Ile - Val - Glu - Leu - Ile - Arg - Gly -

140 145 150
Thr - Gly - Ser - Tyr - Asn - Arg - Ser - Ser - Phe - Glu - Ser - Ser - Ser - Gly - Leu -

155
Val - Trp - Thr - Ser - Gly - Pro - Ala - Thr

FIGURE 1. Amino acid sequence of the coat protein of TMV (wild strain = U1 = vulgare). Positions 85 to 109 and 113 to 122 have been conserved through all mutants and strains of TMV. From Anderer *et al.* (1960), Tsugita *et al.* (1960), and Wittmann-Liebold and Wittmann (1963).

The spontaneous variant flavum, isolated by Melchers (1940) and studied intensively by Aach (1958a,b) and others, induces a strong chlorotic reaction on both the primary inoculated and the systemically invaded leaves of Samsun tobacco. It is not necrotic in java tobacco, retards strongly the growth of Samsun plants and causes necrosis of the primary inoculated leaves. The source and some of the major characteristics of flavum and of a few other variants are presented in Fig. 2. The strain dahlemense (D), isolated by Melchers (1940, 1942), and the strain yellow-tomato atypical mosaic virus (Y-TAMV), isolated by Miller (1953), are now considered to be strains of a different virus species = tomato mosaic virus (see Chapter 9). The green-tomato atypical mosaic virus (G-TAMV), isolated by Knight *et al.* (1962) from Turkish tobacco infected with Y-TAMV, is now considered a separate virus = mild green tobacco mosaic virus (TMGMV) (see Chapter 10). Recently, Wetter (1984) has compared several mild strains of TMGMV with respect to the symptoms they produce on different hosts as well as their serological relationships; his data confirm the view that some variants are almost always present in TMV inocula and indicate that there is no experimental proof for the host-induced mutation that has been hypothesized by many authors (Bawden, 1958; Pelham *et al.*, 1970; Bald *et al.*, 1974; MacNeill and Boxall, 1974).

Data on a few variants of TMV found in a wide variety of host plants and isolated mostly before the era of nitrous acid-induced mutagenesis are summarized in Table II. Several of these have since been identified as separate viruses (see Chapters 9–15).

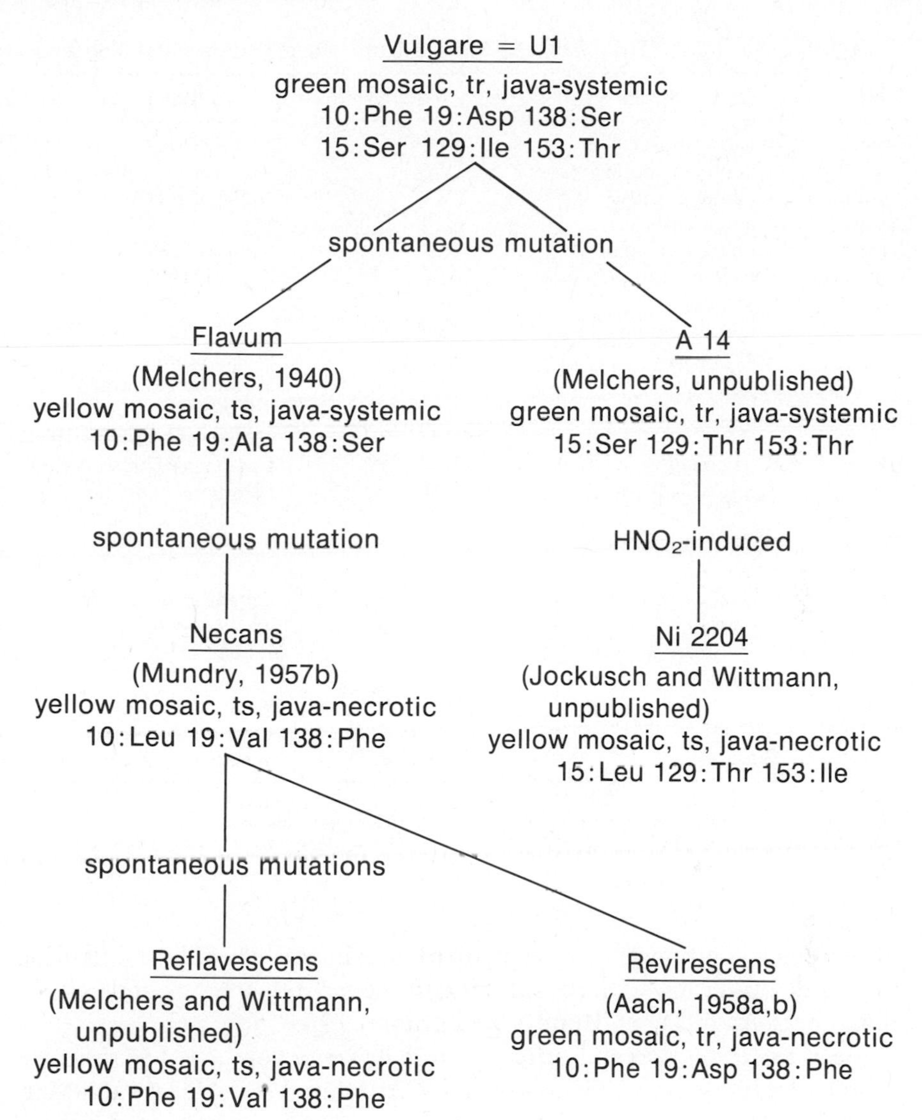

FIGURE 2. Hierarchy and differentiating characters of some of the mutants of TMV. ts = temperature sensitive; tr = resistant or not sensitive to elevated temperature like the ts strains and mutants. Note that there is only one amino acid exchange between revirescens and vulgare.

B. Induced Mutants

Many of the early variants of TMV isolated at elevated temperature or by exposure to ionizing radiation seem to have already been present as spontaneous mutants. Later attempts to increase the rate of spontaneous mutation by UV or X-rays did not give convincing results (Mundry, 1957a,b, 1960). Truly induced mutants were obtained in large number

TABLE II. Strains of TMV Isolated Mostly from Hosts Other Than Tobacco

Host	Strain[a]	Abbreviation	Reference
Tomato	Aucuba mosaic	YA	Bewley (1924)
Cucumber	Cucumber mosaic 3	CV 3	Ainsworth (1935)
Cucumber	Cucumber mosaic 4	CV 4	Ainsworth (1935)
Tomato	Dahlemense	Dahl.	Melchers (1940)
Plantago	Holmes's ribgrass	HR	Holmes (1941)
Sunn hemp	Southern sunn-hemp mosaic	SSM	Capoor (1950)
Orchid	Odontoglossum ringspot	ORSV	Jensen and Gold (1951)
Tobacco	Mild mosaic	U2	Singer *et al.* (1951)
Tobacco	—	U5	Siegel and Wildman (1954)
Tobacco	—	OM	Hiruki and Hidaka (1954)
Cowpea	Cowpea mosaic	CPV	Lister and Thresh (1955)
Cactus	Sammon's Opuntia	SOV	Sammons and Chessin (1961)
Tomato	Yellow-tomato atypical mosaic	Y-TAMV	Knight *et al.* (1962)
Tomato	Green-tomato atypical mosaic	G-TAMV	Knight *et al.* (1962)
Cucumber	Cucumber green mottle mosaic	CGMMV	Inouye *et al.* (1967)
Orchids	—	01 to 07	Kado *et al.* (1968)
Pear tree	—	B-TMV 1 & 2	Opel *et al.* (1969)
Tomato	Various[b]	CV 60, etc.	Wang and Knight (1967)

[a] Several of these have since been identified as separate viruses (see Chapters 9–15).
[b] Wang and Knight (1967) reported on the isolation of 13 different variants from tomato plants and gave them the following abbreviations: CV 60, CV 61, AC 9, HD, SJ, VEN, Aus-II, Dut-I, YLGP, K-1, PTA, PTV, and SAF.

only after the demonstration that nitrous acid can alter the nucleotide bases by deamination without splitting the sugar–phosphate bond (Schuster and Schramm, 1958; Mundry and Gierer, 1958).

Besides HNO_2, several other chemicals have been used to produce mutants: hydroxylamine (Schuster and Wittmann, 1963; Nagata and Martensson, 1968), several alkylating agents, particularly dimethyl sulfate (Fraenkel-Conrat, 1961; Tsugita, 1962; Funatsu and Fraenkel-Conrat, 1964; Singer *et al.*, 1975), fluorouracil (Kramer *et al.*, 1964; Wittmann-Liebold and Wittman, 1965a), *N*-bromo-succinimide (Brammer, 1963; Funatsu and Fraenkel-Conrat, 1964), *N*-methyl-*N*′-nitro-*N*-nitrosoguanidine (Singer and Fraenkel-Conrat, 1967), and some dyes such as proflavine and thiopyronine (Singer and Fraenkel-Conrat, 1965, 1966, 1974). Chemically induced mutants were routinely selected on the basis of one easily detectable altered property, namely the capacity to produce local lesion on java tobacco. An alternative procedure was to inoculate many individuals of a systemic host after diluting the reaction mixture to such an extent that only about half of the plants became infected. This was the method adopted by Siegel *et al.* (1962) for isolating the so-called proteinless mutants PM1 and PM2.

Purification of these proteinless mutants through local lesions is generally difficult, since the poorly protected genomic RNA is quickly inactivated in homogenates of infected plant material. However, they can be transmitted by using a suitable inhibitor of ribonuclease, e.g., bentonite (Fraenkel-Conrat *et al.*, 1961), or by use of an alkaline buffer (Diener, 1962; Sarkar, 1963; Hubert and Bourque, 1981; Sarkar and Smitamana, 1981a), as well as by a prior extraction with water-saturated phenol. Best results are obtained when the lesions are used as soon as they are detectable, e.g., on the second or third day after inoculation.

Among the different criteria chosen to characterize the large number of chemically induced mutants (see Section II.B.2), the structure of the viral coat protein proved to be the most useful and informative. These investigations were stimulated and accelerated by the development of the automatic amino acid analyzer (Spackman *et al.*, 1958).

1. Principle and Proof of Experimental Mutagenesis

When a virus genome is subjected to the action of a mutagenic agent, it can be changed to produce a mutant which will be detected if it retains its ability to infect a host and multiply therein. On the other hand, if its biological activity is lost, the effect of the mutagen was obviously "lethal." How many of the initially active virus particles in a given sample will be inactivated, by receiving one or more "hits," depends on several factors, such as the mode of action of the mutagenic agent and the duration of exposure. With proper controls, one can experimentally determine the fraction of surviving virus as a function of time. The time required to reduce the virus activity to $1/e = 37\%$ of the original value, has been called the "reduced time of treatment" τ (Gierer and Mundry, 1958). In other words, τ is the average time for one lethal conversion of a base per RNA molecule. The relation between the total time of treatment and the reduced time, i.e., the ratio t/τ, is a measure of the average number of "hits." In the case of nitrous acid treatment, it can be looked upon as the average number of deaminations per genomic RNA. Any attempt to achieve more than one mutagenic event per virus particle by increasing the time of exposure or the concentration of a mutagen increases the number of lethal hits. This is clearly illustrated in the finding that TMV mutants with only one amino acid exchange are the commonest among experimentally induced mutants. In 1962 Wittmann reported that among 117 mutants he investigated, 82 had no amino acid exchange, 29 had one each, 6 had two each, and none had 3 or more exchanges.

As already mentioned, the isolation of a variant after treating the virus with a mutagenic agent does not necessarily mean that the variant originated as a result of the action of the mutagen used, nor can one be sure about a possible increase in the mutation rate over that operating under natural conditions. In fact, if during the exposure to a mutagen the starting material, e.g., the wild strain, is inactivated rapidly, the small number of variants already existing could constitute a relatively increased fraction among the survivors, giving the impression of a higher rate of mutation. This would be the case if the mutants are somewhat more resistant to the mutagenic agent and possess a selective advantage over the original strain. Moreover, closely related mutants and strains are known to interfere with each other in their host. It will thus be difficult to obtain the actual ratio between the number of survivors of the original strain and that of the mutants in the test sample. However, by diluting the reaction mixture and the control samples sufficiently, the mutual interference could be avoided and it could be proved that nitrous acid does increase the rate of mutation and that the mutants are intrinsically no more resistant to HNO_2 than the starting strain (Mundry, 1959). Additional convincing proof of mutagenesis by HNO_2 was obtained through detailed analysis of the amino acid sequence of the coat proteins of a large number of mutants. The data fit excellently with the type of conversion of the nucleotide bases in the codons and the corresponding expected amino acid exchange (see also Section III, point 2).

2. Properties of Induced Mutants

The following properties of induced mutants have been studied extensively: type of symptom on tobacco varieties and on a few other host plants; sensitivity to different temperatures; serological and electrophoretic behavior; and the composition and sequence of the coat proteins. The wide range of symptoms, including subtle variations in deformation, discoloration, and mosaic pattern produced by more than 200 nitrous acid-induced mutants and several strains of TMV, have been recorded by Wittmann (1962, 1964). Serological studies have brought to light some very interesting features of the structure and conformation of the coat proteins. It has been found that a single amino acid exchange can be detected by use of a specific antiserum, provided the altered amino acid is present on the surface of the native protein (von Sengbusch, 1965; von Sengbusch and Wittmann, 1965; Van Regenmortel, 1967).

The amino acid sequence has been the most intensively investigated property of the variants of TMV. In Table III the amino acid exchanges in a few selected strains and mutants are presented along with the modification of their codon nucleotides, in order to demonstrate the variety of alterations that have been found.

As can be seen from Table III, not all the alterations in nucleotides are $C \rightarrow U$ or $A \rightarrow I$, indicating that mechanisms other than oxidative deamination are also operating in nature.

Among the temperature-sensitive mutants, some are induced while others are of spontaneous origin. For the isolation of ts mutants of TMV,

TABLE III. Amino Acid Exchanges in the Coat Protein and the Corresponding Possible Nucleotide Alteration in a Few Mutants of TMV

Mutant	Exchange: Amino acid	Exchange: Nucleotide	Position	Reference
Spontaneous				
Flavum	Asp→Ala	GAU→GCU	19	Wittmann *et al.* (1965)
Necans	Phe→Leu	UUU→CUU	10	Wittmann *et al.* (1965)
	Ala→Val	GCU→GUU	19	
	Ser→Phe	UCU→UUU	138	
Reflavescens	Leu→Phe	CUU→UUU	10	Wittmann *et al.* (1965)
Revirescens	Leu→Phe	CUU→UUU	10	Wittmann *et al.* (1965)
	Val→Asp	GUU→GAU	19	
430	Pro→Leu	CCU→CUU	20	Funatsu and Fraenkel-Conrat (1964)
A 14	Ile→Thr	AUU→ACU	129	Wittmann-Liebold *et al.* (1965)
Induced with nitrous acid				
Ni 102	Asp→Gly	GAU→GGU	66	Wittmann-Liebold and Wittmann (1965b)
Ni 118	Pro→Leu	CCC→CUC	20	Wittmann-Liebold *et al.* (1965)
Ni 252	Ser→Phe	UCU→UUU	138	Funatsu and Fraenkel-Conrat (1964)
Ni 1055	Ile→Met	AUA→AUG	21	Wittmann-Liebold and Wittmann (1965b)
Ni 1927	Pro→Leu	CCC→CUC	156	Wittmann-Liebold and Wittmann (1965b)
Ni 2068	Tyr→Cys	UAC→UGC	139	Wittmann-Liebold and Wittmann (1965b)
Ni 2204	Ser→Leu	UCG→UUG	15	Wittmann-Liebold *et al.* (1965)
	Thr→Ile	ACU→AUU	153	
PM5	Arg→Cys	CGU→UGU	112	Hariharasubramanian and Siegel (1969)
321 B	Arg→Gly	AGG→GGG	122	Tsugita (1962)
415	Ile→Val	AUU→GUU	21	Funatsu and Fraenkel-Conrat (1964)
421	Val→Met	GUG→AUG	11	Funatsu and Fraenkel-Conrat (1964)

(*Continued*)

TABLE III. (*continued*)

Mutant	Exchange		Position	Reference
	Amino acid	Nucleotide		
470	Ser→His	UCU→CAU	8	Rombauts and Fraenkel-Conrat (1968)
Induced with fluorouracil				
Fu 41	Thr→Ala	ACU→GCU	28	Wittmann-Liebold and Wittmann (1965a)
Induced with *N*-bromo-succinimide				
218	Pro→Leu	CCU→CUU	20	Funatsu and Fraenkel-Conrat (1964)
414	Ser→Gly	AGU→GGU	65	Funatsu and Fraenkel-Conrat (1964)
Induced with dimethylsulfate				
215	Ser→Phe	UCU→UUU	138	Tsugita (1962)
344	Thr→Ala	ACU→GCU	81	Funatsu and
	Ser→Phe	UCU→UUU	138	Fraenkel-Conrat (1964)

one can take advantage of the slow growth of their lesions at a nonpermissive temperature (Jockusch, 1968). However, this is not always possible (Peters and Murphy, 1975) and an alternative procedure has been described (Dawson and Jones, 1976). The temperature sensitivity of a ts mutant can be due either to an instability of the coat protein (ts-I type; e.g., Ni 118) or to sensitivity to some other function or stage of biosynthesis (ts-II type; e.g., Ni 2519) (Jockusch, 1964, 1966a,b,c, 1968).

C. Apparently Induced but Possibly Spontaneous Mutants

With the exception of most of the mutants isolated after nitrous acid or hydroxylamine treatment, one cannot generally correlate the observed amino acid exchanges with a direct action of the mutagen employed. Even among the so-called HNO_2-induced mutants, the exchange Glu → Asp at position 95 of the coat protein of PM2 (Siegel *et al.*, 1962; Zaitlin and McCaughey, 1965; Wittmann, 1965) seems to have occurred independently of the action of HNO_2. The other exchange in PM2, namely Thr → Ile at position 28, can be explained on the basis of a change from C to U by HNO_2.

The mutants PM1, PM2 as well as DT-1 and DT-2 (Sarkar and Smitamana, 1981a) are conspicuous by their chlorotic reaction on Samsun, necrotic reaction on Xanthi-nc tobacco, and a slow spread from cell to cell. The spontaneous mutants DT-1 and DT-2 were isolated from single lesions produced by PM2 on Xanthi-nc tobacco. Both have a defective insoluble protein, like that of PM1. A spontaneous mutant of DT-1, named DT-1G, is unique in that it produces no coat protein at all (Sarkar and Smitamana, 1981a). It is symptomless on Samsun and java tobacco, produces small necrotic lesions on *N. glutinosa* and Xanthi-nc tobacco, and spreads slowly from cell to cell in Samsun leaves. Among these mutants, one can differentiate between those that have an insoluble coat protein (PM1, DT-1, DT-2) and those whose coat protein is soluble in water (PM2) (Hariharasubramanian and Siegel, 1969; Hariharasubramanian *et al.*, 1973). The defective coat protein of PM2 aggregates in the form of long twisted threads at low pH, and all the PM-type mutants except DT-1G produce inclusion bodies in the infected cells (Zaitlin and Ferris, 1964; Siegel *et al.*, 1966; Kassanis and Turner, 1972).

Certain partially defective mutants have been described: their coat protein denatures easily, the specific infectivity in plant sap is low, and they spread very slowly in systemic hosts (Sehgal, 1973; Bhalla and Sehgal, 1973).

Recently, Wertz *et al.* (1986) have described a mutant (D16y) of dahlemense, which produces a strong chlorosis on Samsun tobacco, retains the java-necrotic character of dahlemense, is sensitive to elevated temperatures, and appears to contain a histidine residue in its coat protein.

III. CONTRIBUTION OF MUTANTS AND STRAINS OF TMV TO VIROLOGY AND MOLECULAR BIOLOGY

TMV was the first virus to be crystallized (Stanley, 1935) and visualized in the electron microscope (Kausche and Ruska, 1939). The basic features of variation, host specificity, selection among viruses as well as chemical composition were evident from some of the very early studies on variants of TMV (McKinney, 1929; Bawden and Pirie, 1937; and others). It was through the study of TMV that RNA (and not only DNA) was shown to be a carrier of hereditary characters (Gierer and Schramm, 1956; Fraenkel-Conrat, 1956; Fraenkel-Conrat and Singer, 1957). Indeed, "TMV has been almost always the first" (Fraenkel-Conrat, 1981).

The following is an attempt to summarize some of the major achievements made through studies of mutants and strains of TMV.

1. TMV was the first virus to be reconstituted *in vitro* from its RNA and protein (Fraenkel-Conrat, 1956). Using different strains of TMV, Fraenkel-Conrat and Singer (1957, 1959) showed that the efficiency of reconstitution *in vitro* was high when the RNA and protein were isolated from the same strain of TMV (homologous combination), indicating a

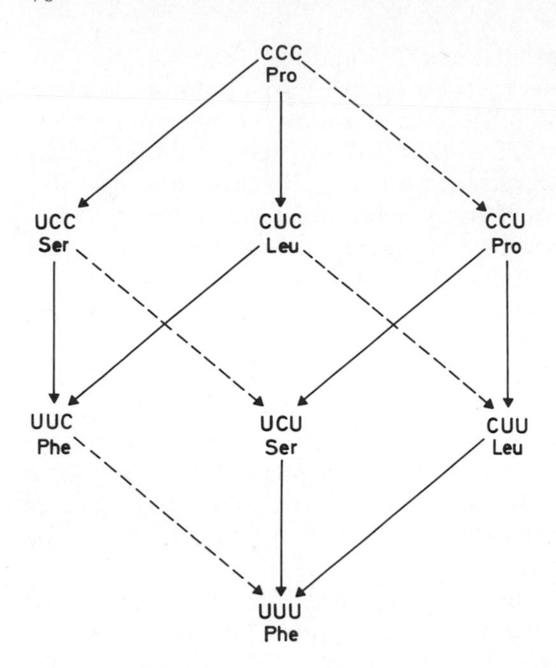

FIGURE 3. One of the eight possible schemes showing the nitrous acid-induced conversion of nucleotide bases in the codons, leading to the corresponding exchanges in amino acids (Wittmann, 1962).

high degree of specificity in the interaction between RNA and protein. These studies offered a very useful tool for experimental investigation of RNA–protein interactions.

2. Probably the most important contribution from the study of mutants and strains of TMV was the elucidation and confirmation of the genetic code. Soon after the codon dictionary was deciphered on the basis of peptide synthesis in the cell-free system of *E. coli* (Nirenberg and Matthaei, 1961), the protein sequence data of TMV mutants were utilized to test the validity of the genetic code (Wittmann, 1962; Tsugita, 1962; Tsugita and Fraenkel-Conrat, 1962; Funatsu and Fraenkel-Conrat, 1964; Wittmann and Wittmann-Liebold, 1966). This would not have been successful if the code were not universal. Moreover, the triplet code was shown to be nonoverlapping, since mutants with two or more neighboring amino acid substitutions would otherwise have been found. The data on TMV mutants also agreed with the concept that the genetic code is degenerate, although there were some restrictions in the pattern of amino acid exchange (cf. Nirenberg *et al.*, 1965). Of particular interest are the mutants produced by HNO_2 treatment. Figure 3 shows one of the eight possible vector diagrams (octets) indicating the direction of induced change in the bases and the corresponding amino acid alteration (Wittmann, 1962). In fact, almost all the amino acid exchanges found in HNO_2-induced mutants of the TMV coat protein have taken place in the direction of the arrows shown, providing a very good proof of the mode of action of HNO_2 as well as the validity of the genetic code. A comparison of the amino acid exchanges of all the TMV strains and mutants analyzed

hitherto (only a few of which are presented in Table III) shows that a host of the changes can be traced back to a modification of only one nucleotide base per codon. In a few other cases, however, more than one nucleotide base has obviously been altered. This is seen by comparing the codons of the different amino acids that have been found in one and the same position of the coat protein molecule of several strains and mutants. This situation does not necessarily speak against the validity of the genetic code, since more than one nucleotide could have been exchanged stepwise or suddenly in the course of evolution.

Attempts to correlate the observed frequency of mutation at definite loci with the probability of alteration of nucleotide bases in the codons have shown that a good qualitative agreement exists between them (Wittmann and Wittmann-Liebold, 1966). However, a better agreement can hardly be expected, since the mutants studied do not necessarily represent a randomized sample of all possible mutations. Depending on the nature and position of the changed amino acid, the conformation and stability of the coat protein may be drastically altered.

3. Comparison of the electrophoretic mobilities of the purified coat proteins and of intact viruses of different strains of TMV led to the understanding that the RNA in a compact structure like the TMV particle does not contribute to the surface charge (Kramer, 1957; Kramer and Wittmann, 1958).

4. It was through a comparison of the electrophoretic mobilities of the mixtures of native proteins of four strains of TMV that the first experimental data in support of a trimer configuration of TMV protein were obtained (Sarkar, 1960). This state of preferred aggregation of TMV protein was postulated by Caspar (1963) on theoretical considerations and crystallographic principles, although the so-called A-protein was at that time considered to be a hexamer (Schramm, 1947); the trimer nature of native TMV protein at neutral pH was confirmed through osmotic pressure measurements (Banerjee and Lauffer, 1966).

5. A comparison of the tryptic peptides of vulgare, dahlemense, flavum, and Holmes's ribgrass strains enabled Wittmann (1960) to arrange the 12 peptides in the proper order and to confirm the total amino acid sequence just determined by the standard methods (Anderer *et al.*, 1960; Tsugita *et al.*, 1960).

6. Among plant viruses, a genomic masking could be demonstrated for the first time using ts mutants of TMV (Schaskolskaya *et al.*, 1968; Sarkar, 1969; Kassanis and Bastow, 1971a,b; Kassanis and Conti, 1971; Dodds and Hamilton, 1972, 1976).

7. Using the truly coat protein-free mutant DT-1G of TMV, it was possible to show that the presence of a viral coat protein is not essential for cross-protection (Sarkar and Smitamana, 1981b). The observed interference may be the result of a competition between the viral genomes for occupying some specific sites in the host cell, as suggested earlier by Bawden and Kassanis (1945).

8. Variants of TMV have found extensive application in the analysis of many phenomena of biological and physicochemical interest such as the size and position of cistrons in the viral genome, the dependence of serological properties on the protein structure, RNA–protein interaction during the initiation and elongation phases of reconstitution *in vitro,* the nature of translation products in cell-free protein-synthesizing systems, the formation of a novel type of nucleoprotein particle of high buoyant density during early stages of TMV biosynthesis, and so on. These themes will be dealt with in other chapters.

ACKNOWLEDGMENTS. Sincere thanks are due to Dr. Dieter Vogel for some of the recent literature on TMV and to Mr. Andreas Dölz for his assistance in the compilation of some valuable data on the structure and temperature sensitivity of virulent strains and mutants.

REFERENCES

Aach, H. G., 1958a, Spektrophotometrische Untersuchungen an Mutanten des Tabakmosaikvirus, *Z. Naturforsch.* **13b:**165.

Aach, H. G., 1958b, Quantitative Aminosäuren-Analysen an Mutanten des Tabakmosaikvirus, *Z. Naturforsch.* **13b:**425.

Ainsworth, G. C., 1935, Mosaic diseases of the cucumber, *Ann. Appl. Biol.* **22:**55.

Anderer, F. A., Uhlig, H., Weber, E., and Schramm, G., 1960, Primary structure of the protein of tobacco mosaic virus, *Nature (London)* **186:**922.

Bald, J. G., Gumpf, D. J., and Heick, J., 1974, Transition from common tobacco mosaic virus to the Nicotiana glauca form, *Virology* **59:**467.

Banerjee, K., and Lauffer, M. A., 1966, Polymerization–depolymerization of tobacco mosaic virus protein. VI. Osmotic pressure studies of early stages of polymerization, *Biochemistry* **5:**1957.

Bawden, F. C., 1950, *Plant Viruses and Virus Diseases,* 3rd ed., Chronica Botanica, Waltham, Mass.

Bawden, F. C., 1958, Reversible changes in strains of tobacco mosaic virus from leguminous plants, *J. Gen. Microbiol.* **18:**751.

Bawden, F. C., and Kassanis, B., 1945, The suppression of one plant virus by another, *Ann. Appl. Biol.* **32:**52.

Bawden, F. C., and Pirie, N. W., 1937, The isolation and some properties of liquid crystalline substances from solanaceous plants infected with three strains of tobacco mosaic virus, *Proc. R. Soc. London Ser. B* **123:**274.

Bewley, W. F., 1924, Mycological investigations, *Chestnut Exp. Res. Stn., Hertfordshire, Annu. Rep.* **9**(1923)**:**66.

Bhalla, R. B., and Sehgal, O. P., 1973, Host range and purification of the nucleic acid of a defective mutant of tobacco mosaic virus, *Phytopathology* **63:**906.

Brammer, K. W., 1963, Chemical modification of viral ribonucleic acid. II. Bromination and iodination, *Biochim. Biophys. Acta* **72:**217.

Capoor, S. P., 1950, A mosaic disease of sunn hemp in Bombay, *Curr. Sci.,* **19:**22.

Caspar, D. L. D., 1963, Assembly and stability of the tobacco mosaic virus particle, *Adv. Protein Chem.* **18:**37.

Dawson, W. O., and Jones, G. E., 1976, A procedure for specifically selecting temperature sensitive mutants of tobacco mosaic virus, *Mol. Gen. Genet.* **145:**307.

Diener, T. O., 1962, Isolation of an infectious, ribonuclease sensitive fraction from tobacco leaves recently inoculated with tobacco mosaic virus, *Virology* **16:**140.

Dodds, J. A., and Hamilton, R. I., 1972, The influence of barley stripe mosaic virus on the replication of tobacco mosaic virus in Hordeum vulgare L., *Virology* **50:**404.

Dodds, J. A., and Hamilton, R. I., 1976, Structural interactions between viruses as a consequence of mixed infections, *Adv. Virus Res.* **20:**33.

Fraenkel-Conrat, H., 1956, The role of the nucleic acid in the reconstitution of active tobacco mosaic virus, *J. Am. Chem. Soc.* **78:**882.

Fraenkel-Conrat, H., 1961, Chemical modification of viral ribonucleic acid. I. Alkylating agents, *Biochim. Biophys. Acta* **49:**169.

Fraenkel-Conrat, H., 1981, Portraits of viruses: Tobacco mosaic virus, *Intervirology* **15:**177.

Fraenkel-Conrat, H., and Singer, B., 1957, Virus reconstitution. II. Combination of protein and nucleic acid from different strains, *Biochim. Biophys. Acta* **24:**540.

Fraenkel-Conrat, H., and Singer, B., 1959, Reconstitution of tobacco mosaic virus. III. Improved method and the use of mixed nucleic acids, *Biochim. Biophys. Acta* **33:**359.

Fraenkel-Conrat, H., Singer, B., and Tsugita, A., 1961, Purification of viral RNA by means of bentonite, *Virology* **14:**54.

Fraser, R. S. S., 1983, Varying effectiveness of the N' gene for resistance to tobacco mosaic virus in tobacco infected with virus strains differing in coat protein properties, *Physiol. Plant Pathol.* **22:**109.

Funatsu, G., and Fraenkel-Conrat, H., 1964, Location of amino acid exchanges in chemically evoked mutants of tobacco mosaic virus. *Biochemistry* **3:**1356.

Garciá-Arenal, F., Palukaitis, P., and Zaitlin, M., 1984, Strains and mutants of tobacco mosaic virus are both found in virus derived from single-lesion-passaged inoculum, *Virology* **132:**131.

Gibbs, A., and Harrison, B., 1976, *Plant Virology: The Principles,* Arnold, London.

Gierer, A., and Mundry, K. W., 1958, Production of mutants of tobacco mosaic virus by chemical alteration of its ribonucleic acid in vitro, *Nature (London)* **182:**1457.

Gierer, A., and Schramm, G., 1956, Die Infektiosität der Nukleinsäure aus Tabakmosaikvirus, *Z. Naturforsch.* **11b:**138.

Hariharasubramanian, V., and Siegel, A., 1969, Characterization of a new defective strain of TMV, *Virology* **37:**203.

Hariharasubramanian, V., Smith, R. C., and Zaitlin, M., 1973, Insoluble coat protein mutants of TMV: Their origin, and characterization of the defective coat proteins, *Virology* **55:**202.

Hennig, B., and Wittmann, H. G., 1972, Tobacco mosaic virus: Mutants and strains, in: *Principles and Techniques in Plant Virology* (C. I. Kado and H. O. Agarwal, eds.), pp. 546–594, Van Nostrand–Reinhold, Princeton, N.J.

Hiruki, C., and Hidaka, Z., 1954, cited by Nozu, Y., and Okada, Y., 1968, Amino acid sequence of a common Japanese strain of tobacco mosaic virus, *J. Mol. Biol.* **35:**643.

Holmes, F. O., 1929, Local lesions in tobacco mosaic, *Bot. Gaz.* **87:**39.

Holmes, F. O., 1934, A masked strain of tobacco mosaic virus, *Phytopathology* **24:**845.

Holmes, F. O., 1941, A distinctive strain of tobacco mosaic virus from Plantago, *Phytopathology* **31:**1089.

Hubert, J. J., and Bourque, D. P., 1981, Improved method for the isolation and propagation of defective tobacco mosaic virus mutants, *Phytopathology* **71:**295.

Inouye, T., Inouye, N., Asatani, M., and Mitsuhata, K., 1967, Studies on cucumber green mottle virus I, *Nogaku Kenkyu* **51:**175.

Ivanowski, D., 1899, Über die Mosaikkrankheit der Tabak pflanze, *Bull.Acad. Imp. Sci. St.-Petersbourg (New Ser.)* **3:**65.

Jauregui-Adell, J., Hindennach, I., and Wittmann, H. G., 1969, Die primäre Proteinstruktur von Stämmen des Tabakmosaikvirus. Teil V: Aminosäuresequenz (1–61) des Proteins des Tabakmosaikvirus-Stammes Holmes rib grass, *Z. Naturforsch.* **24b:**877.

Jensen, D. D., and Gold, A. H., 1951, A virus ring spot of Odontoglossum orchid: Symptoms, transmission and electron microscopy, *Phytopathology* **41:**648.

Jensen, J. H., 1937, Studies on representative strains of tobacco mosaic virus, *Phytopathology* **27**:69.

Jockusch, H., 1964, In vivo und in vitro Verhalten temperatursensitiver Mutanten des Tabakmosaikvirus, *Z. Vererbungsl.* **95**:379.

Jockusch, H., 1966a, Relations between temperature sensitivity, amino acid replacements and quaternary structure of mutant proteins, *Biochem. Biophys. Res. Commun.* **24**:577.

Jockusch, H., 1966b, Temperatursensitive Mutanten des Tabakmosaikvirus. I. In vivo-verhalten, *Z. Vererbungsl.* **98**:320.

Jockusch, H., 1966c, Temperatursensitive Mutanten des Tabakmosaikvirus. II. In vitro-verhalten, *Z. Vererbungsl.* **98**:344.

Jockusch, H., 1968, Two mutants of tobacco mosaic virus temperature-sensitive in two different functions, *Virology* **35**:94.

Kado, C. I., Van Regenmortel, M. H. V., and Knight, C. A., 1968, Studies on some strains of tobacco mosaic virus in orchids. I. Biological, chemical and serological studies, *Virology* **34**:17.

Kassanis, B., and Bastow, C., 1971a, In vivo phenotypic mixing between two strains of tobacco mosaic virus, *J. Gen. Virol.* **10**:95.

Kassanis, B., and Bastow, C., 1971b, Phenotypic mixing between strains of tobacco mosaic virus, *J. Gen. Virol.* **11**:171.

Kassanis, B., and Conti, M., 1971, Defective strains and phenotypic mixing, *J. Gen. Virol.* **13**:361.

Kassanis, B., and Turner, R. H., 1972, Virus inclusions formed by the PM_2 mutant of TMV, *J. Gen. Virol.* **14**:119.

Kausche, G. A., and Ruska, H., 1939, Die Struktur der kristallinen Aggregate des Tabakmosaikvirus-Proteins, *Biochem. Z.* **303**:211.

Knight, C. A., 1963, Chemistry of viruses, in: *Protoplasmatologia, Handbuch der Protoplasmaforschung* (Alfert, M., Bauer, H., and Harding, C. V., eds.), Volume IV, pp. 1–177, Springer-Verlag, Berlin.

Knight, C. A., 1975, Chemistry of viruses, in: *Protoplasmatologia, Handbuch der Protoplasmaforschung* (Alfert, Bauer, and Harding, eds.), Volume IV, 2nd ed., Springer-Verlag, Berlin.

Knight, C. A., Silva, D. M., Dahl, D., and Tsugita, A., 1962, Two distinctive strains of tobacco mosaic virus, *Virology* **16**:236.

Kramer, E., 1957, Elektrophoretische Untersuchungen an Mutanten des Tabakmosaikvirus, *Z. Naturforsch.* **12b**:609.

Kramer, G., and Wittmann, H. G., 1958, Elektrophoretische Untersuchungen der A–proteine dreier Tabakmosaikvirus-Stämme, *Z. Naturforsch.* **13b**:30.

Kramer, G., Wittmann, H. G., and Schuster, H., 1964, Die Erzeugung von Mutanten des Tabakmosaikvirus durch den Einbau von fluorouracil in die Virusnucleinsäure, *Z. Naturforsch.* **19b**:46.

Kunkel, L. O., 1934, Studies on acquired immunity with tobacco and aucuba mosaics, *Phytopathology* **24**:437.

Lister, R. M., and Thresh, J. M., 1955, A mosaic disease of leguminous plants caused by a strain of tobacco mosaic virus, *Nature (London)* **175**:1047.

McKinney, H. H., 1929, Mosaic diseases in the Canary Islands, West Africa and Gibraltar, *J. Agric. Res.* **39**:557.

MacNeill, B. H., and Boxall, M., 1974, The evolution of a pathogenic strain of tobacco mosaic virus in tomato: A host-passage phenomenon, *Can. J. Bot.* **52**:1305.

Matthews, R. E. F., 1981, *Plant Virology*, Academic Press, New York.

Melchers, G., 1940, Die biologische Untersuchung des "Tomatenmosaikvirus Dahlem 1940," *Biol. Zentralbl.* **60**:527.

Melchers, G., 1942, Über einige Mutationen des Tabakmosaikvirus und eine "Parallelmutation" des Tomatenmosaikvirus, *Naturwissenschaften* **30**:48.

Melchers, G., Jockusch, H., and von Sengbusch, P., 1966, A tobacco mutant with a dominant allele for hypersensitivity against some TMV strains, *Phytopathol. Z.* **55**:86.

Miller, P. M., 1953, An apparently new viral disease of tomatoes in Illinois, *Phytopathology* **43:**480.

Mundry, K. W., 1957a, Zur Frage des Einflusses von Röntgen- und UV-Strahlen auf die Mutationsrate des Tabakmosaikvirus nach Behandlung reiner Präparate, *Z. Indukt. Abstamm. Verebungsl.* **88:**115.

Mundry, K. W., 1957b, Die Abhängigkeit des Auftretens neuer Virusstämme von der Kulturtemperatur der Wirtspflanzen, *Z. Indukt. Abstamm. Vererbungsl.* **88:**407.

Mundry, K. W., 1959, The effect of nitrous acid on tobacco mosaic virus: Mutation, not selection, *Virology* **9:**722.

Mundry, K. W., 1960, Mutationsuntersuchungen am Tabakmosaikvirus in vitro. II. Versuche zum Problem der Mutagenese durch UV-Bestrahlung reiner Viruspräparate, *Z. Vererbungsl.* **91:**87.

Mundry, K. W., and Gierer, A., 1958, Die Erzeugung von Mutationen des TMV durch chemische Behandlung seiner Nukleinsäure in vitro, *Z. Vererbungsl.* **89:**614.

Nagata, C., and Martensson, O., 1968, On the mechanism of mutagenic action of hydroxylamine, *J. Theor. Biol.* **19:**133.

Nirenberg, M. W., and Matthaei, J. H., 1961, The dependence of cell-free protein synthesis in E. coli upon naturally occurring or synthetic polyribonucleotides, *Proc. Natl. Acad. Sci. USA* **47:**1588.

Nirenberg, M., Leder, P., Bernfield, M., Brimacombe, R., Trupin, J., Rottmann, F., and O'-Neal, C., 1965, RNA codewords and protein synthesis. VII. On the general nature of the RNA code, *Proc. Natl. Acad. Sci. USA* **53:**1161.

Opel, H., Kegler, H., and Richter, J., 1969, Vorkommen und Charakterisierung von TMV-Stämmen des Kernobstes, *Acta Phytopathol. Acad. Sci. Hung.* **4:**1.

Pelham, J., Fletcher, J. T., and Hawkins, J. H., 1970, The establishment of a new strain of tobacco mosaic virus resulting from the use of resistant varieties, *Ann. Appl. Biol.* **65:**293.

Peters, D. L., and Murphy, T. M., 1975, Selection of temperature-sensitive mutants of tobacco mosaic virus by lesion morphology, *Virology* **65:**595.

Rombauts, W., and Fraenkel-Conrat, H., 1968, Artificial histidine-containing mutants of tobacco mosaic virus, *Biochemistry* **7:**3334.

Sammons, I. M., and Chessin, M., 1961, Cactus virus in the United States, *Nature (London)* **191:**517.

Sarkar, S., 1960, Interaction and mixed aggregation of proteins from tobacco mosaic virus strains, *Z. Naturforsch.* **15b:**778.

Sarkar, S., 1963, Relative infectivity of tobacco mosaic virus and its nucleic acid, *Virology* **20:**185.

Sarkar, S., 1969, Evidence of phenotypic mixing between two strains of tobacco mosaic virus, *Mol. Gen. Genet.* **105:**87.

Sarkar, S., and Smitamana, P., 1981a, A truly coat-protein-free mutant of tobacco mosaic virus, *Naturwissenschaften* **68:**145.

Sarkar, S., and Smitamana, P., 1981b, A proteinless mutant of tobacco mosaic virus: Evidence against the role of a viral coat protein for interference, *Mol. Gen. Genet.* **184:**158.

Schaskolskaya, N. D., Atabekov, J. G., Sacharovskaya, G. N., and Javachia, V. G., 1968, Replication of temperature-sensitive strain of tobacco mosaic virus under nonpermissive conditions in the presence of helper strain, *Biol. Sci. (USSR)* **8:**101.

Schramm, G., 1947, Über die Spaltung des Tabakmosaikvirus und die Wiedervereinigung der Spaltstücke zu höhermolekularen Proteinen. I. Die Spaltungsreaktion, *Z. Naturforsch.* **2b:**112.

Schuster, H., and Schramm, G., 1958, Bestimmung der biologisch wirksamen Einheit in der Ribosenukleinsäure des Tabakmosaikvirus auf chemischen Wege, *Z. Naturforsch.* **13b:**697.

Schuster, H., and Wittmann, H. G., 1963, The inactivating and mutagenic action of hydroxylamine on tobacco mosaic virus ribonucleic acid, *Virology* **19:**421.

Sehgal, O. P., 1973, Biological and physico-chemical properties of an atypical mutant of tobacco mosaic virus, *Mol. Gen. Genet.* **121**:15.

Siegel, A., and Wildman, S. G., 1954, Some natural relationships among strains of tobacco mosaic virus, *Phytopathology* **44**:277.

Siegel, A., Zaitlin, M., and Sehgal, O. P., 1962, The isolation of defective tobacco mosaic virus strains, *Proc. Natl. Acad. Sci. USA* **48**:1845.

Siegel, A., Hills, G. J., and Markham, R., 1966, In vitro and in vivo aggregation of the defective PM2 TMV protein, *J. Mol. Biol.* **19**:140.

Singer, B., and Fraenkel-Conrat, H., 1965, Effects of light in the presence of iron salts on ribonucleic acid and model compounds, *Biochemistry* **4**:226.

Singer, B., and Fraenkel-Conrat, H., 1966, Dye catalyzed photoinactivation of tobacco mosaic virus ribonucleic acid, *Biochemistry* **5**:2446.

Singer, B., and Fraenkel-Conrat, H., 1967, Chemical modification of viral RNA. VI. The action of N-methyl-N'-nitro-N-nitrosoguanidine, *Proc. Natl. Acad. Sci. USA* **58**:234.

Singer, B., and Fraenkel-Conrat, H., 1974, Correlation between amino acid exchanges in coat protein of TMV mutants and the nature of the mutagenesis, *Virology* **60**:485.

Singer, B., Sun, L., and Fraenkel-Conrat, H., 1975, Effects of alkylation of phosphodiesters and of bases on infectivity and stability of tobacco mosaic virus RNA, *Proc. Natl. Acad. Sci. USA* **72**:2232.

Singer, S. J., Bald, J. G., Wildman, S. G., and Owen, R. D., 1951, The detection and isolation of naturally occurring strains of tobacco mosaic virus by electrophoresis, *Science* **114**:463.

Spackman, B. H., Stein, W. H., and Moore, S., 1958, Automatic recording apparatus for use in the chromatography of amino acids, *Anal. Chem.* **30**:1190.

Stanley, W. M., 1935, Isolation of a crystalline protein possessing the properties of the tobacco mosaic virus, *Science* **81**:644.

Takahashi, W. N., 1956, Increasing the sensitivity of the local-lesion method of virus assay, *Phytopathology* **46**:654.

Tsugita, A., 1962, The proteins of mutants of TMV: Composition and structure of chemically evoked mutants of TMV-RNA, *J. Mol. Biol.* **5**:284.

Tsugita, A., and Fraenkel-Conrat, H., 1962, The composition of proteins of chemically evoked mutants of TMV-RNA, *J. Mol. Biol.* **4**:73.

Tsugita, A., Gish, D. T., Young, J., Fraenkel-Conrat, H., Knight, C. A., and Stanley, W. M., 1960, The complete amino acid sequence of the protein of tobacco mosaic virus, *Proc. Natl. Acad. Sci. USA* **46**:1463.

Van Regenmortel, M. H. V., 1967, Serological studies on naturally occurring strains and chemically induced mutants of tobacco mosaic virus, *Virology* **31**:467.

von Sengbusch, P., 1965, Aminosäureaustausche und Tertiärstruktur eines Proteins. Vergleich von Mutanten des Tabakmosaikvirus mit serologischen und physikochemischen Methoden, *Z. Vererbungsl.* **96**:364.

von Sengbusch, P., and Wittmann, H. G., 1965, Serological and physicochemical properties of the wild strains and two mutants of tobacco mosaic virus with the same amino acid exchange in different positions of the protein chain, *Biochem. Biophys. Res. Commun.* **18**:780.

Wang, A. L., and Knight, C. A., 1967, Analysis of protein components of tomato strains of tobacco mosaic virus, *Virology* **31**:101.

Wertz, J. M., Smitamana, P., and Sarkar, S., 1986, Characterisation of a defective mutant of the dahlemense strain of tobacco mosaic virus, *Z. Naturforsch.* (in press).

Wetter, C., 1984, Antigenic relationships between isolates of mild dark-green tobacco mosaic virus, and the problem of host-induced mutation, *Phytopathology* **74**:1308.

Wittmann, H. G., 1959, Vergleich der Proteine des Normalstammes und einer Nitritmutante des Tabakmosaikvirus, *Z. Vererbungsl.* **90**:463.

Wittmann, H. G., 1960, Tryptic peptides within the polypeptide chain of tobacco mosaic virus and a new manner of determining their arrangement, *Virology* **11**:505.

Wittmann, H. G., 1962, Proteinuntersuchungen an Mutanten des Tabakmosaikvirus als Beitrag zum Problem des genetischen Codes, *Z. Vererbungsl.* **93:**491.

Wittmann, H. G., 1964, Proteinanalysen von chemisch induzierten Mutanten des Tabakmosaikvirus, *Z. Vererbungsl.* **95:**333.

Wittmann, H. G., 1965, Die Proteinstruktur der Defektmutante PM2 des Tabakmosaikvirus, *Z. Vererbungsl.* **97:**297.

Wittmann, H. G., and Wittmann-Liebold, B., 1966, Protein chemical studies of the RNA viruses and their mutants, *Cold Spring Harbor Symp. Quant. Biol.* **31:**163.

Wittmann, H. G., Wittmann-Liebold, B., and Jauregui-Adell, J., 1965, Die primäre Proteinstruktur temperatursensitiver Mutanten des Tabakmosaikvirus, *Z. Naturforsch.* **20b:**1224.

Wittmann, H. G., Hindennach, I., and Wittmann-Liebold, B., 1969, Die primäre Proteinstruktur von Stämmen des Tabakmosaikvirus. Teil V: Aminosäuresequenz (Positionen 62–156) des Proteins des Tabakmosaikvirusstammes Holmes rib grass, *Z. Naturforsch.* **24b:**870.

Wittmann-Liebold, B., and Wittmann, H. G., 1963, Die primäre Proteinstruktur von Stämmen des Tabakmosaikvirus, *Z. Vererbungsl.* **94:**427.

Wittmann-Liebold, B., and Wittmann, H. G., 1965a, Lokalisierung von Aminosäureaustauschen bei Spontanmutanten und nach fluorouracil-einbau isolierten Mutanten des Tabakmosaikvirus, *Z. Vererbungsl.* **97:**218.

Wittmann-Liebold, B., and Wittmann, H. G., 1965b, Lokalisierung von Aminosäureaustauschen bei Nitritmutanten des Tabakmosaikvirus, *Z. Vererbungsl.* **97:**305.

Wittmann-Liebold, B., and Wittmann, H. G., 1967, Coat proteins of strains of two RNA viruses: Comparison of their amino acid sequences, *Mol. Gen. Genet.* **100:**358.

Wittmann-Liebold, B., Jauregui-Adell, J., and Wittmann, H. G., 1965, Die primäre Proteinstruktur temperatursensitiver Mutanten des Tabakmosaikvirus. II. Chemisch induzierte Mutanten, *Z. Naturforsch.* **20b:**1235.

Zaitlin, M., and Ferris, W. R., 1964, Unusual aggregation of a nonfunctional tobacco mosaic virus protein, *Science* **143:**1451.

Zaitlin, M., and McCaughey, W. F., 1965, Amino acid composition of a nonfunctional tobacco mosaic virus protein, *Virology* **26:**500.

CHAPTER 4

TOBACCO MOSAIC VIRUS

Antigenic Structure

M. H. V. Van Regenmortel

1. INTRODUCTION

When the primary structure of the TMV protein subunit (TMVP) was elucidated more than 25 years ago (Anderer *et al.*, 1960; Tsugita *et al.*, 1960), it became possible for the first time to analyze the antigenicity of a virus at the molecular level (Anderer, 1963a; Benjamini *et al.*, 1964). Two regions of TMVP were singled out for analysis in these early studies: the C-terminal hexapeptide corresponding to residues 153–158 (Anderer, 1963b), and a tryptic peptide corresponding to residues 93–112 (Benjamini *et al.*, 1965; Young *et al.*, 1966). Several years later, TMV also became the first virus to have its three-dimensional structure elucidated by X-ray crystallography (see Chapter 2) and this made it possible to interpret immunochemical data in terms of the known location of each residue. This led to a rapid increase in our knowledge of the antigenic structure of TMV (Milton and Van Regenmortel, 1979; Altschuh *et al.*, 1983, 1985; Al Moudallal *et al.*, 1982, 1985) with the result that the immunochemical properties of TMV are probably better known than those of any other virus.

Analyzing the antigenic structure of a protein consists mainly in locating its antigenic determinants or epitopes, i.e., in ascertaining which accessible patches on the surface of the molecule are recognized by the binding sites (or paratopes) of specific antibody molecules. As a first approximation, the minimum number of epitopes present on a protein is defined by the number of antibody molecules capable of binding simul-

M. H. V. VAN REGENMORTEL • Institut de Biologie Moléculaire et Cellulaire du CNRS, 67084 Strasbourg Cédex, France.

taneously to the antigen. This number, known as the antigenic valence, was found to be 3–5 for the TMVP subunit (Van Regenmortel and Lelarge, 1973) and 780 for the TMV particle (Van Regenmortel and Hardie, 1976). The fact that the antigenic valence of the virus is smaller than the number of protein subunits of the capsid (i.e., 2130) is due to steric hindrance which limits the number of antibody molecules able to bind simultaneously to the virus.

The classical studies of myoglobin and lysozyme by Atassi and his colleagues (Atassi, 1975; Atassi and Lee, 1978) initially led to the view that proteins contain only a small number of discrete epitopes situated in highly accessible surface locations. Subsequent studies by several other groups altered this initial view of protein antigenicity and eventually established a new paradigm according to which the entire accessible surface of a protein consists of a continuum of overlapping epitopes potentially able to combine with appropriate antibodies (Benjamin *et al.*, 1984). As a result of this new perception of protein antigenicity, investigators no longer attempt to describe the "complete" or "definitive" antigenic structure of a molecule but instead are satisfied to define operationally some of the fragments or regions of the protein that are able to bind to specific antibodies (Van Regenmortel, 1985).

In this chapter, the following types of epitopes will be distinguished (Atassi and Smith, 1978; Van Regenmortel, 1966, 1982): (1) *continuous epitopes* consisting of a linear sequence of residues exposed at the surface of a native protein and possessing distinctive conformational features; (2) *discontinuous epitopes* consisting of residues that are not contiguous in the sequence but are brought together at the surface of the molecule by the folding of the polypeptide chain; (3) *cryptotopes* or hidden epitopes that become antigenically active only after fragmentation, depolymerization, or denaturation of the antigen; and (4) *neotopes* or epitopes that are specific for the quaternary structure of virus particles and are thus absent in the constituent monomeric subunits of the particle. Neotopes owe their existence either to conformational changes of the protein induced by intersubunit bonds or to the juxtaposition of residues from neighboring subunits.

II. METHODS USED TO STUDY TMV ANTIGENICITY

The antigenic properties of TMV have been studied by at least 16 different serological techniques which are listed in an earlier review (Van Regenmortel, 1981). However, only four such techniques, namely precipitation, complement fixation, passive hemagglutination, and enzyme-linked immunosorbent assay (ELISA), have been used for identifying epitopes at the molecular level. These techniques have been used in conjunction with the following analytical approaches (Van Regenmortel and Neurath, 1985; Van Regenmortel, 1985).

A. Fragmentation of the Protein

This method consists in measuring the capacity of natural fragments of TMVP obtained by chemical or enzymatic cleavage to bind to antibodies raised against either the virus or the subunit. Binding of the peptide fragments is usually demonstrated by their capacity to inhibit the reaction between the complete antigen and its specific antibodies. Anderer (1963a) showed that fragments coresponding to residues 18–23, 62–68, 123–134, and 153–158 in TMVP were able to inhibit the precipitation reaction between virus and antibody. Although the best inhibitory activity was observed with the C-terminal peptide 153–158, this peptide, when coupled to a carrier, was not precipitated by TMV antibodies. The same peptide–carrier conjugate, however, was found to react with TMV antibodies when the more sensitive passive hemagglutination test was used (Anderer and Ströbel, 1972a,b). All peptides possessed only a very low antigenic activity and it was assumed that this was due to the fact that they no longer retained the original conformation present in the intact molecule.

Using TMVP subunits and antisera raised against them, Benjamini *et al.* (1964) found that in a mixture of all 12 tryptic peptides of TMVP, only tryptic peptide 8 (residues 93–112) possessed inhibitory activity in the complement fixation technique. Since this peptide completely inhibited the complement fixation reaction between TMVP and anti-TMVP, it was considered to represent the major if not the only epitope of the viral subunit. It was also assumed that since this peptide completely abolished the complement fixation reaction between TMVP and its specific antibodies, the TMVP antisera contained no antibodies specific for other peptides or for any conformational features that would have been destroyed by the tryptic cleavage (Benjamini, 1977). As discussed elsewhere (Van Regenmortel, 1982), such an interpretation of inhibition data obtained by complement fixation is based on the erroneous assumption that inhibition percentages simply reflect the proportion of each antibody population in an antiserum. It is important to realize that a peptide capable of reacting with only 10% of the antibodies present in an antiserum is able to cause complete inhibition of the immunochemical reaction. When in a particular test the percentage of complement fixation is reduced from 80% to 0%, the 100% inhibition pertains only to the antiserum dilution used in the test, and it does not mean that the entire antibody activity of the antiserum has been exhausted by the inhibitor. Repeating the same test using a lower antiserum dilution would reveal the presence of residual active antibody after a so-called complete inhibition. Therefore, the best way to express quantitatively the inhibitory activity of a peptide is to compare the antiserum dilutions that produce the same immunochemical signal, in the presence and in the absence of the inhibitor (Van Regenmortel, 1982).

The inhibitory activity of tryptic peptides of TMVP in complement fixation tests was again examined by Milton and Van Regenmortel (1979). It was found that tryptic peptides 4 (residues 62–68), 8, and 12 (residues 142–158) respectively caused approximately 20, 70, and 80% inhibition of complement fixation in the TMVP anti-TMVP system. Tryptic peptide 1 (residues 1–41) did not inhibit complement fixation in this system but on the contrary enhanced the amount of fixation. This effect was not due to aggregation of the peptide since it was abolished when the TMVP antiserum was first absorbed with TMV in order to remove virus-specific antibodies; after such absorption, peptide 1 was found to possess inhibitory activity. These results provided the first evidence that tryptic peptide 1 possesses two epitopes, one of which is a cryptotope and therefore unreactive with virus-specific antibodies, and the other one an epitope common to both virus and depolymerized subunit. The 41-residue peptide 1 is able to bind two different antibody molecules, one of each specificity, and the resulting complex is therefore able to bind complement. Complement fixation tests are notoriously sensitive to aggregation of the antigen under study. On the one hand, viral subunits tend to form aggregates of various sizes and these may bind variable amounts of complement leading to fluctuating complement fixation curves; on the other hand, aggregation of a peptide may produce a multivalent antigen that will lead to complement binding instead of inhibition of binding.

Problems caused by aggregation of the antigen are eliminated when peptides are tested for activity by inhibition of ELISA. Using this technique, Altschuh and Van Regenmortel (1982) showed that tryptic peptides 3 (residues 47–61) and 6 (residues 72–90) of TMVP also possessed antigenic activity.

In summary, studies with TMVP fragments have led to somewhat discrepant results. Two of the peptides found by Anderer (1963a) to be antigenically active (residues 18–23 and 123–134) were found to be inactive in later work. Furthermore, whereas Benjamini *et al.* (1964) reported that only tryptic peptide 8 possessed inhibitory activity in complement fixation tests, Milton and Van Regenmortel (1979), using the same technique, found four additional antigenic regions in tryptic peptides 1, 4, and 12. It is also worth noting that the latter authors, using three different TMVP antisera, found that the epitopes located in peptides 4 and 12 possessed more activity, on a molar basis, than the epitope located in peptide 8. Finally, by ELISA inhibition tests, antigenic activity could also be demonstrated in tryptic peptides 3 and 6.

B. Studies with Synthetic Peptides

The delineation of epitopes using natural peptide fragments is hampered by the limited number of different cleavage products that can be obtained as well as by the difficulty of obtaining sufficiently pure peptides. When a preparation of an antigenically inactive peptide is contaminated by a small amount of a very active peptide, the activity may erroneously be attributed to the major component. These limitations inherent in the use of natural fragments can be overcome by using synthetic peptides. Advances in solid-phase peptide synthesis (Stewart and Young, 1984) make it possible to obtain peptides of 10–30 residues with considerable ease, and this approach has been extensively used to locate continuous epitopes in TMVP. The antigenic activity of the region 93–112 was further studied in direct binding assays by means of labeled synthetic peptides varying in length from dipeptide to decapeptide and corresponding to residues 103–112 (Young *et al.*, 1967). The shortest peptide showing significant binding to TMVP antibodies was the pentapeptide 108–112 which was initially identified as an epitope of the subunit, although considerably more activity was present in the octapeptide 105–112 (Benjamini *et al.*, 1968). In later work (Morrow *et al.*, 1984), it was found that the entire length of the decapeptide 103–112 was needed for optimal binding to some monoclonal antibodies raised against TMVP, and this decapeptide was therefore considered to represent the epitope. By using a series of synthetic analogues of the decapeptide with substitutions at various locations, it could be shown that not all ten residues were equally important for binding to different monoclonal antibodies (Morrow *et al.*, 1984).

Synthetic peptides were also used to identify which regions in tryptic peptide 1 harbored antigenic activity. By means of ELISA and complement fixation inhibition experiments, it was found that peptides corresponding to residues 1–10, 19–32, and 34–39 were able to bind to TMVP antibodies (Altschuh *et al.*, 1983; Al Moudallal *et al.*, 1985).

C. Cross-Reactivity Studies with Antipeptide Antibodies

Instead of testing the ability of peptide fragments to bind to antibodies raised against TMV or TMVP, another approach consists in using the peptides for immunization. If antipeptide antibodies obtained in this way react with the whole antigen, it is concluded that the immunizing peptide represents an epitope of the intact molecule. Antibodies obtained by immunizing animals with the C-terminal hexa-, penta-, tetra-, or tripeptides of TMVP coupled to bovine serum albumin (BSA) were shown to precipitate the virus and to neutralize its infectivity (Anderer, 1963b; Anderer and Schlumberger, 1965a,b). Antibodies were also prepared against the single terminal residue (threonine) of TMVP coupled to a carrier, and these also precipitated and neutralized the virus. This astonishing result is probably caused by the particularly exposed location of the C-terminus at position 158 which is kept in a favorable orientation by the proline residue at position 156; it was found that the antithreonine antibodies hardly reacted with the terminal threonine of a viral mutant in which only proline 156 is replaced by leucine. These results illustrate the difficulties that are encountered when the immunizing residues are equated with an epitope of the native protein. It would, of course, be incongruous to suggest that the C-terminal threonine corresponds to an epitope.

Peptides corresponding to residues 108–112 and 103–112 of TMVP, after coupling to BSA, were also used to immunize animals, and in both cases the resulting antisera reacted with TMVP (Fearney *et al.*, 1971). Since the immune response of different animals immunized with peptides may vary considerably, it is difficult by this approach to assess the relative contribution of different residues to the structure of an epitope of the native molecule.

D. Cross-Reactivity Studies with TMV Mutants

This approach consists in studying the alterations in antigenic properties of TMVP brought about by modifications of individual amino acid residues of the protein. Modifications resulting from the use of selective chemical reagents (Slobin, 1970; Staab and Anderer, 1976) or from mutational events have been studied. A large number of TMV mutants with one to three amino acid exchanges have been compared with the wild strain by means of precipitation and immunodiffusion tests (von Sengbusch, 1965; Van Regenmortel, 1967). It was found that substitutions occurring at residues 65, 66, 107, 136, 138, 140, 148, and 156 altered the antigenic properties of virus particles, whereas substitutions at positions 20, 21, 25, 33, 46, 59, 63, 81, 97, 99, 126, and 129 did not. However, as discussed previously (Van Regenmortel, 1982), the interpretation of mutant studies is not at all straightforward. It was found, for instance, that exchanges at positions 65 and 140 gave rise to virions that were serologically distinguishable from TMV in precipitation tests, although the corresponding depolymerized subunits were not distinguishable in inhibition assays with a TMVP antiserum. Conversely, the protein subunit of a mutant with an exchange at position 20 was distinguishable from TMVP whereas the corresponding two virions were not distinguishable by precipitation (Milton *et al.*, 1980). Proline substitutions at positions 20 and 156 seemed to affect the antigenic properties by changing the conformation of the polypeptide chain. The exchange Pro 156 → Leu allowed the mutant virion to react with heterospecific antibodies present in TMV antisera. Such heterospecific antibodies, which seem to be elicited in all animals immunized with TMV, are antibodies that are better recognized by another antigen than the one used for immunization (see Van Regenmortel, 1982, 1984a). The presence of these antibodies could be demonstrated by the fact that TMV antisera, after being absorbed with the homologous TMV, were still capable of reacting with the mutant showing a substitution at position 156.

Comparisons between viral proteins possessing a small number of substitutions are most informative when carried out with monoclonal antibodies (McAb). When polyclonal antisera are used in this type of study, those antibodies in the antiserum that recognize an alteration in one epitope could be swamped by the large number of antibodies that continue to react in the normal way with unchanged epitopes of the antigen. When McAbs are used, however, only one epitope is investigated at one time, and the influence of a substitution in the epitope recognized by a McAb is therefore more readily detectable. This was confirmed in studies in which 15 mutants of TMV were compared by means of nine McAbs raised against wild-type virus (Al Moudallal *et al.*, 1982; Briand *et al.*, 1982). All the substitutions which in previous work had been detectable with polyclonal antisera were also recognized by some of the McAbs. The superior discriminatory capacity of McAbs compared with conventional antisera was demonstrated by the finding that three substitutions not detected with antisera in earlier work were distinguished with some McAbs. However, it should be emphasized that many individual McAbs failed to recognize certain exchanges (e.g., at position 156) that were uniformly detectable by all polyclonal antisera.

E. Studies with Monoclonal Antibodies

Some of the advantages of using McAbs for elucidating the antigenic structure of TMV were noted in the previous section. Although it is usually assumed that substitutions that are recognized by McAbs are directly involved in the structure of an epitope, this assumption is not always valid. This is illustrated by the results obtained with TMV (Al Moudallal *et al.*, 1982). Although seven of the substitutions that led to altered binding by certain McAbs are situated near the outer surface of the virus particle, two substitutions at positions 20 and 107 that also altered antibody binding are located 2 and 5 nm away, respectively, from the viral surface. These exchanges, therefore, must have influenced the reactivity of an epitope through long-range conformational changes.

The use of McAbs for localizing epitopes in proteins is hampered by the fact that these antibodies are often specific for conformational features of the native antigen that are absent in peptide fragments (Berzofsky *et al.*, 1982). In a study of 30 McAbs raised against TMVP subunits, more than 80% of the antibodies were unable to recognize any of 18 synthetic peptides representing virtually the entire TMVP polypeptide chain (Al Moudallal *et al.*, 1985). The superior discriminatory capacity of McAbs compared to polyclonal antisera is thus more evident in the type of cross-reactivity studies outlined in Section II.D than in studies using cleavage or synthetic fragments of the antigen. Cross-reactivity studies with McAbs have demonstrated the presence of several discontinuous epitopes in TMV (Van Regenmortel, 1984a; Altschuh *et al.*, 1985), but a complete

TABLE I. Continuous Epitopes of TMVP

Position in sequence	Detected with: Polyclonal antisera	Detected with: Monoclonal antibodies	Reference[a]
1–10	+	−	1, 2
19–32	+	−	2
34–39	+	−	1
47–61	+	−	3
62–68	+	−	4
76–88	+	+	1–3
103–112	+	+	5, 6
115–134	+	+	2
134–146	+	+	2
149–158	+	+	1, 7, 8

[a] 1, Altschuh *et al.* (1983); 2, Al Moudallal *et al.* (1985); 3, Altschuh and Van Regenmortel (1982); 4, Milton and Van Regenmortel (1979); 5, Benjamini *et al.* (1968); 6, Morrow *et al.* (1984); 7, Anderer (1963a); 8, Altschuh *et al.* (1985).

delineation of all the residues constituting any particular discontinuous epitope has not been achieved by this method. Such an aim is probably unrealistic since mutants with substitutions at all the boundary positions of an epitope are unlikely to be available.

III. TYPES OF EPITOPES IDENTIFIED IN TMV AND TMVP

From the preceding account of the different methods used in the antigenic analysis of TMV, it is clear that the picture of antigenicity that emerges in different studies is largely influenced by the analytical approach used.

A. Continuous Epitopes

Linear peptide fragments of TMVP that are able to specifically bind to antibodies raised against the subunit are termed continuous epitopes. It should be realized, however, that any such peptide could represent only a part of a larger discontinuous epitope since it is possible that antibodies directed to a discontinuous epitope could react weakly with subregions of the epitope made up of a few residues in linear sequence (Van Regenmortel, 1985).

In studies ranging over more than 20 years, using techniques such as direct binding in a radioimmunoassay or complement fixation and ELISA inhibition, at least ten continuous epitopes have been identified in TMVP (Table I). The exact boundaries of each epitope have not been

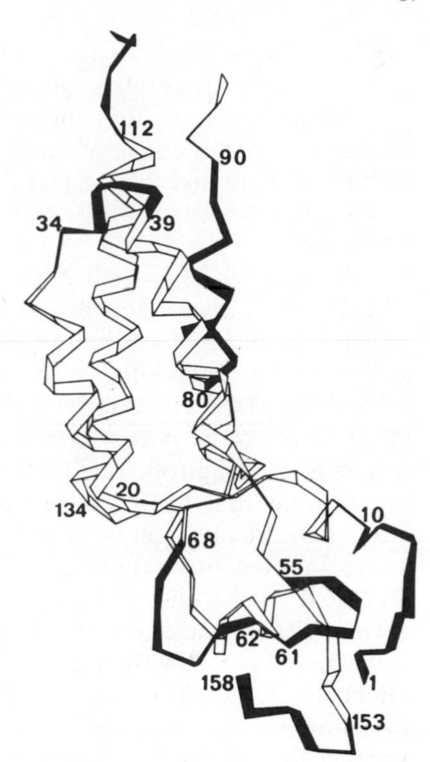

FIGURE 1. Backbone of the TMVP subunit based on crystallographic data of the disk aggregate (see Chapter 2). Residues 94–106 have been omitted because this region has no specific conformation in the disk. The locations of the seven continuous epitopes identified by Altschuh *et al.* (1983) are indicated by heavy lines.

defined in absolute terms, as closely overlapping peptides often possessed similar antigenic activity; furthermore, different antibodies directed against the same general area recognize certain residues preferentially (Benjamini *et al.*, 1968; Morrow *et al.*, 1984).

The activity of the region 1–10 was established in direct binding and in inhibition assays using synthetic peptides corresponding to residues 1–11 (Al Moudallal *et al.*, 1985) and 3–9 (Altschuh *et al.*, 1983). Attempts to localize more precisely the activity found in residues 19–32 failed as none of the shorter peptides 18–25, 22–29, and 27–33 possessed any demonstrable activity either in ELISA binding (using peptides coupled to BSA) or in ELISA inhibition (Al Moudallal *et al.*, 1985). Since the region 19–32 corresponds to an α-helix in TMVP (Fig. 1), it seems likely that the presence of antigenicity only in this longer peptide is linked to the need to stabilize a particular conformation.

Peptide 34–39 possessed antigenic activity only when tested by complement fixation and ELISA inhibition (Altschuh *et al.*, 1983), but not when tested by ELISA in the form of a peptide conjugate (Al Moudallal *et al.*, 1985).

Peptide 47–61, which corresponds to tryptic peptide 3 of TMVP, was shown to possess antigenic activity in ELISA inhibition experiments. By comparing the inhibitory activity of homologous peptides of three tobamoviruses differing in the region 47–61 by only a few residues, the epitope was tentatively assigned to residues 55–61 (Altschuh *et al.*, 1983).

Peptide 62–68, which corresponds to tryptic peptide 4 of TMVP, was active in complement fixation inhibition (Milton and Van Regenmortel, 1979) while the overlapping synthetic peptide 61–74 was antigenically active when tested as a peptide conjugate by ELISA (Al Moudallal *et al.*, 1985).

Peptide 76–88, which closely corresponds to the α-helix 74–88 of TMVP, was active when tested as a peptide conjugate by ELISA, using either a monoclonal antibody raised against TMVP or a polyclonal mouse antiserum (Al Moudallal *et al.*, 1985). In earlier studies based on the comparative inhibitory activity in ELISA of the homologous peptides 72–90 of three tobamoviruses, it was suggested that the region 80–90 corresponds to an epitope of TMVP (Altschuh *et al.*, 1983).

Peptide 103–112 corresponds to the epitope that was exhaustively studied in the classical work of Benjamini and his colleagues (Benjamini, 1977). In a recent study using synthetic peptide analogues of the region 103–112, it was shown that three anti-TMVP McAbs recognizing this epitope nevertheless possessed fine specificities that were very different; this was indicated by the fact that the binding of each McAb was variously affected by substitutions at different positions in the region 103–112 (Morrow *et al.*, 1984). For instance, the fine specificity of some McAbs seemed to indicate that residue 107 was part of the epitope, while the binding of other McAbs indicated that it was not. Such a finding demonstrates that the definition of epitopes in many cases amounts to an operational analysis of the binding specificity of individual immunoglobulin molecules, and that it is meaningless to speak of the intrinsic antigenicity of a residue. Epitopes exist only by virtue of the existence of complementary antibody-combining sites (Van Regenmortel, 1984a, 1985) and it would be futile, in the absence of the relevant antibody molecules, to try to discern the inherent antigenic nature of a portion of a protein molecule.

Peptide 115–134, which corresponds to the α-helix 114–134 of TMVP, was also found to possess antigenic activity, which means that three of the four α-helices found in TMVP harbor antigenic activity (Al Moudallal *et al.*, 1985).

Peptide 134–146 was found to be active both in ELISA binding using a peptide conjugate and in ELISA inhibition. Finally, peptide 149–158 was found to be active in direct ELISA binding as well as in complement fixation and ELISA inhibition (Altschuh *et al.*, 1985; Al Moudallal *et al.*, 1985). Part of this region corresponds to the hexapeptide 153–158 that was extensively studied by Anderer and his colleagues (Anderer, 1963a,b; Anderer and Schlumberger, 1966).

In addition, antigenic activity was also ascribed to the region 90–95 on the basis that a particular McAb raised against TMVP reacted in both ELISA binding and ELISA inhibition with peptide 90–117 but not with

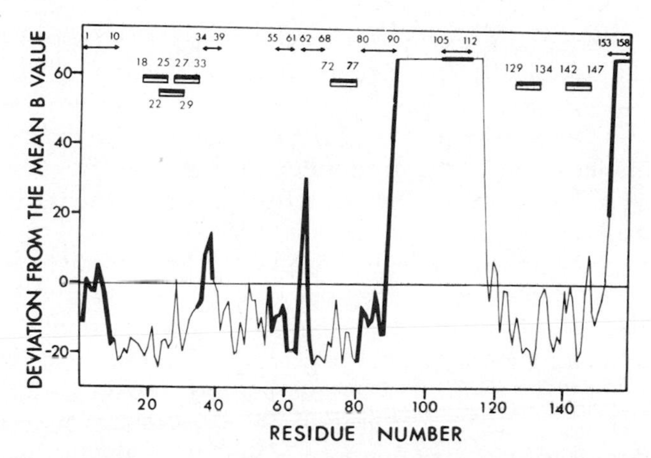

FIGURE 2. Correlation between antigenicity and segmental mobility in TMVP. Deviation from the mean temperature factor of main chain atoms in each residue is plotted against residue number in TMVP. The locations of the seven continuous epitopes identified by Altschuh *et al.* (1983) are indicated by heavy lines and arrows. The locations of antigenically inactive peptides are indicated by boxes. Adapted from Westhof *et al.* (1984) and Al Moudallal *et al.* (1985).

peptide 95–117 (Al Moudallal *et al.*, 1985). However, binding of this McAb to a short peptide corresponding to residues 90–95 has not been demonstrated directly.

In 1984, only seven of the ten continuous epitopes listed in Table I had been identified; their positions in the TMVP polypeptide backbone are indicated in Fig. 1. When E. Westhof and D. Altschuh in the author's laboratory examined the location of these seven epitopes by computer graphics, using the refined coordinates of the viral subunit determined in Klug's laboratory in Cambridge, they noticed that the position of the epitopes agreed closely with regions of high segmental mobility in the protein. Such regions of high mobility correspond to a stretch of residues possessing high atomic temperature factors or B values which represent the mean-square displacement of each atom from a well-defined average position. By plotting B values against residue number, a graphic image of the degree of mobility existing along the polypeptide chain is obtained.

Of these seven continuous epitopes shorter than ten residues, six were found to correspond to peaks of the mobility plot (Fig. 2). The seventh epitope (residues 55–61) was located in a region of low mobility; however, this low mobility could have been due to the intermolecular contacts occurring in this region in the crystal (Westhof *et al.*, 1984).

Since segmental mobility is mostly found in surface projections such as loops and turns, it could be argued that this correlation between mobility and antigenicity is simply a reflection of an underlying correlation with surface accessibility. That this is not the case is shown by the finding that several short peptides (residues 18–25, 22–29, 27–33, 72–77, 129–134, 142–147) situated outside mobile regions possessed no antigenic activity although they correspond to accessible areas of the protein (Westhof *et al.*, 1984; Al Moudallal *et al.*, 1985). It is possible that the mobility of an epitope increases its chances of finding a suitable antibody partner endowed with a sufficient degree of fit. Furthermore, mobility may also improve complementarity by an induced fit phenomenon. Several other instances of a correlation between mobility and antigenicity in different proteins have been reported (Tainer *et al.*, 1985).

It is important to note that the correlation demonstrated by Westhof *et al.* (1984) pertains to mobile regions of 6–8 residues and to epitopes that are themselves 6–10 residues long. Attempts to establish a correlation with much longer antigenic peptides would have had little meaning. In the subsequent study of Al Moudallal *et al.* (1985), three additional continuous epitopes with a length of 13–20 residues were discovered which did not lie within the segments of high mobility in the molecule. Two of these epitopes (residues 19–32 and 115–134) correspond to α-helices and contain within them shorter peptides that were shown to be antigenically inactive (residues 18–25, 22–29, 27–33, 129–134). It seems, therefore, that the use of short peptides (6–8 residues) for probing the antigenic structure favored the identification of epitopes in relatively mobile regions of the protein. When longer peptides were analyzed, however, antigenicity could also be demonstrated in regions of the molecule that are more structured and possess a low mobility. In conclusion, the presence of at least 11 continuous epitopes in TMVP means that virtually the entire polypeptide chain possesses antigenic activity. These results are in line with the current paradigm in protein immunochemistry according to which the entire accessible surface of proteins consists of a continuum of overlapping epitopes that are potentially able to bind to appropriate antibodies (Benjamin *et al.*, 1984).

B. Discontinuous Epitopes

Elucidating the structure of discontinuous epitopes in proteins requires either a crystallographic analysis of antigen–antibody complexes, using McAbs (Amit *et al.*, 1984), or an approach known as surface simulation synthesis, based on the use of synthetic peptides that mimic part of the surface topography of the protein (Lee and Atassi, 1976). Few such investigations have been performed, and evidence for the existence of discontinuous epitopes stems mainly from cross-reactivity studies between related proteins carried out with McAbs (Benjamin *et al.*, 1984; Van Regenmortel, 1984a).

It is widely believed that discontinuous epitopes are more numerous in globular proteins than continuous epitopes. Such a belief partly stems from the observation that when the surface of globular proteins is represented by space filling models, only very small areas of the surface correspond to a linear array of residues in direct peptide linkage.

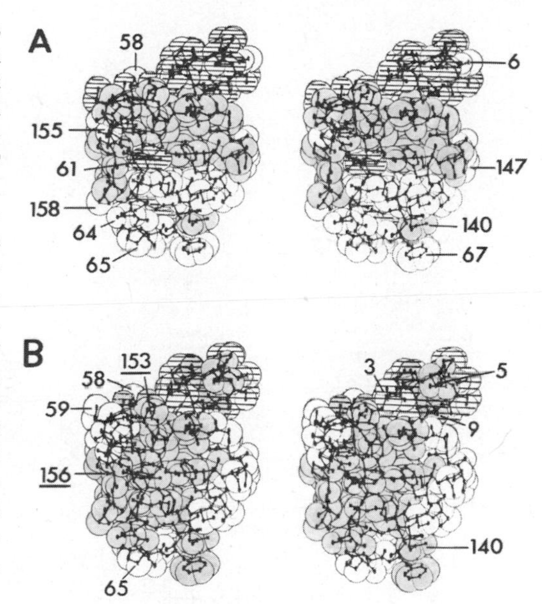

FIGURE 3. Discontinuous epitopes of TMVP. Stereo drawings of the van der Waals surface of a TMV subunit are seen perpendicularly to the helical axis of the virus. The computer graphics used refined coordinates of TMVP obtained from A. C. Bloomer and A. Klug (Cambridge). (A) Surface residues common to TMV and ToMV (see Table II) are indicated by shading (C-terminal residues) and hatching (N-terminal residues and region 55–70). Monoclonal antibody 20 probably binds to residues 66, 67, and 140–143. (B) Residues common to TMV and RMV (see Table II) are indicated by shading. Residues in the region 1–9 that may be part of the epitope are indicated by hatching. Antibody 22 probably recognizes some or all of residues 1, 3, 4, 9, 150, 152, and 153. From Altschuh *et al.* (1985).

In recent studies, a number of discontinuous epitopes have been identified at the surface of TMV particles by means of McAbs prepared against intact virions (Van Regenmortel, 1984a; Altschuh *et al.*, 1985). Immunological cross-reactions between TMV, single point mutants of TMV and several other tobamoviruses were analyzed by means of computer-generated images of those surface residues of the subunit that are exposed at the periphery of the virus particle (Fig. 3). Certain residues of TMVP, originating from distant parts of the sequence but clustered together at the surface, were shown to be implicated in some of the epitopes recognized by certain McAbs. In this type of analysis, it is assumed that the surface residues that differ between TMV and a mutant found to be antigenically dissimilar are directly involved in the structure of an epitope.

Two examples of discontinuous epitopes are illustrated in Fig. 3 in terms of residues present at the outer surface of the viral subunit. The sequence of the external residues 1–10, 55–68, and 140–158 of TMV and of two other tobamoviruses, tomato mosaic virus (ToMV) and ribgrass mosaic virus (RMV), are listed in Table II.

In Fig. 3A, the surface residues that are common to TMV and ToMV have been indicated by shading (C-terminal residues) and hatching (N-terminal residues and region 55–70). Although a large portion of the surface is the same in the two viruses, some McAbs raised against TMV such as antibody 20 did not bind to ToMV (Altschuh *et al.*, 1985). Presumably, these antibodies recognize some of the altered residues 64–67, 141–143, 147, and 158. Since McAb 20 recognized single substitutions at positions 66 and 140 but not at position 65, it seems likely that this

TABLE II. Sequences of Residues 1–10, 55–68, and 140–158 of the Coat Protein of TMV, ToMV, and RMV[a]

	1 10	55 68	140 158
TMV	SYSITTPSQF	SPQVTVRFPDSDFK	NRSSFESSSGLVWTSGPAT
ToMV	SYSITSPSQF	FPQSTVRFPGDVYK	NQNTFESMSGLVWTSAPAS
RMV	SYNITNSNQY	IVAPNQRFPDTGFR	NRAEFE-A-ILPWTTAPAT

[a] From Van Regenmortel (1981).

antibody recognizes a discontinuous epitope comprising residues 67 and 140–143.

In Fig. 3B, the residues common to TMV and RMV have been indicated by shading (see Table II). In spite of the 56% difference in sequence between the two coat proteins, about three-quarters of the surface residues are the same. This probably explains the unusually high degree of serological cross-reactivity between these two viruses (Van Regenmortel, 1975). However, most McAbs raised against TMV did not bind to RMV (Altschuh *et al.*, 1985). One exception was McAb 22 which also reacted with tryptic peptide 1. Accessible residues in the region 1–9 that may be part of the epitope recognized by antibody 22 are indicated by hatching in Fig. 3B. Since a substitution of position 5 did not affect the binding of the antibody, this residue was considered not to be part of the epitope. Furthermore, substitutions at positions 58, 59, 65, and 140 did not affect binding whereas exchanges at positions 153 and 156 did. Taken together, these findings indicate that McAb 22 recognizes some of the residues 1, 3, 4, and 9 as well as some of the residues 150, 152, and 153 situated in the immediate vicinity of the N-terminus.

The existence of many discontinuous epitopes in TMV and TMVP could explain the finding that more than 80% of the McAbs raised against the virus or the subunit were unable to recognize a series of peptides representing the entire polypeptide chain of the coat protein. These results agree with the general finding that McAbs raised against native proteins recognize mainly epitopes that are strongly conformation-dependent (Benjamin *et al.*, 1984; Van Regenmortel, 1984a).

C. Neotopes

The existence of neotopes, i.e., epitopes specific for the quaternary structure of virions, has been demonstrated for all groups of viruses (see Van Regenmortel, 1982). The presence of neotopes on the surface of TMV has been known for many years and was established by showing that TMV antiserum could still react with TMV particles after being depleted by cross-absorption of all antibodies reacting with TMVP subunits (Starlinger, 1955; Aach, 1959; Takahashi and Gold, 1960; Van Regenmortel, 1966, 1982).

Recently, the existence of many neotopes in TMV was confirmed by the finding that out of 18 McAbs raised against TMV, 8 did not react with viral subunits in ELISA (Altschuh *et al.*, 1985). The type of ELISA used in such studies is of crucial importance since it appears that when the virus is directly adsorbed to the plastic surface, the neotope conformation is disrupted and the virions become antigenically similar to monomeric subunits (Al Moudallal *et al.*, 1984; Van Regenmortel, 1984b). These 18 anti-TMV McAbs had been selected for their ability to bind to virions in an ELISA procedure using antibody-coated plates. When these

TABLE III. Reactivity of 18 Anti-TMV Monoclonal Antibodies with Virions and Subunits in Two ELISA Procedures[a]

Clones		ELISA procedure 1[b]		ELISA procedure 2[c]	
		Virions	Subunits	Virions	Subunits
Group A	8 clones	–	–	+	–
Group B	5 clones	–	–	+	+
Group C	5 clones	+	+	+	+

[a] From Altschuh *et al.* (1985).
[b] ELISA procedure 1 uses antigen-coated plates and consists of the following steps: AG, AB^M, anti-M^R, anti-R^G-E [AG = antigen; AB = antibody; M = mouse; R = rabbit; anti-M^R = rabbit anti-mouse globulin; E = enzyme label; G = goat (Al Moudallal *et al.*, 1984)].
[c] ELISA procedure 2 uses antibody-coated plates and consists of the following steps: AB^R, AG, AB^M, anti-M^R-E.

antibodies were tested for their capacity to bind to dissociated viral subunits and to virions in an ELISA procedure using antigen-coated plates, the results shown in Table III were obtained (Altschuh *et al.*, 1985). Group A McAbs were specific for neotopes and did not bind to virus particles on antigen-coated plates. Group B antibodies recognized both virions and subunits but only when they were presented on antibody-coated plates; it appears that adsorption of the antigens to the plastic leads to an alteration of the epitope expressed in both subunits and virions. A third group of McAbs (group C) was able to bind to both types of antigens in both ELISA procedures.

The results in Table III illustrate the ability of McAbs to recognize specific conformational features of viral antigens that are altered when the virions are disrupted or the viral subunits are denatured by attachment to the plastic of microtiter plates.

Three of the McAbs recognizing neotopes of TMV were also shown to be heterospecific antibodies, i.e., they reacted better with several TMV mutants than with the wild-type virus used for immunization (Altschuh *et al.*, 1985).

D. Cryptotopes

The existence of cryptotopes in TMVP subunits, i.e., of epitopes absent in TMV particles, has been a source of controversy for many years (see Van Regenmortel, 1982). Initially, the presence of cryptotope-specific antibodies in TMV or TMVP antisera was demonstrated by absorbing the antisera with virus particles and showing that residual antibodies reacted with dissociated subunits after absorption (Aach, 1959; Van Regenmortel, 1967). The inability of some workers to confirm these results was probably caused by the presence of depolymerized viral subunits in the virus preparations used for absorbing the antisera (Kleczkowski, 1961; Rappaport, 1965). Succinylation of TMV protein has been used to keep the subunits in the monomeric state (Rappaport and Zaitlin, 1970), but since this treatment alters the antigenic specificity of the protein, it is difficult to interpret the results of cross-absorption tests with succinylated protein. When it was shown that the antigenic valence of monomeric TMV subunits was at least three (Van Regenmortel and Lelarge, 1973), the existence of cryptotopes could no longer be contested, since it is impossible to accommodate three antibody molecules on the one extremity of the subunit that is exposed in the virus particle. Finally, when the three-dimensional structure of TMVP was elucidated, it became possible to locate cryptotopes on surfaces of the subunit that are buried inside the virus particle (Milton and Van Regenmortel, 1979).

The epitope identified in the region 103–112 (Benjamini, 1977) is located in a central loop of the polypeptide chain which appears to be in a state of dynamic disorder in the monomeric protein or in disks, but which is ordered in the virus (Jardetsky *et al.*, 1978). This epitope is located in the central hole of the assembled virus particle and is clearly a cryptotope of the virus, although it cannot be excluded that it may be accessible to antibodies at the rod extremities of the TMV particle. These rod extremities represent about 3% of the virion surface, and there is some evidence that they could play a role in certain cross-absorption experiments (Van Regenmortel and Lelarge, 1973).

The epitope identified in the region 34–39 (Altschuh *et al.*, 1983) is located in a hairpin bend between two helices of the polypeptide chain, and is also a cryptotope since it did not react with virus-specific antibodies (Milton and Van Regenmortel, 1979). The location of this region at a radius of 40 Å in the virus particle confirms its cryptotope nature. On similar grounds, it can be concluded that the epitopes identified in the vicinity of residues 19–32, 76–88, 90–95, and 115–134 are also cryptotopes of the virus.

Among the 30 McAbs raised against TMVP, 5 did not react with virions in ELISA using antibody-coated plates and thus were specific for cryptotopes (Table IV). These McAbs were also the only ones that reacted with a number of peptides, namely those corresponding to regions 76–88, 90–117, 115–134, and 134–146 (Al Moudallal *et al.*, 1985). Three of these regions are not accessible at the surface of virus particles, and only some residues of the fourth region (residues 140–146) are potential contact residues at the surface of the virus (see Fig. 1). It is interesting to speculate that the reason why so many of the McAbs reacted with the subunit surface accessible in the virus particle (25 out of 30 clones) is due to the selection procedure used in the fusion experiment; it is indeed possible that in the double-antibody sandwich ELISA procedure used for selection, the primary layer of TMVP antibodies present on the plastic aligned the subunits predominantly in one orientation.

TABLE IV. Specificity of 30 Monoclonal Antibodies Raised against TMV Protein Subunits[a]

	Monomeric subunits	Virions	Peptides
23 clones	+	+	–
5 clones[b]	+	–	+
2 clones[c]	–	+	–

[a] From Al Moudallal *et al.* (1985).
[b] These clones recognize cryptotopes and bind to one of peptides 76–88, 90–117, 115–134, or 134–146.
[c] These clones are heterospecific and recognize neotopes.

IV. DETERMINATION OF THE BINDING CONSTANT OF TMV ANTIBODIES

The quantitative analysis of virus–antibody interactions is usually limited to the determination of antiserum titers and of the smallest quantity of virus detectable in various serological assays. When information is required about the "quality" of viral antibodies, it is possible to calculate their avidity in the form of an affinity constant $K = [AbAg]/[Ab][Ag]$, where Ab represents free antibody, Ag free antigen, and AbAg the antigen–antibody complex at equilibrium. Affinity is a thermodynamic expression of the primary interaction of a single antibody-combining site with an epitope, and strictly speaking it applies only to a monovalent antigen combining with homogeneous antibody molecules (Steward and Steensgaard, 1983). In the case of multivalent antigens and heterogeneous systems, the binding capacity of antibody molecules is expressed in terms of average values that are referred to as avidity. It is essential to include a valency term for the reactants in all equations used for avidity calculations.

A simple transformation of the mass action law (see Van Regenmortel, 1982) leads to the equation

$$f/d = K(s - nf) \tag{1}$$

where f is the ratio of bound antibody to total antigen concentration (mole/liter), d is the free antibody concentration, s is the antigen valence, n is the antibody valence, and K is the equilibrium constant (affinity).

Equation (1) allows the construction of f/d versus f plots known as Scatchard plots from which values of K (as the slope) and of the antigen valence can be derived. The experimental data used to construct such plots are obtained from binding tests performed as follows. A series of dilutions of the virus (1 ml) are mixed with 1 ml of a constant concentration of specific antibody. After ultracentrifugation, the amount of free

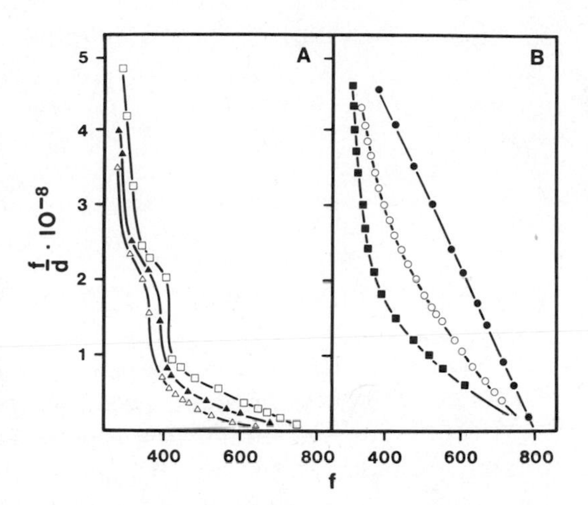

FIGURE 4. Determination of the avidity of TMV antibodies. (A) Plots of f/d versus f representing the interaction between TMV and specific IgG used at 6 mg/ml (□), 1.8 mg/ml (▲), and 0.6 mg/ml (△). The curves extrapolate to the antigenic valence of TMV, $s = 780$. (B) Plots of f/d versus f representing the interaction between TMV and specific Fab fragments used at 1.9 mg/ml (●), 1.18 mg/ml (○), and 0.6 mg/ml (■). The antigenic valence is $s = 800$. From Van Regenmortel (1982).

antibody is determined in the supernatant, and the amount of bound antibody is calculated by subtracting free from total antibody (Van Regenmortel and Hardie, 1976).

Two f/d versus f plots representing binding data between TMV and specific IgG and monovalent Fab fragments, respectively, are shown in *Fig. 4*. The plots obtained with different concentrations of Fab extrapolate to $s = 800$ which is the antigen valence of the virus. The plots obtained with IgG extrapolate to $s = 780$. The similarity in the two s values indicates that in antibody excess, the IgG molecules bind univalently (i.e., $n = 1$). However, the biphasic nature of the curves (Fig. 4A) indicates that bivalent binding of IgG ($n = 2$) occurs at lower antibody–antigen ratios. The flexibility of the IgG molecule makes it possible for the two Fab arms to attach to identical neighboring epitopes on the surface of the virus, a situation corresponding to monogamous bivalent binding. When $n = 2$, the avidity of the antibodies was found to be $K = 24 \times 10^{-7}$ liter/M, whereas when $n = 1$, $K = 7 \times 10^{-7}$ liter/M (Van Regenmortel, 1982). Equation (1) is the method of choice for calculating K in the case of viruses, since the value d is obtained experimentally and the value s can be derived from the plot.

In the case of low-molecular-weight antigens, the following equivalent form of the mass action law is more commonly used for deriving K values:

$$r/c = K(n - sr) \tag{2}$$

where r is the ratio of bound antigen to total antibody and c is the free antigen concentration. However, in the case of viruses it is less suited than equation (1) because values for c and r cannot be obtained experimentally as the surface of virions is likely to be only partly in the bound state. When equation (2) was used to plot binding data between TMV and specific antibody, linear plots were obtained which led to the mistaken belief that TMV antibodies showed homogeneous binding with a single K value (Mamet-Bratley, 1966; Anderer *et al.*, 1971). As discussed previously (Van Regenmortel, 1982), such linear plots do not imply antibody homogeneity and they lead to erroneous K values when incorrect or hypothetical s values are used in the calculations.

V. APPLICATIONS OF SEROLOGY IN TMV STUDIES

Several hundred research papers have been devoted to the antigenic properties of TMV (Rappaport, 1965; Benjamini, 1977; Van Regenmortel, 1982) and it may seem odd that a plant pathogen, unable to induce an immune response in the host it infects, has been the subject of so many immunochemical investigations. One of the reasons for this wide interest, of course, is that many investigators have used TMV simply as a model antigen for investigations in immunology or immunochemistry. However, a knowledge of the antigenic properties of the virus has also been useful for many investigations in virology.

A. Virus Detection

Serological procedures represent the most reliable method for the identification and quantitative assay of TMV. Although a very large number of serological techniques have been used successfully for detecting the virus (Van Regenmortel, 1981), the most useful methods from the practical point of view are immunodiffusion used in conjunction with intragel absorption (Van Regenmortel, 1967; Wetter, 1967; Wetter and Bernard, 1977) and ELISA (Van Regenmortel and Burckard, 1980). A variation of the double-antibody sandwich method of ELISA, using a dissociation step of radiolabeled virus at pH 13, has been shown to be a highly sensitive method for measuring the amount of TMV present around a single infection site on tobacco leaves (Konate and Fritig, 1983). Other techniques such as immunoprecipitation and immunoelectron microscopy have been used successfully for recognizing phenotypically mixed TMV particles (Taliansky *et al.*, 1977) as well as hybrid particles produced during mixed reconstitution experiments (Otsuki *et al.*, 1977).

The advantages and disadvantages of using McAbs for plant virus detection and disease diagnosis have been discussed (Van Regenmortel, 1984b; Sander and Dietzgen, 1984). In general, it is desirable that McAbs should detect the widest possible range of virus strains likely to be encountered in the field. Some of the McAbs raised against TMV were able to discriminate between closely related strains of the virus and were useful mainly because they could differentiate viruses that were not distinguishable previously by conventional antisera (Briand *et al.*, 1982). In this context, it should be stressed that McAbs are epitope-specific rather than virus-specific or strain-specific. In the case of two viruses such as TMV and ToMV that differ in the sequence of their coat proteins by 18%, it is possible to obtain McAbs that are strictly specific for each virus as well as McAbs that will not distinguish between them because they recognize an identical epitope in the two viruses (Briand *et al.*, 1982). In the case of TMV, the main impact of hybridoma research until now has been to provide superior analytical tools for studying epitopes at the molecular level rather than the production of better diagnostic reagents (Dietzgen and Sander, 1982; Dietzgen, 1983; Altschuh *et al.*, 1985).

B. Virus Classification

The usefulness of serology for classifying viruses is based on the finding that viruses that are serologically related always share most of their other properties. In plant virology, serological cross-reactions have been found only between the members of the 30-odd virus groups that are presently recognized; the alleged cross-reaction between TMV and the isometric cocksfoot mild mosaic virus (Bercks and Querfurth, 1971) could not be confirmed in later work (Altschuh and Van Regenmortel, 1983).

Some authors have questioned the validity of using antigenic properties for classifying viruses, on the grounds that the antigenically relevant part of viral coat proteins corresponds to only a very small percentage of the total coding capacity of the viral genome. However, successful classifications have often been developed on the basis of very few parameters, as shown for instance by the success achieved in reconstructing phylogenetic trees of higher organisms on the basis of the degree of sequence similarity found in a small number of their proteins (Wilson *et al.*, 1977).

In the tobamovirus group, the degree of antigenic similarity among members has been shown to agree closely with the extent of sequence similarity in their coat proteins (Van Regenmortel, 1975). This correlation between sequence difference and antigenic distance between tobamoviruses is shown in Fig. 5. A close correlation has also been demonstrated between similarities based on the amino acid composition of tobamovirus coat proteins and those based on antigenic relationship (Paul *et al.*, 1980; Gibbs, 1980). The extent of antigenic similarity between these viruses was expressed by a serological differentiation index (SDI) which corresponds to the average number of twofold dilution steps separating homologous from heterologous precipitin titers (Van Regenmortel and von Wechmar, 1970). Recently, it was shown that reciprocal ELISA experiments could also be used to obtain reliable SDI values for expressing the degree of cross-reactivity between viruses (Jaegle and Van Regenmortel, 1985). In Fig. 5, the data corresponding to the cross-reactivity between TMV and RMV (SDI = 2.1) show the least agreement since these two

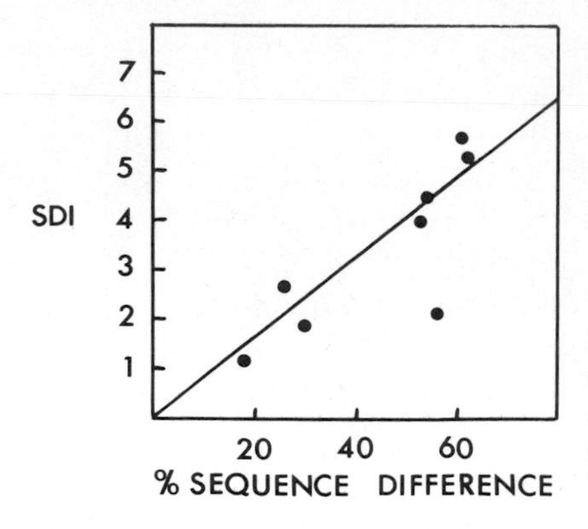

FIGURE 5. Correlation between sequence difference in the coat protein and antigenic difference as measured by serological differentiation index (SDI) in the tobamovirus group. Data from Van Regenmortel (1975).

viruses have a 56% difference in the sequence of their coat proteins. However, as discussed in Section III.B, a considerable proportion (about 75%) of the residues which can be expected to have an antigenic role (i.e., the surface residues shown in Fig. 3B) are identical in TMV and RMV. The poor correlation thus stems from the fact that the extent of chemical similarity between the two viruses is higher in the surface residues than over the entire polypeptide chain.

When a large number of related viruses are examined serologically, it is always found that they lie on a continuum of decreasing antigenic relatedness (Van Regenmortel, 1982). In the tobamovirus group, for instance, it has not been possible to draw a precise borderline between clusters of virus strains and different virus species. In fact, the difficulty in defining virus species, using antigenic or any other criteria, has made many plant virologists reluctant to use the Linnean terms *species* and *genus* for entities such as viruses that do not reproduce by sexual means (Milne, 1984). Instead of these terms, *virus* and *group* are commonly used.

It cannot be denied, however, that there is a need for a term that would denote a taxonomic entity made up of closely related viruses that seem to form a cluster to a human observer. To use the word *virus* for individual isolates or strains as well as for a taxonomic entity comprising all known strains of a "species" is confusing; it is clear that in this context, using the term *species* is helpful.

Since there are no cross-reactions between the members of separate plant virus groups, these groups are serologically homogeneous. The simplest way, therefore, to build up a uniform classification encompassing all viruses would be to consider the recognized clusters of plant virus strains as species and the established groups of plant viruses as genera.

REFERENCES

Aach, H. G., 1959, Serologische Untersuchungen zur Struktur des Tabakmosaikvirus, *Biochim. Biophys. Acta* **32:**140.

Al Moudallal, Z., Briand, J. P., and Van Regenmortel, M. H. V., 1982, Monoclonal antibodies as probes of the antigenic structure of tobacco mosaic virus, *EMBO J.* **1:**1005.

Al Moudallal, Z., Altschuh, D., Briand, J. P., and Van Regenmortel, M. H. V., 1984, Comparative sensitivity of different ELISA procedures for detecting monoclonal antibodies, *J. Immunol. Methods* **68:**35.

Al Moudallal, Z., Briand, J. P., and Van Regenmortel, M. H. V., 1985, The major part of the polypeptide chain of tobacco mosaic virus protein is antigenic, *EMBO J.* **4:**1231.

Altschuh, D., and Van Regenmortel, M. H. V., 1982, Localization of antigenic determinants of a viral protein by inhibition of enzyme immunosorbent assay (ELISA) with tryptic peptides, *J. Immunol. Methods* **50:**99.

Altschuh, D., and Van Regenmortel, M. H. V., 1983, Refutation by ELISA of the alleged antigenic relationship between an isometric and a rod-shaped plant virus, *Int. J. Microbiol.* **1:**13.

Altschuh, D., Hartmann, D., Reinbolt, J., and Van Regenmortel, M. H. V., 1983, Immunochemical studies of tobacco mosaic virus. V. Localization of four epitopes in the protein subunit by inhibition tests with synthetic peptides and cleavage peptides from three strains, *Mol. Immunol.* **20:**271.

Altschuh, D., Al Moudallal, Z., Briand, J. P., and Van Regenmortel, M. H. V., 1985, Immunochemical studies of tobacco mosaic virus. VI. Attempts to localize viral epitopes with monoclonal antibodies, *Mol. Immunol.* **22:**329.

Amit, A. G., Mariuzza, R. A., Phillips, S. E. V., and Poljak, R. J., 1984, Three-dimensional structure of an antigen–antibody complex at 6 Å resolution, *Nature (London)* **313:**156.

Anderer, F. A., 1963a, Versuche zur Bestimmung der serologisch determinanten Gruppen des Tabakmosaikvirus, *Z. Naturforsch.* **18b:**1010.

Anderer, F. A., 1963b, Preparation and properties of an artificial antigen immunologically related to tobacco mosaic virus, *Biochim. Biophys. Acta* **71:**246.

Anderer, F. A., and Schlumberger, H. D., 1965a, Properties of different artificial antigens immunologically related to tobacco mosaic virus, *Biochim. Biophys. Acta* **97:**503.

Anderer, F. A., and Schlumberger, H. D., 1965b, Kreuzreaktionen von Antiseren gegen heterologe terminale Aminosäuresequenzen und zwei Stammen des Tabakmosaikvirus, *Z. Naturforsch.* **20b:**564.

Anderer, F. A., and Schlumberger, H. D., 1966, Cross-reactions of antisera against the terminal amino acid and dipeptide of tobacco mosaic virus, *Biochim. Biophys. Acta* **115:**222.

Anderer, F. A., and Ströbel, G,. 1972a, Recognition of conjugated and native peptide determinants. I. Conformational and sequential specificities of rabbit antibodies versus tobacco mosaic virus, *Eur. J. Immunol.* **2:**274.

Anderer, F. A., and Ströbel, G., 1972b, Recognition of conjugated and native peptide determinants. II. Enhancement of antibodies with sequential specificity in anti-TMV sera by preimmunization with conjugated peptide antigens, *Eur. J. Immunol.* **2:**278.

Anderer, F. A., Uhlig, H., Weber, E., and Schramm, G., 1960, Primary structure of the protein of tobacco mosaic virus, *Nature (London)* **186:**922.

Anderer, F. A., Koch, M. A., and Hirschle, S. D., 1971, Tobacco mosaic virus specific immunoglobulins from horse serum. II. Structural specificity and association constants, *Eur. J. Immunol.* **1:**81.

Atassi, M. Z., 1975, Antigenic structure of myoglobin: The complete immunochemical anatomy of a protein and conclusions relating to antigenic structures of proteins, *Immunochemistry* **12:**423.

Atassi, M. Z., and Lee, C., 1978, The precise and entire antigenic structure of native lysozyme, *Biochem. J.* **171:**429.

Atassi, M. Z., and Smith, J. A., 1978, A proposal for the nomenclature of antigenic sites in peptides and proteins, *Immunochemistry* **15:**609.

Benjamin, D. C., Berzofsky, J. A., East, I. J., Gurd, F. R. N., Hannum, C., Leach, S. J., Margoliash, E., Michael, J. G., Miller, A., Prager, E. M., Reichlin, M., Sercarz, E. E., Smith-

Gill, S. J., Todd, P. A., and Wilson, A. C., 1984, The antigenic structure of proteins: A reappraisal, *Annu. Rev. Immunol.* **2**:67.

Benjamini, E., 1977, Immunochemistry of the tobacco mosaic virus protein, in: *Immunochemistry of Proteins* (M. Z. Atassi, ed.), Volume 2, pp. 265–310, Plenum Press, New York.

Benjamini, E., Young, J. D., Shimizu, M., and Leung, C. Y., 1964, Immunochemical studies on the tobacco mosaic virus protein. I. The immunological relationship of the tryptic peptides of tobacco mosaic virus protein to the whole protein, *Biochemistry* **3**:1115.

Benjamini, E., Young, J. D., Peterson, W. J., Leung, C. Y., and Shimizu, M., 1965, Immunochemical studies on the tobacco mosaic virus protein. II. The specific binding of a tryptic peptide of the protein with antibodies to the whole protein, *Biochemistry* **4**:2081.

Benjamini, E.,, Shimizu, M., Young, J. D., and Leung, C. Y., 1968, Immunochemical studies on the tobacco mosaic virus protein. VI. Characterization of antibody populations following immunization with tobacco mosaic virus protein, *Biochemistry* **7**:1253.

Bercks, R., and Querfurth, G., 1971, Serologische Beziehungen zwischen einem gestreckten (tobacco mosaic) und einem isometrischen (cocksfoot mild mosaic) Virus, *Phytopathol. Z.* **72**:354.

Berzofsky, J. A., Buckenmeyer, G. K., Hicks, G., Gurd, F. R. N., Feldmann, R. J., and Minna, J., 1982, Topographic antigenic determinants recognized by monoclonal antibodies to sperm whale myoglobin, *J. Biol. Chem.* **257**:3189.

Briand, J. P., Al Moudallal, Z., and Van Regenmortel, M. H. V,. 1982, Serological differentiation of tobamoviruses by means of monoclonal antibodies, *J. Virol. Methods* **5**:293.

Dietzgen, R. G., 1983, Doctoral dissertation, University of Tübingen.

Dietzgen, R. G., and Sander, E., 1982, Monoclonal antibodies against a plant virus, *Arch. Virol.* **74**:197.

Fearney, F. J., Leung, C. Y., Young, J. D., and Benjamini, E., 1971, The specificity of antibodies to a peptide determinant of the tobacco mosaic virus protein induced by immunization with the peptide conjugate, *Biochim. Biophys. Acta* **243**:509.

Gibbs, A., 1980, How ancient are the tobamoviruses?, *Intervirology* **14**:101.

Jaegle, M., and Van Regenmortel, M. H. V., 1985, Use of ELISA for measuring the extent of serological cross-reactivity between plant viruses, *J. Virol. Methods* **11**:189.

Jardetsky, O., Akasaka, K., Vogel, D., Morris, S., and Holmes, K. C., 1978, Unusual segmental flexibility in a region of tobacco mosaic virus coat protein, *Nature (London)* **273**:564.

Kleczkowski, A., 1961, Serological behaviour of tobacco mosaic virus and of its protein fragments, *Immunochemistry* **4**:130.

Konate, G., and Fritig, B., 1983, Extension of the ELISA method to the measurement of the specific radioactivity of viruses in crude cellular extracts, *J. Virol. Methods* **6**:347.

Lee, C. L., and Atassi, M. Z., 1976, Delineation of the third antigenic site of lysozyme by application of a novel "surface-simulation" synthetic approach directly linking the conformationally adjacent residues forming the site, *Biochem. J.* **159**:89.

Mamet-Bratley, M. D., 1966, Evidence concerning homogeneity of the combining sites of purified antibody, *Immunochemistry* **3**:155.

Milne, R. G., 1984, The species problem in plant virology, *Microbiol. Sci.* **1**:113.

Milton, De L. R. C., and Van Regenmortel, M. H. V., 1979, Immunochemical studies of tobacco mosaic virus. III. Demonstration of five antigenic regions in the protein subunit, *Mol. Immunol.* **16**:179.

Milton, De L. R. C., Milton, S. C. F., von Wechmar, M. B., and Van Regenmortel, M. H. V., 1980, Immunochemical studies of tobacco mosaic virus. IV. Influence of single amino acid exchanges on the antigenic activity of mutant coat proteins and peptides, *Mol. Immunol.* **17**:1205.

Morrow, P. R., Rennick, D. M.,, Leung, C. Y., and Benjamini, E., 1984, The antibody response to a single antigenic determinant of the tobacco mosaic virus protein: Analysis using monoclonal antibodies, mutant proteins and synthetic peptides, *Mol. Immunol.* **21**:301.

Otsuki, Y., Takebe, I., Ohno, T., Fukuda, M., and Okada, Y., 1977, Reconstitution of tobacco mosaic virus rods occurs bidirectionally from an internal initiation region: Demonstration by electron microscopic serology, *Proc. Natl. Acad. Sci. USA* **74**:1913.

Paul, H. L., Gibbs, A., and Wittmann-Liebold, B., 1980, The relationships of certain tymoviruses assessed from the amino acid composition of their coat proteins, *Intervirology* **13**:99.

Rappaport, I., 1965, The antigenic structure of tobacco mosaic virus, *Adv. Virus Res.* **11**:223.

Rappaport, I., and Zaitlin, M., 1970, Conformational changes in the antigenic determinant of tobacco mosaic virus protein resulting from polymerization of the subunits, *Virology* **41**:208.

Sander, E., and Dietzgen, R. G., 1984, Monoclonal antibodies against plant viruses, *Adv. Virus Res.* **29**:131.

Slobin, L. I., 1970, Antigenic properties of a homogeneous tobacco mosaic virus–hapten conjugate, *Nature (London)* **225**:698.

Staab, H. J., and Anderer, F. A., 1976, Structure and immunogenic behaviour of methylated tobacco mosaic virus, *Biochim. Biophys. Acta* **427**:453.

Starlinger, P., 1955, Vergleich der serologischen Spezifität des Tabakmosaikvirus mit Nucleinsäure*freien und-haltigen Abbauprodukten des Virus, Z. Naturforsch.* **10b**:339.

Steward, M. W., and Steensgaard, J, 1983, *Antibody Affinity: Thermodynamic Aspects and Biological Significance*, CRC Press, Boca Raton, Fla.

Steward, J. M., and Young, J. D.,, 1984, *Solid Phase Peptide Synthesis*, 2nd ed., Pierce Chemical Corp., Rockford, Ill.

Tainer, J. A., Getzoff, E. D., Paterson, Y., Olson, A. J., and Lerner, R. A., 1985, The atomic mobility component of protein antigenicity, *Annu. Rev. Immunol.* **3**:501.

Takahashi, W. N., and Gold, A. H., 1960, Serological studies with X protein, tobacco mosaic virus, polymerized X protein, and virus reconstituted from nucleic acid and X protein, *Virology* **10**:449.

Taliansky, M. E., Atabekov, T. I., and Atabekov, J. G., 1977, The formation of phenotypically mixed particles upon mixed assembly of some tobacco mosaic virus (TMV) strains, *Virology* **76**:701.

Tsugita, A., Gish, D. T., Young, J., Fraenkel-Conrat, H., Knight, C. A., and Stanley, W. M., 1960, The complete amino acid sequence of the protein of tobacco mosaic virus, *Proc. Natl. Acad. Sci. USA* **46**:1463.

Van Regenmortel, M. H. V., 1966, Plant virus serology, *Adv. Virus Res.* **12**:207.

Van Regenmortel, M. H. V., 1967, Serological studies on naturally occurring strains and chemically induced mutants of tobacco mosaic virus, *Virology* **31**:467.

Van Regenmortel, M. H. V., 1975, Antigenic relationships between strains of tobacco mosaic virus, *Virology* **64**:415.

Van Regenmortel, M. H. V., 1981, Tobamoviruses, in: *Handbook of Plant Virus Infections and Comparative Diagnosis* (E. Kurstak, ed.), pp. 541–564, Elsevier/North-Holland, Amsterdam.

Van Regenmortel, M. H. V., 1982, *Serology and Immunochemistry of Plant Viruses*, Academic Press, New York.

Van Regenmortel, M. H. V., 1984a, Molecular dissection of antigens with monoclonal antibodies, in: *Hybridoma Technology in Agricultural and Veterinary Research* (N. J. Stern and H. R. Gamble, eds.), pp. 43–82, Rowman & Allanheld, Totowa, N.J.

Van Regenmortel, M. H. V., 1984b, Monoclonal antibodies in plant virology, *Microbiol Sci.* **1**:73.

Van Regenmortel, M. H. V., 1985, Operational aspects of the definition of epitopes in peptides and proteins, in: *Proceedings of the Symposium on Synthetic Peptides in Biology and Medicine*, Elsevier, Amsterdam (in press).

Van Regenmortel, M. H. V., and Burckard, J., 1980, Detection of a wide spectrum of tobacco mosaic virus strains by indirect enzyme immunosorbent assays (ELISA), *Virology* **106**:327.

Van Regenmortel, M. H. V., and Hardie, G., 1976, Immunochemical studies of tobacco

mosaic virus. II. Univalent and monogamous bivalent binding of IgG antibody, *Immunochemistry* **13**:503.
Van Regenmortel, M. H. V., and Lelarge, N., 1973, The antigenic specificity of different states of aggregation of tobacco mosaic virus protein, *Virology* **52**:89.
Van Regenmortel, M. H. V., and Neurath, A. R., 1985, Structure of viral antigens, in: *Immunochemistry of Viruses: The Basis for Serodiagnosis and Vaccines* (M. H. V. Van Regenmortel and A. R. Neurath, eds.), pp. 1–12, Elsevier, Amsterdam.
Van Regenmortel, M. H. V., and von Wechmar, M. B., 1970, A reexamination of the serological relationship between tobacco mosaic virus and cucumber mosaic virus, *Virology* **41**:330.
von Sengbusch, P., 1965, Aminosäureaustausche und Tertiärstruktur eines Proteins. Vergleich von Mutanten des Tabakmosaikvirus mit serologischen und physikochemischen Methoden, *Z. Vererbungsl.* **96**:364.
Westhof, E., Altschuh, D., Moras, D., Bloomer, A. C., Mondragon, A., Klug, A., and Van Regenmortel, M. H. V., 1984, Correlation between segmental mobility and the location of antigenic determinants in proteins, *Nature (London)* **311**:123.
Wetter, C., 1967, Immunodiffusion of tobacco mosaic virus and its interaction with agar, *Virology* **31**:498.
Wetter, C., and Bernard, M., 1977, Identifizierung, Reinigung, und serologischer Nachweis von Tabakmosaikvirus und Para-Tabakmosaikvirus aus Zigaretten, *Phytopathol. Z.* **90**:257.
Wilson, A. C., Carlson, S. S., and White, T. J., 1977, Biochemical evolution, *Annu. Rev. Biochem.* **46**:573.
Young, J. D., Benjamini, E., Shimizu, M., and Leung, C. Y., 1966, Immunochemical studies on the tobacco mosaic virus protein. III. The degradation of an immunologically active tryptic peptide of tobacco mosaic virus protein and the reactivity of the degradation products with antibodies to the whole protein, *Biochemistry* **5**:1481.
Young, J. D., Benjamini, E., Stewart, J. M., and Leung, C. Y., 1967, Immunochemical studies on tobacco mosaic virus protein. V. The solid-phase synthesis of peptides of an antigenically active decapeptide of tobacco mosaic virus protein and the reaction of these peptides with antibodies to the whole protein, *Biochemistry* **6**:1455.

CHAPTER 5

TOBACCO MOSAIC VIRUS

Infectivity and Replication

PETER PALUKAITIS AND MILTON ZAITLIN

I. INTRODUCTION

Tobacco mosaic was first characterized as an infectious disease in 1886 by Mayer and the virus has been the subject of intensive investigation since the 1930s. Early studies focused on chemistry and structure, and much of our fundamental knowledge of the chemical and physical properties of simple RNA-containing viruses derives from TMV. While the large amounts of easily extracted, easily purified virus available in high yield facilitated these studies, the complex nature of the virus–host system has slowed our understanding of the details of TMV replication (and of all other plant viruses, for that matter); for instance, when a leaf is inoculated with virus, initially only a very small proportion of the cells become infected. These initial centers serve as foci for the subsequent rounds of replication in adjacent cells. Thus, early events in the few initially infected cells must be detected against an overwhelming background of uninfected cells. Further difficulties arise from the relatively slow rate of virus replication when compared with animal or bacterial viruses and, most troublesome, the fact that plant virus infection does not shut down host protein and nucleic acid synthesis. Thus, it is sometimes not easy to distinguish virus-directed proteins and nucleic acids from those of the host. Fortunately, recently developed technologies such as molecular hybridization and Western blotting having solved some of these problems.

PETER PALUKAITIS AND MILTON ZAITLIN • Department of Plant Pathology, Cornell University, Ithaca, New York 14853.

The development of leaf protoplast technology (pioneered by Takebe and Otsuki, 1968) has been very useful in overcoming some of these obstacles, and much of our knowledge of the temporal events in TMV replication derives from protoplast studies. It should be appreciated, however, that isolated protoplasts may not always be equivalent to cells in a leaf; they suffer osmotic shock during their preparation and have been shown to exhibit aberrations in protein synthesis when compared with cells in a leaf (Fleck *et al.*, 1983). Although important, isolated protoplasts have not proven to be as useful to the plant virologist as tissue culture cells have been to animal virologists.

We present here a description of the TMV replication process, centering our discussion around a putative life cycle, which is an updated version of one presented earlier by one of us (Zaitlin, 1977). As the reader will quickly ascertain, many details of TMV replication are still to be resolved. We present here a life cycle in which we have incorporated concepts generated with TMV, but also some shown with other viruses which we feel might be applicable to TMV.

When preparing this chapter, it became obvious to us that we still have large gaps in our knowledge. Many uncertainties relate to fundamental processes, the understanding of which have been addressed for decades. For instance, we are still not certain as to the intracellular site of replication of the viral RNA, we do not know the method of entry of the infectious agent into the cell, nor have we identified the exact types of cells into which the virus first enters. The form in which the infectious principle moves in the plant is a matter of conjecture, and the function(s) of the amorphous inclusion bodies seen in infected cells is not known. Other uncertainties will be apparent in our description below.

Our discussion is centered on the so-called common strain of the virus, although most of the principles considered by us should apply to the tobamoviruses discussed in the following chapters. The common strain has a number of pseudonyms, namely vulgare, ordinary, type, wild type, and U1. The Japanese common strain, OM, has two amino acid replacements in the coat protein (Nozu *et al.*, 1970), but is apparently otherwise very similar to the common strain described here.

II. INFECTION AND PATHOGENICITY

A. Viral Ingress into the Plant

For infection by TMV, a cell must suffer a small, healable wound. TMV is rare among plant viruses in that it has no known vectors; it is not transmitted through the seed or pollen; but only by mechanical means. Man and his machines are the principal transmitting agents.

The TMV infection process was a subject of intensive studies in the

1950s and 1960s, but has received little attention in recent years. It is generally conceded that successful infection of each infectible site on a leaf is initiated by a single virion or RNA molecule; however, many infectious units (up to 10^6/site) must normally be applied to the leaf to produce infection of a site, although by careful inoculation methods virion numbers as low as 450 have been successfully utilized (Walker and Pirone, 1972). In our laboratory, we did not get consistent infections when we employed fewer than 3000 virions in inoculations using volumes of 1 μl (Garciá-Arenal *et al.*, 1984). Virions other than the infecting one seem to get into the infection site at inoculation, but they can only replicate on rare occasions (Zaitlin *et al.*, 1977; Garciá-Arenal *et al.*, 1984). In a classical study, Rappaport and Wu (1962) demonstrated that when they used a mixed inoculum of TMV strains U1 and VM, the *postinoculation* selective inactivation of strain VM by ultraviolet light at infectible sites increased the number of U1 infections, indicating that more than one particle was at the infectible site but the noninfecting ones only became functional when the original, infecting particle was inactivated.

No exclusion of one strain by another seems to be extant in protoplasts, as one can get mixedly infected cells if the viruses are introduced into the cells at the same time or with a short interval between individual inoculations (Otsuki and Takebe, 1978).

With protoplasts, one must still add large numbers of particles or RNA molecules to the medium to secure efficient infection, but as with infection of leaves, based on "one-hit" inactivation by ultraviolet light, the infection of the cell is believed to result from a single particle (Takebe, 1977); the mixed strain example given above being slightly contradictory.

B. Site of Viral Entry into the Plant and the Cell

Common sense and previous studies (Fry and Matthews, 1963) suggest that when a leaf is inoculated mechanically with a virus, the epidermal cells will be the first to be infected. Surprisingly, some recent data of ours (Sulzinski and Zaitlin, 1982) and others (Calvete and Wieringa-Brants, 1984) suggest that the first cells to be infected are those of the mesophyll. If this is the cell of entry, then virus would have to pass through the epidermal cells without infecting them and enter the mesophyll where the actual infection would occur.

Exactly how the virus enters the cells in a leaf is not known. A prime candidate in recent years has been the ectodesmata—protoplasmic channels in the epidermal cell walls adjacent to the cuticle. (Of course, if the epidermal cells are not the prime sites of infection, then this discussion is superfluous.) Mechanical disruption of the cell wall would be another route of entry (Matthews, 1981). The evidence favoring ectodesmata as sites of entry reflects the correlation seen between environmental conditions and various perturbations which concomitantly affect infection and ectodesmata numbers (Brants, 1964; Thomas and Fulton, 1968). Virus particles were not seen in ectodesmata (Merkens *et al.*, 1972), which apparently prompted the suggestion that the virus is uncoated outside of the cell (de Zoeten, 1981; de Zoeten and Gaard, 1984). We consider this an unlikely hypothesis, particularly in the light of recent evidence showing that ribosomes (within the cell) are the likely agents for uncoating (see Section III.C.1). It seems more probable to us, and to Wilson (1985), that the virus is forced by physical means through the cell wall, and either penetrates the plasmalemma, which can then heal, or alternatively, once inside the wall pinocytosis can convey the particle through the plasmalemma. Infection of protoplasts *in vitro* normally requires some perturbation to the cell membrane, giving support to the wound hypothesis. However, the exact method whereby the infecting particle enters the naked protoplast is not settled. Both endocytosis and damaged areas on the plasmalemma are suggested (reviewed by Takebe, 1977). Furthermore, in *in vitro* infection particles must be positively charged relative to the plasma membrane for infection to take place; such a condition can be satisfied when the virus has a high isoelectric point (pI), or by adding a polycation such as poly-L-ornithine (Watts and King, 1984). TMV virions have a low pI and usually require polycations.

C. Virus Movement from Cell to Cell

Once inside a cell, for the infection to be productive, the virus must move to adjacent cells. There seems to be no argument to the postulate that the infectious principle moves through the protoplasmic bridges which interconnect plant cells—the plasmodesmata. Indeed, TMV virions have been seen in these structures (Weintraub *et al.*, 1976). From that observation, one must conclude that virions can move from cell to cell, but the fact that the agent can spread after infection by TMV mutants possessing a nonfunctional coat protein, indicates that unencapsidated RNA can also be transported.

It is also now apparent that the movement of the infectious agent to adjacent cells is not a passive process; rather, the virus actively potentiates its own movement. As pointed out by Gunning and Overall (1983), many isometric viruses and ribosomes are of similar size, yet the viruses pass into plasmodesmata and the ribosomes do not. The active role of the virus in the transport process was most forcefully demonstrated by the work of Nishiguchi *et al.* (1978, 1980). They isolated a temperature-sensitive strain of tomato mosaic virus (LS-1), which at a restrictive temperature was unable to move in its host. Replication was not impaired at that temperature; the defect was correlated with a slight modification of the virus-coded 30K protein (Leonard and Zaitlin, 1982). Subsequent determination of the RNA sequence coding for this protein indicated that, when compared with its parent strain L, LS-1 had a single base change which substituted a serine for a proline residue (Ohno *et al.*, 1983). Although these studies are only correlative (there could be undetected changes elsewhere in the viral genome), they seem to implicate the 30K protein in transport.

Further evidence for the role of the virus in mediating cell to cell movement is found in the work of Atabekov and his colleagues (reviewed in Atabekov and Dorokhov, 1984). They show that movement-competent viruses can help movement-deficient viruses; even unrelated viruses can help one another (Taliansky *et al.*, 1982). Moreover, viruses normally confined to the phloem can invade other tissues in the presence of a helper virus (Carr and Kim, 1983), suggesting that the virus's contribution to movement can overcome the normal barriers to movement within a plant.

On the other hand, the plant can apparently play a role in virus movement too, as there is genetic variation in segregating populations of a given plant species for the degree of virus movement (see, e.g., Kuhn *et al.*, 1981). In the extreme case with TMV, there are some plant species in which TMV replication is restricted to those cells into which it is first delivered during inoculation; it cannot move from those cells (Sulzinski and Zaitlin, 1982). The reason for this phenomenon is not known; however, one can generate a scenario in which those particular plants can inactivate or destroy the virus's "movement protein."

D. Long-Distance Virus Movement

TMV can move long distances in its host, and at a fast rate. Capoor (1949) recorded movement rates of about 8 cm/hr. Furthermore, classical studies by Samuel (1934) showed that virus can move through the stem without replicating there; the most probable tissue is the phloem (Bennett, 1940). The form in which the infectious principle moves (i.e., virions, RNA, replicative complexes) is not known with certainty. We favor the virion as the agent, although de Zoeten (1981) suggests that the agent might be a ". . . membrane-associated virus-specific RNA [ds RNA, (−) RNA, or (+) RNA]." Our reason for favoring the virion stems from the aberrant movement seen in coat protein-defective TMV mutants (Siegel *et al.*, 1962). These nitrous acid-induced mutants do not have functional coat proteins and in the plant the infectious agent exists as free RNA. The infectious principle does not show long-distance movement; rather, the infection spreads by slow cell to cell movement. As the viral RNA replicates in an apparently normal manner, the RNA would be expected to generate normal replicative structures, but no virions. If replicative structures were the form in which the infection moved over long distances, one would expect these mutants to move in the same manner as normal TMV strains. However, they do not; somehow the moving entity does not enter the phloem, or perhaps it is inactivated there.

E. Symptom Determinants

In general, the symptoms induced by common strain TMV on various plant species and cultivars are of two types: a systemic mosaic, or a hypersensitive necrotic local lesion. In tobacco, the hypersensitive response, which in effect renders the plant resistant to systemic disease, is determined by a single, dominant gene (the N gene). There are alleles of the N gene in some plants in which some TMV strains develop systemic infections and others local lesions. The hypersensitive response is temperature dependent; at temperatures of about 32°C or higher, the virus invades the plant systemically. The exact temperature at which the systemic reaction develops is somewhat dependent on the genetic background which accompanies the N gene (McKinney and Clayton, 1945).

The molecular bases for any TMV-induced symptoms have not been elucidated. The approximate region of the genome which seems to determine whether TMV will induce local lesions has been found to be about 25% from the 3′-end (Kado and Knight, 1966). (The latter reference actually stated that the region is 25% from the 5′-end but the ends of the RNA were misidentified.)

A number of "pathogenesis-related" proteins (PR-proteins) are engendered in TMV plants exhibiting the hypersensitive response (for a recent review, see Gianinazzi, 1984). These proteins, sometimes called "b proteins," are also induced by agents other than viruses and thus it is unlikely that they are the substances responsible for hypersensitivity; rather, they are produced as a consequence of the hypersensitive reaction. Furthermore, they are found in the intercellular portions of the leaf and are also present in uninfected plants (Parent and Asselin, 1984). Hooft van Huijsduijnen *et al.* (1985) have shown that mRNAs coding for several PR-proteins are stimulated by viral infection suggesting *de novo* synthesis. The amino acid sequence of one of these proteins has been determined (Lucas *et al.*, 1985).

The nucleotide sequence differences between an "attenuated" and a closely related "virulent strain" of a tobamovirus have been determined in a first step to attempt to identify the specific sequences which influence symptomatology (Nishiguchi *et al.*, 1985). Between the virulent strain L and the attenuated strain $L_{II}A$, ten base changes were found, dispersed over the genome. Seven occurred in the third base which did not influence the coding for amino acids, while three did result in amino acid substitutions. In our view, symptoms may not necessarily be determined by virus-coded proteins; nucleic acid interactions may be involved (Zaitlin, 1983), so that any of the above replacements could be symptom determinants. Nishiguchi *et al.* (1985) favor an amino acid-altering change in the 126/183K protein coding region as being important. Obviously, more work will resolve this issue and will identify those sequences which interact with host components to cause disease. How that interaction comes about and what host processes are affected first is a complete mystery.]

III. REPLICATION

A. Introduction

Over the past 15 years, a great deal of information has accumulated concerning the molecular aspects of the replication of TMV. While we still do not know all the details of the "life-cycle" of the virus, by combining a series of individual snapshots of various aspects of TMV replication with those from other plant viruses, we can put together a speculative picture of the life cycle as we now see it (Fig. 1). However, prior to describing this life cycle, it is necessary to understand the organization, the mode of expression, and the function of the genes of TMV.

B. Gene Organization

1. Gene Expression

The genome of TMV contains at least five genes (Fig. 2), the products of which are referred to by the molecular weights of the proteins they encode. Since TMV RNA itself is an mRNA, it is referred to as a (+) sense RNA, or (+) RNA.

The 5′-proximal gene of TMV RNA encodes two coinitiated proteins, the 126K and the 183K protein, which are synthesized by translation of the virion RNA (Beachy *et al.*, 1976; Hunter *et al.*, 1976; Beachy and Zaitlin, 1977; Pelham, 1978). The three genes located internally in the genome are not expressed from the genomic RNA, but rather derive from "subgenomic RNAs" generated during the course of infection (Beachy *et al.*, 1976; Hunter *et al.*, 1976; Beachy and Zaitlin, 1977; Palukaitis *et al.*, 1983; Sulzinski *et al.*, 1985). Although all of these subgenomic RNAs are 3′ coterminal with the genomic RNA (Palukaitis *et al.*, 1983; Sulzinski *et al.*, 1985), each acts only as an mRNA for the translation of one gene located near the 5′-end of that particular subgenomic RNA. Thus, the I_1 mRNA (Fig. 2) translation product is the 54K protein (Sulzinski *et al.*, 1985), the I_2 mRNA translates into the 30K protein (Beachy and Zaitlin, 1977), and the LMC encodes the 17.5K coat protein (Hunter *et al.*, 1976).

The genomic RNA and the LMC have been shown to be capped at their 5′-ends by 7-MeG (Keith and Fraenkel-Conrat, 1975; Zimmern, 1975; Guilley *et al.*, 1979); indirect tests indicate that the I_2 mRNA is not capped (Hunter *et al.*, 1983; Joshi *et al.*, 1983); the I_1 mRNA has not been analyzed for cap structures.

The mechanism for the generation of the subgenomic RNAs of TMV

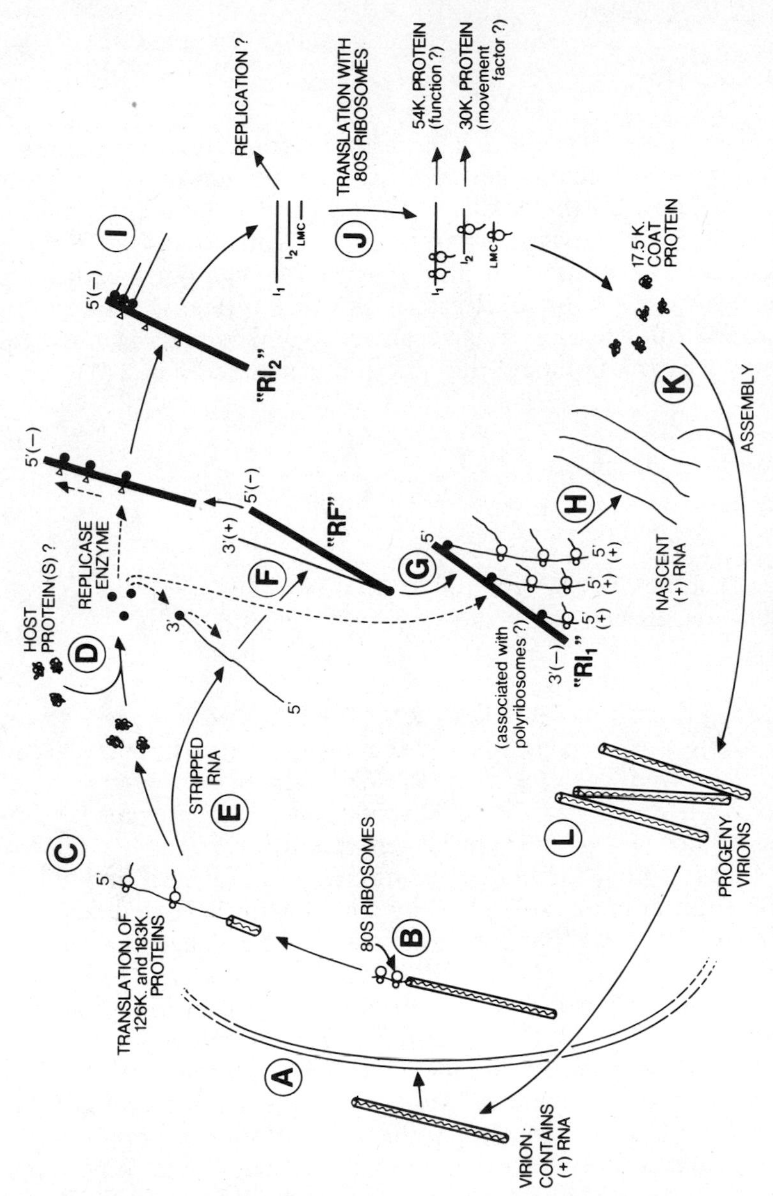

FIGURE 1. The "life cycle" of tobacco mosaic virus. See text for explanation.

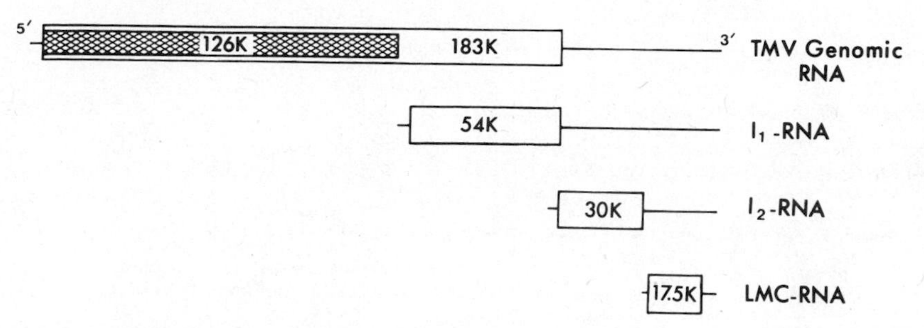

FIGURE 2. The genetic map of tobacco mosaic virus, showing the five virus-coded proteins and three subgenomic RNAs. See text for explanation. From Sulzinski *et al.* (1985).

is unknown, although the one postulated for brome mosaic virus (BMV) may also apply to TMV (see Section III.C.3).

From the complete 6395-nucleotide sequence of TMV (Fig. 3) (Goelet *et al.*, 1982), it is possible to determine the actual size and the amino acid sequence of the various TMV-specific proteins described above. The locations within the genomic RNA of the 5′-ends of each of the three subgenomic RNAs have also been determined (arrows in Fig. 3): I_1 RNA begins at residue 3405 (Sulzinski *et al.*, 1985), I_2 RNA begins at residue 4838 (Watanabe *et al.*, 1984b), and LMC begins at residue 5703 (Guilley *et al.*, 1979; Goelet *et al.*, 1982). Thus, the sizes of the subgenomic RNAs are 2991 residues (I_1), 2058 residues (I_2), and 693 residues (LMC).

The viral RNA (Fig. 3) consists of: (1) a 5′-end untranslated region of 68 nucleotides; (2) the 126K gene and a readthrough of the UAG termination signal (residues 3417–3419) by plant suppressor Tyr-tRNAs (Beier *et al.*, 1984) to produce a protein of 183K (terminated by a UAA at residues 4917–4919); (3) the 54K gene, probably beginning at residue 3495 (Sulzinski *et al.*, 1985) and coincident with the C-terminal one-third of the 183K protein; (4) the 30K gene, beginning at residue 4903 and overlapping the 54K and 183K genes by five amino acids, albeit in a different reading frame; (5) a five-nucleotide internal nontranslated region (5707–5711) that includes the UAA termination signal of the 30K gene; (6) the 17.5K gene (beginning at residue 5712 and terminating with a UGA codon at residues 6189–6191); and (7) a 3′-end untranslated region of 204 nucleotides.

The 3′-terminus of TMV RNA can be aminoacylated *in vitro* by histidine (Öberg and Philipson, 1972). The 3′-end of the RNA also can be folded into a tRNA-like tertiary structure (Rietveld *et al.*, 1984), the function of which remains unknown. Numerous models have been put forward to explain the function of this 3′-end tRNA-like structure (Hall, 1979; Haenni *et al.*, 1982; Florentz *et al.*, 1984), the newest and most intriguing of which is as follows: the 3′-end of the RNA can base-pair with the 5′-end of the genomic RNA, burying the initiation site for protein synthesis in the process. Subsequently, interacting with tRNA ami-

10 20 30 40 50 60 70
GUAUUUUUAC AACAAUUACC AACAACAACA AACAACAAAC AACAUUACAA UUACUAUUUA CAAUUACAAU

80 90 100 110 120 130 140
GGCAUACACA CAGACAGCUA CCACAUCAGC UUUGCUGGAC ACUGUCCGAG GAAACAACUC CUUGGUCAAU

150 160 170 180 190 200 210
GAUCUAGCAA AGCGUCGUCU UUACGACACA GCGGUUGAAG AGUUUAACGC UCGUGACCGC AGGCCCAAGG

220 230 240 250 260 270 280
UGAACUUUUC AAAAGUAAUA AGCGAGGAGC AGACGCUUAU UGCUACCCGG GCGUAUCCAG AAUUCCAAAU

290 300 310 320 330 340 350
UACAUUUUAU AACACGCAAA AUGCCGUGCA UUCGCUUGCA GGUGGAUUGC GAUCUUUAGA ACUGGAAUAU

360 370 380 390 400 410 420
CUGAUGAUGC AAAUUCCCUA CGGAUCAUUG ACUUAUGACA UAGGCGGGAA UUUUGCAUCG CAUCUGUUCA

430 440 450 460 470 480 490
AGGGACGAGC AUAUGUACAC UGCUGCAUGC CCAACCUGGA CGUUCGAGAC AUCAUGCGGC ACGAAGGCCA

500 510 520 530 540 550 560
GAAAGACAGU AUUGAACUAU ACCUUUCUAG GCUAGAGAGA GGGGGGAAAA CAGUCCCCAA CUUCCAAAAG

570 580 590 600 610 620 630
GAAGCAUUUG ACAGAUACGC AGAAAUUCCU GAAGACGCUG UCUGUCACAA UACUUUCCAG ACAAUGCGAC

640 650 660 670 680 690 700
AUCAGCCGAU GCAGCAAUCA GGCAGAGUGU AUGCCAUUGC GCUACACAGC AUAUAUGACA UACCAGCCGA

710 720 730 740 750 760 770
UGAGUUCGGG GCGGCACUCU UGAGGAAAAA UGUCCAUACG UGCUAUGCCG CUUUCCACUU CUCUGAGAAC

780 790 800 810 820 830 840
CUGCUUCUUG AAGAUUCAUA CGUCAAUUUG GACGAAAUCA ACGCGUGUUU UUCGCGCGAU GGAGACAAGU

850 860 870 880 890 900 910
UGACCUUUUC UUUUGCAUCA GAGAGUACUC UUAAUUAUUG UCAUAGUUAU UCUAAUAUUC UUAAGUAUGU

920 930 940 950 960 970 980
GUGCAAAACU UACUUCCCGG CCUCUAAUAG AGAGGUUUAC AUGAAGGAGU UUUUAGUCAC CAGAGUUAAU

990 1000 1010 1020 1030 1040 1050
ACCUGGUUUU GUAAGUUUUC UAGAAUAGAU ACUUUUCUUU UGUACAAAGG UGUGGCCCAU AAAAGUGUAG

1060 1070 1080 1090 1100 1110 1120
AUAGUGAGCA GUUUUAUACU GCAAUGGAAG ACGCAUGGCA UUACAAAAAG ACUCUUGCAA UGUGCAACAG

1130 1140 1150 1160 1170 1180 1190
CGAGAGAAUC CUCCUUGAGG AUUCAUCAUC AGUCAAUUAC UGGUUUCCCA AAAUGAGGGA UAUGGUCAUC

1200 1210 1220 1230 1240 1250 1260
GUACCAUUAU UCGACAUUUC UUUGGAGACU AGUAAGAGGA CGCGCAAGGA AGUCUUAGUG UCCAAGGAUU

1270 1280 1290 1300 1310 1320 1330
UCGUGUUUAC AGUGCUUAAC CACAUUCGAA CAUACCAGGC GAAAGCUCUU ACAUACGCAA AUGUUUUGUC

1340 1350 1360 1370 1380 1390 1400
CUUUGUCGAA UCGAUUCGAU CGAGGGUAAU CAUUAACGGU GUGACAGCGA GGUCCGAAUG GGAUGUGGAC

1410 1420 1430 1440 1450 1460 1470
AAAUCUUUGU UACAAUCCUU GUCCAUGACG UUUUACCUGC AUACUAAGCU UGCCGUUCUA AAGGAUGACU

1480 1490 1500 1510 1520 1530 1540
UACUGAUUAG CAAGUUUAGU CUCGGUUCGA AAACGGUGUG CCAGCAUGUG UGGGAUGAGA UUUCGCUGGC

1550 1560 1570 1580 1590 1600 1610
GUUUGGGAAC GCAUUUCCCU CCGUGAAAGA GAGGCUCUUG AACAGGAAAC UUAUCAGAGU GGCAGGCGAC

FIGURE 3. The nucleotide sequence of the common strain of tobacco mosaic virus as determined by Goelet *et al.* (1982). Arrows indicate the 5′-ends of the three subgenomic RNAs and boxes indicate the positions of gene initiation and termination sites. See text for details.

```
      1620       1630       1640       1650       1660       1670       1680
GCAUUAGAGA UCAGGGUGCC UGAUCUAUAU GUGACCUUCC ACGACAGAUU AGUGACUGAG UACAAGGCCU

      1690       1700       1710       1720       1730       1740       1750
CUGUGGACAU GCCUGCGCUU GACAUUAGGA AGAAGAUGGA AGAAACGGAA GUGAUGUACA AUGCACUUUC

      1760       1770       1780       1790       1800       1810       1820
AGAGUUAUCG GUGUUAAGGG AGUCUGACAA AUUCGAUGUU GAUGUUUUUU CCCAGAUGUG CCAAUCUUUG

      1830       1840       1850       1860       1870       1880       1890
GAAGUUGACC CAAUGACGGC AGCGAAGGUU AUAGUCGCGG UCAUGAGCAA UGAGAGCGGU CUGACUCUCA

      1900       1910       1920       1930       1940       1950       1960
CAUUUGAACG ACCUACUGAG GCGAAUGUUG CGCUAGCUUU ACAGGAUCAA GAGAAGGCUU CAGAAGGUGC

      1970       1980       1990       2000       2010       2020       2030
UUUGGUAGUU ACCUCAAGAG AAGUUGAAGA ACCGUCCAUG AAGGGUUCGA UGGCCAGAGG AGAGUUACAA

      2040       2050       2060       2070       2080       2090       2100
UUAGCUGGUC UUGCUGGAGA UCAUCCGGAG UCGUCCUAUU CUAAGAACGA GGAGAUAGAG UCUUUAGAGC

      2110       2120       2130       2140       2150       2160       2170
AGUUUCAUAU GGCAACGGCA GAUUCGUUAA UUCGUAAGCA GAUGAGCUCG AUUGUGUACA CGGGUCCGAU

      2180       2190       2200       2210       2220       2230       2240
UAAAGUUCAG CAAAUGAAAA ACUUUAUCGA UAGCCUGGUA GCAUCACUAU CUGCUGCGGU GUCGAAUCUC

      2250       2260       2270       2280       2290       2300       2310
GUCAAGAUCC UCAAAGAUAC AGCUGCUAUU GACCUUGAAA CCCGUCAAAA GUUUGGAGUC UUGGAUGUUG

      2320       2330       2340       2350       2360       2370       2380
CAUCUAGGAA GUGGUUAAUC AAACCAACGG CCAAGAGUCA UGCAUGGGGU GUUGUUGAAA CCCACGCGAG

      2390       2400       2410       2420       2430       2440       2450
GAAGUAUCAU GUGGCGCUUU UGGAAUAUGA UGAGCAGGGU GUGGUGACAU GCGAUGAUUG GAGAAGAGUA

      2460       2470       2480       2490       2500       2510       2520
GCUGUCAGCU CUGAGUCUGU UGUUUAUUCC GACAUGGCGA AACUCAGAAC UCUGCGCAGA CUGCUUCGAA

      2530       2540       2550       2560       2570       2580       2590
ACGGAGAACC GCAUGUCAGU AGCGCAAAGG UUGUUCUUGU GGACGGAGUU CCGGGCUGUG GGAAAACCAA

      2600       2610       2620       2630       2640       2650       2660
AGAAAUUCUU UCCAGGGUUA AUUUUGAUGA AGAUCUAAUU UUAGUACCUG GGAAGCAAGC CGCGGAAAUG

      2670       2680       2690       2700       2710       2720       2730
AUCAGAAGAC GUGCGAAUUC CUCAGGGAUU AUUGUGGCCA CGAAGGACAA CGUUAAAACC GUUGAUUCUU

      2740       2750       2760       2770       2780       2790       2800
UCAUGAUGAA UUUUGGGAAA AGCACACGCU GUCAGUUCAA GAGGUUAUUC AUUGAUGAAG GGUUGAUGUU

      2810       2820       2830       2840       2850       2860       2870
GCAUACUGGU UGUGUUAAUU UUCUUGUGGC GAUGUCAUUG UGCGAAAUUG CAUAUGUUUA CGGAGACACA

      2880       2890       2900       2910       2920       2930       2940
CAGCAGAUUC CAUACAUCAA UAGAGUUUCA GGAUUCCCGU ACCCCGCCCA UUUUGCCAAA UUGGAAGUUG

      2950       2960       2970       2980       2990       3000       3010
ACGAGGUGGA GACACGCAGA ACUACUCUCC GUUGUCCAGC CGAUGUCACA CAUUAUCUGA ACAGGAGAUA

      3020       3030       3040       3050       3060       3070       3080
UGAGGGCUUU GUCAUGAGCA CUUCUUCGGU UAAAAAGUCU GUUUCGCAGG AGAUGGUCGG CGGAGCCGCC

      3090       3100       3110       3120       3130       3140       3150
GUGAUCAAUC CGAUCUCAAA ACCCUUGCAU GGCAAGAUCC UGACUUUUAC CCAAUCGGAU AAAGAAGCUC

      3160       3170       3180       3190       3200       3210       3220
UGCUUUCAAG AGGGUAUUCA GAUGUUCACA CUGUGCAUGA AGUGCAAGGC GAGACAUACU CUGAUGUUUC

      3230       3240       3250       3260       3270       3280       3290
ACUAGUUAGG UUAACCCCUA CACCAGUCUC CAUCAUUGCA GGAGACAGCC CACAUGUUUU GGUCGCAUUG

      3300       3310       3320       3330       3340       3350       3360
UCAAGGCACA CCUGUUCGCU CAAGUACUAC ACUGUUGUUA UGGAUCCUUU AGUUAGUAUC AUUAGAGAUC
```

FIGURE 3. *(continued)*

```
      3370       3380       3390       3400       3410       3420       3430
UAGAGAAACU UAGCUCGUAC UUGUUAGAUA UGUAUAAGGU CGAUGCAGGA ACACAAUAGC AAUUACAGAU

      3440       3450       3460       3470       3480       3490       3500
UGACUCGGUG UUCAAAGGUU CCAAUCUUUU UGUUGCAGCG CCAAAGACUG GUGAUAUUUC UGAUAUGCAG

      3510       3520       3530       3540       3550       3560       3570
UUUUACUAUG AUAAGUGUCU CCCAGGCAAC AGCACCAUGA UGAAUAAUUU UGAUGCUGUU ACCAUGAGGU

      3580       3590       3600       3610       3620       3630       3640
UGACUGACAU UUCAUUGAAU GUCAAAGAUU GCAUAUUGGA UAUGUCUAAG UCUGUUGCUG CGCCUAAGGA

      3650       3660       3670       3680       3690       3700       3710
UCAAAUCAAA CCACUAAUAC CUAUGGUACG AACGGCGGCA GAAAUGCCAC GCCAGACUGG ACUAUUGGAA

      3720       3730       3740       3750       3760       3770       3780
AAUUUAGUGG CGAUGAUUAA AAGGAACUUU AACGCACCCG AGUUGUCUGG CAUCAUUGAU AUUGAAAAUA

      3790       3800       3810       3820       3830       3840       3850
CUGCAUCUUU AGUUGUAGAU AAGUUUUUUG AUAGUUAUUU GCUUAAAGAA AAAAGAAAAC CAAAUAAAAA

      3860       3870       3880       3890       3900       3910       3920
UGUUUCUUUG UUCAGUAGAG AGUCUCUCAA UAGAUGGUUA GAAAAGCAGG AACAGGUAAC AAUAGGCCAG

      3930       3940       3950       3960       3970       3980       3990
CUCGCAGAUU UUGAUUUUGU AGAUUUGCCA GCAGUUGAUC AGUACAGACA CAUGAUUAAA GCACAACCCA

      4000       4010       4020       4030       4040       4050       4060
AGCAAAAAUU GGACACUUCA AUCCAAACGG AGUACCCGGC UUUGCAGACG AUUGUGUACC AUUCAAAAAA

      4070       4080       4090       4100       4110       4120       4130
GAUCAAUGCA AUAUUUGGCC CGUUGUUUAG UGAGCUUACU AGGCAAUUAC UGGACAGUGU UGAUUCGAGC

      4140       4150       4160       4170       4180       4190       4200
AGAUUUUUGU UUUUCACAAG AAAGACACCA GCGCAGAUUG AGGAUUUCUU CGGAGAUCUC GACAGUCAUG

      4210       4220       4230       4240       4250       4260       4270
UGCCGAUGGA UGUCUUGGAG CUGGAUAUAU CAAAAUACGA CAAAUCUCAG AAUGAAUUCC ACUGUGCAGU

      4280       4290       4300       4310       4320       4330       4340
AGAAUACGAG AUCUGGCGAA GAUUGGGUUU UGAAGACUUC UUGGGAGAAG UUUGGAAACA AGGGCAUAGA

      4350       4360       4370       4380       4390       4400       4410
AAGACCACCC UCAAGGAUUA UACCGCAGGU AUAAAAACUU GCAUCUGGUA UCAAAGAAAG AGCGGGGACG

      4420       4430       4440       4450       4460       4470       4480
UCACGACGUU CAUUGGAAAC ACUGUGAUCA UUGCUGCAUG UUUGGCCUCG AUGCUUCCGA UGGAGAAAAU

      4490       4500       4510       4520       4530       4540       4550
AAUCAAAGGA GCCUUUUGCG GUGACGAUAG UCUGCUGUAC UUUCCAAAGG GUUGUGAGUU UCCGGAUGUG

      4560       4570       4580       4590       4600       4610       4620
CAACACUCCG CGAAUCUUAU GUGGAAUUUU GAAGCAAAAC UGUUUAAAAA ACAGUAUGGA UACUUUUGCG

      4630       4640       4650       4660       4670       4680       4690
GAAGAUAUGU AAUACAUCAC GACAGAGGAU GCAUUGUGUA UUACGAUCCC CUAAAGUUGA UCUCGAAACU

      4700       4710       4720       4730       4740       4750       4760
UGGUGCUAAA CACAUCAAGG AUUGGGAACA CUUGGAGGAG UUCAGAAGGU CUCUUUGUGA UGUUGCUGUU

      4770       4780       4790       4800       4810       4820       4830
UCGUUGAACA AUUGUGCGUA UUACACACAG UUGGACGACG CUGUAUGGGA GGUUCAUAAG ACCGCCCCUC

      4840       4850       4860       4870       4880       4890       4900
CAGGUUCGUU UGUUUAUAAA AGUCUGGUGA AGUAUUUGUC UGAUAAAGUU CUUUUUAGAA GUUUGUUUAU

      4910       4920       4930       4940       4950       4960       4970
AGAUGGCUCU AGUUGUUAAA GGAAAAGUGA AUAUCAAUGA GUUUAUCGAC CUGACAAAAA UGGAGAAGAU

      4980       4990       5000       5010       5020       5030       5040
CUUACCGUCG AUGUUUACCC CUGUAAAGAG UGUUAUGUGU UCCAAAGUUG AUAAAAUAAU GGUUCAUGAG
```

FIGURE 3. *(continued)*

```
      5050       5060       5070       5080       5090       5100       5110
AAUGAGUCAU UGUCAGAGGU GAACCUUCUU AAAGGAGUUA AGCUUAUUGA UAGUGGAUAC GUCUGUUUAG

      5120       5130       5140       5150       5160       5170       5180
CCGGUUUGGU CGUCACGGGC GAGUGGAACU UGCCUGACAA UUGCAGAGGA GGUGUGAGCG UGUGUCUGGU

      5190       5200       5210       5220       5230       5240       5250
GGACAAAAGG AUGGAAAGAG CCGACGAGGC CACUCUCGGA UCUUACUACA CAGCAGCUGC AAAGAAAAGA

      5260       5270       5280       5290       5300       5310       5320
UUUCAGUUCA AGGUCGUUCC CAAUUAUGCU AUAACCACCC AGGACGCGAU GAAAAACGUC UGGCAAGUUU

      5330       5340       5350       5360       5370       5380       5390
UAGUUAAUAU UAGAAAUGUG AAGAUGUCAG CGGGUUUCUG UCCGCUUUCU CUGGAGUUUG UGUCGGUGUG

      5400       5410       5420       5430       5440       5450       5460
UAUUGUUUAU AGAAAUAAUA UAAAAUUAGG UUUGAGAGAG AAGAUUACAA ACGUGAGAGA CGGAGGGCCC

      5470       5480       5490       5500       5510       5520       5530
AUGGAACUUA CAGAAGAAGU CGUUGAUGAG UUCAUGGAAG AUGUCCCUAU GUCGAUCAGG CUUGCAAAGU

      5540       5550       5560       5570       5580       5590       5600
UUCGAUCUCG AACCGGAAAA AAGAGUGAUG UCCGCAAAGG GAAAAAUAGU AGUAAUGAUC GGUCAGUGCC

      5610       5620       5630       5640       5650       5660       5670
GAACAAGAAC UAUAGAAAUG UUAAGGAUUU UGGAGGAAUG AGUUUUAAAA AGAAUAAUUU AAUCGAUGAU

      5680       5690       5700       5710       5720       5730       5740
GAUUCGGAGG CUACUGUCGC CGAAUCGGAU UCGUUUUAAA UAUGUCUUAC AGUAUCACUA CUCCAUCUCA

      5750       5760       5770       5780       5790       5800       5810
GUUCGUGUUC UUGUCAUCAG CGUGGGCCGA CCCAAUAGAG UUAAUUAAUU UAUGUACUAA UGCCUUAGGA

      5820       5830       5840       5850       5860       5870       5880
AAUCAGUUUC AAACACAACA AGCUCGAACU GUCGUUCAAA GACAAUUCAG UGAGGUGUGG AAACCUUCAC

      5890       5900       5910       5920       5930       5940       5950
CACAAGUAAC UGUUAGGUUC CCUGACAGUG ACUUUAAGGU GUACAGGUAC AAUGCGGUAU UAGACCCGCU

      5960       5970       5980       5990       6000       6010       6020
AGUCACAGCA CUGUUAGGUG CAUUCGACAC UAGAAAUAGA AUAAUAGAAG UUGAAAAUCA GGCGAACCCC

      6030       6040       6050       6060       6070       6080       6090
ACGACUGCCG AAACGUUAGA UGCUACUCGU AGAGUAGACG ACGCAACGGU GGCCAUAAGG AGCGCGAUAA

      6100       6110       6120       6130       6140       6150       6160
AUAAUUUAAU AGUAGAAUUG AUCAGAGGAA CCGGAUCUUA UAAUCGGAGC UCUUUCGAGA GCUCUUCUGG

      6170       6180       6190       6200       6210       6220       6230
UUUGGUUUGG ACCUCUGGUC CUGCAACUUG AGGUAGUCAA GAUGCAUAAU AAAUAACGGA UUGUGUCCGU

      6240       6250       6260       6270       6280       6290       6300
AAUCACACGU GGUGCGUACG AUAACGCAUA GUGUUUUUCC CUCCACUUAA AUCGAAGGGU UGUGUCUUGG

      6310       6320       6330       6340       6350       6360       6370
AUCGCGCGGG UCAAAUGUAU AUGGUUCAUA UACAUCCGCA GGCACGUAAU AAAGCGAGGG GUUCGAAUCC

      6380       6390
CCCCGUUACC CCCGGUAGGG GCCCA
```

FIGURE 3. *(continued)*

noacylase, the 3′-end of the RNA changes its secondary and tertiary structure, frees the 5′-end of the RNA for interaction with ribosomes, and permits the expression of the genes on the 5′-end of the RNA. Working in reverse, such a scenario would result in the inhibition of protein synthesis from newly synthesized genomic RNA, and could be considered a regulatory mechanism of gene expression.

2. Gene Products and Function

The only gene product with an assigned function is the 17.5K viral coat protein. It exists in two forms in virions: 2130 copies of the coat protein per virion (Caspar, 1963) and approximately one copy of a 26.5K hybrid protein of unknown function (Asselin and Zaitlin, 1978), which appears to be a fusion protein involving the coat protein and another non-coat protein molecule of either host or viral origin (Collmer *et al.*, 1983).

Although three of the other four virus-coded proteins have been observed *in vivo* (Sakai and Takebe, 1972, 1974; Zaitlin and Hariharasubramanian, 1972; Paterson and Knight, 1975; Scalla *et al.*, 1978; Joshi *et al.*, 1983; Kiberstis *et al.*, 1983; Ooshika *et al.*, 1984; Watanabe *et al.*, 1984a), we only have circumstantial evidence of their functions.

The 30K protein is phosphorylated (T. Hunter, personal communication) and is believed to be involved in potentiating the cell to cell movement of the virus. This effect was described in detail in Section II.C.

The 126K and 183K proteins are both found in partially purified preparations of the viral replicase (Zaitlin *et al.*, 1973; Young and Zaitlin, unpublished results). Although it has not been shown that either or both of these proteins are subunits of the replicase, their appearance early after infection of tobacco protoplasts by TMV (Sakai and Takebe, 1974; Paterson and Knight, 1975; Siegel *et al.*, 1978; Ogawa and Sakai, 1984), their location at the 5′-end of the RNA (permitting their preferential translation upon invasion of the cell by TMV RNA), and their protein-sequence homology to the gene products of RNAs 1 and 2 of several tricornaviruses (Cornelissen and Bol, 1984; Haseloff *et al.*, 1984; Rezaian *et al.*, 1984), known to be required for the formation of replicative RNA forms (Kiberstis *et al.*, 1981; Rao and Francki, 1981; Nassuth and Bol, 1983; Sarachu *et al.*, 1983), strongly suggest that the 126K and 183K proteins are both part of the replicase. Recently, Evans *et al.* (1985) employed a photoaffinity-labeling procedure which enabled them to analyze the proteins extracted from TMV-infected plants for nucleotide binding ability. They found that the 126K protein could bind nucleotides and suggested it was a component of the replicase. Although they did not specifically discuss it as such, their gel figures show a nucleotide-binding high-molecular-weight protein in extracts of virus-infected plants which is most probably the 183K protein. This observation does not necessarily implicate the 183K protein as a component of the replicase, since the sequence of the already implicated 126K protein is entirely contained within the 183K protein (Fig. 3), but it also does not rule it out.

The 54K protein has yet to be found in cells; however, the homology between this protein and the protein products encoded by RNA 2 of the tricornaviruses, suggests that the 54K protein may also be involved in replication. Possible roles for this protein are: (1) modulating the switch-over in RNA synthesis from (−) to (+) strands (see Fig. 1); (2) altering the replicase to synthesize subgenomic RNAs; and/or (3) as a cap-binding protein that is involved in the initiation of (+) TMV RNA synthesis on a (−) RNA template. Roles (1) and (3) may be identical.

None of the nonvirion, virus-coded proteins is believed to be involved in assembly, since the process occurs spontaneously *in vitro* involving only the RNA and the coat protein (Fraenkel-Conrat and Williams, 1955).

The only other known functional region of the RNA is the cluster of nucleotides (within the 30K gene) comprising the viral RNA origin of encapsidation: nucleotides 5420–5546 (Jonard *et al.*, 1977; Zimmern, 1977; Goelet *et al.*, 1982).

C. TMV Life Cycle

The putative life cycle of TMV is shown in Fig. 1. Although many of the depicted structures have been isolated in TMV or other viruses, several are inferred to exist as intermediates but have yet to be shown to exist *in vivo*.

1. Disassembly

After the virus has entered the cell (A in Fig. 1; see Section II.B), the virus particle must disassemble. Since TMV particles are very stable, until recently it was difficult to understand how this took place. Based on *in vitro* studies by Wilson (1984a,b, 1985) and Wilson and Watkins (1985), we can postulate a mechanism as follows (B in Fig. 1): (1) the virus particle swells (the cause of this swelling is not known) and 80 S cytoplasmic ribosomes bind to the 5′-end of the viral RNA, protruding out from one end of the virion (Wu *et al.*, 1984); (2) the ribosomes move along the viral RNA, removing the coat protein from the RNA; and (3) the ribosomes translate the first 5′-proximal gene, producing the 126K and 183K proteins (C in Fig. 1). Although the above scenario has been shown to occur *in vitro* (Wilson, 1984a,b, 1985), there is as yet no evidence for its existence *in vivo*.

2. Formation of Viral Replicase and Replication Complexes

The 126K and 183K proteins translated from the disassembled viral RNA probably associate with one or more host proteins to form the active viral replicase (D in Fig. 1). Although we have no evidence for the involvement of host proteins in the TMV replicase, there is at least one host protein associated with the replicase of turnip yellow mosaic virus (Mouches *et al.*, 1984). While numerous host proteins are found in the partially purified replicase preparations from both TMV and cowpea mosaic virus (Young and Zaitlin, unpublished observations; Dorssers *et al.*, 1984), it is not known if any are *functionally* associated with the enzyme.

After translation by the ribosomes, the disassembled TMV RNA (E in Fig. 1) binds a replicase molecule at its 3′-end and becomes a template for the synthesis of a complementary (−) strand (F in Fig. 1). This structure is referred to as the "replicative form" (RF), the two strands of which are believed not to be base-paired *in vivo* (Garnier *et al.*, 1980; Richards *et al.*, 1984); however, upon extraction of the RNA from tissue and the removal of proteins (replicase molecules separating the strands?), the RF is isolated as a double-stranded RNA structure (Thatch and Thatch, 1973).

Traditionally, it has been suggested that the RF leads to the production of a "replicative intermediate" (RI; G in Fig. 1). The RI is defined as a molecule that contains a number of (+) strands associated with a fewer number of (−) strands of the same RNA (Kamen, 1975). Upon extraction, the RI is known to be an RNA structure containing single-stranded as well as double-stranded (+) sequences, but a near-completely base-paired (−) strand (Nilsson-Tillgren, 1970; Jackson *et al.*, 1971). What causes the switchover from (−) RNA synthesis (i.e., RF formation) to (+) RNA synthesis is unknown; however, it may involve one of the viral translation products interacting with the viral replicase.

Because the above models for RFs and RIs have ignored the production of subgenomic RNAs, we have considered the possibility that there actually may be two functionally distinct RIs: those that are involved in the production of new genomic (+) RNA (RI_1; G in Fig. 1) and those that are involved in the production of subgenomic RNAs (RI_2; I in Fig. 1). The temporal relationship of the two RIs in the replication process is unclear; however, they are probably generated simultaneously, with the number of molecules of RI_1, greatly outnumbering those of RI_2. The nature of the RNAs produced during both TMV RF and RI synthesis have been investigated *in vitro* with a partially-purified replicase enzyme. The results are consistent with the structures shown for the RF and RIs in Fig. 1 (Young and Zaitlin, 1986).

RI_1 structures are presumably produced when the original replicase molecule at the 3′-end of the (−) strand [as well as several other replicase molecules which subsequently bind to the 3′-end of the (−) TMV RNA] of the RF initiates the synthesis of (+) TMV RNA. The resultant structure now has multiple (+) strands on a single (−) strand (G in Fig. 1).

It is also possible that RIs could be formed very early during the life cycle that contain one (+) strand and multiple (−) strands and then each of these progeny (−) strands goes on to produce new RFs and the RI_1 structures; however, at the earliest time postinoculation that RNA synthesis can be measured, only RIs with (+) strands in excess over (−) strands can be detected (Aoki and Takebe, 1975). Perhaps both multiple (−) stranded and multiple (+) stranded RIs are present at such early stages, but the multiple (+) stranded RIs (RI_1) again vastly outnumber the multiple (−) stranded ones. Since the generation of multiple (−) stranded RIs is merely an amplification step increasing the number of molecules of RF going to RI_1, and for the sake of simplicity, we have chosen to omit this from Fig. 1.

The events described above relating to the formation of RI_1 complexes (G in Fig. 1) are thought to occur on membranous structures (Ralph *et al.*, 1971; Nilsson-Tillgren *et al.*, 1974). Since ribosomes have also been found associated with the RI_1 (Beachy and Zaitlin, 1975), it is quite probable that translation of the (+) TMV RNA occurs even before the 3′-end of that molecule has been completely synthesized. After the synthesis of (+) TMV RNA has been completed and the RNA is released from the RI_1 structures (H in Fig. 1), there is reason to believe that very little of this RNA becomes associated with polyribosomes for further *in vivo* translation (Aoki and Takebe, 1975; Palukaitis *et al.*, 1983). Rather, most of these nascent viral RNA molecules are encapsidated to form virus particles (K in Fig. 1).

As of late 1984, the intracellular site of TMV RNA replication was considered to be cytoplasmic membranes (Ralph *et al.*, 1971; Nilsson-Tillgren *et al.*, 1974). Nilsson-Tillgren *et al.* had partitioned leaf homogenates into fractions enriched in nuclei, chloroplasts, mitochondria, and membranes. Each fraction was characterized by measurement of the amount of marker substances characteristic of that fraction and for the amount of TMV (−) strand; they found that viral (−) sequences were grossly enriched in the membrane fraction. Recently, however, two groups have suggested that viral RNA may be synthesized in the nucleus (Watanabe and Okada, 1984; van Telgen *et al.*, 1985). Watanabe and Okada based their conclusions on the finding of replicase activity as well as the 126K and 183K proteins in that fraction; van Telgen *et al.* found a 126K protein in a chromatin fraction. Our (unpublished) observations do not agree with these two recent studies; this subject needs clarification.

3. Production and Expression of Subgenomic RNAs

The mechanism for the generation of subgenomic RNAs of TMV has not been determined; however, the presence of double-stranded forms of the three subgenomic mRNAs of TMV in infected tissues (Zelcer *et al.*, 1981; Palukaitis *et al.*, 1983) suggests the possibility of the replication of subgenomic mRNAs via their own RFs and RIs. We consider this improbable, since only the genomic TMV RNA is required for infectivity and new subgenomic RNAs are generated from infection by purified genomic TMV RNA; it is unlikely that there are two mechanisms for the generation of subgenomic RNAs. Moreover, among the tricornaviruses, there is evidence that the subgenomic RNA is not replicated (Nassuth and Bol, 1983). Thus, the generation of RFs of the subgenomic mRNAs may either be a method of regulating their expression, or an unavoidable consequence as a result of a 3′-terminus common to both the genomic and subgenomic RNAs; the 3′-terminus is presumably involved in replicase binding upon initiation. Since the 3′-ends of the newly synthesized (−) strands of the subgenomic RNAs do not contain the recognition sequence for (+) subgenomic RNA synthesis, "replication" of the subgenomic RNAs would terminate at the RF stage.

Experiments carried out with BMV have shown that BMV replicase can bind internally on the (−) strand of BMV RNA 3, initiate the synthesis of and generate a copy of (+) BMV RNA 4, a subgenomic RNA of BMV RNA 3 (Miller *et al.*, 1985). We are postulating that a similar mechanism occurs with TMV (I in Fig. 1).

After the formation of the TMV RF (F in Fig. 1), other molecules of replicase bind internally onto the (−) TMV RNA strand and begin the synthesis of (+) TMV subgenomic RNAs. The RI structure that is produced is referred to as RI_2 (I in Fig. 1). The replicase molecules complete the synthesis of the subgenomic RNAs, which are then translated on both membrane-bound and free polyribosomes (J in Fig. 1; Ogawa *et al.*, 1983). Once again, it is conceivable that translation of the subgenomic RNAs also begins before the completion of synthesis or the release of the subgenomic RNAs from RI_2 structures.

Of course, it is entirely possible that RI_1 and RI_2 are really part of the same RI complex with the replicase molecules initiating the synthesis of both (+) genomic RNAs and (+) subgenomic RNAs on the same (−) TMV RNA molecule. The switchover by replicase molecules from genomic RNA synthesis to subgenomic RNA synthesis may involve either modulation of the viral replicase by either host or virus-coded proteins, or competition between "promoter sites" on the (−) TMV RNA molecule for replicase binding, or a combination of both effects.

As stated above, the subgenomic RNAs are probably also binding ribosomes and being translated while they are still part of the RI_2 complex. Evidence for this postulate and for the low level of translation of (+) genomic RNA after release from RI_1, comes from several independent studies:

1. Time course studies in protoplasts showed the early synthesis of 126K/183K proteins, which reached a plateau at about the time the synthesis of coat protein became exponential. At later stages of infection, the rate of synthesis of 126K/183K proteins usually declined (Paterson and Knight, 1975; Siegel *et al.*, 1978; Ogawa and Sakai, 1984).

2. Most of the LMC found associated with polyribosomes was present on free polyribosomes and only a small fraction was found associated with membrane-bound polyribosomes (Ogawa *et al.*, 1983).

3. Very little nonencapsidated genomic RNA was found either in infected tissues (Palukaitis *et al.*, 1983) or in protoplasts "late" in infection (Aoki and Takebe, 1975).

4. Temperature-sensitive mutants of TMV RNA synthesis have been isolated (Dawson and White, 1979). Studies with such mutants have shown the following: (i) Mutations leading to temperature-sensitive functions that result in the shutoff of the production of single-stranded (+) RNA synthesis (both genomic and LMC), do not affect the production of RFs or RIs, and also do not affect the synthesis of the 126K, 183K, or

17.5K proteins (the only ones analyzed); and (ii) other temperature-sensitive mutations which shut off single-stranded RNA, RF, and RI synthesis also result in the inhibition of virus-specific protein synthesis. Thus, protein synthesis is closely regulated with RF/RI production, and a model that suggests translation occurring on RI complexes would be consistent with this observation.

On the other hand, results obtained from protoplast studies indicate that some subgenomic RNAs are either only transiently expressed (I_2 mRNA: Joshi *et al.*, 1983; Kiberstis *et al.*, 1983; Watanabe *et al.*, 1984a) or expressed later during replication (LMC: Sakai and Takebe, 1974; Paterson and Knight, 1975; Siegel *et al.*, 1978; Ogawa and Sakai, 1984). Thus, they are not coordinately regulated. Furthermore, LMC is found preferentially associated with free and not membrane-bound polyribosomes (J in Fig. 1; Ogawa *et al.*, 1983). Clearly, we still have much to learn about the regulatory mechanisms involved in the generation and expression of the subgenomic RNAs of TMV.

4. Late Gene Functions

In Section III.B.2, the functions of the various TMV gene products were described. In this subsection, the functions of the products of two genes (the 30K gene and the 17.5K coat protein gene) will be described in relationship to the TMV life cycle.

The mRNA (I_2 mRNA) for the 30K gene was shown to be transiently synthesized in protoplasts and the expression of the 30K gene paralleled the synthesis of I_2 mRNA (Watanabe *et al.*, 1984a). Since I_2 mRNA is found encapsidated in virions (Beachy and Zaitlin, 1977), the transient expression of the 30K gene is not due to a rapid turnover of the I_2 mRNA *per se*. Rather, the synthesis of the I_2 mRNA appears to be inhibited by some unknown regulatory mechanism, while the preexisting pool of I_2 mRNA is encapsidated into less-than-full-length virus particles.

The 30K gene product first appears in protoplasts and in plants at the same time as the 17.5K coat protein mRNA (LMC) (Watanabe *et al.*, 1984a). However, unlike the I_2 mRNA, the LMC continues to be synthesized and translated to produce coat protein throughout the infection process, as coat protein is required for the encapsidation of the newly synthesized virion RNA (K in Fig. 1). Since coat protein production is considered a "late (albeit continuous) gene function," the 30K gene product appears also to have a late gene function; the limited amount of 30K gene product synthesized suggests that the 30K protein has an enzymatic function. In Section II.C, evidence was presented that this function was to promote the cell to cell movement of the virus; the temporal link of the synthesis of 30K protein with the production of coat protein [the latter is linked temporally to virus assembly (Sakai and Takebe, 1974; Paterson and Knight, 1975; Ogawa and Sakai, 1984)] is not at variance with such a role. The mechanism by which the 30K protein promotes the cell to cell movement of the virus has not been elucidated.

Encapsidation of TMV RNA into virions was described in Chapter 2. The process occurs spontaneously (Fraenkel-Conrat and Williams, 1955) and does not require the presence of other (e.g., scaffolding) proteins.

5. The End and the Beginning of the Life Cycle

Following encapsidation, the virus particles are present in such a high localized concentration that they crystallize. The types of crystals and their intracellular locations are described in Chapter 7.

Since the virus disassembles so readily upon entering the cell (B in Fig. 1), why does it remain as a stable particle at this stage of the life cycle? One possibility may be the change in the intracellular environment that has taken place as a result of infection. For example: (1) the ribosomes may not be free to "attack" TMV particles because they are still translating LMC—a very efficient and stable mRNA (Watanabe *et al.*, 1984a); (2) the ratio of virus particles to free ribosomes is much higher than at the beginning of the life cycle; (3) the initial event in disassembly (equivalent to the *in vitro* swelling of TMV particles) may have occurred under conditions no longer present in the cell or in a subcellular location different from the sites of virus accumulation; and (4) the crystallization of virions removes a large pool of such particles from interacting with ribosomes.

Some of the noncrystalline virus particles pass through plasmodesmata to adjacent cells (L in Fig. 1), where they once again begin the life cycle (A and B in Fig. 1). Other particles find their way into the vascular system (Section II.D), where they are translocated to the plant meristem and begin the life cycle anew in other tissues.

D. Cross-Protection versus Replication

Infection of a plant with a mild strain (little or no symptom production) of a virus can "cross-protect" the plant against disease incited by either a severe strain of the same virus or a "closely related" virus (reviewed by Hamilton, 1980). This phenomenon is called cross-protection and usually results in little or no replication of the second inoculated virus (referred to as the "challenge" strain).

Various models have been proposed to explain the mechanism of cross-protection (reviewed by Hamilton, 1980). Of course, we favor our hypothesis (Palukaitis and Zaitlin, 1984) that cross-protection occurs by the following mechanism: the (−) strand of the RF (Fig. 1), or of a multiple (−) stranded RI of the challenge virus, is "captured" (hydrogen bonded) by (+) RNA molecules of the already established (initial) virus. The formation of such double-stranded structures would prevent the challenge virus RNA from being further replicated and expressed. Isolated TMV RF is known to be noninfectious, unless heat-denatured prior to inoculation (Jackson *et al.*, 1971). Recently, the first steps necessary to test this hypothesis have been taken; a clone containing the coat protein gene of TMV has been inserted into the tobacco genome. This gene was expressed as (+) TMV RNA and low levels of both the mRNA and TMV coat protein were produced (Bevan *et al.*, 1985). It remains to be determined if this level of transcription is sufficient to engender cross protection. Full-length clones of TMV have recently been generated (Beck *et al.*, 1985) which should facilitate this type of study.

This hypothesis is called "negative strand capture," and there is evidence that similar mechanisms may also occur in both prokaryotes and eukaryotes as a method of translational control of mRNAs (Izant and Weintraub, 1984; Mizuno *et al.*, 1984; Pestka *et al.*, 1984; Melton, 1985; Rosenberg *et al.*, 1985). In these cases, however, the (−) strand RNA is transcribed from the antisense strand of the same gene and by hydrogen bonding to (a part of or all of) the (+) mRNA, prevents the translation of the mRNA; the (−) strand RNA is referred to here as "antisense RNA."

The above model for cross-protection would also predict the hydrogen bonding of nascent (+) viral RNA to the (−) RNA of the RF, resulting in the inhibition of further (+) RNA synthesis from the RF. There is some evidence from protoplast studies that such a phenomenon may actually occur. RF and RI synthesis do indeed reach a plateau early in infection (Aoki and Takebe, 1975; Ogawa and Sakai, 1984), and the synthesis of such RNAs does not begin anew until late in infection (albeit at reduced levels), when most of the (+) RNA has been encapsidated (Aoki and Takebe, 1975; Gonda and Symons, 1979). In plants, this phenomenon would be difficult to observe because of cell to cell movement and the infection of neighboring cells. Nevertheless, at some stage during the life cycle in each cell, "self-protection" is likely to occur, resulting in either termination or at least a reduction in the rate of replication.

Finally, the apparent sequence complementarity between the 5′-end of the (+) genomic RNA and its 3′-end (Section III.B.1), which has been postulated to inhibit the translation of the genomic RNA, may also alter the balance of synthesis of RNA from RI_1 to RI_2, as well as play a role in initiating self-protection and/or cross-protection. Clearly, the world's most studied virus has yet to reveal all of its mysteries.

ACKNOWLEDGMENTS. We wish to thank Drs. R. I. B. Francki and N. D. Young for their careful consideration of the manuscript, and Dr. P. J. G. Butler for permission to reprint the complete sequence of TMV RNA. Our studies described herein were supported in part by Grant 84-09881 from the National Science Foundation and Grant 83-00181 from the Competitive Grants Program of the United States Department of Agriculture.

REFERENCES

Aoki, S., and Takebe, I., 1975, Replication of tobacco mosaic virus RNA in tobacco mesophyll protoplasts inoculated *in vitro, Virology* **65:**343.

Asselin, A., and Zaitlin, M., 1978, Characterization of a second protein associated with virions of tobacco mosaic virus, *Virology* **91:**173.

Atabekov, J. G., and Dorokhov, Y. L., 1984, Plant virus-specific transport function and resistance of plants to viruses, *Adv. Virus. Res.* **29:**313.

Beachy, R. N., and Zaitlin, M., 1975, Replication of tobacco mosaic virus. VI. Replicative intermediate and TMV-RNA-related RNAs associated with polyribosomes, *Virology* **63:**84.

Beachy, R. N., and Zaitlin, M., 1977, Characterization and *in vitro* translation of the RNAs from less-than-full-length, virus-related nucleoprotein rods present in tobacco mosaic virus preparations, *Virology* **81:**160.

Beachy, R. N., Zaitlin, M., Bruening, G., and Israel, H. W., 1976, A genetic map for the cowpea strain of TMV, *Virology* **73:**498.

Beck, D. L., Knorr, D., Grantham, G., and Dawson, W. O., 1985, Synthesis of full-length cDNA clones of TMV, *Phytopathology* (Abstract) **75:**1334.

Beier, H., Barciszewska, M., Krupp, G., Mitnacht, R., and Gross, H. J., 1984, UAG read-through during TMV RNA translation: Isolation and sequence of two tRNAs ^{Tyr} with suppressor activity from tobacco plants, *EMBO J.* **3:**351.

Bennett, C. W., 1940, The relation of viruses to plant tissues, *Bot. Rev.* **6:**427.

Bevan, M. W., Mason, S. E., and Goelet, P., 1985, Expression of tobacco mosaic virus coat protein by a cauliflower mosaic virus promoter in plants transformed by *Agrobacterium, EMBO J.* **4:**1921.

Brants, D. H., 1964, The susceptibility of tobacco and bean leaves to tobacco mosaic virus infection in relation to the condition of ectodesmata, *Virology* **23:**588.

Calvete, J. S., and Wieringa-Brants, D. H., 1984, Infection and necrosis of cowpea mesophyll cells by tobacco necrosis virus and two strains of tobacco mosaic virus, *Neth. J. Plant Pathol.* **90:**71.

Capoor, S. P., 1949, The movement of tobacco mosaic viruses and potato virus X through tomato plants, *Ann. Appl. Biol.* **36:**307.

Carr, R. J., and Kim, K. S., 1983, Evidence that bean golden mosaic virus invades non-phloem tissue in double infections with tobacco mosaic virus, *J. Gen. Virol.* **64:**2489.

Caspar, D. L. D., 1963, Assembly and stability of the tobacco mosaic virus particle, *Adv. Protein Chem.* **18:**37.

Collmer, C. W., Vogt, V. M., and Zaitlin, M., 1983, H-protein, a minor protein of TMV virions, contains sequences of the viral coat protein, *Virology* **126:**429.

Cornelissen, B. J. C., and Bol, J. F., 1984, Homology between the proteins encoded by tobacco mosaic virus and two tricornaviruses, *Plant Mol. Biol.* **3:**379.

Dawson, W. O., and White, J. L., 1979, A temperature-sensitive mutant of tobacco mosaic virus deficient in synthesis of single-stranded RNA, *Virology* **93:**104.

de Zoeten, G. A., 1981, Early events in plant virus infection, in: *Plant Diseases and Vectors: Ecology and Epidemiology* (K. Maramorosch and K. F. Harris, eds.), pp. 221–239, Academic Press, New York.

de Zoeten, G. A., and Gaard, G., 1984, The presence of viral antigen in the apoplast of systemically virus-infected plants, *Virus Res.* **1:**713.

Dorssers, L., van der Krol, S., van der Meer, J., van Kammen, A., and Zabel, P., 1984, Purification of cowpea mosaic virus RNA replication complex: Identification of a virus-encoded 110,000-dalton polypeptide responsible for RNA chain elongation, *Proc. Natl. Acad. Sci. USA* **81:**1951.

Evans, R. K., Haley, B. E., and Roth, D. A., 1985, Photoaffinity labeling of a viral induced protein from tobacco: Characterization of nucleotide-binding properties, *J. Biol. Chem.* 260:7800.

Fleck, J., Durr, A., and Hirth, L., 1983, Gene expression in freshly isolated protoplasts from *Nicotiana sylvestris,* in: *Proceedings 6th International Protoplast Symposium,* pp. 240–241.

Florentz, C., Briand, J. P., and Giege, R., 1984, Possible functional role of viral tRNA-like structures, *FEBS Lett.* **176:**295.

Fraenkel-Conrat, H., and Williams, R. C., 1955, Reconstitution of active tobacco mosaic virus from its inactive protein and nucleic acid components, *Proc. Natl. Acad. Sci. USA* **41:**690.

Fry, P. R., and Matthews, R. E. F., 1963, Timing of some early events following inoculation with tobacco mosaic virus, *Virology* **19:**461.

Garciá-Arenal, F., Palukaitis, P., and Zaitlin, M., 1984, Strains and mutants of tobacco mosaic virus are both found in virus derived from single-lesion-passaged inoculum, *Virology* **132:**131.

Garnier, M., Mamoun, R., and Bové, J. M., 1980, TYMV RNA replication *in vivo:* Replicative intermediate is mainly single stranded, *Virology* **104:**357.

Gianinazzi, S., 1984, Genetic and molecular aspects of resistance induced by infections or chemicals, in: *Plant–Microbe Interactions: Molecular and Genetic Perspectives* (T. Kosuge and E. W. Nester, eds.), Volume 1, pp. 321–342, Macmillan Co., New York.

Goelet, P., Lomonossoff, G. P., Butler, P. J. G., Akam, M. E., Gait, M. J., and Karn, J., 1982, Nucleotide sequence of tobacco mosaic virus RNA, *Proc. Natl. Acad. Sci. USA* **79:**5818.

Gonda, T. J., and Symons, R. H., 1979, Cucumber mosaic virus replication in cowpea protoplasts: Time course of virus, coat protein and RNA synthesis, *J. Gen. Virol.* **45:**723.

Guilley, H., Jonard, G., Kukla, B., and Richards, K. E., 1979, Sequence of 1000 nucleotides at the 3′ end of tobacco mosaic virus RNA, *Nucleic Acids Res.* **6:**1287.

Gunning, B. E. S., and Overall, R. L., 1983, Plasmodesmata and cell-to-cell transport in plants, *Bioscience* **33:**260.

Haenni, A. L., Joshi, S., and Chapeville, F., 1982, tRNA like structures in genomes of RNA viruses, *Prog. Nucleic Acid Res. Mol. Biol.* **27:**85.

Hall, T. C., 1979, Transfer RNA-like structures in viral genomes, *Int. Rev. Cytol.* **60:**1.

Hamilton, R. I., 1980, Defenses triggered by previous invaders: Viruses, in: *Plant Disease: An Advanced Treatise* (J. G. Horsfall and E. B. Cowling, eds.), Volume V, pp. 279–303, Academic Press, New York.

Haseloff, J., Goelet, P., Zimmern, D., Ahlquist, P., Dasgupta, R., and Kaesberg, P., 1984, Striking similarities in amino acid sequence among nonstructural proteins encoded by RNA viruses that have dissimilar genomic organization, *Proc. Natl. Acad. Sci. USA* **81:**4358.

Hooft van Huijsduijnen, R. A. M., Cornelissen, B. J. C., van Loon, L. C., van Boom, J. H., Tromp, M., and Bol, J. F., 1985, Virus-induced synthesis of messenger RNAs for precursors of pathogenesis-related proteins in tobacco, *EMBO J.* **4:**2167.

Hunter, T. R., Hunt, T., Knowland, J., and Zimmern, D., 1976, Messenger RNA for the coat protein of tobacco mosaic virus, *Nature* (*London*) **260:**759.

Hunter, T., Jackson, R., and Zimmern, D., 1983, Multiple proteins and subgenomic mRNAs may be derived from a single open reading frame on tobacco mosaic virus RNA, *Nucleic Acids Res.* **11:**801.

Izant, J. G., and Weintraub, H., 1984, Inhibition of thymidine kinase gene expression by anti-sense RNA: A molecular approach to genetic analysis, *Cell* **36:**1007.

Jackson, A. O., Mitchell, D. M., and Siegel, A., 1971, Replication of tobacco mosaic virus. I. Isolation and characterization of double-stranded forms of ribonucleic acid, *Virology* **45:**182.

Jonard, G., Richards, K. E., Guilley, H., and Hirth, L., 1977, Sequence from the assembly nucleation region of TMV RNA, *Cell* **11:**483.

Joshi, S., Pleij, C. W. A., Haenni, H. L., Chapeville, F., and Bosch, L., 1983, Properties of the tobacco mosaic virus intermediate length RNA-2 and its translation, *Virology* **127:**100.

Kado, C. I., and Knight, C. A., 1966, Location of a local lesion gene in tobacco mosaic virus RNA, *Proc. Natl. Acad. Sci. USA* **55**:1276.

Kamen, R. I., 1975, Structures and function of the QB replicase, in: *RNA Phages* (N. D. Zinder, ed.), pp. 203–234, Cold Spring Harbor Laboratory, Cold Spring Harbor, N.Y.

Keith, J., and Fraenkel-Conrat, H., 1975, Tobacco mosaic virus RNA carries 5′-terminal triphosphorylated guanosine blocked by 5′-linked 7-methylguanosine, *FEBS Lett.* **57**:31.

Kiberstis, P. A., Loesch-Fries, L. S., and Hall, T. C., 1981, Viral protein synthesis in barley protoplasts inoculated with native and fractionated brome mosaic virus RNA, *Virology* **112**:804.

Kiberstis, P. A., Pessi, A., Atherton, E., Jackson, R., Hunter, T., and Zimmern, D., 1983, Analysis of *in vitro* and *in vivo* products of the TMV 30kDa open reading frame using antisera raised against a synthetic peptide, *FEBS Lett.* **164**:355.

Kuhn, C. W., Wyatt, S. D., and Brantley, B. B., 1981, Genetic control of symptoms, movement, and virus accumulation in cowpea plants infected with cowpea chlorotic mottle virus, *Phytopathology* **71**:1310.

Leonard, D. A., and Zaitlin, M., 1982, A temperature sensitive strain of tobacco mosaic virus defective in cell-to-cell movement generates an altered viral-coded protein, *Virology* **117**:416.

Lucas, J., Camacho Henriquez, A., Lottspeich, F., Henschen, A., and Sänger, H. L., 1985, Amino acid sequence of the 'pathogenesis-related' leaf protein p14 from viroid-infected tomato reveals a new type of structurally unfamiliar proteins, *EMBO J.* **4**:2745.

McKinney, H. H., and Clayton, E. E., 1945, Genotype and temperature in relation to symptoms caused in *Nicotiana* by the mosaic virus, *J. Hered.* **36**:323.

Matthews, R. E. F., 1981, *Plant Virology*, 2nd ed., Academic Press, New York.

Mayer, A., 1886, Über die Mosaikkrankheit des Tabaks, *Landwirtsch. Vers. Stn.* **32**:451. Reprinted in English as Phytopathological Classics No. 7, American Phytopathological Society, Minneapolis.

Melton, D. A., 1985, Injected anti-sense RNAs specifically block messenger RNA translation *in vivo*, *Proc. Natl. Acad. Sci. USA* **82**:144.

Merkens, W. S. W., de Zoeten, G. A., and Gaard, G., 1972, Observations on ectodesmata and the virus infection process, *J. Ultrastruct. Res.* **41**:397.

Miller, W. A., Dreher, T. W., and Hall, T. C., 1985, Synthesis of brome mosaic virus subgenomic RNA *in vitro* by internal initiation on (−) sense genomic RNA, *Nature (London)* **313**:68.

Mizuno, T., Chou, M.-Y., and Inouye, M., 1984, A unique mechanism regulating gene expression: Translational inhibition by a complementary RNA transcript (micRNA), *Proc. Natl. Acad. Sci. USA* **81**:1966.

Mouches, C., Candresse, T., and Bové, J. M., 1984, Turnip yellow mosaic virus RNA-replicase contains host and virus-encoded subunits, *Virology* **134**:78.

Nassuth, A., and Bol, J. F., 1983, Altered balance of the synthesis of plus- and minus-strand RNAs induced by RNAs 1 and 2 of alfalfa mosaic virus in the absence of RNA 3, *Virology* **124**:75.

Nilsson-Tillgren, T., 1970, Studies on the biosynthesis of TMV. III. Isolation and characterization of the replicative form and the replicative intermediate RNA, *Mol. Gen. Genet.* **109**:246.

Nilsson-Tillgren, T., Kielland-Brandt, M. C., and Bekke, B., 1974, Studies on the biosynthesis of tobacco mosaic virus. VI. On the subcellular localization of double-stranded viral RNA, *Mol. Gen. Genet.* **128**:157.

Nishiguchi, M., Motoyoshi, F., and Oshima, N., 1978, Behavior of a temperature sensitive strain of tobacco mosaic virus in tomato leaves and protoplasts, *J. Gen. Virol.* **39**:53.

Nishiguchi, M., Motoyoshi, F., and Oshima, N., 1980, Further investigation of a temperature-sensitive strain of tobacco mosaic virus: Its behavior in tomato leaf epidermis, *J. Gen. Virol.* **46**:497.

Nishiguchi, M., Kikuchi, S., Kiho, Y., Ohno, T., Meshi, T., and Okada, Y., 1985, Molecular basis of plant viral virulence: The complete nucleotide sequence of an attenuated strain of tobacco mosaic virus, *Nucleic Acids Res.* **13:**5585.

Nozu, Y., Ohno, T., and Okada, Y., 1970, Amino acid sequences of some common Japanese strains of tobacco mosaic virus, *J. Biochem.* (*Tokyo*) **68:**39.

Öberg, B., and Philipson, L., 1972, Binding of histidine to tobacco mosaic virus RNA, *Biochem. Biophys. Res. Commun.* **48:**927.

Ogawa, M., and Sakai, F., 1984, A messenger RNA for tobacco mosaic virus coat protein in infected tobacco mesophyll protoplasts. II. Time course of its synthesis, *Phytopathol. Z.* **109:**193.

Ogawa, M., Sakai, F., and Takebe, I., 1983, A messenger RNA for tobacco mosaic virus coat protein in infected tobacco mesophyll protoplasts, *Phytopathol. Z.* **107:**146.

Ohno, T., Takamatsu, N., Meshi, T., Okada, Y., Nishiguichi, M., and Kiho, Y., 1983, Single amino acid substitution in 30K protein of TMV defective in virus transport function, *Virology* **131:**255.

Ooshika, I., Watanabe, Y., Meshi, T., Okada, Y., Igano, K., Inouye, K., and Yoshida, N., 1984, Identification of the 30K protein of TMV by immunoprecipitation with antibodies directed against a synthetic peptide, *Virology* **132:**71.

Otsuki, Y., and Takebe, I., 1978, Production of mixedly coated particles in tobacco mesophyll protoplasts doubly infected by strains of tobacco mosaic virus, *Virology* **84:**162.

Palukaitis, P., and Zaitlin, M., 1984, A model to explain the "crossprotection" phenomenon shown by plant viruses and viroids, in: *Plant–Microbe Interactions: Molecular and Genetic Perspectives* (T. Kosuge and E. W. Nester, eds.), Volume I, pp. 420–429, Macmillan Co., New York.

Palukaitis, P., Garciá-Arenal, F., Sulzinski, M. A., and Zaitlin, M., 1983, Replication of tobacco mosaic virus. VII. Further characterization of single- and double-stranded virus-related RNAs from TMV-infected plants, *Virology* **131:**533.

Parent, J.-G., and Asselin, A., 1984, Detection of pathogenesis-related proteins (PR or b) and of other proteins in the intercellular fluid of hypersensitive plants infected with tobacco mosaic virus, *Can. J. Bot.* **62:**564.

Paterson, R., and Knight, C. A., 1975, Protein synthesis in tobacco protoplasts infected with tobacco mosaic virus, *Virology* **64:**10.

Pelham, H. R. B., 1978, Leaky UAG termination codon in tobacco mosaic virus RNA, *Nature* (*London*) **272:**469.

Pestka, S., Daugherty, B. L., Jung, V., Hotto, K., and Pestka, R. K., 1984, Anti-mRNA: Specific inhibition of translation of single mRNA molecules, *Proc. Natl. Acad. Sci. USA* **81:**7525.

Ralph, R. K., Bullivant, S., and Wojcik, S. Y., 1971, Cytoplasmic membranes, a possible site of tobacco mosaic virus RNA replication, *Virology* **43:**713.

Rao, A. L. N., and Francki, R. I. B., 1981, Comparative studies on tomato aspermy and cucumber mosaic viruses. VI. Partial compatibility of genome segments from the two viruses, *Virology* **114:**573.

Rappaport, I., and Wu, J.-H., 1962, Release of inhibited virus infection following irradiation with ultraviolet light, *Virology* **17:**411.

Rezaian, M. A., Williams, R. H., Gordon, K. H., Gould, A. R., and Symons, R. H., 1984, Nucleotide sequence of cucumber mosaic virus RNA2 reveals a translation product significantly homologous to corresponding proteins of other viruses, *Eur. J. Biochem.* **143:**277.

Richards, O. C., Martin, S. C., Jense, H. G., and Ehrenfeld, E., 1984, Structure of poliovirus replicative intermediate RNA, *J. Mol. Biol.* **173:**325.

Rietveld, K., Linschooten, K., Pleij, C. W. A., and Bosch, L., 1984, The three-dimensional folding of the tRNA-like structure of tobacco mosaic virus RNA: A new building principle applied twice, *EMBO J.* **3:**2613.

Rosenberg, U. B., Preiss, A., Seifert, E., Jäckle, H., and Knipple, D. C., 1985, Production of phencopies by *Krüppel* antisense RNA injection into *Drosophila* embryos, *Nature* (*London*) **313:**703.

Sakai, F., and Takebe, I., 1972, A non-coat protein synthesized in tobacco mesophyll protoplasts infected by tobacco mosaic virus, *Mol. Gen. Genet.* **118**:93.
Sakai, F., and Takebe, I., 1974, Protein synthesis in tobacco mesophyll protoplasts induced by tobacco mosaic virus infection, *Virology* **62**:426.
Samuel, G., 1934, The movement of tobacco mosaic virus within the plant, *Ann. Appl. Biol.* **21**:90.
Sarachu, A. N., Nassuth, A., Roosien, J., Van Vloten-Doting, L., and Bol, J. F., 1983, Replication of temperature-sensitive mutants of alfalfa mosaic virus in protoplasts, *Virology* **125**:64.
Scalla, R., Romaine, P., Asselin, A., Rigaud, J., and Zaitlin, M., 1978, An *in vivo* study of a non-structural polypeptide synthesized upon TMV infection and its identification with a polypeptide synthesized *in vitro* from TMV RNA, *Virology* **91**:182.
Siegel, A., Zaitlin, M., and Sehgal, O. P., 1962, The isolation of defective tobacco mosaic virus strains, *Proc. Natl. Acad. Sci. USA* **48**:1845.
Siegel, A., Hari, V., and Kolacz, K., 1978, The effect of tobacco mosaic virus infection on host and virus-specific protein synthesis in protoplasts, *Virology* **85**:494.
Sulzinski, M. A., and Zaitlin, M., 1982, Tobacco mosaic virus replication in resistant and susceptible plants: In some resistant species virus is confined to a small number of initially infected cells, *Virology* **121**:12.
Sulzinski, M. A., Gabard, K., Palukaitis, P., and Zaitlin, M., 1985, Replication of tobacco mosaic virus. VIII. Characterization of a third subgenomic TMV RNA, *Virology* **145**:132.
Takebe, I., 1977, Protoplasts in the study of plant virus replication, *Comprehensive Virology* **11**:237.
Takebe, I., and Otsuki, Y., 1968, Isolation of tobacco mesophyll cells in intact and active state, *Plant Cell Physiol.* **9**:115.
Taliansky, M. E., Malyshenko, S. I., Pshennikova, E. S., and Atabekov, J. G., 1982, Plant virus-specific transport function. II. A factor controlling virus host range, *Virology* **122**:327.
Thatch, S. S., and Thatch, R. E., 1973, Mechanism of viral replication. I. Structure of replication complexes of R17 bacteriophage, *J. Mol. Biol.* **81**:367.
Thomas, P. E., and Fulton, R. W., 1968, Correlation of ectodesmata number with nonspecific resistance to initial virus infection, *Virology* **34**:459.
van Telgen, H. J., Goldbach, R. W., and van Loon, L. C., 1985, The 126,000 molecular weight protein of tobacco mosaic virus is associated with host chromatin in mosaic-diseased tobacco plants, *Virology* **143**:612.
Walker, H. L., and Pirone, T. P., 1972, Number of TMV particles required to infect locally or systemically susceptible tobacco cultivars, *J. Gen. Virol.* **17**:241.
Watanabe, Y., and Okada, Y., 1984, Paper presented at the Sixth International Congress of Virology, Sendai, Japan.
Watanabe, Y., Emori, Y., Ooshika, I., Meshi, T., Ohno, T., and Okada, Y., 1984a, Synthesis of TMV-specific RNAs and proteins at the early stage of infection in tobacco protoplasts: Transient expression of the 30K protein and its mRNA, *Virology* **133**:18.
Watanabe, Y., Meshi, T., and Okada, Y., 1984b. The initiation site for transcription of the TMV 30-kDa protein messenger RNA, *FEBS Lett.* **173**:247.
Watts, J. W., and King, J. M., 1984, The effect of charge on infection of tobacco protoplasts by bromoviruses, *J. Gen. Virol.* **65**:1709.
Weintraub, M., Ragetli, H. W. J., and Leung, E., 1976, Elongated virus particles in plasmodesmata, *J. Ultrastruct. Res.* **56**:351.
Wilson, T. M. A., 1984a, Cotranslational disassembly of tobacco mosaic virus *in vitro*, *Virology* **137**:255.
Wilson, T. M. A., 1984b, Cotranslational disassembly increases the efficiency of expression of TMV RNA in wheat germ cell-free extracts, *Virology* **138**:353.
Wilson, T. M. A., 1985, Nucleocapsid disassembly and early gene expression by positive-strand RNA viruses, *J. Gen. Virol.* **66**:1201.
Wilson, T. M. A., and Watkins, P. A. C., 1985, Cotranslational disassembly of a cowpea

strain (C_c) of TMV: Evidence that viral RNA–protein interactions at the assembly origin block ribosome translocation *in vitro, Virology* **145**:346.

Wu, A.-Z., Dai, R.-M., Shen, X.-R., and Sun, Y.-K., 1984, The location and function of the 5′-cap structure of the RNA of tobacco mosaic virus in the virion, in: *Abstracts of the Sixth International Congress of Virology, Sendai, Japan,* p. 231.

Young, N. D., and Zaitlin, M., 1986, An analysis of tobacco mosaic virus replicative structures synthesized *in vitro, Plant Molecular Biology,* in press.

Zaitlin, M., 1977, Replication of plant viruses—An overview, in: *Beltsville Symposia in Agricultural Research I. Virology in Agriculture* (J. A. Romberger, ed.), pp. 33–46, Alenheld, Osmun, Montclair, N.J.

Zaitlin, M., 1983, Functions of the translation products of RNA plant viruses, *Plant Mol. Biol. Rep.* **1**:111.

Zaitlin, M., and Hariharasubramanian, V., 1972, A gel electrophoretic analysis of proteins from plants infected with tobacco mosaic and potato spindle tuber viruses, *Virology* **47**:296.

Zaitlin, M., Duda, C. T., and Petti, M. A., 1973, Replication of tobacco mosaic virus. V. Properties of the bound and solubilized replicase, *Virology* **53**:300.

Zaitlin, M., Beachy, R. N., and Bruening, G., 1977, Lack of molecular hybridization between RNAs of two strains of TMV: A reconsideration of the criteria for strain relationships, *Virology* **82**:237.

Zelcer, A., Weaber, K. F., Balázs, E., and Zaitlin, M., 1981, The detection and characterization of viral-related double-stranded RNAs in tobacco mosaic virus-infected plants, *Virology* **113**:417.

Zimmern, D., 1975, The 5′ end group of tobacco mosaic virus RNA is $m^7G^{5'}ppp^{5'}Gp$, *Nucleic Acids Res.* **2**:1189.

Zimmern, D., 1977, The nucleotide sequence at the origin for assembly on tobacco mosaic virus RNA, *Cell* **11**:463.

Appearance

CHAPTER 6

TOBACCO MOSAIC VIRUS

Epidemiology and Control

G. V. GOODING, JR.

I. INTRODUCTION

Tobacco mosaic is an endemic disease of *Nicotiana tabacum* L. throughout the world. It was recognized as a problem in The Netherlands before 1879 when Adolf Mayer initiated his research into a disease of unknown etiology. He later named the disease tobacco mosaic and demonstrated the infectious nature of the causative agent (Mayer, 1886).

The disease usually is not lethal but results in significant losses because of its high incidence in the population.

II. THE DISEASE

A. Etiology

TMV is the most thoroughly characterized viral pathogen of plants. Its properties are described in other chapters of this book. Numerous variants of TMV have been reported but the predominant strain currently found in tobacco is probably the same that caused the disease described by Mayer (Ford and Tolin, 1983).

The symptoms on individually infected plants in the field are remarkably similar but plants occasionally are observed that differ in symptom severity or in which the mosaic pattern tends to be more yellow or

G. V. GOODING, JR. • Department of Plant Pathology, North Carolina State University, Raleigh, North Carolina 27695.

white than the typical green. The most definitive study of field strains of TMV or related viruses on tobacco was done on burley tobacco in Kentucky (Valleau *et al.*, 1954; Johnson and Valleau, 1946). Valleau *et al.* (1954) reported that naturally occurring field isolates of TMV on burley tobacco could be divided into four major groups based on symptomatology: (1) mild mosaic, (2) yellow mosaic, (3) white mosaic, and (4) a type which caused leaf burning. A fifth type of symptom, consisting of severe stunting, leaf distortion and necrosis, occasionally was observed in the field. These symptoms were caused by the plantago strain of TMV (Holmes, 1941) which Valleau (1952b) postulated was the origin of the strains now found on tobacco. The strains of TMV delineated by Valleau *et al.* were based on isolates collected over a number of years but the relative incidence of the different strains was not reported. Gooding (1969) reported that 98 randomly collected isolates of TMV from flue-cured tobacco caused similar symptoms when inoculated into two cultivars of tobacco under greenhouse conditions. Field studies, which might have revealed greater differences than those found in the greenhouse, were not conducted.

Any naturally infected tobacco plant undoubtedly contains many more or less closely related variants and these can be separated (Jensen, 1933, 1936; McKinney, 1935). The symptoms on individual plants in a field are relatively uniform, however. This indicates that the population is predominantly that of the common strain. Even if a single virion of the common strain were introduced into a plant, the population would soon consist of numerous variants due to the high mutation rate of the virus (Garciá-Arenal *et al.*, 1984; Holland *et al.*, 1982; Hennig and Wittmann, 1972).

The resistance of TMV to inactivation and its ease of mechanical transmission are critical factors in its survival without an efficient vector and in the epidemiology of the disease. The virus survives at least 50 years *in vitro* (Silber and Burk, 1965; Johnson and Valleau, 1935) and its resistance to chemical inactivation was reported early in this century (Allard, 1918).

Although TMV is highly resistant to physical and chemical inactivation in comparison to most plant viruses, it is eventually degraded in nature by microorganisms and its elements returned to the biomass (Alexander, 1964). The effectiveness of various species of fungi and bacteria in inactivating TMV was demonstrated in early studies in which fungi were generally more effective inactivators than bacteria (Johnson and Hoggan, 1937). The rate of inactivation of TMV in the soil is correlated with soil type, pH, temperature, and moisture (Cheo, 1980; Lehman, 1936; Hoggan and Johnson, 1936) and these factors are related to the degradation of substrate by microorganisms (Alexander, 1964). These variables also explain reports of TMV survival in the soil ranging from weeks to months (Cheo and Nickoloff, 1981; Avigliano, 1972; Huang and Huang, 1960; Uozumi and Nishimura, 1971).

Mechanical transmission of TMV is very efficient and a single particle (Boxall and MacNeill, 1974) or, at most, a few (Walker and Pirone, 1972) are capable of infecting tobacco.

B. Symptoms

The most characteristic macroscopic symptom of mosaic in tobacco consists of light and dark green areas on leaves that form the characteristic mosaic pattern. Less common symptoms are leaf distortion and vein clearing on young leaves and line patterns and necrotic areas on old leaves (Fig. 1). The latter symptoms are thought to be related to strains of the virus and/or water stress (Valleau *et al.*, 1954). As leaves approach maturity, the symptoms fade to the extent that the disease may not be recognizable on a mature plant in the field.

Microscopic signs and symptoms include the appearance of crystals, paracrystals, and X-bodies in the cytoplasm (Esau, 1968).

C. Identification

A simple, accurate technique for identifying TMV is essential for epidemiological studies. Cucumber mosaic virus usually is the only virus infecting tobacco in the field that causes symptoms which an experienced observer might confuse with common mosaic. Thus, in areas of tobacco production where CMV rarely, if ever, occurs, tobacco mosaic can be identified with a high degree of accuracy on symptoms alone. This is the situation in the flue-cured (Virginia-type) tobacco production area in the United States, but in the areas of burley tobacco production in the United States and in countries such as Japan, Switzerland, and Taiwan, CMV may be prevalent (Lucas, 1975; Johnson, 1933).

There are many techniques available for identifying and/or detecting TMV (Noordam, 1973) and new ones are constantly being developed (Sela *et al.*, 1984). Serology is the most universally applicable procedure for virus identification and many different techniques are available (Van Regenmortel, 1975, 1982). The agar-gel double-diffusion technique is the simplest procedure for those tobacco viruses for which it is applicable (Gooding, 1975a) because checks against nonspecific reaction are in the same assay system as the virus and its homologous antibody.

D. Effects on the Host

The most obvious effects of TMV on young plants are mosaic symptoms and a distinct reduction in growth compared to noninfected plants. Plants which become infected as they approach maturity develop mosaic

FIGURE 1. Symptoms of tobacco mosaic on flue-cured tobacco. Top photographs show symptoms on young leaves. The leaf distortion seen in the upper right photograph commonly occurs on plants infected when they are in the three- to five-leaf stage. The lower photographs show symptoms on old leaves. The leaf at the lower left has line patterns along the veins; the leaf at the lower right shows mosaic burn.

symptoms but little or no stunting is evident. The effect of the time of infection on flue-cured tobacco growth as measured by dry leaf weight was recently reported by Johnson *et al.* (1983) and others (e.g., Reilly, 1983; Dialbo and Mulchi, 1981; Chaplin, 1964; Wolf and Moss, 1932; Valleau and Johnson, 1927b). Based on their results, significant yield repression occurs if plants become infected within 7 weeks of transplanting, but little or no repression if infection occurs 9 weeks or later after transplanting.

The host's metabolism and subsequent chemical composition are greatly affected by TMV infection. The primary effect of the virus on chemical constituents is to increase total nitrogen and decrease reducing sugars and starch (Reilly, 1983; Spurr and Chaplin, 1974; Harman *et al.*, 1970).

E. Economic Importance

Tobacco mosaic occurs in every country that produces tobacco (Akehurst, 1981; Lucas, 1975). Annual losses of 1% of the crop as estimated by Lucas (1975) for the United States may also occur in other areas where susceptible cultivars are grown.

Losses are very difficult to accurately determine for most diseases (Bos, 1982) and tobacco mosaic is no exception. Johnson *et al.* (1983) developed regression models for estimating losses in flue-cured tobacco that give reliable data under current cultural and market conditions. Yield was significantly reduced and related to time of infection but the quality of the cured leaf was not significantly decreased. In contrast, the quality of cigar-wrapper tobacco leaves is drastically affected by TMV because the tensile strength of the leaf tissue is reduced, rendering it unacceptable for use as wrappers which are of premium value.

III. EPIDEMIOLOGY

A. Sources of Inoculum

1. Tobacco Plants or Tissue

a. Plants

The usual disease progression for mosaic is that a few plants (<1%) in a field become infected (primary infection) and the virus is transmitted from these plants to the remainder of the field (secondary spread) in a polycyclic manner (Thresh, 1983; Vanderplank, 1982). Infected tobacco plants in one field can also serve as sources of inoculum for plants in another field. Thus, infected tobacco plants are the main source of inoculum of TMV for the development of epidemics.

b. Debris from the Previous Crop

Tobacco is produced for its leaves. When they are marketed, much of the TMV that was in infected plants is removed from the farm on which the disease occurred. However, a significant amount of TMV remains in leaf debris or in stalk and root tissue, and these constitute sources of inoculum for the next crop.

Residual dried leaf tissue is commonly found in barns and other buildings where tobacco is handled. Such tissue that contains TMV may be brought into contact with the next tobacco crop by man, wind, or other means. Perhaps the most important source of the virus where monoculture is practiced is in the soil, however.

Tobacco growers interviewed by Mayer (1886) believed that mosaic was associated with the soil and Beijerinck (1898) presented evidence of its soilborne nature by demonstrating that tobacco became infected when grown in pots from which diseased plants had been removed. The importance of soilborne inoculum in primary infection was established by Johnson and Ogden (1929) and Lehman (1934) in the United States and its worldwide importance is now recognized (e.g., Park and Kim, 1980; Van Winckel and Andronova, 1962; Huang and Huang, 1960; Bruyns-Haylett, 1959).

c. Tobacco Products

TMV remains infectious for many years in dried tobacco leaves and so it is not surprising that products made from these leaves can serve as sources of inoculum (Wetter and Bernard, 1977; Gooding, 1969; Kumuro and Iwaki, 1968; Limasset, 1957; Johnson and Ogden, 1929). Considerable infectious virus even survives the heat-curing process used for Virginia-type tobacco (Harman *et al.*, 1971). Valleau and Johnson (1927a) considered chewing tobacco made from TMV-infected natural leaf an important source of inoculum, but products from processed leaf unimportant.

2. Hosts Other Than Tobacco

Holmes (1946) reported infection of 199 different plant species with TMV. The role of weeds as inoculum sources of viruses is well known (Duffus, 1971), but tobacco and other solanaceous crops, and especially their fruit, are usually much more important as inoculum sources of TMV than weeds.

3. Contaminated Equipment

Equipment used in cultivating and harvesting the tobacco crop becomes coated with a film of gum derived from contact with plant tissue.

TMV can remain active in this material for several months as a source of inoculum.

4. Water

TMV presumably leaches from infected crop debris into water in the soil. Should TMV-infested water that flows into ponds, streams, or rivers be used for irrigating the crop, it would be a source of inoculum. Tosic and Tosic (1984) demonstrated such sources in Hungary by isolating TMV from the Danube and Sava rivers.

5. Seed

Although TMV has been detected in tobacco seed (Benoit and Maury, 1976), seedborne virus is not considered an important source of inoculum (Zaitlin and Israel, 1975; Gooding, 1969).

B. Transmission

1. Vertebrates

Man is the primary transmitter of TMV either directly with his hands and/or body or indirectly with the various types of equipment used in cultivating the crop. A rather unusual type of transmission occurs at transplanting when the host is brought into contact with the inoculum rather than vice versa. It is unlikely that TMV persists in soil water (Cheo and Nickoloff, 1981; Hoggan and Johnson, 1936). It may overwinter bound to sand (Lefler and Knott, 1974) or clay colloids (Miyamoto, 1959), but it definitely survives in living or dead crop debris (Gooding, 1969; Lehman, 1934; Johnson and Ogden, 1929). Gooding (1969) reported that infection occurs at transplanting when plants are physically brought in contact with soilborne virus. Plants whose leaves and stem were protected from soil contact by polyethylene (only the roots were exposed) were symptomless 6 weeks after transplanting whereas approximately 1.5% of the unprotected plants had symptoms. Pieces of root or stalk debris are usually found contiguous to leaf or stalk tissue of plants that develop symptoms 2–3 weeks after transplanting. This does not prove a cause–effect relationship but suggests that much, if not all, infection that occurs at transplanting is due to plants being forced into contact with infected debris.

Animals other than man that come in contact with diseased and then healthy plants are potential transmitters of TMV. Transmission by this means is probably rare because tobacco fields are not usually frequented by animals other than man. Transmission by animals could be important where they are used in the cultivation of the crop but they currently are seldom employed for such purposes.

2. Invertebrates

Various invertebrates have been reported to transmit TMV but their role in the development of epidemics has not been evaluated.

Although most plant viruses are efficiently transmitted by aphids, TMV transmission by aphids is very inefficient (Hoggan, 1934; Lojek and Orlob, 1972). Transmission apparently results from the injury of cells by aphid tarsi (Bradley and Harris, 1972) and not by their mouthparts. Pirone and Shaw (1973) were able to obtain stylet transmission by using poly-L-ornithine in acquisition inoculum, however. That a highly infectious and stable virus such as TMV is not readily transmitted by aphids is an anomaly discussed by Harris and Maramorsch (1977).

The ease with which TMV is transmitted mechanically suggests that chewing insects should be efficient transmitters, but the potato flea beetle (Orlob, 1963), the tobacco flea beetle (Mercer *et al.*, 1982), and a grasshopper, *Melanoplus differentialis* (Walters, 1952), only inefficiently transmitted TMV from tobacco to tobacco. Mercer *et al.* obtained no transmission from TMV-infected horse nettle (*Solanum carolinense* L.) to tobacco with the eggplant flea beetle (*Epitrix fuscula*). Lack of transmission may have resulted from inefficient feeding by *E. fuscula* on tobacco or may have been related to inhibition of infection by regurgitant as described for three beetles by Gergerich *et al.* (1983).

Several species of nematodes failed to transmit TMV from roots of infected to roots of noninfected tobacco plants (Gooding and Barker, unpublished findings). The importance of root infection (Fulton, 1941) and the role that nematodes or other organisms may have as transmitters of the virus to roots apparently is unknown, based on the literature reviewed.

The relative inefficiency of transmission of TMV by insects does not, of course, mean that they could not be responsible for extensive spread of the virus where high insect populations occur. The question has not been addressed on tobacco because of the apparently overwhelming role of man in dissemination of the virus. Randles *et al.* (1981) reported spread of TMV by an unknown mechanism in the wild tobacco species, *Nicotiana glauca* L., in Australia. Elucidation of the mechanism of transmission of TMV in this wild species should prove to be interesting because man is unlikely to be involved.

C. Environmental Factors

1. Geographic Area of Production

The importance of the different sources of inoculum of TMV is dependent primarily on the flora and the climatic zone in which the crop is produced. The flora is important as it relates to weed and crop hosts of the virus other than tobacco. Climate is the most important factor permitting or restricting the continuous (year-round) production of tobacco and other solanaceous crops.

In regions where crop production is restricted to the warm months, the disease usually progresses from a few early infections to other plants in the field in a polycyclic manner if factors are conducive to mechanical spread by cultivation equipment. By the time harvest is completed, essentially all of the population may be infected.

In production areas where tobacco and other solanaceous crops are grown continuously in overlapping sequence, spread from outside sources can lead to a high incidence of disease even though there is little or no secondary spread in a field. This can also occur in temperate zones if a grower has several fields and transmits the virus from one to another on contaminated equipment.

2. Cultural Practices

Cultural practices resulting in sequential plant contact have a profound effect on the spread of TMV. Some cultural practices that relate to sequential plant contact are dependent and others independent of the type of tobacco produced.

Tobacco-type-dependent practices primarily involve the manner in which the crop is harvested. Two extremes are cigar-wrapper tobacco where individual leaves are removed as they mature versus burley tobacco which is usually harvested once by cutting the entire plant. Individual plants of cigar-wrapper-type tobacco are thus handled many times during harvest compared to burley tobacco. Tobacco-type-dependent practices usually cannot be altered by the grower to reduce the opportunity for virus spread.

Tobacco-type-independent practices affecting the development of epidemics are usually under grower control. Practices favoring serial contact of plants favor virus spread, and thus a high incidence of virus.

3. Meteorological Factors

The primary meteorological factor affecting the mosaic disease is water. Mechanical transmission between plants occurs more readily if the leaves are turgid than if they are wilted. The effect of mosaic on the tobacco plant is greater if infected plants are exposed to water stress, however.

IV. CONTROL

A. Resistance

No plant disease has been more effectively controlled by the use of resistant cultivars than tobacco mosaic. The well-known hypersensitive reaction of *Nicotiana glutinosa* L. to TMV that leads to the formation of local lesions was reported by Holmes (1929). The virus is restricted to these local lesions within temperatures of normal tobacco production but may move systemically at higher temperatures (Kassanis, 1952). Mosaic resistance in *N. glutinosa* is monogenic and dominant (Holmes, 1938). Valleau (1952a) incorporated this resistance into burley tobacco cultivars in the 1940s and most of the burley cultivars utilized today contain the "N" factor from *N. glutinosa.* The once widespread and economically important mosaic disease of burley tobacco (Lucas, 1975: Valleau *et al.*, 1954) essentially has been eliminated. Strains of TMV that "break" this resistance may eventually become established in the crop, but to date no such strain has been reported. A strain that "breaks" TMV-resistance in tomatoes has been reported but the mechanism of resistance is different (Pelham *et al.*, 1970; Dawson, 1967).

Virginia (flue-cured-type) tobacco comprises a significant proportion of the total tobacco produced worldwide. Cultivars with "N" factor resistance have been developed but constitute only a small percentage of the crop because their monetary value, based on purchase preferences in regard to leaf color, is less than susceptible cultivars (Johnson and Main, 1983). Although the market value of mosaic-resistant flue-cured tobacco cultivars is less than susceptible cultivars, the quality of the leaf is as suitable for the manufacture of tobacco products as susceptible cultivars (Chaplin, personal communication). Thus, in situations where the manufacturer of tobacco products controls the cultivars produced by the grower, mosaic can be easily controlled by planting resistant cultivars.

Cultivars of other tobacco types have been developed that contain resistance derived from *N. glutinosa* (Stavely, 1979; Lucas, 1975; Reyno and Balico, 1964; Clayton, 1936) but they too have not found the universal grower acceptance that the burley tobacco cultivars have.

The dramatic success in the control of TMV in burley tobacco has led to intensive efforts to incorporate resistance from *N. glutinosa* into agronomically acceptable cultivars of other tobacco types. Anticipation of the impending success of these efforts has discouraged the search for other means of control.

B. Tolerance

Nolla and Moller (1933) reported that the Colombian tobacco cultivar Ambalema is tolerant to infection by TMV. Tolerance in Ambalema is linked to undesirable agronomic characteristics, however, and attempts to develop cultivars from it have been discontinued (Valleau, 1952a). Tolerance also has been identified in other cultivars of *N. tabacum* (Chaplin and Gooding, 1969; Burk and Heggestad, 1966; Lucas, 1975). Apparently, no efforts currently are under way to incorporate tolerance to TMV in tobacco cultivars. This probably is due to the continued expectation of success with *N. glutinosa*-type resistance. Also, the use of asymptomatic tolerant cultivars could constitute an unrecognized source of inoculum for sensitive cultivars.

C. Prevention of Primary Infection

The potential sources of TMV are so numerous that sanitation would seem a hopeless strategy for controlling mosaic.

This is not the situation in regions where winter temperatures kill all plants except perennial species. The simple practice of crop rotation, which eliminates primary infection from the soil, will usually decrease disease incidence to insignificant levels (Gooding and Todd, 1976).

Primary infection is much more difficult to control in regions of continuous culture of solanaceous crops. Crop production in these areas usually involves much manual labor. Workers moving between fields are very efficient transmitters of the virus via their hands or clothing (Broadbent and Fletcher, 1963).

1. Elimination of Soilborne Inoculum

The simplest way to eliminate soilborne inoculum is to alternate tobacco with a nonhost crop and control weed hosts. Should rotation with a nonhost crop not be possible, susceptible cultivars can be rotated with resistant cultivars.

No methods have been reported to eliminate TMV from the soil where susceptible tobacco cultivars are used in monoculture. Lehman (1934) reported that enough TMV survived in the soil to infect a small percentage of transplants even where the stalks and roots of the previous crop had been removed. Gooding and Lucas (1969) identified some herbicides that, when sprayed on the stalks after harvest, would kill all stalk and root tissue within 3 months of application. Even with this treatment, primary infection was not eliminated. Approximately 0.1% of the plants transplanted to treated plots were infected from soilborne inoculum. This was not surprising because dead roots collected at the time of transplanting, 6 months after the previous crop residue had been disced into the soil, contained approximately 10% as much active virus as living roots (Gooding, 1969). Although the physical removal of stalks and roots or their killing with herbicides will not eliminate soilborne inoculum, the number of infected transplants will be reduced from about 1% to 0.1%, thus reducing the number of plants that require rogueing.

Chemical soil fumigants evaluated as inactivators of TMV (Broadbent *et al.*, 1965; McKeen, 1962) have not been found effective. Conversely, some have increased the survival of TMV (Broadbent *et al.*, 1965), presumably by killing microorganisms that denature the virus. Moist heat has been reported effective in inactivating TMV (Broadbent *et al.*, 1965; Van Winckel and Geypens, 1965), however.

2. Elimination of Tobacco Products as a Source of Inoculum

Tobacco products can be eliminated as a source of inoculum by preventing their use by workers. This concept is obviously simple but difficult to enforce. The more realistic approach of heating tobacco products to inactivate the virus and providing them to workers is now being recommended in North Carolina. This procedure was recommended to growers by Thornberry *et al.* in 1937 as a result of extensive studies on the inactivation of TMV in tobacco products by dry heat, but apparently has not been utilized. Thornberry *et al.* (1937) recommended heating natural leaves for 30 min at 150°C. Gooding (unpublished findings) found this inadequate for manufactured products which were wrapped in aluminum foil to reduce drying. Therefore, it is recommended that heating be done for 1 hr at 150°C. Cartons of cigarettes thus treated resulted in a usable product in informal farm tests.

3. Control of Inoculum from Weeds and Solanaceous Crops

TMV overwinters in various perennial weed hosts in temperate zones (Duffus, 1971) and this source of inoculum can be eliminated by various means. The incentive to adopt this approach and its feasibility will depend on the losses due to mosaic on a given farm and the number and abundance of overwintering hosts. Horse nettle (*Solanum carolinense*) apparently is the only weed species in which TMV overwinters in the flue-cured areas of tobacco production in the southeastern United States (Gooding, 1969). Tobacco etch virus and other tobacco pathogens also overwinter in this weed so that eradication measures are fully justified.

Tobacco produced in climates permitting year-round cultivation of solanaceous crops is subject to primary infection from large crop reservoirs of TMV in addition to weed hosts. It would be an unusual situation where elimination of these sources of inoculum would be feasible. Primary infection from these sources may be reduced by requiring workers to wash their hands prior to working in the tobacco crop, however.

Many types of soap and other agents such as tribasic sodium phosphate have been recommended to aid in removing TMV from workers' hands (Nakajima *et al.*, 1983; Krausz and Fortnum, 1982; Broadbent, 1976; Reilly, 1979; Valleau *et al.*, 1954). No significant difference in removing or inactivating TMV from contaminated hands was found among several brands of abrasive or nonabrasive soaps, anionic or nonionic detergents, or tribasic sodium phosphate (3%) (Gooding, unpublished findings). All significantly reduced the quantity of virus surviving compared with washing in water alone. No practical means of removing or inactivating all virus from the hands has been reported. The primary problem is the small pieces of tissue that become lodged under the fingernails when working with plants (Broadbent, 1976).

4. Destruction of Inoculum on Equipment

Higher concentrations of the chemicals used to remove or inactivate TMV on workers' hands or more caustic chemicals can be used on equipment.

Mechanical clipping of transplants in the plant bed to increase uniformity or delay transplanting is becoming popular (Stephenson *et al.*, 1984; Miner *et al.*, 1983). Clipping is usually done with a conventional lawnmower modified to allow for greater adjustments in height above the cutting surface. The contamination of these mowers with TMV from weed sources or from TMV-infected plants and subsequent transmission to virus-free plant beds must be prevented. Stephenson *et al.* (1984) recommended cleaning debris from mowers and then spraying them with 0.5% sodium hypochlorite between each operation to prevent the spread of tobacco pathogens. Sodium hypochlorite is the most effective inactivator of TMV readily available (Gooding, 1975b) and is undoubtedly the chemical of choice. It was effective at 1.5% but not at 0.5% concentration in eliminating all infectious virus from mowers and other equipment (Gooding, unpublished findings). An alternative is to steam clean the equipment if facilities are available.

D. Prevention of Secondary Spread

1. Prevention of Spread by the Use of Chemicals

Removal of TMV from or inactivation of TMV on workers' hands or equipment is effective in reducing primary infection in a bed or field but of limited value in preventing secondary spread. The removal or inactivation treatment would have to be repeated between each plant contacted, to be of meaningful value, and this would be impractical.

The most feasible approach to preventing secondary spread by chemicals is to coat plants with a chemical that will prevent infection. Milk is the most widely used product for this purpose, and Lucas (1975) was instrumental in adapting this long-known inhibitor of infection (Chester, 1934) for use on tobacco. Milk reduces but does not eliminate secondary spread of the virus even at rates [120 g powdered milk/1000 ml water (Reilly and Komm, 1983)] that approach phytotoxicity. The use of milk to reduce secondary spread of TMV on tobacco is feasible in countries with a milk surplus but not in countries with shortages. In these countries, the use of other inhibitors should be considered.

There are numerous reports of TMV inhibitors from plants, animals, or microorganisms (Dawson, 1984; Cheo, 1980; Lucas, 1975; Bawden,

1954). Species of higher plants that contain inhibitors are likely to occur in all areas where tobacco is produced and these could replace milk in reducing the secondary spread of TMV. Murthy *et al.* (1981) reported this approach in India where the spread in Virginia tobacco was significantly reduced using a 1:100 dilution of juice from *Basella alba* L. sprayed on plants in the field. *Phytolacca americana* L. (Wyatt and Shepherd, 1969) and other *Phytolacca* species (Kassanis and Kleczkowski, 1948) contain a highly active inhibitor(s) of TMV infection. *P. americana* is a relatively large, succulent plant from which crude juice can be easily extracted. Its value in preventing spread of TMV should be investigated.

Economics discourage the identification and introduction of new chemicals for preventing spread of TMV, but more work with known inhibitory compounds, such as Dodine (*n*-dodecylguanidine acetate) and Glyodin (2-heptadecyl-2-imidazoline acetate) (Chou and Rodgers, 1973), may be worth pursuing.

Secondary spread of TMV occurs both in the plant bed and in the field. The most critical spread occurs in the plant bed or at transplanting. On farms where mosaic occurs in the plant bed, plants should be treated with a substance to reduce secondary spread before each operation that involves handling plants.

2. Rogueing

The value of removing infected plants from the field before each cultivation to prevent spread of TMV is well established (Lehman, 1934a; Wolf, 1933). Rogueing should be started 3–4 weeks after transplanting when symptoms are distinct enough to identify infected plants.

3. Cultural Practices

The most effective way of preventing secondary spread of TMV is to adhere to cultural practices that minimize plant contact by workers' hands or equipment.

Extreme care should be taken to prevent spread of TMV in the plant bed. Historically, most spread in the plant bed occurred when workers removed weeds by hand. This should be avoided by using another means of weed control. Beds should be no wider than 1 m so that transplants can be removed without having to walk on the beds.

Field operations should also be designed to minimize contact with plants; e.g., all cultivations should be done in a manner that prevents plant contact by equipment.

E. Chemical

A panacea for the control of TMV, as for other viral diseases, would obviously be a chemotherapeutic that could be used after infection occurs. This possibility has received much attention in the past (Yun and Hirai, 1968) and studies continue (Dawson, 1984; White *et al.*, 1983), but no suitable chemical has been released for grower usage.

A chemical that selectively kills or causes a reaction making infected plants readily identifiable (Schuster, 1971) would be very valuable in eliminating or locating sources of inoculum.

F. Cross-Protection and Interference

Cross-protection or induced resistance has been utilized commercially to reduce losses in tomatoes from TMV (Broadbent, 1976; Rast, 1972, 1975). Cross-protection has also been shown to reduce yield losses on tobacco under experimental conditions (Gooding, 1981).

Cross-protection was not recommended for commercial use by Gooding (1981) for several reasons. The mild strain used, Holmes's masked strain (Holmes, 1934), decreased yield by 8.0% and few farmers experience such losses. More importantly, there is no reliable method for predicting the incidence of mosaic in a given field to indicate where the use of a mild strain would be justified. Nevertheless, cross-protection is worthy of further investigation, especially for use on farms or in areas restricted to monoculture where mosaic is a chronic, severe problem. A strain of TMV may be selected or induced (Yarwood, 1979; Rast, 1972) with less adverse effects on the host than Holmes's masked strain.

Simultaneous inoculation of a plant with a mild and a severe strain of a virus usually results in a reaction intermediate in severity between either strain alone (Gibbs and Harrison, 1976). This type of interference may offer another approach to the control of mosaic in tobacco. The usual course of the disease involves infection of a few plants and secondary spread from these plants to the remainder of the plants in a field. Inoculation of only a portion of the plants in a field would provide inoculum of a mild strain that could interfere with the effects of a more severe strain if conditions facilitate secondary spread. Should conditions be unfavorable for secondary spread, yield repression by the mild strain would only occur on a portion of the plants. For example, if the mild strain caused a 10% yield repression but only 10% of the plants in a field were inoculated, the cost of insurance would be only 1% of the potential yield.

Elucidation of the molecular basis for interference between mild and severe strains of TMV should enhance utilization of this phenomenon in virus disease control (Kiko and Nishigucki, 1984).

V. DISCUSSION

Tobacco mosaic is known to have been a problem to growers since the latter part of the 19th century. The most effective control for mosaic is the use of resistant cultivars, and where they have been utilized, mosaic has been eliminated. Mosaic can be controlled using other strategies but these are not understood or are unacceptable to tobacco growers on whose crop the disease persists. Losses to mosaic in susceptible cultivars may be reduced in the future by using techniques such as cross-protection, interference, or chemotherapeutics. The disappearance of the disease on susceptible cultivars is more likely to occur as a result of discontinuance of the culture of the crop than it is by control of the disease, however.

REFERENCES

Akehurst, B. C., 1981, *Tobacco* 2nd ed., Longman Press, New York.

Alexander, M., 1964, Biochemical ecology of soil microorganisms, *Annu. Rev. Microbiol.* **18**:217–252.

Allard, H. A., 1918, Effects of various salts, acids, germicides, etc. on the infectivity of the virus causing the mosaic disease of tobacco, *J. Agric. Res.* **13**:619–637.

Avigliano, M., 1972, Contribution to the study of tobacco mosaic virus transmission through the soil, *Tobacco* **76**:21–26.

Bawden, F. C., 1954, Inhibitors and plant viruses, *Adv. Virus Res.* **2**:31–57.

Beijerinck, M. W., 1898, Über ein contagium virum fluidum als sorzaak van de vlekziekte der tabaksbladen, *Verh. K. Akad. Wet. Amsterdam* **65**:3–21. English translation by James Johnson, 1942, Phytopathological Classics No. 7.

Benoit, M. A., and Maury, Y., 1976, Location of tobacco mosaic virus in tobacco seed, *Planter* **52**:392–396.

Bos, L., 1982, Crop losses caused by viruses, *Crop Proc.* **1**:263–282.

Boxall, M., and MacNeill, B. H., 1974, Local lesions as sources of biologically pure strains of tobacco mosaic virus, *Can. J. Bot.* **52**:23–25.

Bradley, R. H. E., and Harris, K. F., 1972, Aphids can inoculate plants with tobacco mosaic virus by clawing, *Virology* **50**:615–618.

Broadbent, L., 1976, Epidemiology and control of tomato mosaic virus, *Annu. Rev. Phytopathol.* **14**:75–96.

Broadbent, L., and Fletcher, J. T., 1963, Epidemiology of tomato mosaic. IV. Persistence of virus on clothing and greenhouse structures, *Ann. Appl. Biol.* **52**:233–241.

Broadbent, L., Read, W. H., and Last, F. T., 1965, The epidemiology of tomato mosaic. X. Persistence of TMV-infected debris in soil, and the effects of partial soil sterilization, *Ann. Appl. Biol.* **55**:471–483.

Bruyns-Haylett, J. P., 1959, Influence of mosaic-infected tobacco crop residues in the soil on the incidence of primary mosaic infection in the subsequent tobacco crop, *J. Agric. Sci.* **2**:331–341.

Burk, L. G., and Heggestad, H. E., 1966, The genus Nicotiana: A source of resistance to diseases of cultivated tobacco, *Econ. Bot.* **20**:76–88.

Chaplin, J. F., 1964, Effects of tobacco mosaic on flue-cured tobacco resistant and susceptible varieties, *S. C. Agric. Exp. Stn. Bull.* **513**.

Chaplin, J. F., and Gooding, G. V., Jr., 1969, Reactions of diverse Nicotiana tobacco germplasm to tobacco mosaic virus, *Tob. Sci.* **13**:130–135.

Cheo, P. C., 1980, Antiviral factors in soil, *Soil Sci.* **44**:62–67.

Cheo, P. C., and Nickoloff, J. A., 1981, Rate of tobacco mosaic virus degradation in a field plot after repeated application of the virus, *Soil Sci.* **131**:284–289.

Chester, K. S., 1934, Specific quantitative neutralization of the viruses of tobacco mosaic, tobacco ringspot and cucumber mosaic by immune sera, *Phytopathology* **24**:1180–1202.

Chou, H. S., and Rodgers, E. M., 1973, Efficacy of Dodine and Glyodin as foliar protectants for control of tobacco mosaic virus, *Phytopathology* **63**:1428–1429.

Clayton, E. E., 1936, Control of tobacco diseases through resistance, *Phytopathology* **26:**239–244.

Dawson, J. R., 1967, The adaption of tomato mosaic virus to resistant tomato plants, *Ann. Appl. Biol.* **60:**209–214.

Dawson, W. O., 1984, Effects of animal antiviral chemicals on plant viruses, *Phytopathology* **74:**211–213.

Dialbo, I., and Mulchi, C. L., 1981, Influence of time of injection by TMV and TEV on agronomic, chemical and physical properties of tobacco, *Tob. Sci.* **25:**25–29.

Duffus, J. E., 1971, Role of weeds in the incidence of virus diseases, *Annu. Rev. Phytopathol.* **9:**319–340.

Esau, K., 1968, *Viruses in Plant Hosts,* University of Wisconsin Press, Madison.

Ford, R. H., and Tolin, S. A., 1983, Genetic stability of tobacco mosaic virus in nature. *Tob. Sci.* **27:**14–17.

Fulton, R. W., 1941, The behaviour of certain viruses in plant roots, *Phytopathology* **31:**575–598.

Garciá-Arenal, F., Palukaitis, P., and Zaitlin, M., 1984, Strains and mutants of tobacco mosaic virus are both found in virus derived from single-lesion-passaged inoculum, *Virology* **132:**131–137.

Gergerich, R. C., Scott, H. A., and Fulton, J. P., 1983, Regurgitant as a determinant of specificity in the transmission of plant viruses by beetles, *Phytopathology* **72:**936–938.

Gibbs, A., and Harrison, B., 1976, *Plant Virology: The Principles,* Wiley, New York.

Gooding, G. V., Jr., 1969, Epidemiology of tobacco mosaic on flue-cured tobacco in North Carolina, *N.C. State Univ. Tech. Bull.* **195.**

Gooding, G. V., Jr., 1975a, Serological identification of tobacco viruses, *Tob. Sci.* **19:**135–139.

Gooding, G. V., Jr., 1975b, Inactivation of tobacco mosaic virus on tomato seed with trisodium orthophosphate and sodium hypochlorite, *Plant Dis. Rep.* **59:**770–772.

Gooding, G. V., Jr., 1981, Control of tobacco mosaic virus on flue-cured tobacco by cross protection, *Tob. Sci.* **25:**40–41.

Gooding, G. V., Jr., and Lucas, G. B., 1969, Tobacco stalk and root destruction with herbicides and their effects on tobacco mosaic virus, *Plant Dis. Rep.* **53:**174–178.

Gooding, G. V., Jr., and Todd, F. A., 1976, Soil-borne tobacco mosaic virus as an inoculum source for flue-cured tobacco, *Tob. Sci.* **20:**140–142.

Harman, G. E., Gooding, G. V., Jr., and Hebert, T. T., 1970, Effect of tobacco mosaic virus infection on some chemical constituents of flue-cured tobacco, *Tob. Sci.* **14:**138–140.

Harman, G. E., Gooding, G. V., Jr., and Hebert, T. T., 1971, State and ineffectivity of tobacco mosaic virus in flue-cured tobacco tissue, *Phytopathology* **61:**1032–1033.

Harris, K. F., and Maramorsch, K., 1977, *Aphids as Virus Vectors,* Academic Press, New York.

Hennig, B., and Wittmann, H. G., 1972, Tobacco mosaic virus: Mutants and strains, in: *Principles and Techniques in Plant Virology* (I. Kado and H. O. Agrawal, eds.), pp. 546–594, Van Nostrand–Reinhold, Princeton, N.J.

Hoggan, I. A., 1934, Transmissibility by aphids of the tobacco mosaic virus from different hosts, *J. Agric. Res.* **49:**1135–1142.

Hoggan, I. A., and Johnson, J., 1936, Behaviour of the ordinary tobacco mosaic virus in the soil, *J. Agric. Res.* **52:**271.

Holland, J., Spindler, K., Norodyski, F., Grabau, E., Nichol, S., and Vonde Pal, S., 1982, Rapid evolution of RNA genomes, *Science* **215:**1577–1585.

Holmes, F. O., 1929, Local lesions in tobacco mosaic, *Bot. Gaz.* **87:**39–55.

Holmes, F. O., 1934, A masked strain of tobacco mosaic virus, *Phytopathology* **24:**845–873.

Holmes, F. O., 1938, Inheritance of resistance to tobacco-mosaic disease in tobacco, *Phytopathology* **28:**553–561.

Holmes, F. O., 1941, A distinctive strain of tobacco mosaic virus from plantago, *Phytopathology* **31:**1089–1098.

Holmes, F. O., 1946, A comparison of the experimental host ranges of tobacco-etch and tobacco mosaic viruses, *Phytopathology* **36:**643–657.

Huang, T. L., and Huang, C. S., 1960, Some preliminary studies on the use of tobacco stalks relating to soil survival of tobacco common mosaic virus in farm practice, *Taiwan Tob. Wine Monop. Bur. Tob. Res. Inst. Annu. Rep.* **1960:**158–168.

Jensen, J. H., 1933, Isolation of yellow-mosaic viruses from plants infected with tobacco mosaic virus, *Phytopathology* **23:**964–974.

Jensen, J. H., 1936, Studies on the origin of yellow-mosaic viruses, *Phytopathology* **26:**266–277.

Johnson, C. S., and Main, C. E., 1983, Yield/quality trade-offs of tobacco mosaic virus resistant tobacco cultivars in relation to disease management, *Plant Dis.* **67:**886–890.

Johnson, C. S., Main, C. E., and Gooding, G. V., Jr., 1983, Crop loss assessment for flue-cured tobacco infected with tobacco mosaic virus, *Plant Dis.* **67:**881–885.

Johnson, E. M., and Valleau, W. D., 1935, Mosaic from tobacco one to fifty-two years old, *Ky. Agric. Exp. Stn. Bull.* **361**.

Johnson, E. M., and Valleau, W. D., 1946, Field strains of tobacco mosaic virus, *Phytopathology* **36:**112–116.

Johnson, J., 1933, Cucumber mosaic on tobacco in Wisconsin, *Phytopathology* **23:**311.

Johnson, J., and Hoggan, I. A., 1937, Inactivation of the ordinary tobacco mosaic virus by microorganisms, *Phytopathology* **27:**1014–1027.

Johnson, J., and Ogden, W. B., 1929, The overwintering of the tobacco mosaic virus, *Wis. Agric. Exp. Stn. Bull.* **95**.

Kassanis, B., 1952, Some effects of high temperature on the susceptibility of plants to infection with viruses, *Ann. Appl. Biol.* **39:**358–369.

Kassanis, B., and Kleczkowski, A., 1948, The isolation and some properties of a virus-inhibiting protein from *Phytolacca esculenta, J. Gen. Microbiol.* **2:**143–153.

Kiko, Y., and Nishigucki, M., 1984, Unique nature of an attenuated strain of tobacco mosaic virus: Autoregulation, *Microbiol. Immunol.* **28:**589–599.

Krausz, J. P., and Fortnum, B. A., 1982, Alternative control for tobacco mosaic virus, *Tob. Sci.* **26:**124–125.

Kumuro, Y., and Iwaki, M., 1968, Strains of tobacco mosaic virus contained in cigarettes, with special reference to the infection source for mosaic disease of tomato, *Ann. Phytopathol. Soc. Jpn.* **34:**98–102.

Lefler, E., and Knott, J., 1974, Virus retention and survival in sand, in: *Virus Survival in Water and Waste Water Systems* (J. F. Malina and B. P. Sagik, eds.), pp. 84–91, Center for Research in Water Resources, Austin.

Lehman, S. G., 1934, Contaminated soil and cultural practices as related to occurrence and spread of tobacco mosaic, *N.C. Agric. Exp. Stn. Bull.* **46:**43.

Lehman, S. G., 1936, Inactivation of the virus of common mosaic by drying and by freezing in soil, *J. Elisha Mitchell Sci. Soc.* **52:**159.

Limasset, M. P., 1957, Observations relatives au role des terreaux et des tobacs manufactures dans la transmission dua virus de la mosique du tobac, *Rev. Int. Tab.* **33:**77–79.

Lojek, J. S., and Orlob, G. B., 1972, Transmission of tobacco mosaic virus by *Myzus persicae, J. Gen. Virol.* **17:**125–127.

Lucas, G. B., 1975, *Diseases of Tobacco,* 3rd ed., Biol. Consult. Assoc., Raleigh, N.C.

McKeen, C. D., 1962, The destruction of viruses by soil fumigants, *Phytopathology* **52:**742.

McKinney, H. H., 1935, Evidence of virus mutation in the common mosaic of tobacco, *J. Agric. Res.* **31:**951–981.

Mayer, A., 1886, Über die Mosaikkrankheit des Tabaks, *Landwirtsch. Vers. Stn.* **32:**451–467. English translation by James Johnson, 1942, Phytopathological Classics No. 7.

Mercer, D. W., Gooding, G. V., Jr., and Kennedy, G. G., 1982, Transmission of tobacco mosaic virus by *Epitrix hirtipennis* (Melsheimer), *Tob. Sci.* **26:**129–132.

Miner, G. S., Cundiff, J. S., and Miles, J. D., 1983, Effects of clipping flue-cured tobacco plantbeds on transplant production efficiency and uniformity, *Tob. Sci.* **27:**70–74.

Miyamoto, Y., 1959, Further evidence for the longevity of soil-borne plant viruses absorbed on soil particles, *Virology* **9**:290–291.

Murthy, N. S., Nagarajan, K., and Sastry, A. B., 1981, Effect of prophylactic sprays of leaf extracts on the infection of tobacco by tobacco mosaic virus, *Indian J. Agric. Sci.* **51**:792–795.

Nakajima, T., Teraoka, T., Shigematsu, T., and Kasugai, H., 1983, Inhibitory activity of surfactants and polyelectrolytes against TMV infection, *Nippon Noyaku Gakkaishi* **8**:499–503.

Nolla, J. A. B., and Moller, W. J., 1933, A variety of tobacco resistant to ordinary tobacco mosaic, *J. Dep. Agric. P.R.* **17**.

Noordam, D., 1973, Identification of Plant Viruses: Methods and Experiments, Centre for Agric. Publ. and Doc., Wageningen.

Orlob, G. B., 1963, Reappraisal of transmission of tobacco mosaic virus by insects, *Phytopathology* **53**:822–830.

Park, E. K., and Kim, J. J., 1980, Studies on infection sources of tobacco mosaic virus (TMV) in tobacco fields, *Korean Soc. Tob. Sci. J.* **2**:53–60.

Pelham, J., Fletcher, J. T., and Hawkins, J. H., 1970, The establishment of a new strain of tobacco mosaic virus resulting from the use of resistant varieties of tomato, *Ann. Appl. Biol.* **65**:293–297.

Pirone, T. P., and Shaw, J. G., 1973, Aphid stylet transmission of poly-L-ornithine treated tobacco mosaic virus, *Virology* **53**:274–276.

Randles, J. W., Palukaitis, P., and Davis, C., 1981, Natural distribution, spread, and variation in the tobacco mosaic virus infecting *Nicotiana glauca* in Australia, *Ann. Appl. Biol.* **98**:109–119.

Rast, A. T. B., 1972, M11-16, an artificial symptomless mutant of tobacco mosaic virus for seedling inoculation of tomato crops, *Neth. J. Plant Pathol.* **78**:110–112.

Rast, A. T. B., 1975, Variability of tobacco mosaic virus in relation to control of tomato mosaic in glasshouse tomato crops by resistance breeding and cross production, *Agric. Res. Rep.* **834**.

Reilly, J. J., 1979, Chemical control of tobacco mosaic virus, *Tob. Sci.* **23**:97–99.

Reilly, J. J., 1983, Effects of sequential virus infections on flue-cured tobacco, *Tob. Sci.* **27**:147–151.

Reilly, J. J., and Komm, D. A., 1983, Incidence of tobacco mosaic virus as affected by various concentrations of milk, *Tob. Sci.* **27**:62–63.

Reyno, C. C., and Balico, G., 1964, Comparative resistance of different tobacco hybrids to mosaic disease, *Philipp. J. Agric.* **26**:155–162.

Schuster, G., 1971, Selective killing of virus-diseased plants by certain chemicals, *Zesz. Probl. Postepow Nauk Roln.* **111**:183–196.

Sela, J., Reichman, M., and Weisbach, A., 1984, Comparison of dot molecular hybridization and enzyme-linked immunosorbent assay for detecting tobacco mosaic virus in plant tissue and protoplasts, *Phytopathology* **74**:385–389.

Silber, G., and Burk, L. G., 1965, Ineffectivity of tobacco mosaic virus stored for fifty years in extracted 'unpreserved' plant juice, *Nature* (*London*) **206**:740–741.

Spurr, H. W., Jr., and Chaplin, J. F., 1974, Tobacco mosaic virus concentration, total nitrogen and altered enzyme activities of field-grown tobacco cultivars differing in mosaic symptom severity, *Tob. Sci.* **18**:134–136.

Stavely, J. R., 1979, Disease resistance, in: *Nicotiana Procedure for Experimental Use* (R. D. Durbin, ed.), pp. 87–110, *U.S. Dep. Agric. Tech. Bull.* **1586**.

Stephenson, M. G., Miles, J. D., Saina, T. P., and Wilson, W. H., Jr., 1984, Clipping effects on transplant yield and field performance of flue-cured tobacco, *Tob. Sci.* **28**:55–58.

Thornberry, H. H., Valleau, W. D., and Johnson, E. M., 1937, Inactivation of tobacco-mosaic virus in cured tobacco leaves by dry heat, *Phytopathology* **27**:129–134.

Thresh, J. M., 1983, Progress curves for plant virus diseases, *Adv. Appl. Biol.* **8**:1–85.

Tosic, M., and Tosic, D., 1984, Occurrence of tobacco mosaic virus in water of the Danube and Sava rivers, *Phytopathol. Z.* **110**:200–202.

Uozumi, T., and Nishimura, N., 1971, Studies on the soil infection of tobacco mosaic disease, *Morioka. Jpn. Tob. Exp. Stn. Bull.* **6**:41–64.

Valleau, W. D., 1952a, Breeding tobacco for disease resistance, *Econ. Bot.* **6**:69–102.

Valleau, W. D., 1952b, The evolution of susceptibility to tobacco mosaic in Nicotianae and the origin of the tobacco mosaic virus, *Phytopathology* **42**:40–42.

Valleau, W. D., and Johnson, E. M., 1927a, Commercial tobaccos and cured leaf as a source of mosaic disease in tobacco, *Phytopathology* **17**:513–522.

Valleau, W. D., and Johnson, E. M., 1927b, The effect of a strain of tobacco mosaic on the yield and quality of burley tobacco, *Phytopathology* **17**:523–527.

Valleau, W. D., Johnson, E. M., and Diachu, M. S., 1954, Tobacco diseases, *Univ. Ky. Coop. Ext. Serv. Circ.* **522-A**.

Vanderplank, J. E., 1982, *Host–Pathogen Interactions in Plant Disease,* Academic Press, New York.

Van Regenmortel, M. H. V., 1975, Antigenic relationship between strains of tobacco mosaic virus, *Virology* **64**:415–420.

Van Regenmortel, M. H. V., 1982, *Serology and Immunochemistry of Plant Viruses,* Academic Press, New York.

Van Winckel, A., and Andronova, A. V., 1962, On the problem of the role of the soil in the spread of tobacco mosaic virus, *Akad. Nauk. SSSR Inst. Genet. Tr., Moscow* **29**:389–403.

Van Winckel, A., and Geypens, M., 1965, Inactivity of TMV in the soil, *Parasitica* **21**:124–138.

Walker, H. L., and Pirone, T. P., 1972, Number of TMV particles required to infect locally or systemically susceptible tobacco cultivars, *J. Gen. Virol.* **17**:241–243.

Walters, H. J., 1952, Some relationships of three plant viruses to the differential grasshopper, *Melanoplus differentialis* (Thos.), *Phytopathology* **42**:355–362.

Wetter, C., and Bernard, M., 1977, Identification, purification, and serological differentiation of tobacco mosaic and paratobacco mosaic virus from cigarettes, *Phytopathol. Z.* **90**:257–267.

White, R. F., Antoniw, J. F., Carr, J. P., and Woods, R. D., 1983, Effects of aspirin and polyacrylic acid on the multiplication and spread of TMV in different cultivars of tobacco with and without the N-gene, *Phytopathol. Z.* **107**:224–232.

Wolf, F. A., 1933, Rogueing as a means of control of tobacco mosaic, *Phytopathology* **23**:831–833.

Wolf, F. A., and Moss, E. G., 1932, Effect of mosaic of flue-cured tobacco on yield and quantity, *Phytopathology* **22**:834–836.

Wyatt, S. D., and Shepherd, R. J., 1969, Isolation and characterization of a virus inhibitor from *Phytolacca americana, Phytopathology* **59**:1787–1794.

Yarwood, C. E., 1979, Host passage effects with plant viruses, *Adv. Virus Res.* **24**:169–190.

Yun, T. G., and Hirai, T., 1968, Studies on the chemotherapy for plant virus diseases. XII. Methods of applying antiviral chemicals against infected plants, *Ann. Phytopathol. Soc. Jpn.* **34**:109–113.

CHAPTER 7

TOBACCO MOSAIC VIRUS

Cytopathological Effects

J. R. EDWARDSON AND R. G. CHRISTIE

I. INTRODUCTION

Inclusion bodies were listed as one of the criteria for virus classification by the Plant Virus Subcommittee of the International Committee on Taxonomy of Viruses (Harrison *et al.*, 1971), although they were not used in that classification scheme. This paper did, however, establish the tobacco mosaic virus group which was designated by the acronym tobamoviruses. Fenner (1976) was the first to incorporate inclusions in the classification of plant viruses. Gibbs (1977) listed, among the main characteristics of the tobamovirus group, cytoplasmic inclusions composed of virus particles in elongated paracrystalline arrays, in plate-shaped crystalline form, and aggregated cell constituents with or without virus particles forming X-bodies. Martelli and Russo (1977) stated that tobamovirus members induce inclusions of virus particles aggregated in crystalline arrays appearing as plates with hexagonal or rounded shapes. Edwardson and Christie (1978) suggested cytoplasmic crystalline inclusions as a main characteristic of the group and proposed them as diagnostic for infection by tobamoviruses. Matthews (1979, 1982) listed, as the main characteristics of the tobamovirus group, virus-induced viroplasms, and virus particles often forming large crystalline arrays, visible by light microscopy. We assume Matthews uses the term *viroplasm* in place of the term *X-body.* The morphology and structure of these and other inclusions induced by tobamoviruses are well depicted in the fol-

J. R. EDWARDSON AND R. G. CHRISTIE • Plant Virus Laboratory, Agronomy Department, University of Florida, Gainesville, Florida 32611.

lowing: Warmke and Christie (1967), Esau (1968), Warmke (1974), Christie and Edwardson (1977).

Currently, the tobamovirus group consists of ten members and six possible members with common tobacco mosaic virus (TMV) as the type member (Matthews, 1982). All members of the group have been studied cytologically except Sammon's Opuntia and *Chara australis* viruses. Most of the cytological studies of tobamoviruses, 105 of 174 reports, have been devoted to TMV and its strains. Although this survey is incomplete, it provides material for drawing some conclusions about the usefulness of inclusions for classification and diagnosis of tobamoviruses based on their composition and structure. Since ten of the tobamoviruses have been reported to induce cytoplasmic crystalline inclusions of virus particles, stacked or in monolayers, this type of inclusion can be considered one of the main characteristics of the group (Edwardson and Christie, 1978; Matthews, 1979).

Several additional inclusion types have been reported to be associated with tobamoviruses; however, they have been reported in less than half the members of the group. Therefore, these inclusions cannot be considered main characteristics at this time.

II. CRYSTALLINE INCLUSIONS

The crystalline virus aggregates of the tobamovirus group are located in the cytoplasm. These inclusions were first reported in TMV-infected tobacco cells by Ivanowski (1903) who described them as consisting of waxlike material. The crystals were depicted in polar (top) and lateral (side) views in living cells. The crystals in polar views were regular in outline; many were hexagonal. Ivanowski also reported formation of striations across the short axes of the crystals in lateral view when they were treated with Flemming's solution. The hexagonal outline of the crystals in polar views was described by Goldstein (1924, 1926). Goldstein reported faint striations of the crystals *in vivo.* The inclusions exhibited pronounced striations (paracrystal formation) when treated with Flemming's solution. She also reported that strong acids or bases disrupted the crystals into formless masses. Crystals were also reorganized into paracrystalline arrays when they were touched with a microneedle (Sheffield, 1939).

When Wilkins *et al.* (1950) examined TMV crystals in lateral views in isolated tobacco hair cells, they observed longitudinal light and dark bands in polarized light. The width of the bands was approximately 300 nm. Wilkins *et al.* interpreted this to indicate that crystals were composed of stacked plates of virus particles with the long axes of the particles parallel to each other and to the lateral face of the plates. The particles were slightly tilted with respect to particles in adjacent plates. Thin-section and freeze-etching studies (such as Steere, 1957; Warmke and Edwardson, 1966; Esau, 1968; Willison and Cocking, 1969) have confirmed the interpretations given by Wilkins *et al.* (1950). Steere and Williams (1953), using freeze-drying techniques to extract hexagonal stacked plate crystalline inclusions from tobacco cells, showed that they were composed of virus particles which retained infectivity after dispersal in water and inoculation to healthy plants. Small aggregates of parallel particles with ends aligned in hexagonal close packing increase in size and become monolayers (i.e., three-dimensional crystalline inclusions) (Warmke and Edwardson, 1966). Stacked plate hexagonal crystalline inclusions contain two or more stacked monolayers (Warmke and Edwardson, 1966). In lateral views, portions of these complex crystals may exhibit a herringbone pattern where particles in adjacent plates are tilted in opposite directions (Warmke and Christie, 1967), while other particles in the same or other plates may be perpendicular to the polar faces of the crystal (Warmke and Christie, 1967), or the particles in adjacent plates may be tilted at the same angle in the same direction (Esau, 1968). Willison (1976) determined that TMV stacked plate crystalline inclusions always contained ice in frozen sucrose-treated cells; the relative proportions of virus and ice (55:45) were determined stereometrically. Interparticle spacing within monolayers was found to be 17.6 nm. In addition to intercrystalline water and virus particles, stacked plate crystals may also contain endoplasmic reticulum, mitochondria, plastids, and ribosomes (Warmke and Edwardson, 1966; Esau, 1968).

Stacked plate crystalline inclusions have been reported in tissues infected with the flavum (Kolehmainen *et al.*, 1965), the mutant Ni 118 at 20°C (Kassanis and Milne, 1971), pepper (Feldman and Oremianer, 1972), TD (Brcak and Pozdena, 1976), U1 (Bald and Solberg, 1961), U1C (Woods and Eck, 1948), U1D, and U1SB strains (Granett and Shalla, 1970a). Tissues infected with the following strains have not been reported to contain crystalline inclusions: English PM_2, German PM_2 (Kassanis and Turner, 1972), Kazahkstan (Goldin *et al.*, 1967), and U2–U5 (Bald and Solberg, 1961). However, Granett and Shalla (1970a) observed hexagonal crystals usually in the form of monolayers, in about 1% of cells infected with U5. A majority of the tobamoviruses which have been studied cytologically (10/14) induce crystalline inclusions usually in the form of stacked plates: tobacco mosaic, cucumber green mottle mosaic, cucumber 4, frangipani mosaic, Odontoglossum ringspot, ribgrass mosaic, sunn-hemp mosaic, tomato mosaic, potato moptop, soilborne wheat mosaic. However, the inclusions induced by tobamoviruses other than TMV are usually not hexagonal but contain rounded plates in polar (top) view, each with successively smaller diameters than the central plate so that in lateral views the plates resemble stair steps on either side of the central plate.

The monolayer and stacked plate crystalline inclusions induced by tobamoviruses stain green in the calcomine orange–Luxol brilliant green stain, and magenta in Azure A after heating (60°C, 1 min) (Christie and Edwardson, 1977).

Crystalline inclusions of virus particles are also main characteristics of other virus groups (Fenner, 1976; Matthews, 1979, 1982). The crystalline inclusions of the tobamoviruses can be distinguished from crystalline inclusions induced by viruses in other groups by differences in staining reactions in Azure A in the light microscope and by differences in particle morphology and inclusion structure in thin sections.

III. PARACRYSTALS

Paracrystalline inclusions of TMV particles are elongated aggregates, without evidence of regularity in the direction of length. The particles are packed in hexagonal two-dimensional regularity at right angles to the particles' length (Warmke and Edwardson, 1966). There appears to be little or no restriction on repetition of particles along the length of these inclusions. These inclusions are located in the vacuoles.

Ivanowski (1903) first reported transformation of crystalline inclusions into paracrystals in his descriptions of striations occurring across the short axes of crystalline inclusions where they were oriented in lateral views. Goldstein (1924, 1926) was the first to describe paracrystals (needlelike crystals or raphides) extending through most of the long axes of palisade and mesophyll cells. Aggregation of TMV particles in a paracrystalline array may extend the entire length of the cell, and in many cases the aggregates double back to form loops or figure eights (Goldin, 1963). Paracrystalline inclusions have been detected in U2–U4 where crystalline inclusions have not been reported (Bald and Solberg, 1961). Granett and Shalla (1970a) reported few hexagonal crystal inclusions in U5-infected cells (ca. 1%) which were usually in the form of monolayers, while most of the cells (73%) contained paracrystals. Paracrystalline inclusions of virus particles have also been reported to be associated with other tobamoviruses: Odontoglossum ringspot (Christie and Edwardson, 1977), potato moptop (Harrison, 1974), sunn-hemp mosaic (Sulochana and Solomon, 1970), tomato mosaic (Warmke, 1968), soilborne wheat mosaic (Hibino *et al.*, 1974b).

Paracrystalline inclusions of virus particles are considered to be a main characteristic of the carlaviruses (Matthews, 1982) and they have been reported to be induced by some viruses in certain other groups such as the clostero-, como-, and potexvirus groups. The paracrystalline inclusions of virus particles induced by tobamoviruses can be distinguished from paracrystals induced by viruses in other groups by staining reactions of the inclusions in Azure A. Tobamovirus paracrystalline inclusions do not register magenta coloration until heat is applied to the tissues; paracrystalline inclusions induced by other viruses are stained without the application of heat.

IV. ANGLED-LAYER AGGREGATES

Cytoplasmic angled-layer aggregate inclusions are induced by a pepper strain (Herold and Munz, 1967), the U5 strain (Shalla, 1968), and a necrotic strain of TMV (Vela and Rubio-Huertos, 1972). Angled-layer aggregate inclusions are gently twisted and elongate elements composed of many layers of aligned and parallel virus particles oriented across the long axes of the aggregates; one layer is flat against the next and is rotated at an angle of just less than 60° (or just over its supplement, 120°). In cross sections the inclusions appear as three-way cross-hatched figures, and in longitudinal sections as parallel lines interspersed with rows of dots. This type of inclusion has been described in detail by Warmke (1967, 1968, 1969, 1974) in tobacco infected with the aucuba strain of tomato mosaic virus. They have also been observed in beet necrotic yellow vein infections (Putz and Vuittenez, 1980), in peanut clump virus infections (Thouvenel and Fauquet, 1981; Reddy *et al.*, 1983), and in Odontoglossum ringspot virus infections (Inouye, 1983). Angled-layer aggregate inclusions do not stain with Azure A prior to the application of heat.

V. X-BODIES

Cytoplasmic coarse to fine granular bodies either vacuolate or nonvacuolate, rounded or amoeboid in outline were first reported by Ivanowski (1903) in TMV-infected tobacco cells. Goldstein (1924) termed these inclusions *X-bodies* and stated that they were bounded by a plasma membrane. Although Goldstein (1926) depicted X-bodies as being limited by membranes, later studies have shown no evidence of these inclusions being bounded by membranes. X-bodies were found to be resistant to treatment with formalin, HCl, and HNO_3 and soluble in KOH and ethanol (Smith, 1926). Unfortunately, the term *X-body* has been used to describe inclusions of very different morphologies and compositions. It has been proposed (Warmke, 1969; Esau and Hoefert, 1971) that the term be confined to those structures induced by the tobamoviruses.

X-bodies are cytoplasmic inclusions, induced by tobamoviruses, that consist of aggregates of virus-induced broad filaments, aggregated virus within electron-lucent areas, and host components, principally ER and ribosomes and occasionally dictyosomes; "vacuoles" within the inclusions appear to be regions in which cytoplasmic components have degenerated (Esau and Cronshaw, 1967a; Esau, 1968). The proteinaceous filaments of the X-body have a diameter of about 28.3 nm (Esau and Cronshaw, 1967b), while the diameter of TMV particles is about 18 nm

(Esau, 1968). The filaments are more electron-opaque than the coat of TMV particles (Esau, 1968). Esau (1968) states that a parallel might be drawn between the formation of the thick filaments (X-tubules) from granular X-material which appears early in X-body formation, and the polymerization into rods of spherical subunits of X-protein (Takahashi and Ishii, 1953). However, Esau (1968) does not equate X-material and X-protein. Granett and Shalla (1970b) found no consistent differences between amounts of extracted X-protein from TMV U1(SB) and TMV U1(D) although the SB strain induced more X-bodies than the D strain. TMV coat protein has been shown to be serologically unrelated to the protein in X-tubules (Granett and Shalla, 1970b).

Although Bald and Solberg (1961) were unable to detect X-bodies in tobacco infected with U2–U5, Granett and Shalla (1970a) reported X-bodies in about 1% of tobacco cells infected with the U5 strain. Feldman and Oremianer (1972) did not detect X-bodies in tobacco infected with a pepper strain of TMV. Since these authors also reported the presence of somewhat rounded crystalline inclusions, they may have examined a tobamovirus other than common TMV. X-bodies have also been reported in cells infected with sunn-hemp mosaic (Resconich, 1961) and tomato mosaic viruses (Warmke, 1969). Although Inaba and Saito (1968) reported X-bodies in light microscopy of soilborne wheat mosaic virus-infected *Poa annua* L. and wheat, electron microscopy of cytoplasmic inclusions of infected rye and wheat cells did not reveal typical X-body ultrastructure (Hibino *et al.*, 1974a,b). Matthews (1979, 1982) lists X-bodies under the term *viroplasms* as a main characteristic of the group. However, since most tobamoviruses (11/14) have not been reported to induce X-bodies, this type of inclusion should not be considered as a main characteristic of the group.

Scattered and aggregated virus particles have been observed within X-bodies. Smith and Schlegel (1965) used actinomycin D to suppress host RNA syntheses and found ^{3}H-uridine-labeled TMV RNA first in nuclei; later some label passed into the cytoplasm and still later some label accumulated in X-bodies in tobacco. Although double-stranded RNA, complementary viral RNA, and virus-induced RNA replicase have been associated with membranes, none of the membrane fractions have been demonstrated to be derived from TMV-induced X-bodies.

X-bodies stain olive green in the calcomine orange–Luxol brilliant green stain and red in Azure A (Christie and Edwardson, 1977). The X-bodies are stained by Azure A in the absence of heat, indicating a concentration of ribonucleoprotein with structure differing from that of TMV particles.

VI. VIRUS AGGREGATES IN CHLOROPLASTS

Shalla (1964) used light microscopy to study the development of chloroplast appendages in epidermal cells of TMV-infected tomato leaves. The appendages extended and contracted. After contraction the chloroplasts often enveloped ground cytoplasm and mitochondria. In thin sections, loosely aggregated virus particles sometimes associated with cytoplasmic ribosomes and ER were enveloped in the chloroplasts. Aggregated TMV particles occurred in tomato (Shalla, 1964) and tobacco (Milne, 1966; Scalla *et al.*, 1969a) chloroplast vacuoles, and TMV-U1 particles occurred in tobacco chloroplast vacuoles (Shalla, 1968).

A different type of association between TMV particles and chloroplasts has been reported by Esau and Cronshaw (1967a), Pratt (1969), and Honda and Matsui (1971). Since the closely aggregated particles were not associated with cytoplasmic membranes or enclosed in plastid membranes, the particles were assumed to be synthesized within the plastids. A similar condition in TMV U5-infected tobacco was described by Shalla (1968), in U1- and U5-infected tobacco by Granett and Shalla (1970a), and in a TMV-necrotic strain infecting *Digitalis thlaspi* by Vela and Rubio-Huertos (1972). Etiolation of leaf cells increased this type of particle–plastid association (Pratt, 1969; Honda and Matsui, 1971). Both types of association were reported by Honda and Matsui (1971) in the same leaf tissue of TMV-infected tobacco.

Autoradiographic studies in U5-infected tobacco leaf tissue by Betto *et al.* (1972) do not support the assumption that virus synthesis takes place within chloroplasts, since [^{3}H]uridine incorporation in plastids was practically absent, while incorporation in the cytoplasm was high. Shalla *et al.* (1975) isolated virus particles from chloroplasts of tobacco infected with the U5 strain. These particles (ca. 100 nm in length) exhibited minimal infectivity and the authors concluded that the particles derived from chloroplasts were noninfective TMV.

Chloroplast inclusions have been reported to be induced by only one other tobamovirus, a thermosensitive strain of tomato mosaic virus (Boudon and Meignoz, 1970). Here, inclusions of homogeneous appearance, less electron-opaque than osmiophilic bodies, occur in chloroplasts, sometimes bounded by membranes, sometimes not.

VII. OTHER TYPES OF CYTOPLASMIC INCLUSIONS

In tobacco infected with the flavum strain (producing insoluble coat protein), long hollow tubes with outer diameters ca. 19 nm and core diameters 6 to 9 nm are reported either oriented in groups between ER or parallel in large groups in the cytoplasm. These tubes seem to have lower affinities for OsO_4 and uranyl ions than do virus particles (Kolehmainen *et al.*, 1965). These authors also described relatively large areas of cytoplasm as being filled with electron-opaque globules, of various sizes, that tended to fuse into large masses. The globules were presumed to be composed of lipid.

The Ni 118 mutant of TMV in infected tobacco grown at 35°C produced insoluble coat protein and few complete virus particles, but at 20°C it produced many particles. Large electron-opaque inclusions were observed in plants grown at 35°C (Kassanis and Milne, 1971); these authors assumed the inclusions to consist of insoluble defective Ni 118 coat protein.

Kassanis and Turner (1972) compared two TMV mutants, German PM_2 which produced a soluble defective protein unable to coat its RNA, and English PM_2 which produced insoluble coat protein. Long fibers helically twisted in the form of figure eights, probably consisting only of coat protein, were observed in German PM_2-infected tobacco and not in English PM_2-infected cells. Bald (1964b) described similar inclusions in PM_2-infected cells. The width of the individual components of the filaments is ca. 19–22 nm (Kassanis and Turner, 1972). Cells infected with English PM_2 contained large electron-opaque globules similar to those described by Kolehmainen *et al.* (1965). However, Kassanis and Turner (1972) assumed that these inclusions consisted of insoluble coat protein and not lipids, since they did not stain as intensely as lipid droplets of chloroplasts. However, Liu and Boyle (1972) described osmiophilic globules in the cytoplasm of yellow-strain-infected cells which stained as intensely as lipid droplets in chloroplasts. German PM_2 and type TMV did not induce this type of inclusion.

Long electron-opaque hollow filaments have been reported in the cytoplasm in thin sections of leaf cells infected with TMV U1 (Shalla, 1964; Milne, 1966; Esau and Cronshaw, 1967b), TMV flavum (Kolehmainen *et al.*, 1965), and German PM_2 (Kassanis and Turner, 1972), whereas the English PM_2 strain did not induce such filaments. Liu and Boyle (1972) also reported long filaments in infections with the U1 strain but did not detect them in infections with a yellow strain of TMV. While Shalla (1964) and Milne (1966) have suggested that the filaments are developmental forms of the virus particles, Kolehmainen *et al.* (1965) proposed that they were composed of coat protein polymerized in a different way from that in complete particles. Kassanis and Turner (1972) pointed out that this latter proposal is supported by the appearance of filaments in infections with the German PM_2 strain (soluble coat protein) and the absence of filaments in infections with the English PM_2 strain (insoluble coat protein). Shalla's (1964) and Milne's (1966) proposition might be tested by staining with Azure A. The staining of the filaments after heating would indicate they were nucleoprotein. If they did not stain, then Kolehmainen and colleagues' (1965) proposition would appear to be more plausible.

In tobacco cells infected with the Kazakhstan strain of TMV, Miličić *et al.* (1979) observed filaments aggregated into cytoplasmic inclusions. The filaments, termed *stretched threads,* were difficult to stain in thin sections. They had a diameter of ca. 9 nm and were often aggregated parallel to each other. The composition of the filaments is unknown. Cytoplasmic inclusions of convuluted tubules 20–50 nm in diameter have been reported in soilborne wheat mosaic virus infections (Hibino *et al.*, 1974a,b). In potato moptop virus infections, aggregates of tubules ca. 18–22 nm in diameter and often branched have been described in the cytoplasm and vacuoles (Fraser, 1976). The composition of the threads and tubules is unknown.

VIII. NUCLEAR INCLUSIONS

Square or oblong plate inclusions of unknown composition have been reported in 11 solanaceous species infected with TMV (Hoggan, 1927). Crystalline and paracrystalline inclusions of TMV strain 1C particles were reported in nuclei of living tomato, pepper, tobacco, and *Nicotiana paniculata* (Woods and Eck, 1948). Acridine orange staining indicated an accumulation of RNA in nuclei of TMV-infected tomato cells (Hirai and Wildman, 1963). Hibino and Matsui (1964) applied [^{3}H]uridine to TMV-infected tobacco and observed developed silver grains over nuclei and ground cytoplasm indicating the presence of TMV RNA in those regions. Using fluorescent antibody specific for TMV coat protein, Hirai and Hirai (1964) observed that nuclei first show the presence of newly synthesized TMV protein in infected tomato hair cells. Singh and Hildebrandt (1966) employed still and time-lapse motion pictures in their study of TMV-infected tobacco callus cells. These authors observed paracrystals and small crystals moving into and out of nuclei.

The Kazakhstan strain of TMV in living cells and callus tissues of tobacco has been reported to induce crystalline and paracrystalline inclusions in nuclei (Goldin, 1963). In examinations of living and stained tobacco cells infected with TMV U1, Bald (1964a) described viral RNA originating in nucleoli and passing into the cytoplasm, whereas TMV U5 was apparently synthesized in the nucleus or near its perimeter.

In thin-section investigations of TMV-infected tobacco, aggregates of virus have been shown to occur in the nuclear matrix in resting stages (Esau and Cronshaw, 1967a), while TMV particles and aggregates of these were observed among chromosomes in dividing nuclei (Esau, 1968; Esau and Gill, 1969). However, the particles and aggregates were usually outside the nuclear envelope when division was completed. Aggregates of virus particles in the nuclear matrix of TMV-infected tobacco have also been described by Honda and Matsui (1969), Granett and Shalla (1970a), and Langenberg and Schroeder (1973). Membranes have not been observed to be associated with these aggregates. Some of these aggregates may represent portions of crystalline monolayers of virus particles.

Light and electron-lucent areas have been reported in nuclei of TMV-infected plants. These areas have been reported to contain no virus particles (Gianinazzi and Schneider, 1979), a few particles (Da Graca and Martin, 1976), masses of incomplete particles (Scalla *et al.*, 1969b,c), masses of complete particles (Honda and Matsui, 1971; Scalla *et al.*,

1969b), and crystalline aggregates of particles (Scalla *et al.*, 1969b). The intranuclear masses and crystalline aggregates of particles were not associated with membranes. Electron-lucent areas have also been reported in nuclei of tobacco cells infected with the Kazakhstan strain. These areas may contain masses of virus particles or aggregates of stretched threads ca. 9 nm in diameter (Miličić *et al.*, 1979).

None of the other members or possible members of the tobamovirus group have been reported to induce crystalline, paracrystalline, or particle aggregate nuclear inclusions. Homogeneous nuclear inclusions more electron-lucent than nucleoli or chromatin have been described in tobacco infected with a temperature-sensitive strain of tomato mosaic virus (Boudon and Meignoz, 1970).

IX. CONCLUDING REMARKS

Tobamoviruses have been reported to induce several types of inclusions. At present, only one type, the crystalline cytoplasmic inclusion, is a main characteristic of the group. However, this situation may change when more cytological information becomes available. Problems involving sampling probably account for some of the reported discontinuity of inclusion types encountered within this group. Some of the sampling factors would include unequal distribution of inclusions in tissues, destruction or obliteration of some inclusions by certain fixatives, and confining studies to the electron microscope, a notoriously poor sampling device for sectioned material.

Cytology of inclusions was not the primary objective of many of the reports reviewed here, and some studies were confined to a single type of inclusion. Only single reports have been devoted to four tobamoviruses: cucumber virus 4 (Brcak and Hrsel, 1961), frangipani mosaic (Varma and Gibbs, 1978), *Nicotiana velutina* mosaic (Randles, 1978), and U2 tobacco mosaic (Bald and Solberg, 1961). None of these investigations mentioned X-body detection, and Bald and Solberg (1961) stated that U2 through U5 did not induce X-bodies. However, Granett and Shalla (1970a) observed X-bodies in cells infected with the U5 strain.

Randles's (1978) statement that no intracellular changes were observed in tissues infected with *N. velutina* mosaic virus should not be interpreted as indicating that this virus does not induce inclusions since all viruses that have been studied intensively have been shown to induce inclusions (Edwardson and Christie, 1978).

The functions of the various inclusions induced by tobamoviruses are unknown. Procedures utilized in purification of cytoplasmic inclusions induced by potyviruses (Hiebert *et al.*, 1984; de Mejia *et al.*, 1985a,b) may be applicable to X-body purification. Here, light microscopy of products arising from various purification steps is essential for maintaining the purity of the preparations. Serological and electrophoretic analyses of purified X-body components may reveal the accumulation of nonstructural TMV protein and may shed some light on the function(s) of TMV-induced X-body inclusions.

Crystalline inclusions induced by tobamoviruses may be disrupted and transformed into paracrystals or masses of particles by fixatives, salts, or mechanical stresses. In whatever form tobamoviruses particle aggregates are encountered, they can be distinguished from inclusions induced by viruses in the other groups. Only tobamovirus particle aggregate inclusions require the application of heat (60°C, 1 min) in order for the Azure A stain to register its characteristic magenta color in ribonucleoprotein-containing inclusions. This type of staining reaction is thus diagnostic for tobamovirus infections (Christie and Edwardson, 1977).

REFERENCES

Bald, J. G., 1964a, Cytological evidence for the production of plant virus ribonucleic acid in the nucleus, *Virology* **22**:377.

Bald, J. G., 1964b, Symptoms and cytology of living cells infected with defective mutants of tobacco mosaic virus, *Virology* **22**:388.

Bald, J. G., and Solberg, R. A., 1961, Apparent release of tobacco mosaic virus in living infected cells, *Nature* (*London*) **190**:651.

Betto, E., Bassi, M., Favali, M. A., and Conti, G. G., 1972, An electron microscopic and autoradiographic study of tobacco leaves infected with the U5 strain of tobacco mosaic virus, *Phytopathol. Z.* **75**:193.

Boudon, E., and Meignoz, R., 1970, Etude au microscope electronique de Tabacs infectes par une souche aucuba thermosensible du virus de la Mosaique du Tabac (V.M.T.). Absence de virions a 33°C, inclusions anormales et alterations cellulaires, *C.R. Acad. Sci. Ser. D* **270**:3003.

Brcak, J., and Hrsel, I., 1961, The electron microscopic examination of inclusions of cucumber virus 4, *Biol. Plant.* **3**:132.

Brcak, J., and Pozdena, J., 1976, Some virus and virus-like diseases of tobacco, tomato, papaya, and rubber tree in Vietnam and Cambodia, *Biol. Plant.* **18**:290.

Christie, R. G., and Edwardson, J. R., 1977, Light and electron microscopy of plant virus inclusions, *Fla. Agric. Exp. Stn. Monogr.* **9**.

Da Graca, J. V., and Martin, M. M., 1976, An electron microscope study of hypersensitive tobacco infected with tobacco mosaic virus at 32° C, *Physiol. Plant Pathol.* **8**:215.

de Mejia, M. V. G., Hiebert, E., and Purcifull, D. E., 1985a, Isolation and partial characterization of the amorphous, cytoplasmic inclusions associated with infections caused by two potyviruses, *Virology* **142**:24.

de Mejia, M. V. G., Hiebert, E., Purcifull, D. E., Thornberry, D. W., and Pirone, T. P., 1985b, Identification of potyviral amorphous inclusion protein as a nonstructural virus-specific protein related to helper component, *Virology* **142**:34.

Edwardson, J. R., and Christie, R. G., 1978, Use of virus-induced inclusions in classification and diagnosis, *Annu. Rev. Phytopathol.* **16**:31.

Esau, K., 1968, *Viruses in Plant Hosts* University of Wisconsin Press, Madison.

Esau, K., and Cronshaw, J., 1967a, Relation of tobacco mosaic virus to the host cells, *J. Cell Biol.* **33**:665.

Esau, K., and Cronshaw, J., 1967b, Tubular components in cells of healthy and tobacco mosaic virus-infected *Nicotiana*, *Virology* **33**:26.

Esau, K., and Gill, R. H., 1969, Tobacco mosaic virus in dividing mesophyll cells of Nicotiana, *Virology* **38**:464.

Esau, K., and Hoefert, L. L., 1971, Cytology of beet yellows infection in *Tetragonia*. III. Conformations of virus in infected cells, *Protoplasma* **73**:51.

Feldman, J. M., and Oremianer, S., 1972, An unusual strain of tobacco mosaic virus from pepper, *Phytopathol. Z.* **75**:250.

Fenner, F., 1976, Classification and nomenclature of viruses, *Intervirology* **7**:1.

Fraser, T. W., 1976, Mop-top tubules: The ulstrastructure of unusual tubular elements associated with two different leaf symptoms of potato mop-top virus infected potato leaves, *Protoplasma* **90**:15.

Gianinazzi, S., and Schneider, C., 1979, Differential reaction to tobacco mosaic virus infection in Samsun 'nn' tobacco plants. II. A comparative ultrastructural study of the virus induced necroses with those of Xanthi-nc, *Phytopathol. Z.* **96**:313.

Gibbs, A. J., 1977, Tobamovirus group, *CMI/AAB Descriptions of Plant Viruses No. 184.*

Goldin, M. I., 1963, *Viral Inclusions in the Plant Cell and the Nature of Viruses*, Akad. Nauk USSR, Moscow.

Goldin, M. I., Agoyeva, N. V., and Tumanova, V. A., 1967, *In vivo* investigations on viral inclusions of the Kazakhstan strain of tobacco mosaic virus in tissue cultures and free living cells, *Acta Virol.* **11**:462.

Goldstein, B., 1924, Cytological study of living cells of tobacco plants affected with mosaic disease, *Bull. Torrey Bot. Club* **51**:261.

Goldstein, B., 1926, A cytological study of the leaves and growing points of healthy and mosaic diseased tobacco plants, *Bull. Torrey Bot. Club* **53**:299.

Granett, A. L., and Shalla, T. A., 1970a, Discrepancies in the intracellular behavior of three strains of tobacco mosaic virus, two of which are serologically indistinguishable, *Phytopathology* **60**:419.

Granett, A. L., and Shalla, T. A., 1970b, The relation of tobacco mosaic virus X-protein to amorphous cellular inclusions (X-bodies), *Phytopathology* **60**:426.

Harrison, B. D., 1974, Potato mop-top virus, *CMI/AAB Descriptions of Plant Viruses No. 138.*

Harrison, B. D., Finch, J. T., Gibbs, A. J., Hollings, M., Shepherd, R. J., Valenta, V., and Wetter, C., 1971, Sixteen groups of plant viruses, *Virology* **45**:356.

Herold, F., and Munz, K., 1967, An unusual kind of crystalline inclusion of tobacco mosaic virus, *J. Gen. Virol.* **1**:375.

Hibino, H., and Matsui, C., 1964, Electron microscopic autoradiography of leaf cells infected with tobacco mosaic virus. I. Uridine-H^3 uptake, *Virology* **24**:102.

Hibino, H., Tsuchizaki, T., and Saito, Y., 1974a, Comparative electron microscopy of cytoplasmic inclusions induced by 9 isolates of soil-borne wheat mosaic virus, *Virology* **57**:510.

Hibino, H., Tsuchizaki, T., and Saito, Y., 1974b, Electron microscopy of inclusion developments in the leaf cells infected with soil-borne wheat mosaic virus, *Virology* **57**:522.

Hiebert, E., Purcifull, D. E., and Christie, R. G., 1984, Purification and immunological analyses of plant viral inclusion bodies, in: *Methods in Virology*, Volume 8 (K. Maramorosch and H. Koprowski, eds.), pp. 225–280, Academic Press, New York.

Hirai, T., and Hirai, A., 1964, Tobacco mosaic virus: Cytological evidence of synthesis in the nucleus, *Science* **145**:589.

Hirai, T., and Wildman, S. G., 1963, Cytological and cytochemical observations on the early stage of infection of tomato hair cells by tobacco mosaic virus, *Plant Cell Physiol.* **4**:265.

Hoggan, I. A., 1927, Cytological studies on virus diseases of solanaceous plants, *J. Agric. Res.* **35**:651.

Honda, Y., and Matsui, C., 1969, Occurrence of tobacco mosaic virus within the nucleus, *Virology* **39**:593.

Honda, Y., and Matsui, C., 1971, Distribution of tobacco mosaic virus in etiolated tobacco leaf cells infected with two viruses, *Phytopathology* **61**:759.

Inaba, T., and Saito, Y., 1968, Soil-borne wheat mosaic virus from *Poa annua* L., a new host, *Ann. Phytopathol. Soc. Jpn.* **34**:40.

Inouye, N., 1983, Host range and properties of a strain of Odontoglossum ringspot virus in Japan, *Nogaku Kenkyu* **60:**53.

Ivanowski, D., 1903, Über die Mosaikkrankheit der Tabakspflanze, *Z. Pflanzenkr. (Pflanzenpathol.) Pflanzenschutz* **13:**1.

Kassanis, B., and Milne, R. G., 1971, An unusual inclusion in plants infected with a tobacco mosaic virus mutant, *J. Gen. Virol.* **11:**193.

Kassanis, B., and Turner, R. H., 1972, Virus inclusions formed by the PM_2 mutant of TMV, *J. Gen. Virol.* **14:**119.

Kolehmainen, L., Zech, H., and von Wettstein, D., 1965, The structure of cells during tobacco mosaic virus reproduction, *J. Cell Biol.* **25:**77.

Langenberg, W. G., and Schroeder, H. F., 1973, Effects of preparative procedures on the preservation of tobacco mosaic virus inclusions, *Phytopathology* **63:**1003.

Liu, K. C., and Boyle, J. S., 1972, Intracellular morphology of two tobacco mosaic virus strains in, and cytological responses of, systemically susceptible potato plants, *Phytopathology* **62:**1303.

Martelli, G. P., and Russo, M., 1977, Plant virus inclusion bodies, *Adv. Virus Res.* **21:**175.

Matthews, R. E. F., 1979, Classification and nomenclature of viruses, *Intervirology* **12:**132.

Matthews, R. E. F., 1982, Classification and nomenclature of viruses, *Intervirology* **17:**4.

Miličić, D., Wrischer, M., Brcak, J., and Juretić, N., 1979, Intercellular changes induced by the defective Kazakhstan strain of tobacco mosaic virus, *Acta Bot. Croat.* **38:**1.

Milne, R. G., 1966, Multiplication of tobacco mosaic virus in tobacco leaf palisade cells, *Virology* **28:**79.

Pratt, M. T., 1969, The incidence of tobacco mosaic virus in plastids of etiolated tobacco leaves, *Virology* **39:**344.

Putz, C., and Vuittenez, A., 1980, The intracellular location of beet necrotic yellow vein virus, *J. Gen. Virol.* **50:**201.

Randles, J. W., 1978, Nicotiana velutina mosaic virus, *CMI/AAB Descriptions of Plant Viruses No. 189.*

Reddy, D. V. R., Rajeshwari, R., Iizuka, N., Lesemann, D. E., Nolt, B. L., and Goto, T., 1983, The occurrence of Indian peanut clump, a soilborne virus disease of groundnuts (*Arachis hypogea*) in India, *Ann. Appl. Biol.* **102:**305.

Resconich, E. C., 1961, Interpretation of the forms of inclusions in bean systemically infected with tobacco mosaic virus. *Virology* **15:**16.

Scalla, R., Meignoz, R., Noirot-Timothee, C., and Martin, C., 1969a, Etude au microscope electronique des premiers states des alterations des chloroplastes chez le Nicotiana Xanthi n.C. au cours de la reaction hypersensible au virus de la mosaique du Tabac, *C.R. Acad. Sci. Ser. D.* **268:**527.

Scalla, R., Martin, C., Lhoste, B., and Meignoz, R., 1969b, Modification de la structure nucleaire chez le Tabac infecte a 32° C par le virus de la mosaique du Tabac, *C.R. Acad. Sci. Ser. D* **269:**466.

Scalla, R., Martin, C., Meignoz, R., and Lhoste, B., 1969c, Modification de la structure nucleaire chez le Tabac infecte a 32° C par le virus de la mosaique du Tabac. Nicotiana tabacum var. Samsun, *C.R. Acad. Sci. Ser. D* **269:**586.

Shalla, T. A., 1964, Assembly and aggregation of tobacco mosaic virus in tomato leaflets, *J. Cell Biol.* **21:**253.

Shalla, T. A., 1968, Virus particles in chloroplasts of plants infected with the U5 strain of tobacco mosaic virus, *Virology* **35:**194.

Shalla, T. A., Peterson, L. J., and Giunchedi, L., 1975, Partial characterization of virus-like particles in chloroplasts of plants infected with the U-5 strain of TMV, *Virology* **66:**94.

Sheffield, F. M. L., 1939, Micrurgical studies on virus infected plants, *Proc. R. Soc. London Ser. B* **126:**529.

Singh, M., and Hildebrandt, A. C., 1966, Movements of tobacco mosaic virus inclusion bodies within tobacco callus cells, *Virology* **30:**134.

Smith, F. F., 1926, Some cytological and physiological studies of mosaic diseases and leaf variegations, *Ann. Mo. Bot. Gard.* **13:**425.

Smith, S. H., and Schlegel, D. E., 1965, The incorporation of ribonucleic acid precursors in healthy and virus-infected plant cells, *Virology* **26:**180.

Steere, R. L., 1957, Electron microscopy of structural detail in frozen biological specimens, *J. Biophys. Biochem. Cytol.* **3:**45.

Steere, R. L., and Williams, R. C., 1953, Identification of crystalline inclusion bodies extracted intact from plant cells infected with tobacco mosaic virus, *Am. J. Bot.* **40:**81.

Sulochana, C. B., and Solomon, J. J., 1970, Local lesion formation in *Cyamopsis tetragonoloba* Taub. with DEMV and other plant viruses, *Proc. Indian Acad. Sci. Sect. B* **71:**56.

Takahashi, W. N., and Ishii, M., 1953, A macromolecular protein associated with tobacco mosaic virus infection: Its isolation and properties, *Am. J. Bot.* **40:**85.

Thouvenel, J. C., and Fauquet, C., 1981, Peanut clump virus, *CMI/AAB Descriptions of Plant Viruses No. 235.*

Varma, A., and Gibbs, A. J., 1978, Frangipani mosaic virus, *CMI/AAB Descriptions of Plant Viruses No. 196.*

Vela, A, and Rubio-Huertos, M., 1972, Ultraestructura de hojas de *Digitalis thlaspi* infectadas espontaneamente con dos virus, *Microbiol. Esp.* **25:**1.

Warmke, H. E., 1967, Aucuba strain of tobacco mosaic virus: An unusual aggregate, *Science* **156:**262.

Warmke, H. E., 1968, Fine structure of inclusions formed by the aucuba strain of tobacco mosaic virus, *Virology* **34:**149.

Warmke, H. E., 1969, A reinterpretation of amorphous inclusions of tobacco mosaic virus, *Virology* **39:**695.

Warmke, H. E., 1974, Direction of rotation of aucuba (TMV) angled-layer aggregates, *Virology* **59:**591.

Warmke, H. E., and Christie, R. G., 1967, Use of dilute osmium tetroxide for preservation of three-dimensional crystals of tobacco mosaic virus, *Virology* **32:**534.

Warmke, H. E., and Edwardson, J. R., 1966, Electron microscopy of crystalline inclusions of tobacco mosaic virus in leaf tissue, *Virology* **30:**45.

Wilkins, M. H. F., Stokes, A. R., Seeds, W. E., and Oster, G., 1950, Tobacco mosaic virus crystals and three-dimensional microscopic vision, *Nature (London)* **166:**127.

Willison, J. H. M., 1976, The hexagonal lattice spacing of intracellular crystalline tobacco mosaic virus, *J. Ultrastruct. Res.* **54:**176.

Willison, J. H. M., and Cocking, E. C., 1969, Freeze-etching observations of tobacco leaves infected with tobacco mosaic virus, *J. Gen. Virol.* **4:**229.

Woods, M. W., and Eck, R. V., 1948, Nuclear inclusions produced by a strain of tobacco mosaic virus, *Phytopathology* **38:**852.

CHAPTER 8

Tobamovirus Classification

ADRIAN GIBBS

The tobamoviruses are one of the best known and most fully reviewed groups of plant viruses (Gibbs, 1977; Shikata, 1977; Van Regenmortel, 1981); nevertheless, another review of their classification is justified by the increasing flood of molecular sequence data which is providing new insights into their classification and, possibly, into their evolution.

Early attempts to classify and name plant viruses by their host and symptoms were unsatisfactory (McKinney, 1944). However, the practice, advocated by Bawden (1941), of grouping plant viruses on characters such as the morphology, chemical composition, and serological specificity of their particles, and on their cross-protection behavior, soon resulted in a stable and useful classification. The most widely used criteria were serological tests, pioneered by Chester (1937), and cross-protection tests (McKinney, 1929; Salaman, 1933; Thung, 1931). They established the tobamoviruses as a discrete group of which there are now more than a dozen definitive members (Matthews, 1982). These share the following characteristics:

Infective particles are straight tubes with a modal length, for most members, of about 300 × 18 nm and a sedimentation coefficient of around 190 S. Each particle is constructed of ca. 2000 protein subunits of a single protein species (relative molecular weight 1.8K) arranged as a helix (pitch ca. 2.3 nm) enclosing the genome, which is a single molecule of single-stranded RNA (relative molecular weight ca. 2000K, ca. 5% of the particle weight). This RNA has a 5′ methyl guanosine terminal "cap," and an amino acid accepting ability at the 3′-terminus. It codes for at least four proteins including the coat protein, which is translated from a subgenomic mRNA. Infectivity in sap survives heating to 90°C and storage for many years. Particles occur in sap at concentrations up to 10 g/liter.

ADRIAN GIBBS • Research School of Biological Sciences, Australian National University, Canberra, Australian Capital Territory, 2601, Australia.

Most tobamoviruses occur naturally in one or a few angiosperm species, but may be transmitted experimentally to a very wide range of such species, and cause mottles and mosaics. They are transmitted in nature by contact between plants or from contaminated soil, and are sometimes carried on seed, but no efficient and specific vectors are known. They are readily transmitted by sap inoculation. Tobamovirus particles are found in the cytoplasm, and sometimes the chloroplasts, of cells of all tissues (except perhaps embryos) where they may form elongated (paracrystalline) or plate-shaped (crystalline) inclusions, or, together with various cell constituents, they may form amorphous inclusions.

There are other viruses that share some of these properties and these are reviewed elsewhere in this volume. Closest perhaps is the *Chara australis* virus, originally incorrectly called *Chara corallina* virus. This differs most notably from other tobamoviruses in having particles with a modal length of 532 nm and an algal host, but no known angiosperm host.

More distant are several viruses with fungal vectors and particles of two or more modal lengths, some of which are now grouped together as the furoviruses.

I. RELATIONSHIPS AMONG THE DEFINITIVE TOBAMOVIRUSES

Many different characters are available for assessing the relatedness of the definitive tobamoviruses; however, I shall concentrate primarily on the available molecular information, particularly sequences, as such data have been found to be very useful for assessing relationships of organisms. I will then attempt to correlate the relationships indicated by such data with those indicated by other criteria.

A. Coat Protein

1. Amino Acid Composition

Early analyses of the amino acid composition of tobamovirus particles showed that they had a specific composition, which differed between isolates. The relatedness of different tobamoviruses indicated by these data correlated with that indicated by other criteria, and clearly distinguished them from other viruses (Hennig and Wittmann, 1972; Knight, 1964; Siegel and Wildman, 1954; Tremaine and Argyle, 1970; Tsugita and Fraenkel-Conrat, 1960, 1962; Van Regenmortel, 1967; Wittmann, 1960).

Tsugita (1962) was the first to propose a classification of tobamovirus variants and strains based on the composition of their coat proteins. He defined four clusters on such characters as the methionine and histidine content, and N-terminal amino acids of the proteins, though as this clustering was based on so few characters it is not surprising that it misclassified some unusual histidine-containing mutants of the type strain analyzed later (Rombauts and Fraenkel-Conrat, 1968). However, a more stable and predictive classification of the tobamoviruses, which confirmed the classifications of Tsugita (1962) and Van Regenmortel (1967), was computed using estimates of the molar amounts of all amino acids in the coat protein (Gibbs, 1969; Gibbs and Harrison, 1976); the full membership of the clusters in that classification, and the sources of the data, were given by Gibbs (1977).

2. Nucleotide Sequence

Recently, there has been a great increase in information of the nucleotide sequences not only of the coat protein genes of tobamoviruses, but also of some of their other genes; indeed, the entire genome of some isolates has been sequenced. So far the amino acid sequences of the coat proteins of more than ten tobamovirus isolates have been determined, some on several occasions, some directly and others via the nucleotide sequence of the genome or coat protein mRNA. These sequences are readily aligned over most of their length. Table I shows the amino acid sequences of the coat proteins of:

- TMV-type strain
- Tomato mosaic (ToMV; syn. dahlemense)
- Tobacco mild green mosaic (TMGMV; syn. U2 or U5 strain)
- Odontoglossum ringspot (ORSV; syn. orchid strain)
- Ribgrass mosaic (RMV; syn. Holmes's ribgrass strain)
- Sunn-hemp mosaic (SHMV; syn. cowpea or bean strain)
- Cucumber green mottle mosaic (CGMMV; syn. CV3)

The relatedness of these sequences (Table II) may be used to calculate a dendrogram (Fig. 1). The branching pattern of this dendrogram is quite unambiguous as shown by the similarity of the homology estimates contributing to each branching point. For example, the standard error of the ten estimates contributing to the primary branching point, which has a mean homology of 33.8%, is 1.29%, whereas the next nearest branch point is determined by four estimates and has a mean homology of 43.75% with a standard error of 1.94% (i.e., they differ by more than five times the standard error). It is noteworthy that the data derived from the ORSV sequence contribute more of the variance than the other sequences. This may indicate that this sequence is in need of revision; it was determined by conventional techniques some time ago, the data have never been published and possibly involved assignment of amino acids by analogy.

The four coat protein gene sequences that have been reported (TMV-

TABLE I. The Aligned Amino Acid Sequences of Tobamovirus Coat Proteins[a,b,c]

	N-termini			C-termini
TMV-type	SYSITTPSQFVFLSSAWADPIELINLCTNALGNQFQTQQARTVVQRQFSE	VWKPSPQVTVRFPD-SD-FKVYRYNAVLDPLVTALLGAFDTRNRIIEVEN	QANPTTAETLDATRRVDDATVAIRSAINNLIVELIRGTGSYNRSSFESSS	GLVWTSGPAT
ToMV	SYSITSPSQFVFLSSVWADPIELLNVCTSSLGNQFQTQQARTTVQQQFSE	VWKPFPQSTVRFPG-DV-YKVYRYNAVLDPLITALLGTFDTRNRIIEVEN	QQSPTTAETLDATRRVDDATVAIRSAINNLVNELVRGTGLYNQNTFESMS	GLVWTSAPAS
TMGMV	PYTINSPSQFVYLSSAYADPVELINLCTNALGNQFQTQQARTTVQQQFAD	AWKPSPVMTVRFPA-SD-FYVYRYNSTLDPLITALLNSFDTRNRIIZVBB	ZPAPNTTVPIBTZZRVDDATVAIRASINNLANELVRGTGMFNQAGFETAS	GLVWTTTPAT
ORSV	SYSITTPSZLBYLSSAWABPKZLIBLCTBALGBSFZTZBARTTVQQQFAD	VWTPSPQLTVRFPAGAGYFRVYRYBFILBPLITPLMGTFDTRNRIIZVZB	ZPBPTTAZTLBTTRRVDDATVAIRSAINNLLNELVRGTGMYBZSTFZVMG	---WTSSLST
RMV	SYNITNSNQYQYFAAVWAEPTPMLNQCVSALSQSYQTQAGRDTVRQQFAN	LLSTIVAPNQRFPD-TG-FRVYVNSAVIKPLYEALMKSFDTRNRIIQTEE	QSRPSASQVANATQRVDDATVAIRSQIQLLLNELSNHGGYMNRAEFE-A-	ILPWTTAPAT
SHMV	AYSIPTPSQLVYFTENYADYIPFVNRLINARSNSFQTQSGRDELREILIK	SQVSVVSPISRFPA-EPAYYIYLRDPSISTVYTALLQSTDTRNRVIEVEN	STDVTTAEQLNAVRRTDDASTAIHNNLEQLLSLLTNGTGVFNRTSFESAS	GLWLVTTPTRTA
CGMMV	AYNPITPSKLIAFSASYVPVRTLLNFLVASQGTAFQTQAGRDSFRESLSA	LPSSVVDINSRFPD-AG-FYAFLNGPVLRPIFVSLLSSTDTRNRVIEVVD	PSNPTTAESLNAVKRTDDASTAARAEIDNLIESISKGFDVYDRASFEAAF	SVVWSEATTSKA

[a] Data from Altschuh *et al.* (1981), Meshi *et al.* (1983), Ohno *et al.* (1984), and from the National Biochemical Research Foundation amino acid data base, 1985.

[b] Other tobamovirus coat proteins that have been sequenced, but are not listed here, include the "O," "Kokubu," and "OM" isolates, which differ from the TMV-type sequence by one, one, and two amino acids, respectively (i.e., $< 1\%$).

[c] The sequences were aligned by the "diagram" method of Gibbs and McIntyre (1970), and one or two spaces inserted in some of them at positions 65 and 68, and additional spaces inserted near the C-termini of the HRV and ORSV sequences to maximize homology; there are several alternative ways of inserting additional gaps to maximize the homology of residues 151–162, but whether these are meaningful is uncertain, and none were included.

type, ToMV, SHMV, and CGMMV) confirm that the relationships found in the amino acid sequences mirror similar, but not linearly related, nucleotide sequence homologies. For example, the TMV-type and ToMV coat protein genes have a nucleotide homology of about 75% (Takamatsu *et al.*, 1983) and yield proteins with an amino acid homology of 82.3%, whereas the TMV-type and CGMMV genes have nucleotide and amino acid homologies of 45.8 and 36.7%, respectively.

TABLE II. Homologies of the Coat Protein Sequences in Table I

TMV-type	ToMV	TMGMV	ORSV	RMV	SHMV	CGMMV
	82.5%	68.7%	60.6%	45.6%	41.4%	37.0%
		66.9%	61.9%	46.9%	38.9%	35.2%
			60.0%	44.4%	38.3%	31.5%
				38.1%	30.3%	28.4%
					34.0%	35.8%
						43.8%

3. More Composition Data

The relatedness of tobamovirus coat proteins assessed from their amino acid sequences has been shown (Gibbs, 1980) to correlate closely with that assessed from their amino acid compositions. Thus, it is possible to extend the classification based on sequence comparisons by computing one based on amino acid compositions. This was done using data for clusters of individual tobamoviruses from various sources. Compositions expressed as molar percentages were used to compute classifications; molar percentages were used because (1) there is clear evidence (Fauquet *et al.*, 1985; Gibbs and McIntyre, 1970) that such data are more reliable than the number of residues estimated from them, and (2) it is likely that the sizes of the coat proteins of different tobamoviruses are similar.

First, a classification was computed from all the data. This grouped most replicate analyses for individual tobamoviruses, and so mean compositions were calculated for these clusters of isolates (Table III). Different metrics (i.e., methods of computing the relatedness of pairs of individuals) were then compared and found to give different dendrogram patterns; however, the "Canberra" metric (Lance and Williams, 1967) produced a dendrogram (Fig. 2) similar to that of the sequence classification, with SHMV and CGMMV in a primary branch separate from the others.

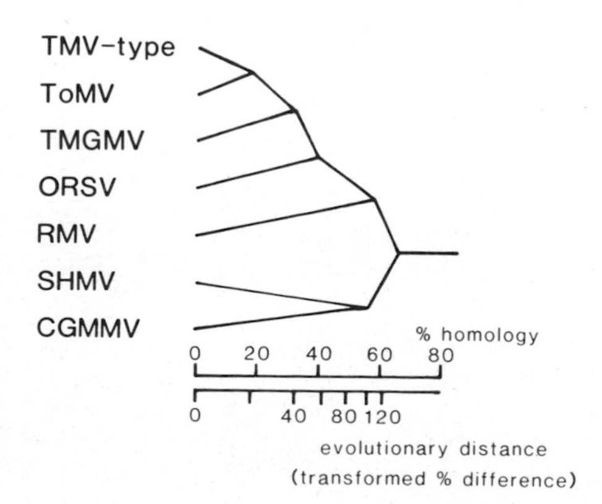

FIGURE 1. Dendrogram calculated from the coat protein sequence homologies in Table I by the simple centroid sorting method of Gibbs and Fenner (1984). "Evolutionary distance" is a measure of the actual minimum percentage amino acid change that occurred during evolution but is not now seen because of reversion (Dayhoff, 1972).

TABLE III. Composition of the Tobamovirus Coat Proteins Classified in Fig. 2[a,b]

Virus	Amino acid composition (mole%)																	
	Ala	Arg	Asx	Cys	Glx	Gly	His	Ile	Leu	Lys	Met	Phe	Pro	Ser	Thr	Trp	Tyr	Val
TMV-type	8.94	7.27	11.49	0.63	10.01	3.69	0	5.38	7.57	1.37	0	5.05	5.05	9.90	10.22	1.90	2.53	8.95
ToMV	6.98	5.71	11.20	0.63	12.05	3.81	0	4.44	8.25	1.27	0.63	5.07	5.07	9.72	10.57	1.90	3.17	9.51
PMMV	10.76	5.70	11.39	0.63	11.39	5.06	0	3.16	10.13	1.27	0.63	4.43	3.80	6.33	12.66	1.27	2.53	8.86
SOV	9.81	6.54	9.81	0.82	9.81	4.09	0	5.72	9.81	3.27	0.82	4.91	4.91	7.36	8.99	1.90	4.09	7.36
TMGMV	11.18	5.06	13.92	0.63	10.13	2.74	0	5.06	6.96	0.63	1.27	5.06	6.33	6.33	12.03	1.27	3.80	7.59
ORSV	7.01	6.37	12.74	0.64	9.55	4.46	0	5.10	8.92	0.64	1.91	4.46	5.73	7.64	13.38	1.91	3.82	5.73
RMV	11.54	6.09	10.58	0.64	14.42	2.56	0.64	4.81	7.37	1.28	1.92	3.53	5.77	8.98	7.69	1.60	4.17	6.41
CRMV	13.46	6.41	9.62	0.64	13.46	2.56	0.64	4.49	9.62	1.92	1.92	3.85	5.13	7.69	5.77	1.92	4.49	6.41
RMV-CW	11.04	7.15	9.08	0.65	14.60	1.94	0.65	4.87	9.08	1.62	2.60	3.90	4.87	9.09	7.60	1.30	3.90	6.17
RMV-Y	12.18	5.77	8.97	1.92	16.03	1.92	0.64	4.49	7.69	1.28	1.28	3.21	7.05	8.97	7.05	1.28	3.85	6.41
UMMV	11.39	6.96	6.96	0.63	15.19	3.16	1.27	4.43	8.86	1.90	0.63	4.43	6.33	9.49	6.96	1.27	3.16	6.96
RMV-Ly	10.76	6.96	10.13	0.63	13.29	1.90	0.63	4.43	7.59	1.27	2.53	3.16	5.70	10.13	8.23	1.90	4.43	6.33
ToSNV	12.84	6.76	10.14	0.68	14.86	2.03	0.68	2.70	9.46	1.35	2.03	3.38	4.73	10.14	7.43	1.35	4.05	5.41
CGMMV	12.37	6.02	11.58	0	6.36	3.18	0	4.11	7.95	2.37	0	6.83	5.40	14.30	7.62	0.64	2.54	8.74
CV3-Jap	13.12	5.00	12.50	0	6.25	5.63	0.63	4.38	11.25	2.50	0	5.63	3.75	15.00	6.25	1.25	2.50	4.38
SHMV	7.40	7.40	11.11	0	10.49	1.85	0.61	6.17	9.26	0.61	0	3.70	4.94	11.11	12.35	0.61	4.94	7.40
FMV	8.86	6.96	10.76	0.63	10.13	5.70	0.63	6.96	8.23	2.53	0	4.43	2.53	8.86	8.23	3.16	3.16	8.23
CAV	8.30	4.71	14.60	0	8.77	6.89	0.59	3.59	6.95	5.89	1.77	8.36	5.30	9.01	8.36	0	2.24	4.71

[a] Data from Fauquet *et al.* (1985), Gibbs (1977), Gibbs and Harrison (1976), Gibbs *et al.* (1982), Shen *et al.* (1982), Wetter *et al.* (1984), and chapters in this volume.

[b] In addition to the virus acronyms given in the text; CV3-Jap is the Japanese strain of cucumber virus 3, CRMV is Chinese rape mosaic virus (syn. YMV-15), FMV is frangipani mosaic virus, PMMV is pepper mild mottle virus, RMV is ribgrass mosaic type strain and strain M, RMV-CW is RMV strains C and W, RMV-Ly is RMV strain Ly, RMV-Y is RMV strain Y, ToSNV is tomato stripe necrosis virus, and UMMV is Ullucus mild mottle virus.

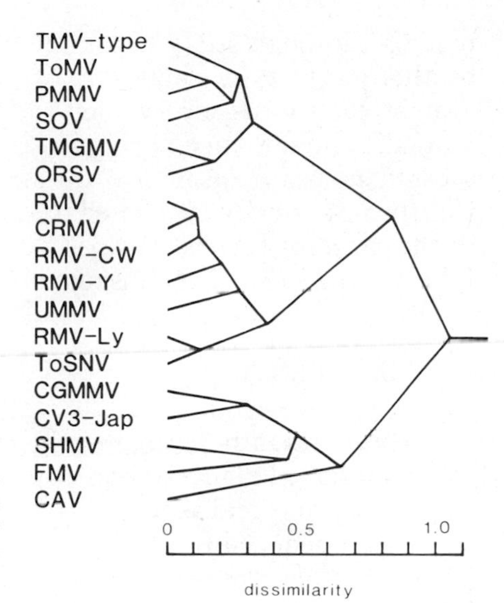

FIGURE 2. Dendrogram computed from the amino acid compositions of tobamovirus coat proteins using the TAXON library of SIRONET; "Canberra" metric, "flexible" sorting strategy.

The "Canberra" metric is

$$\Sigma_s \left[(x_{ik} - x_{jk}) / (x_{ik} + x_{jk}) \right]$$

where x_{ik} and x_{jk} are the molar percentages for proteins i and j for the kth of s amino acids.

Some of the data used in this classification, such as those for the type strain, ToMV, and TMGMV, are from very large numbers of replicate analyses, and hence are more reliably established than those of some of the other viruses; it would be useful to have confirmatory data for individual viruses in the RMV cluster, for frangipani mosaic (FMV), and for *Chara australis* virus (CAV).

This classification suggests that the naming of the cucumber tobamoviruses needs to be revised. The CGMMV (watermelon strain) protein, recently sequenced by Meshi *et al.* (1983), clustered closely with the proteins of Berkeley CV3 and Berkeley and Czechoslovakian isolates of CV4, whereas the Japanese CV3 protein (Tung and Knight, 1972) is clearly distinct though related.

A recent principal coordinates analysis of data on the amino acid composition of 134 plant virus coat proteins by Fauquet *et al.* (1985) confirms the value of using composition data for classification. They found that most of the virus coat proteins fell into groups that correlated closely with the accepted groups of plant viruses (Matthews, 1982).

4. Serological Tests

Even though serological tests were one of the first tests to be used for grouping and comparing viruses, only Van Regenmortel (1975) has made a thorough study of the relationships of a group of tobamoviruses by this method. By using a large number of rabbits to prepare antisera against each virus, he avoided the variability of response of individual animals, and obtained consistent serological differences, which were expressed as "serological differentiation indices." These indices are significantly statistically correlated ($p < 0.001$) with the sequence similarities of the proteins (Gibbs, 1980). The use of serological data for classifying tobamoviruses has been discussed by Van Regenmortel (1982, 1985).

B. "30K" Protein

This protein is thought to be involved in the transport of the virus within infected plants (Ohno *et al.*, 1983).

The amino acid sequences of the 30K proteins of four tobamoviruses have been deduced from nucleotide sequences (Takamatsu *et al.*, 1983). Their relationships are similar to those of the coat proteins of the same viruses. For example, the TMV-type and ToMV proteins are 77% homologous with most of the differences in one region near the C-termini, whereas both are only about 22% homologous to the SHMV protein.

C. 3′ Noncoding Region of the Genome

The 3′ noncoding region of the tobamovirus genome includes the part which forms a tRNA-like amino acid acceptor, which for TMV-type, ToMV, TMGMV, and RMV may be specifically charged with histidine (Carriquiry and Litvak, 1974), but that of SHMV with valine (Beachy *et al.*, 1976). These regions of the TMV-type and ToMV genomes are of 210 and 205 nucleotides, respectively, and over 85% homologous (Takamatsu *et al.*, 1983), whereas that of CGMMV is only 175 nucleotides long though the 3′-terminal 156 nucleotides are about 70% homologous (Meshi *et al.*, 1983). The SHMV 3′ noncoding region is 209 nucleotides long, and shows no significant homology with that of the other tobamoviruses; however, Meshi *et al.* (1981) found that it shows considerable homology (Tables IV and V) with the 3′ noncoding region of turnip yellow mosaic virus (TYMV) (Briand *et al.*, 1977), which, like SHMV, specifically "accepts" valine. This suggests that this part of the SHMV genome was acquired by recombination with a tymovirus; probably not TYMV itself as SHMV is a virus of tropical legumes, whereas TYMV is a virus of cool temperate crucifers.

D. The Genome

Genome sequence homology has been assessed directly by nucleic acid hybridization tests. Palukaitis and Symons (1980) were the first to

TABLE IV. Homology[a] of the 3′-Terminal 150 Nucleotides of the Noncoding Regions of the Genomes of TMV-type, ToMV, SHMV, CGMMV, and Turnip Yellow Mosaic Virus

TMV-type	ToMV	SHMV	CGMMV	TYMV
	28.67	1.51	12.66	2.82
		0.02	11.91	2.10
			2.18	6.89
				0.96

[a] The homologies were assessed by the National Biomedical Research Foundation ALIGN program, a version of the algorithm of Needleman and Wunsch (1970). The homology of each pair of sequences is given as the number of standard deviations between the observed alignment score and the mean of the scores for 50 randomized sequences of the same compositions. Gap penalty 5.

use the genomic RNAs and cDNAs prepared from them for this purpose. They confirmed by this method that the recognized members of the tobamovirus group were distinct, and only demonstrated significant cross-hybridization between TMV-type and ToMV isolates. Van De Walle and Siegel (1982) obtained hybridization results with broader specificity for five tobamoviruses using a similar method to prepare the cDNA, and J. Blok and A. Mackenzie (personal communication) obtained similar results for four tobamoviruses using poly(A) tailing to prime the cDNA synthesis. Table VI summarizes these latter two sets of data, which agree closely with one another (correlation coefficient 0.953, 5 *df*, $p < 0.001$). They also correlate well with the coat protein similarities in Table II; correlation coefficients of 0.786 (8 *df*, $p = 0.01$–0.001) and 0.614 (5 *df*, p = ca. 0.15), respectively.

TABLE V. The Aligned[a] 3′-Terminal Noncoding Sequence of the Genomes of SHMV (Top Line), Turnip Yellow Mosaic Virus (Middle Line), and the Nucleotides They Share (Bottom Line)

```
5'-termini
---------TTAAGAGTCCACGCAAATCGAACTCTAGAACTTATGAACAG
CUCUCUCGAUGCACUCUCCGCUCAUCACGGACACUUCCACCUAAGUUC--
            A    CC C CA   CG AC C    AC  A G  C

TCATGGTTTCCATGCCGTAAAGTTCATAACCGCGAAGTCGCGGCGCCGTC
UCGAUCUUUAAAAUCGUUAGCUCGCCAGUUAGCGAGGUCU-------GUC
 C         A  C   A     C      GCGA G C        G C

AAGACACGACGGTGAGTGGGGAGCATTACCCCCCCAAAACCCTGGGGATA
CCCACACGACAGAUAAUCGGGUGCAACUCCCGCCCCUCUUCCGAGGGUCA
   ACACGAC G  A   GGG GCA   CCC CCC     CC  GGG  A

3'-termini
CAGGGCCCA
UCGGAACCA
  GG  CCA
```

[a] These sequences were aligned by the ALIGN program used to estimate the homologies given in Table IV.

TABLE VI. Hybridization Distance (%) of Tobamovirus Genome RNAs Assessed by Reciprocal Hybridization Tests with cDNA Probes

Set A[a]

TMV-type	ToMV	TMGMV	SHMV	CGMMV
	86.75	97.5	100.25	100
		99	100	100
			97.5	102
				Not tested

[a] Van De Walle and Siegel (1982).

Set B[a]

TMV-type	ToMV	TMGMV	SHMV
	46	84	96.5
		84.5	81.5
			76

[a] J. Blok and A. Mackenzie (personal communication).

The 3′-terminal quarter of the genomes of five tobamoviruses have been sequenced, and their structure (i.e., the positions of the functional parts) determined (Takamatsu *et al.*, 1983) (see Fig. 1 of Okada, this volume). It can be seen that they fall into two classes that correspond to two primary divisions of the coat protein sequence classification (Fig. 1). In the genomes of TMV-type, OM (a strain of TMV-type), and ToMV, the particle assembly site is located in the 30K protein gene and the 30K and coat protein genes do not overlap; in contrast, in the genomes of SHMV and CGMMV the particle assembly site is in the coat protein gene, and the 30K and coat protein genes overlap.

In the 1940s to 1960s, the base ratios of many tobamovirus genome RNAs were analyzed. Computer classification of these data shows no significant clustering of replicate analyses of the same isolate, nor correlation with the groupings obtained from the coat protein composition and sequence studies, though it confirmed the observation of Knight (1954) that the base ratio of cucumber tobamoviruses seems to be distinct from that of all the others, including that of SHMV.

E. Host Range

In the coat protein classifications in Figs. 1 and 2, there seems to be a clustering of tobamoviruses with natural hosts in the Solanaceae, and this may reflect a correlation between the taxonomies of the viruses and their hosts, as has been reported for the tymoviruses (Guy *et al.*, 1984). However, the "solanaceous cluster" includes many natural hosts that are

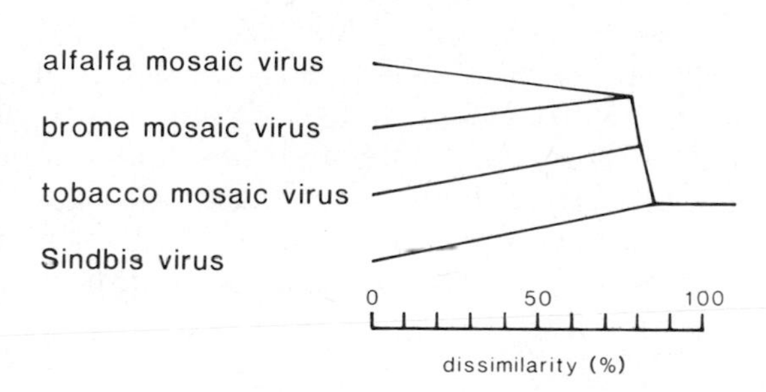

FIGURE 3. Dendrogram calculated from the sequence homologies in Table II of Ahlquist *et al.* (1985). They compared three regions of the genome, and I have "normalized" their three similarity matrices to contribute equally to the single matrix used to calculate the dendrogram; thus, the branching points of the dendrogram at 78.3, 81.7, and 85.75% are relative, not absolute, values.

natives of the Americas; Holmes (1951) reported that more plants from the Americas, than from elsewhere, have genes for hypersensitivity to TMV, and suggested that this indicated a "New World" origin for the virus.

II. RELATIONSHIPS BETWEEN TOBAMOVIRUSES AND OTHER VIRUSES

The recent spate of nucleotide and amino acid sequence data has given surprising substance to earlier speculation about the origins and relationships of different groups of plant viruses. For example, Cornelissen and Bol (1984) have shown that there are clear sequence homologies between the 5′ half of the TMV-type genome and the RNA 1 and RNA 2 genome segments of alfalfa and brome mosaic viruses. These homologies are convincing evidence of a common origin of these genes. Even more surprisingly, this "supergroup" has been shown recently by Ahlquist *et al.* (1985) to include Sindbis alphavirus, a virus of mosquitoes and vertebrates. The sequence homologies between these four virus genomes are around 10–30%, near the limit of detection; Fig. 3 is a dendrogram calculated from estimates from three regions of these genomes.

Other viruses are known to be members of this supergroup including tobacco streak virus (Cornelissen *et al.*, 1984) and cucumber mosaic virus (Rezaian *et al.*, 1984). It is likely that further viruses will be included when more genomes have been sequenced; indeed, Fauquet *et al.* (1985) noted that their ordination based on coat protein composition may provide clues on the membership of such supergroups.

Whether the genes of the tobamoviruses and related viruses are related to other genes or represent a unique class is unknown. McLachlan *et al.* (1980) suggested that TMV coat protein originated by the dimerization of a smaller structure, but there is perhaps insufficient surviving historical information in the primary structure of these proteins to indicate their origins and this will possibly only be clarified when the secondary and tertiary structure of many proteins are known.

Also of interest is the timing of evolutionary events. If the relationships shown in the classifications in Figs. 1–3 reflect the phylogenetic relationships of these viral genes, when did the "branchings" occur? I have suggested (Gibbs, 1980) that the magnitude of the sequence differences among tobamovirus coat proteins suggest that they are at least as old as flowering plants, and thus the common ancestor of the tobamoviruses, "tricornaviruses," and alphaviruses was probably one of those that received the instruction to "Be fruitful and multiply, and replenish the earth, and subdue it; and have dominion over the fish of the sea . . ." (Genesis 1,28).

III. CONCLUSIONS

The definitive tobamoviruses are probably a distinct group of viruses with a common ancestor. They seem to have "speciated" in a simple hierarchical manner. All measures of their relatedness correlate with that shown by the sequences (and composition) of their coat proteins, except that the 3′-terminal noncoding region of the genome of SHMV was probably acquired by recombination with the same region of a tymovirus genome.

REFERENCES

Ahlquist, P., Strauss, E. G., Rice, C. M., Strauss, J. H., Haseloff, J., and Zimmern, D., 1985, Sindbis virus proteins nsP1 and nsP2 contain homology to nonstructural proteins from several RNA plant viruses, *J. Virol.* **53:**536.

Altschuh, D., Reinbolt, J., and Van Regenmortel, M. H. V., 1981, Sequence and antigenic activity of the region 93 to 113 of the coat protein of strain U2 of tobacco mosaic virus, *J. Gen. Virol.* **52:**363.

Bawden, F. C., 1941, *Plant Viruses and Virus Diseases*, 1st ed., Chronica Botanica, Waltham, Mass.

Beachy, R. N., Zaitlin, M., Bruening, G., and Israel, H. W., 1976, A genetic map for the cowpea strain of TMV, *Virology* **73:**498.

Briand, J.-P., Jonard, G., Guilley, H., Richards, K., and Hirth, L., 1977, Nucleotide sequence (n = 159) of the amino-acid-accepting 3′-OH extremity of turnip-yellow-mosaic-virus RNA and the last portion of its coat-protein cistron, *Eur. J. Biochem.* **72:**453.

Carriquiry, E., and Litvak, S., 1974, Further studies on the enzymatic amino-acylation of TMV-RNA by histidine, *FEBS Lett.* **38:**287.

Chester, K. S., 1937, Serological studies of plant viruses, *Phytopathology* **27:**903.

Cornelissen, B. J. C., and Bol, J. F., 1984, Homology between the proteins encoded by tobacco mosaic virus and two tricornaviruses, *Plant Mol. Biol.* **3:**379.

Cornelissen, B. J. C., Janssen, H., Zuidema, D., and Bol, J. F., 1984, Complete nucleotide sequence of tobacco streak virus RNA3, *Nucleic Acids Res.* **12:**1253.

Dayhoff, M. O. (ed.), 1972, *Atlas of Protein Sequence and Structure*, Volume 5, National Biomedical Research Foundation, Washington, D.C.

Fauquet, C., Dejardin, J., and Thouvenel, J.-C., 1985, Evidence that the amino acid composition of the particle proteins of plant viruses is characteristic of the virus group, *Intervirology* (in press).

Gibbs, A., 1969, Plant virus classification, *Adv. Virus Res.* **14:**263.
Gibbs, A. J., 1977, Tobamovirus group, *CMI/AAB Descriptions of Plant Viruses No. 184.*
Gibbs, A., 1980, How ancient are the tobamoviruses?, *Intervirology* **14:**101.
Gibbs, A., and Fenner, F., 1984, Methods for comparing sequence data such as restriction endonuclease maps or nucleotide sequences of viral nucleic acid molecules, *J. Virol. Methods.* **9:**317.
Gibbs, A., and Harrison, B., 1976, *Plant Virology: The Principles,* Arnold, London.
Gibbs, A. J., and McIntyre, G. A., 1970, A method for assessing the size of a protein from its composition: Its use in evaluating data on the size of the protein subunits of plant virus particles, *J. Gen. Virol.* **9:**51.
Gibbs, A., Tien, P., Kang, L.-Y., Tian, Y.-C., and Randles, J., 1982, Classification of several tobamoviruses isolated in China on the basis of the amino acid composition of their virion proteins, *Intervirology* **18:**160.
Guy, P. L., Dale, J. L., Adena, M. A., and Gibbs, A. J., 1984, A taxonomic study of the host ranges of tymoviruses, *Plant Pathol.* **33:**337.
Hennig, B., and Wittmann, H. G., 1972, Tobacco mosaic virus: Mutants and strains, in: *Principles and Techniques in Plant Virology* (C. I. Kado and H. O. Agrawal, eds.), pp. 546–594, Van Nostrand—Reinhold, Princeton, N.J.
Holmes, F. O., 1951, Indications of a New-World origin of tobacco-mosaic virus, *Phytopathology* **41:**341.
Knight, C. A., 1954, The chemical constitution of viruses, *Adv. Virus Res.* **2:**153.
Knight, C. A., 1964, Structural biochemistry of plant viruses, in: *Plant Virology* (M. K. Corbett and H. D. Sisler, eds.), pp. 292–314, University of Florida Press, Gainesville.
Lance, G. N., and Williams, W. T., 1967, Mixed-data classificatory programs. I. Agglomerative systems, *Aust. Comput. J.* **1:**15.
McKinney, H. H., 1929, Mosaic diseases in the Canary Islands, West Africa and Gibraltar, *J. Agric. Res.* **39:**557.
McKinney, H. H., 1944, Genera of the plant viruses, *J. Wash. Acad. Sci.* **34:**139.
McLachlan, A. D., Bloomer, A. C., and Butler, P. J. G., 1980, Structural repeats and evolution of tobacco mosaic virus coat protein and RNA, *J. Mol. Biol.* **136:**203.
Matthews, R. E. F., 1982, Classification and nomenclature of viruses, *Intervirology* **17:**1.
Meshi, T., Ohno, T., Iba, H., and Okada, Y., 1981, Nucleotide sequence of a cloned cDNA copy of TMV (cowpea strain) RNA, including the assembly origin, the coat protein cistron, and the 3′ non-coding region, *Mol. Gen. Genet.* **184:**20.
Meshi, T., Kiyama, R., Ohno, T., and Okada, Y., 1983, Nucleotide sequence of the coat protein cistron and the 3′ noncoding region of cucumber green mottle mosaic virus (watermelon strain) RNA, *Virology* **127:**54.
Needleman, S. B., and Wunsch, C. D., 1970, A general method applicable to the search for similarities in the amino acid sequence of two proteins, *J. Mol. Biol.* **48:**443.
Ohno, T., Takamatsu, N., Meshi, T., Okada, Y., Nishiguchi, M., and Kiho, Y., 1983, Single amino acid substitution in 30K protein of TMV defective in virus transport function, *Virology* **131:**255.
Ohno, T., Aoyagi, M., Yamanashi, Y., Saito, H., Ikawa, S., Meshi, T., and Okada, Y., 1984, Nucleotide sequence of the tobacco mosaic virus (tomato strain) genome and comparison with the common strain genome, *J. Biochem.* (*Tokyo*) **96:**1915.
Palukaitis, P., and Symons, R. H., 1980, Nucleotide sequence homology of thirteen tobamovirus RNAs determined by hybridization analysis with complementary DNA, *Virology* **107:**354.
Rezaian, M. A., Williams, R. H. V., Gordon, F. H. J., and Symons, R. H., 1984, The primary structure of the four RNAs of cucumber mosaic virus, in: *6th Int. Congr. Virol.* Abstr. P42-6.
Rombauts, W., and Fraenkel-Conrat, H., 1968, Artificial histidine-containing mutants of tobacco mosaic virus, *Biochemistry* **7:**3334.
Salaman, R. N., 1933, Protective inoculation against a plant virus, *Nature* (*London*) **131:**468.

Shen, X.-R., Wu, A.-Z., Dai, R.-M., Ching, E.-D., and Sun, Y.-K., 1982, Study on the tomato stripe necrosis virus, *Acta Biochim. Biophys. Sin.* **14**:257.

Shikata, E., 1977, Tobamovirus (tobacco mosaic virus) group, in: *The Atlas of Insect and Plant Viruses* (K. Maramorosch, ed.), pp. 237–255, Academic Press, New York.

Siegel, A., and Wildman, S. G., 1954, Some natural relationships among strains of tobacco mosaic virus, *Phytopathology* **44**:277.

Takamatsu, N., Ohno, T., Meshi, T., and Okada, Y., 1983, Molecular cloning and nucleotide sequence of the 30K and the coat protein cistron of TMV (tomato strain) genome, *Nucleic Acids Res.* **11**:3767.

Thung, T. H., 1931, Smetstof en plantencel bij enkele virusziekten van de Tabaksplant, *Handel. 6 Ned.-Indisch. Natuurwetensch. Congr. Bandoeng. Java* pp. 450–463.

Tremaine, J. H., and Argyle, E., 1970, Cluster analysis of viral proteins, *Phytopathology* **60**:654.

Tsugita, A., 1962, The proteins of mutants of TMV: Classification of spontaneous and chemically evoked strains, *J. Mol. Biol.* **5**:293.

Tsugita, A., and Fraenkel-Conrat, H., 1960, The composition of proteins of chemically evoked mutants of tobacco mosaic virus, *Proc. Natl. Acad. Sci. USA* **46**:636.

Tsugita, A., and Fraenkel-Conrat, H., 1962, The composition of proteins of chemically evoked mutants of TMV, *J. Mol. Biol.* **4**:73.

Tung, J.-S., and Knight, C. A., 1972, The coat protein subunits of cucumber viruses 3 and 4 and a comparison of methods for determining their molecular weights, *Virology* **48**:574.

Van De Walle, M. J., and Siegel, A., 1982, Relationships between strains of tobacco mosaic virus and other selected plant viruses, *Phytopathology* **72**:390.

Van Regenmortel, M. H. V., 1967, Serological studies on naturally occurring strains and chemically induced mutants of tobacco mosaic virus, *Virology* **31**:467.

Van Regenmortel, M. H. V., 1975, Antigenic relationships between strains of tobacco mosaic virus, *Virology* **64**:415.

Van Regenmortel, M. H. V., 1981, Tobamoviruses, in: *Handbook of Plant Virus Infections and Comparative Diagnosis* (E. Kustak, ed.), pp. 541–564, Elsevier/North-Holland, Amsterdam.

Van Regenmortel, M. H. V., 1982, *Serology and Immunochemistry of Plant Viruses*, Academic Press, New York.

Van Regenmortel, M. H. V., 1985, Antigenic structure of plant viruses, in: *Immunochemistry of Viruses: The Basis for Serodiagnosis and Vaccines* (M. H. V. Van Regenmortel and A. R. Neurath, eds.), pp. 467–478, Elsevier, Amsterdam.

Wetter, C., Conti, M., Altschuh, D., Tabillion, R., and Van Regenmortel, M. H. V., 1984, Pepper mild mottle virus, a tobamovirus infecting pepper cultivars in Sicily, *Phytopathology* **74**:405.

Wittmann, H. G., 1960, Composition of the tryptic peptides of chemically induced and spontaneous mutants of tobacco mosaic virus, *Virology* **12**:609.

CHAPTER 9

Tomato Mosaic Virus

ALAN A. BRUNT

I. INTRODUCTION

Tomato mosaic virus (ToMV), which has occurred in tomato (*Lycopersicon esculentum*) in the United States (Clinton, 1909; Allard, 1916) and the Netherlands (Westerdijk, 1910) for over 75 years, now probably occurs wherever tomato crops are grown (Broadbent, 1976; Hollings and Huttinga, 1976). Because of its close serological relationship to tobacco mosaic virus (TMV), ToMV is still often considered to be a strain of TMV (e.g., Van Regenmortel, 1975, 1981; Van De Walle and Siegel, 1976; Dawson *et al.*, 1979; Kukla *et al.*, 1979; Fraser and Loughlin, 1980; Fraser *et al.*, 1980; Burgyán and Gaborjanyi, 1984). However, because the two can be readily distinguished by differences in host range, serological affinities, and protein compositions, ToMV has during the past 15 years been increasingly recognized as a distinct virus (e.g., Harrison *et al.*, 1971; Fenner, 1976; Hollings and Huttinga, 1976; Gibbs, 1977). I briefly review here its natural occurrence, major properties, affinities, epidemiology, and control.

II. THE VIRUS AND ITS STRAINS

The major properties of ToMV and TMV are so similar, and their serological relationship so close, that ToMV hardly justifies recognition as a distinct virus (Hamilton *et al.*, 1981). Nevertheless, it has since 1971 been increasingly accepted as a distinct virus, mainly because it fails to protect tomato plants from subsequent infection by TMV (e.g., MacNeill and Ismen, 1960; Rast, 1975) and is otherwise biologically distinct; more-

ALAN A. BRUNT • Glasshouse Crops Research Institute, Littlehampton, West Sussex BN17 6LP, England.

over, it differs in its serological affinities (see Section IV) and the composition and sequence of amino acids in its coat protein subunits (see Section III).

Extensive surveys in several countries have established that TMV occurs comparatively rarely in tomato crops (MacNeill, 1962; Broadbent, 1962; Komuro *et al.*, 1966; van Winckel, 1967a,b; Rast, 1975), presumably because it replicates and spreads within plants much slower than ToMV (Komuro *et al.*, 1966; Jensen, 1969; Tomaru *et al.*, 1970). The two viruses were initially distinguished by the symptoms induced in some differential hosts (Termohlen and van Dorst, 1959; Broadbent, 1962; Komuro *et al.*, 1966; Wang and Knight, 1967; Rast, 1973; Hollings and Huttinga, 1976); thus, unlike TMV which induces systemic leaf chlorosis, ToMV induces only necrotic lesions in inoculated leaves of *Datura stramonium, Nicotiana rustica, N. sylvestris, N. tabacum* cv. White Burley (Dutch A type), *Petunia hybrida, P. nyctagyniflora,* and *Physalis ixocarpa*; conversely, in *Solanum giganteum* ToMV induces conspicuous systemic infection whereas TMV induces only local lesions. In addition, although typical isolates of ToMV fail to infect *Phaseolus vulgaris*, TMV induces red necrotic lesions in inoculated primary leaves. However, tests with a wide range of isolates have shown that, in some hosts, the Dahlemense and Ohio strains I, II, and III induce symptoms more like those usually attributable to TMV (Hollings and Huttinga, 1976; Dawson *et al.*, 1979). Nevertheless, both ToMV and TMV are readily distinguishable from other tobamoviruses (Gibbs, 1977; Wetter, 1984).

Although tomato is by far its most important host, ToMV also occurs naturally in other species. It is one of at least five tobamoviruses infecting peppers (*Capsicum annuum* and *C. frutescens*) in widely scattered geographical locations; it is known to occur in crops in the United States (Miller and Thornberry, 1958; Knight *et al.*, 1962; Wetter, 1984), the United Kingdom (Fletcher, 1963), Italy (Conti and Masenga, 1977), France (Van Regenmortel, in Wetter, 1984), and Hungary (Csillery *et al.*, 1983); due to past confusion with other tobamoviruses, it may also occur elsewhere. ToMV is also known to infect potato (*Solanum tuberosum*) in Hungary (Juretić *et al.*, 1977; Horvath *et al.*, 1978), cherry (*Prunus avium* and *P. cerasus*), pear (*Pyrus communis*), apple (*Malus sylvestris*), and grape (*Vitis vinifera*) in the United States (Gilmer and Wilks, 1967a,b, 1968; Kirkpatrick and Lindner, 1964), and to induce chlorotic leaf spotting in pear (*Pyrus calleryana*) in the Democratic Republic of Germany (Opel *et al.*, 1969). It can also induce severe stunting, distortion, and necrosis of *Chenopodium murale* in the United States (Bald and Paulus, 1963).

Most isolates of ToMV are serologically closely related or even indistinguishable, have similar physicochemical properties, and differ little in the composition and sequences of their RNAs and proteins (Wang and Knight, 1967; Mosch *et al.*, 1973; Dawson *et al.*, 1975; Hollings and Huttinga, 1976). Nevertheless, many strains have been differentiated by the symptoms they induce in tomato and/or other species, and by their abilities to infect tomato genotypes containing genes for resistance (McRitchie and Alexander, 1963; Cirulli and Alexander, 1969; Pelham, 1972).

A. Symptomatologically Distinct Strains

Although the development of symptoms in tomato is dependent on environmental conditions (mainly light intensity, temperature, and day length), duration of infection, and age and relative tolerance of the cultivar (e.g., Ainsworth and Selman, 1936; Broadbent and Cooper, 1964; Rast, 1967, 1975; Paludan, 1968; Fletcher and McNeill, 1971; Crill *et al.*, 1973), strains have been differentiated by the symptoms they induce in intolerant cultivars. The better-known symptomatologically distinct strains include:

1. Aucuba or Yellow Mosaic

This characteristically induces discoloration of fruits and the development of bright yellow mottling, distortion, and necrosis of leaves which, at maturity, may be almost white. This long-recognized strain occurs sporadically, and often only infects a few plants within a crop (Bewley, 1923; Henderson Smith, 1928; Kunkel, 1934).

2. Common Mosaic

This very commonly occurring strain induces a light and dark green leaf mosaic. Malformed leaves ("fern leaves") also commonly occur during early stages of infection, especially on plants grown under low light conditions (Rast, 1975).

3. Corky Ringspot

This strain, first detected in 1982 in cvs. Jackpot Blazer and Valerie in California, induces mild to moderately severe leaf chlorosis and corky rings and lines on small and unevenly ripened fruits (Mayhew *et al.*, 1984).

4. Crusty Fruit

Some isolates inducing yellow ringspot leaf symptoms in the Netherlands also induce on fruits the development of complete or incomplete rings of superficial corky tissue which, due to differential growth rates of normal and affected areas, cause fruits to burst (Rast, 1975).

5. Dahlemense

This isolate, which has been studied extensively, induces yellow leaf mottling in tomato (Melchers *et al.*, 1940).

6. Enation

Initially designated "tomato enation mosaic virus" (Ainsworth, 1937), this strain characteristically induces at all seasons enations on abaxial leaf surfaces, especially those with "fern leaf" symptoms. It also causes abnormal development and setting of flowers and thus heavy yield losses (Rast, 1967).

7. Black Fleck

This occurs only very rarely in Britain and, in addition to chlorosis, induces black necrotic flecks on leaves (Hollings, 1957).

8. M11–16

This is best known for its former widespread use in the mild strain protection of nonresistant tomato cultivars. Obtained in 1968 as a nitrous acid mutant of a virulent Dutch strain, it induces virtually symptomless infection in inoculated tomato plants which are then often efficiently protected from the effects of more severe isolates (Rast, 1972).

9. Rosette

This unusually virulent strain causes severe distortion and stunting of infected plants which, when very severe, resembles that caused by herbicides (Price and Fenne, 1951).

10. Summer Necrosis

This strain, which occurs rarely in the Netherlands, can when temperatures exceed 30°C induce leaf chlorosis as well as necrosis of petioles and stems. The necrosis is sometimes lethal, but surviving plants remain stunted and have leaves with chlorotic rings; fruits may also be malformed (Rast, 1975).

11. Yellow Atypical Mosaic

This isolate, studied extensively in the United States, was separated by Knight *et al.* (1962) from a mixture of two tobamoviruses isolated from naturally infected tomato (Miller and Thornberry, 1958).

12. Yellow Ringspot

This strain, which occurs commonly in the Netherlands, sometimes induces yellow ringspots in the leaves of tomato and some related species (Rast, 1965, 1975). Leaf symptoms, however, are usually mild, fruits are

TABLE I. Nomenclature of Virus Strains (or Pathotypes) According to Their Interaction with Host Resistance Genes[a,b]

ToMV strain	Resistance genes in host								
	0	Tm-1/+	Tm-1/Tm-1	Tm-2/+	Tm-2/Tm-2	$Tm\text{-}2^2/+$	$Tm\text{-}2^2/Tm\text{-}2^2$	Tm-1/+ Tm-2/+	Tm-1/+ $Tm\text{-}2/Tm\text{-}2^2$
0	S	T	T	R*	R	R*	R	R	R
1	S	S	S	R*	R	R*	R	R*	R
2	S	T	T	S	S	R*	R	R	R
1.2	S	S	S	S	S	R*	R	S	R*
1.2^2	S	S	S	R	R	S	S	R	R
$1.2.2^2$	S	S	S	S	S	S	S	S	S

[a] Data mainly from Hall and Bowes (1981).
[b] T = tolerance reaction, mild mosaic symptoms, little or no effect on growth. S = normal susceptible reaction. R = normal resistance reaction, no symptoms. R* = resistance reaction; a systemic hypersensitive necrotic reaction *may* occur.

rarely affected, and yield small, if any losses. Affected plants are often restricted to small areas of glasshouses where temperatures are abnormally high (Rast, 1965, 1975).

13. Winter Necrosis

This strain initially induces leaf symptoms similar to those of the common mosaic strain; from late autumn to early spring, however, it can cause necrosis of leaf bases, resulting in subsequent wilting and death of leaves. Immature fruits may develop irregular necrotic blotches (Rast, 1975).

B. Strains (or Pathotypes) Differentiated by Host Resistance Genes

Classification of isolates by their ability to induce symptoms in *Lycopersicon* spp. (Clayberg *et al.*, 1960) or in near-isogenic lines of cv. Craigella with individual or combinations of the resistance genes Tm-1, Tm-2, or Tm-2^2 (Pelham, 1972) has proved of particular interest to breeders. Unfortunately, their classification by such criteria is not correlated with serological affinities, reactions of tomato and other indicator species, or physicochemical properties (Rast, 1975; Hollings and Huttinga, 1976; Dawson *et al.*, 1979). Although recognized by breeders as strains, such pathogenically distinct isolates are analogous to fungal *formae speciales* or races and are perhaps best considered to be pathotypes. For convenience, however, the earlier terminology is used here, and the designation of strains according to their ability to overcome different resistance genes is shown in Table I; thus, strain 0 is unable to overcome any of the resistance genes, strain 1 can overcome gene Tm-1, strain 2 gene Tm-2, and strain 2^2 gene Tm-2^2. Very rarely occurring strains such as 1.2 can overcome genes Tm-1 and Tm-2 when present in the homozygous or heterozygous state, strain 1.2^2 overcomes genes Tm-1 and Tm-2^2 together and separately, and strain 1.2.2^2 overcomes all genes. Of the strains described previously from Ohio (McRitchie and Alexander, 1963; Cirulli and Alexander, 1969; Alexander, 1971), strains I and II are equivalent to Pelham (1972) strain 0, strain III to Pelham strain 1, strain IV to Pelham strain 2, and strain V to Pelham strain 1.2 (Hollings and Huttinga, 1976; Dawson *et al.*, 1975).

III. PROPERTIES AND COMPOSITION

A. Properties

1. In Sap and Leaf Tissue

Like morphologically similar viruses, ToMV is very stable *in vitro*. In tomato leaf extracts, the virus is usually still infective after 10 min at

85 but not 90°C, after dilution to 10^{-5} but not 10^{-6}, after many months at laboratory temperature (18–24°C) and after several years at 2°C (Hollings and Huttinga, 1976), and in desiccated tomato leaf tissue after 24 years at room temperature (Caldwell, 1959). Losses or changes in virulence that are liable to occur in stored sap (Rast, 1975) can be prevented by preserving cultures as lyophilized sap (Hollings and Huttinga, 1976).

2. In Purified Preparations

Like other tobamoviruses, ToMV has rod-shaped particles mostly measuring 18 × 300 nm (Caspar and Holmes, 1969) with an isoelectric precipitation point of pH 4.5–4.6 (Oster, 1951) and an electrophoretic mobility of 0.97 (μm/sec)/(V/cm) (Kramer and Wittmann, 1958). In homogeneous preparations they sediment as a single component with a buoyant density in cesium chloride of 1.325 g/cm^3 and a sedimentation coefficient ($s^0_{20,w}$) of 177–190 S (Mosch *et al.*, 1973; Hollings and Huttinga, 1976); some particles aggregate end-to-end during purification so that preparations also often contain dimers and trimers with sedimentation coefficients of 240 and 330 S, respectively (Kassanis *et al.*, 1972).

Purified virus preparations, with maximum and minimum absorption at 262 and 249 nm and with A_{260}/A_{280} and A_{max}/A_{min} ratios of 1.17 and 1.10, respectively, are typical of a nucleoprotein contain ca. 5% nucleic acid.

The structural order of RNA and protein in particles of the Dahlemense strain, as shown by X-ray diffraction, is essentially similar to that of tobacco mild green mosaic virus (TMGMV), cucumber green mottle mosaic virus (CGMMV), and sunn-hemp mosaic virus (SHMV) (Franklin, 1956; Holmes and Franklin, 1958; Caspar, 1963). Thus, particles have a maximum diameter of 18 nm and a central hollow core ca. 4 nm in diameter with the RNA helix 3–5 nm from the axis enclosed by coat protein folded in coaxial cylindrical order (Caspar, 1963). Unlike TMV, in which the protein subunits form a uniform surface structure, ToMV (Dahlemense strain) has groups of four subunits in closer proximity and interstices between adjacent groups (Caspar, 1963).

B. Composition

1. Protein

Like that of similar viruses, the protein of ToMV is composed of a single polypeptide species with a molecular weight of 17,640 when determined by amino acid analysis (Wittmann-Liebold and Wittmann, 1967) and ca. 21,000 when estimated by polyacrylamide gel electrophoresis (R. J. Barton, personal communication).

The amino acid composition of many ToMV isolates has been determined (Table II). That of the strain previously described as "tomato yellow atypical mosaic virus" (YTAMV) was initially found to be similar to that of TMV; however, it contains one methionine residue and has C-terminal serine, whereas that of TMV contains no methionine and has C-terminal threonine (Knight *et al.*, 1962). The protein subunit composition of many other ToMV strains has since been shown to be remarkably similar to that of YTAMV. Thus, the Dahlemense strain differs by only one amino acid residue (Tsugita, 1962; Wittmann-Liebold and Wittmann, 1963); moreover, of 13 isolates from widely different geographical locations (United States, Australia, Netherlands, Japan, Germany, Philippines, and South Africa), five had compositions identical to that of YTAMV, one differed by only one exchange, and another differed by two exchanges (Wang and Knight, 1967). Similarly, the composition of the polypeptide of a Japanese mild isolate differed little from that of the typical strain from which it was derived (Oshima *et al.*, 1976; Nishiguchi *et al.* 1984).

Dawson *et al.* (1975) later compared the composition of Alexander's (McRitchie and Alexander, 1963; Cirulli and Alexander, 1969) five strains of ToMV which, unlike those studied previously by Wang and Knight (1967), were selected for their ability to overcome specific host genes for resistance to infection. The results obtained are confusing in that the composition of only one strain (Alexander strain IV) was similar to that of ToMV; of the remaining four strains, two (Alexander strains I and II, both Pelham strain 0) had compositions similar to TMV, and those of the other two (Alexander strains III and V) were significantly different from both ToMV and TMV. The reported compositions of strains III and IV are essentially similar to those recorded previously by Mosch *et al.* (1973). In further studies, Dawson *et al.* (1979) demonstrated that the differing responses of a wide range of ToMV strains to various resistance genes in tomato were not correlated with differences in their amino acid compositions.

The sequence of amino acids in the polypeptide of ToMV (Dahlemense strain), first determined almost 20 years ago (Wittmann and Wittmann-Liebold, 1967), differs from that of TMV in only 30 (18%) of its 158 positions (Table III); this suggests that the specifying genes must differ by at least 36 (7.6%) of the 474 nucleotides necessary to code for coat protein (Van De Walle and Siegel, 1976). Polypeptide mapping after treatment with carboxypeptidase A indicated that the protein subunits of 13 ToMV isolates from widely scattered geographical locations had similar amino acid sequences; unlike that of TMV, those of all ToMV strains contained serine instead of threonine at the C-termini (Wang and Knight, 1967).

In addition to coat protein, ToMV RNA encodes three nonstructural proteins of M_r 30,000 (30K), 130,000 (130K), and 165,000 (165K), respectively, the four proteins accounting for ca. 85% of its coding capacity (Leonard and Zaitlin, 1982). The coat protein and the two larger non-

TABLE II. The Amino Acid Compositions of Tomato Mosaic Virus Strains and Tobacco Mosaic Virus

		ToMV strains							
	TMV[a]	Dahlemense[b]	TAMV[c]	Field[d] isolates	Alexander I Pelham 0	Alexander II Pelham 0	Alexander III Pelham 1	Alexander IV Pelham 2	Alexander V
Lys	2	2	2	2	2	2	2	2	3
His	0	0	0	0	0	0	0	0	0
Arg	11	9	9	9	11	11	10	9	9
Asp	18	17	18	18–19	18	18	19	18	18
Thr	16	17	17	16–17	16	16	17	16	17
Ser	16	16	15	15–16	16	15	15	16	15
Glu	16	19	19	19	16	16	17	19	17
Pro	8	8	8	8	8	8	7	8	8
Gly	6	6	6	6	6	6	6	6	6
Ala	14	11	11	11–12	14	15	15	11	15
Cys	1	1	1	1	1	1	1	1	1
Val	14	15	15	14–15	14	14	14	15	14
Met	0	1	1	1	0	0	0	1	0
Ile	9	7	7	7	9	9	8	7	8
Leu	12	13	13	13	12	12	12	13	12
Tyr	4	5	5	5	4	4	4	5	4
Phe	8	8	8	8	8	8	8	8	8
Trp	3	3	3	3	3	3	3	3	3
Total	158	158	158	158	158	158	158	158	158

Data from: [a] Anderer (1983), [b] Wittmann-Liebold and Wittmann (1963), [c] Tsugita (1962), [d] Wang and Knight (1967), [e] Dawson *et al.* (1975).

TABLE III. Amino Acid Sequences[a] of Tomato Mosaic Virus (Dahlemense Strain) and Tobacco Mosaic Virus (Vulgare Strain)

1	AcSer	Tyr	Ser	Ile	Thr	Ser	Pro	Ser	Gln	Phe	Val	Phe	Leu	Ser	Ser	Val	Trp	Ala	Asp	Pro
						Thr										Ala				
21	–Ile	Glu	Leu	Leu	Asn	Val	Cys	Thr	Ser	Ser	Leu	Gly	Asn	Gln	Phe	Gln	Thr	Gln	Gln	Ala
				Ile		Leu			Asn	Ala										
41	–Arg	Thr	Thr	Val	Gln	Gln	Gln	Phe	Ser	Glu	Val	Trp	Lys	Pro	Phe	Pro	Gln	Ser	Thr	Val
			Val			Arg				Gln					Ser			Val		
61	–Arg	Phe	Pro	Gly	Asp	Val	Tyr	Lys	Val	Tyr	Arg	Tyr	Asn	Ala	Val	Leu	Asp	Pro	Leu	Ile
				Asp	Ser	Asp	Phe													Val
81	–Thr	Ala	Leu	Leu	Gly	Thr	Phe	Asp	Thr	Arg	Asn	Arg	Ile	Ile	Glu	Val	Glu	Asn	Gln	Gln
						Ala														Ala
101	–Ser	Pro	Thr	Thr	Ala	Glu	Thr	Leu	Asp	Ala	Thr	Arg	Arg	Val	Asp	Asp	Ala	Thr	Val	Ala
	Asn																			
121	Ile	Arg	Ser	Ala	Ile	Asn	Asn	Leu	Val	Asn	Glu	Leu	Val	Arg	Gly	Thr	Gly	Leu	Tyr	Asn
									Ile	Val			Ile					Ser		
141	–Gln	Asn	Thr	Phe	Glu	Ser	Met	Ser	Gly	Leu	Val	Trp	Thr	Ser	Ala	Pro	Ala	Ser		
	Arg	Ser	Ser												Gly			Thr		

[a] Upper sequence, ToMV; lower sequence, TMV showing positional differences only. Data from Wittmann-Liebold and Wittmann (1967).

structural proteins induced by a Japanese isolate of ToMV are closely similar to those of a mutant strain which is not translocated from cell to cell at 32°C. The 30K protein of the mutant, however, differs slightly from that of the parent strain, which provides further evidence that the protein is implicated in virus translocation (Leonard and Zaitlin, 1982).

The sequences of amino acids in the 30K protein of ToMV and other tobamoviruses have been determined and shown to have highly conserved common sequences (Meshi *et al.*, 1982a,b, 1983b; Ohno *et al.*, 1983), one of which contains the amino acid substitution (proline for serine) of the parental and mutant ToMV strains (Ohno *et al.*, 1983). The predicted structure of the 30K protein indicates that a conserved proline residue occurs in the β-turn of its structure (Ohno *et al.*, 1983). This substitution of proline for serine may result in decreased rigidity of the mutant protein, thus possibly explaining its temperature sensitivity and its presumed function in virus transport (Ohno *et al.*, 1983).

2. Nucleic Acid

ToMV particles each contain a molecule of ssRNA of 6384 nucleotides (total M_r ca. 2.2×10^6) enclosed within the helically arranged coat protein subunits. Base ratio analyses indicated that, in containing relatively high amounts of adenine (28%) and guanine (30%) and relatively low amounts of cytosine (19%) and uracil (23%), ToMV RNA differed little from that of TMV (Knight *et al.*, 1962; Mosch *et al.*, 1973). It has long been known that ToMV RNA contains adenine at its 3′ and 5′-termini (Sugiyama and Fraenkel-Conrat, 1963) and, like TMGMV RNA, has long guanosine-free sequences (Kukla *et al.*, 1979).

The sequence of 1614 and 278 nucleotides at the 3′- and 5′-termini of the RNA of ToMV strain L were determined by Takamatsu *et al.* (1983). Like those of TMV, the cistrons for the coat and 30K proteins are located at residues 203–682 and 685–1479 from the 3′-terminus; the RNAs of the two viruses are also homologous in the 3′ noncoding region, the assembly origin, and the 5′ flanking region of the 30K protein cistron (Takamatsu *et al.*, 1983). The assembly origin of both viruses is located 800–1000 nucleotides from the 3′-end of their genomic RNAs (Fukuda *et al.*, 1980, 1981).

The sequence of the 6384 nucleotides in ToMV RNA was recently completed with the sequencing of the remaining 4492 nucleotides (Ohno *et al.*, 1984). The cistrons for the 130K, 180K, 30K, and coat proteins were shown to be located at residues 72–3442, 72–4922, 4906–5700, and 5703–6182, respectively.

Although the nucleotide sequences of ToMV and TMV show ca. 80% homology, they differ in the regions coding for the N-terminal portion of the 130K/180K proteins and the C-terminal portion of the 30K protein; the latter results in the 30K proteins of ToMV and TMV having 263 and 267 amino acids, respectively (Ohno *et al.*, 1984).

TABLE IV. Serological Differentiation Indices[a] between ToMV and Four Other Tobamoviruses[b]

	Antiserum to				
Virus	ToMV	TMV	TMGMV	RMV	CGMMV(CV_4)
ToMV	0	1.2	1.8	4.3	7.0
TMV	1.1	0	2.8	2.1	7.5
TMGMV	1.9	2.5	0	3.8	6.7
RMV	3.7	2.1	5.1	0	NT
CV_4	6.1	3.8	5.8	4.6	0

[a] The difference between homologous and heterologous titers in tube precipitin tests expressed as negative $\log_2$.
[b] Data from Van Regenmortel (1975). Each value obtained from a minimum of 40 serum samples collected from at least five rabbits.

IV. AFFINITIES

A. Serological Relationships

ToMV has long been known to be closely related to, but serologically distinguishable from TMV (Chesters, 1936; Van Regenmortel, 1967; Van Regenmortel and von Wechmar, 1974; Hollings and Huttinga, 1976). More recently, however, the extent of the relationship between ToMV and other tobamoviruses has been carefully reexamined using 40 sequential samples of antiserum to each virus from at least five rabbits (Van Regenmortel, 1975). The serological differentiation indices (SDI) between ToMV and TMV, TMGMV, ribgrass mosaic virus (RMV), and CGMMV, expressed as the mean of the values obtained in homologous and heterologous reactions (Table IV), confirm that ToMV is serologically closely related to TMV and TMGMV, but only very distantly to RMV and CGMMV. The extent of the relationship is closely correlated with differences in the amino acid sequence of the individual viruses (Van Regenmortel, 1975). Thus, the sequence of ToMV differs from those of TMV, TMGMV, and RMV by 18, 30, and 50%, respectively (Hennig and Wittmann, 1972) and from that of CGMMV to a larger extent (Kurachi *et al.*, 1972).

The relationship between ToMV and three other tobamoviruses has also been determined quantitatively by immunosorbent electron microscopy (Nicolaieff and Van Regenmortel, 1980). Electron microscope grids coated with diluted TMV antiserum trapped per unit area almost equal numbers of TMV and ToMV particles from preparations of each containing 0.5 μg virus/ml, but far fewer particles of RMV and TMGMV. The efficacy of trapping is thus clearly correlated with the degree of affinity between TMV and the three viruses.

Similarly, enzyme-linked immunosorbent assays (ELISA) showed that ToMV and TMV are closely related although only distantly so to RMV, CGMMV, and SHMV (Van Regenmortel and Burckard, 1980). When calculated from such assays, the SDI values between ToMV and TMV, TMGMV, RMV, and CGMMV are similar to those obtained in tube precipitin tests (Jaegle and Van Regenmortel, 1985).

Monoclonal antibodies to TMV, although failing to bind with tobamoviruses differing by SDI values of two or more, can in indirect ELISA differentiate TMV from its strains and from ToMV, TMGMV, RMV, and odontoglossum ringspot virus (ORSV). Thus, of nine antibody clones obtained, four failed to react with ToMV (e.g., clone 20), two reacted equally with TMV and ToMV (e.g., clone 19), and three reacted better with TMV than ToMV (e.g., clone 21); thus, the two closely related viruses can be distinguished by their differential reactions with seven clones (Briand *et al.*, 1982).

Despite the close serological relationship between ToMV and TMV, prior inoculation of tomato plants with TMV fails to prevent subsequent infection by ToMV (Broadbent and Winsor, 1964); conversely, ToMV fails to protect tobacco plants from subsequent infection by TMV (Jensen, 1969). However, mild strains of ToMV, although not preventing infection by more virulent strains, often prevent the development of severe symptoms (Boyle and Bergman, 1967; Rast, 1972).

B. Molecular Hybridization Analyses

More recently, the degree of relationship between ToMV and other tobamoviruses has been investigated by molecular hybridization techniques. Competitive hybridization procedures initially failed to detect any homology between the RNA of ToMV and those of TMV, TMGMV, SHMV, CGMMV, and RMV (Van De Walle and Siegel, 1976) or, incidentally, between TMV and SHMV (Zaitlin *et al.*, 1977). These failures, however, are now attributed to deficiencies of the earlier used procedure (Palukaitis and Symons, 1980).

In more sensitive hybridization analyses using cDNA techniques, ToMV RNA showed 14% homology with TMV cDNA under stringent conditions (in 0.12 M NaCl) for hybridization and S1 nuclease assay, and 35% under less stringent conditions (in 0.56 M NaCl); in similar tests, ToMV RNA failed to hybridize with cDNAs of TMGMV, FMV, SHMV, or CGMMV (Palukaitis and Symons, 1980). ToMV cDNA initially failed to hybridize with TMV RNA, but under less stringent conditions it showed 31% homology with TMV RNA, while TMV cDNA showed 47% homology with ToMV RNA; these results indicate that the RNAs have many short sequences in common and, due to instability of the hybrids, that the viruses are only distantly related. Under these conditions, ToMV cDNA showed no significant homology with RNA of SHMV or CGMMV, although some with that of TMGMV, FMV, and ORSV (Palukaitis and Symons, 1980).

In similar tests, RNA from two strains of ToMV later showed ca.

15% homology with TMV cDNA; similar results were obtained in reciprocal tests using ToMV cDNA, but neither ToMV cDNA nor TMV cDNA showed any homology with RNA from TMGMV, SHMV, CGMMV, and probably also RMV (Van De Walle and Siegel, 1982).

C. Amino Acid Analyses

The affinities of individual tobamoviruses, estimated from the amino acid composition of their coat proteins, is correlated with the degree of relationship determined by other methods (Gibbs, 1977, 1980). The percentage difference between the sequences of the coat proteins of seven tobamoviruses allow them to be arranged in a single dendrogram which probably reflects the phylogeny of their proteins (Gibbs, 1977) (see chapter 8).

V. INTRACELLULAR OCCURRENCE

Simple infectivity assays have shown that, with the possible exception of embryos and meristems, all parts of ToMV-infected plants contain virus (Taylor *et al.*, 1961; Broadbent, 1965c).

ToMV, like other tobamoviruses (Christie and Edwardson, 1977), induces the formation of several types of intracellular inclusions which can be readily detected by either optical (Henderson Smith, 1930; Sheffield, 1931; Kassanis and Sheffield, 1941; Miličić and Juretić, 1971) or electron microscopy (Warmke, 1967, 1968, 1969; Allen, 1969). So-called amorphous inclusions (Henderson Smith, 1930; Sheffield, 1931) are usually present during early stages of infection and often appear to be vacuolated (Henderson Smith, 1930; Sheffield, 1931). Electron microscopy, however, indicates that such inclusions consist of massive accumulations of angled-layer aggregates (Warmke, 1967, 1968, 1969). Individual aggregates contain layers of 12–16 particles, adjacent layers being at ca. 60° to each other. In longitudinal section, such aggregates appear as alternating layers of particles in transverse and longitudinal section, and in other planes as cross-hatched structures. Critical examination of serial sections of inclusions has shown that there is a 6.5° shift of particles in successive sections, results indicating that the inclusion is in the form of a right-handed helix (Warmke, 1974). Although angled-layer aggregates were initially considered to be characteristic of ToMV (Warmke, 1968), they are also frequently induced by pepper mild mottle virus (Herold and Munz, 1967; Wetter *et al.*, 1984).

Hexagonal crystalline inclusions, although readily disrupted, can with suitable preparative techniques also be detected by optical microscopy (Sheffield, 1931; Kassanis and Sheffield, 1941). Electron microscopy has shown that these inclusions consist of sheets of particles which are aligned closely side by side, and end to end, to appear at higher magnification as "herringbone" structures (Warmke, 1968, 1969).

Other inclusions, previously described as spikes, spindles, and needles (Kassanis and Sheffield, 1941), are now considered to be bundles of particles in paracrystalline array (Warmke, 1968, 1969).

X-bodies, which also occur in ToMV-infected cells, consist of comparatively few virus particles entrapped in aggregated cellular components such as ribosomes and protein (Warmke, 1969).

VI. EPIDEMIOLOGY AND CONTROL

A. Epidemiology

The epidemiology of ToMV has been intensively investigated in numerous countries and the main factors affecting virus spread have been clearly established (Broadbent, 1961, 1976). Virus can remain infective for many months on the testae of seed collected from infected plants, and transmitted mechanically therefrom to very young seedlings during transplantation (Broadbent, 1965c; van Winckel, 1965, 1968). The virus can also persist for many months on glasshouse structures, and in debris of infected plants in soils from which it is eradicated with great difficulty by sterilization (Johnson and Ogden, 1929; Jones and Burnett, 1935; Nitzany, 1960; Broadbent, 1965a; Broadbent *et al.*, 1965; Fletcher, 1969). It is unlikely that weeds act as sources of infection of ToMV (Broadbent, 1976), but virus has been detected occasionally in water used for irrigating crops in the Netherlands (van Dorst, 1970). From such primary foci of infection, the virus can be subsequently spread within crops by insects (Bradley and Harris, 1972; Harris and Bradley, 1973), small mammals, birds (Broadbent, 1965b), and, most importantly, by horticultural workers on contaminated hands, clothing, and tools during routine cultural operations (Broadbent, 1976).

B. Control

Because ToMV can reduce yields of intolerant cultivars by 20% or more (Broadbent, 1976), considerable efforts have been made to control infection in commercial crops. The use of virus-free seed, growth of plants in sterilized compost contained in plastic bags ("grow bags"), and application of strict hygiene can often reduce, but not prevent, infection (Broadbent, 1976). However, the severity of infection in nonresistant cultivars has sometimes been minimized by mild strain protection and, more recently, effectively controlled by the widespread use of virus-resistant cultivars.

1. Mild Strain Protection

Before genotypes with effective resistance to ToMV became generally available, virus-induced losses were sometimes reduced by inoculating young seedlings with mild strains of virus. Such attenuated strains were obtained initially from infected plants subjected to prolonged heat treatment (Oshima *et al.*, 1965; Komochi *et al.*, 1966; Paludan, 1968, 1973; Oshima, 1975, 1981), but the widely used strain M11–16 was obtained in 1968 as a nitrous acid mutant of a virulent Dutch strain (Rast, 1972, 1975). M11–16 temporarily checks the growth of inoculated tomato seedlings and so delays flowering and fruit set, but these adverse effects are readily overcome by advancing the date on which seed is sown (Rast, 1975). This strain has been used successfully in several countries, including the Netherlands, France, United Kingdom, Belgium, New Zealand and the USA (Migliori *et al.*, 1972; Rast, 1975; Fletcher and Rowe, 1975; Zimmerman-Gries and Pelowsky, 1975; Vanderveken and Coutissi, 1975; Mossop and Proctor, 1975; Channon *et al.*, 1978; Fletcher, 1978; Ahoonmanesh and Shalla, 1981). When widely used between 1971 and 1973 in the Netherlands, the average yields of early tomato crops increased by 15% (Rast, 1975); trials in the United Kingdom also indicated that "protected" crops yielded 7% more (Upstone, 1974; Channon *et al.*, 1978). Nevertheless, because M11–16 was derived from strain 1 (Pelham, 1966), it protects inoculated tomato plants only against naturally occurring isolates of this strain; moreover, it fails to protect inoculated plants grown at 25–30°C and, under some environmental conditions, readily reverts to its virulent form (Thomas, 1974; Rast, 1975).

2. Use of Resistant Cultivars

Although problems of ToMV infection can be alleviated by strict hygiene and/or mild strain protection, ToMV is now undoubtedly best controlled by the use of resistant cultivars. The use of genetic resistance to control ToMV has been well reviewed (Pelham, 1966, 1972; Rast, 1975; Hall, 1980; Hall and Bowes, 1981).

Several sources of genetic resistance have been recognized (Walter, 1956, 1967; Pelham, 1966; Alexander, 1971; Fletcher, 1973), but three major genes (designated Tm-1, Tm-2, and Tm-2^2) have been most used in breeding programs. The earlier use of cultivars containing genes Tm-1 and/or Tm-2 resulted in the rapid emergence or increased incidence of strains which were able to overcome resistance (Pelham *et al.*, 1970; MacNeill and Fletcher, 1971; Fletcher and Butler, 1975; Hall, 1978, 1979). Thus, although the commonly occurring isolates (strain 0) were unable to induce symptoms in cultivars containing any of the resistance genes, those of strain 1 were able to infect cultivars containing the Tm-1 genes and those of strain 1.2 cultivars possessing genes Tm-1 and/or Tm-2. The strain (2^2) which is able to overcome gene Tm-2^2, by contrast, has been found only very rarely in northwestern Europe and then only in plants within unusually hot glasshouses (Rast, 1975; Hall, 1980). After a decade or so of widespread use, the gene Tm-2^2 in homozygous cultivars is still highly effective in controlling ToMV; as an added safeguard, it is recommended that it be present together with genes Tm-1 and Tm-2 (Pelham *et al.*, 1970; Thomas, 1973). The origin and exploitation of genes Tm-2 and Tm-2^2 have been fully discussed by Hall (1980).

Little is known about the molecular basis for genetic resistance to ToMV. It has long been known that, although the Tm-1 gene suppresses the development of symptoms attributable to strain 0 in resistant cultivars, it does not prevent detectable virus multiplication (Clayberg, 1960; Alexander, 1962; Pécaut, 1962, 1964; Dawson, 1965, 1967; Phillip *et al.*, 1965; Pelham, 1966, 1972). It was uncertain whether Tm-1 inhibits virus multiplication (Pelham, 1972; Arroya and Selman, 1977; Motoyoshi and Oshima, 1977), stimulates its replication (Pelham, 1972), or has no effect (Clayberg, 1960; Pelham, 1972) until it was demonstrated that it both delayed multiplication and inhibited the accumulation of viral RNA and coat protein (Fraser and Loughlin, 1980, 1982). The resistance gene products, however, have yet to be identified and characterized. It was later shown that although Tm-1 failed to suppress symptoms of strain 1, it still partly inhibited its multiplication, results further suggesting that the gene has independent effects on symptom suppression and virus multiplication (Fraser *et al.*, 1980). The action of the gene Tm-2^2 has yet to be further investigated. However, although ToMV multiplies slowly in leaf disks of cv. Craigella homozygous for Tm-2^2, it readily infects and multiplies within leaf mesophyll protoplasts (Motoyoshi and Oshima, 1975, 1977); these results suggest that Tm-2^2, like Tm-2, possibly inhibits virus multiplication in epidermal cells and/or prevents cell-to-cell movement (Hall, 1980). Although the molecular basis for resistance has yet to be determined, strain 2^2 failed to develop in scions of cultivars containing Tm-2^2 when grafted onto infected susceptible cultivars (Hall, 1978, 1979), and probably did not occur naturally before the use of resistant cultivars (Rast, 1975; Hall, 1980). Moreover, because naturally occurring inoculum of ToMV has continued to fall with the greater use of resistant cultivars and "grow bags" and the decreased use of mild strain protection, there are good prospects that resistance conferred by Tm-2^2 will be durable.

REFERENCES

Ahoonmanesh, A., and Shalla, T. A., 1981, Feasibility of cross-protection for control of tomato mosaic virus in fresh market field-grown tomatoes, *Plant Dis.* **65**:56.

Ainsworth, G. C., 1937, 'Enation mosaic' of tomato caused by a virus of the tobacco virus 1 type, *Ann. Appl. Biol.* **24**:545.

Ainsworth, G. C., and Selman, I. W., 1936, Some effects of tobacco mosaic virus on the growth of seedling tomato plants, *Ann. Appl. Biol.* **23**:89.

Alexander, L. J., 1962, Strains of TMV on tomato in the Netherlands and in Ohio, U.S.A., *Meded. Landbouwhogesch. Opzoekingsstn. Staat Gent* **72:**1020.

Alexander, L. J., 1971, Host–pathogen dynamics of tobacco mosaic virus on tomato, *Phytopathology* **61:**611.

Allard, H. A., 1916, The mosaic disease of tomatoes and petunias, *Phytopathology* **6:**328.

Allen, A. S., 1969, Cytopathology of inclusions caused by six strains of tobacco mosaic virus in leaf cells of resistant and susceptible *Lycopersicon esculentum* Mill, *Diss. Abstr. B* **29:**3633.

Anderer, F. A., 1963, Recent studies on the structure of tobacco mosaic virus, *Adv. Protein Chem.* **18:**1.

Anderer, F. A., Uhlig, H., Weber, E., and Schramm, G., 1960, Primary structure of the protein of tobacco mosaic virus, *Nature (London)* **186:**922.

Arroya, A., and Selman, I. W., 1977, The effects of rootstock and scion on tobacco mosaic virus infection in susceptible, tolerant and immune cultivars of tomato, *Ann. Appl. Biol.* **85:**249.

Bald, J. G., and Paulus, A. O., 1963, A characteristic form of tobacco mosaic virus in tomato and *Chenopodium murale, Phytopathology* **53:**627.

Bewley, W. F., 1923, *Diseases of Glasshouse Plants,* Ernest Benn, London.

Boyle, J. S., and Bergman, E. L., 1967, Factors affecting incidence and severity of internal browning of tomato induced by tobacco mosaic virus, *Phytopathology* **57:**354.

Bradley, R. H. E., and Harris, K. F., 1972, Aphids can inoculate plants with tobacco mosaic virus by clawing, *Virology* **50:**615.

Briand, J. P., Al Moudallal, Z., and Van Regenmortel, M. H. V., 1982, Serological differentiation of tobamoviruses by means of monoclonal antibodies, *J. Virol. Methods* **5:**293.

Broadbent, L., 1961, The epidemiology of tomato mosaic, 1. A review of the literature, *Rep. Glasshouse Crops Res. Inst. 1960* **96.**

Broadbent, L., 1962, The epidemiology of tomato mosaic. 2. Smoking tobacco as a source of virus, *Ann. Appl. Biol.* **50:**461.

Broadbent, L., 1965a, The epidemiology of tomato mosaic. 8. Virus infection through tomato roots, *Ann. Appl. Biol.* **55:**57.

Broadbent, L., 1965b, The epidemiology of tomato mosaic. 9. Transmission of TMV by birds, *Ann. Appl. Biol.* **55:**67.

Broadbent, L., 1965c, The epidemiology of tomato mosaic. 11. Seed-transmission of TMV, *Ann. Appl. Biol.* **56:**177.

Broadbent, L., 1976, Epidemiology and control of tomato mosaic virus, *Annu. Rev. Phytopathol.* **14:**75.

Broadbent, L., and Cooper, A. J., 1964, The epidemiology of tomato mosaic. 6. The influence of tomato mosaic virus on root growth and the annual pattern of fruit production, *Ann. Appl. Biol.* **54:**31.

Broadbent, L., and Winsor, G. W., 1964, The epidemiology of tomato mosaic. 5. The effect on TMV-infected plants of nutrient foliar sprays and of steaming the soil, *Ann. Appl. Biol.* **54:**23.

Broadbent, L., Read, W. H., and Last, F. T., 1965, The epidemiology of tomato mosaic. 10. Persistence of TMV-infected debris in soil and the effects of soil partial sterilization, *Ann. Appl. Biol.* **55:**471.

Burgyan, J., and Gaborjanyi, R., 1984, Cross-protection and multiplication of mild and severe strains of TMV in tomato plants, *Phytopathol. Z.* **110:**156.

Caldwell, J., 1959, Persistence of tomato aucuba mosaic virus in dried leaf tissue, *Nature (London)* **183:**1142.

Caspar, D. L. D., 1963, Radial density distribution in the tobacco mosaic virus particle, *Nature (London)* **177:**928.

Caspar, D. L. D., and Holmes, K. C., 1969, Structure of dahlemense strain of tobacco mosaic virus: A periodically deformed helix, *J. Mol. Biol.* **46:**99.

Channon, A. G., Cheffins, N. J., Hitchon, G. M., and Barker, J., 1978, The effect of inoculation with an attenuated mutant strain of tobacco mosaic virus on the growth and yield of early glasshouse tomato crops, *Ann. Appl. Biol.* **88:**121.

Chester, K. S., 1936, Separation and analysis of virus strains by means of precipitin tests, *Phytopathology* **26:**778.

Christie, R. G., and Edwardson, J. R., 1977, Light and electron microscopy of plant virus inclusions, *Fla Agric. Exp. Stn. Monogr.* **9:**11.

Cirulli, M., and Alexander, L. J., 1969, Influence of temperature and strain of tobacco mosaic virus on resistance in a tomato breeding line derived from *Lycopersicon peruvianum*, *Phytopathology* **59:**1287.

Clayberg, C. D., 1960, Relative resistance of Tm_1 and Tm_2 to tobacco mosaic virus, *Tomato Genetics Co-operative Report* **10:**13.

Clayberg, C. D., Butler, L., Rick, C. M., and Young, P. A., 1960, Second list of known genes in the tomato, *Heredity* **51:**167.

Clinton, G. P., 1909, Tomato calico, lima bean, string bean and musk melon chlorosis: Peach yellows, tobacco and tomato mosaic, *Rep. Conn. Agric. Exp. Stn. 1907–1908* p. 854.

Conti, M., and Masenga, V., 1977, Identification and prevalence of pepper viruses in northwest Italy, *Phytopathol. Z.* **90:**212.

Crill, P., Burgis, D. S., Jones, J. P., and Strobel, J. W., 1973, Effect of tomato mosaic virus on yield of fresh-market, machine-harvest type tomatoes, *Plant Dis. Rep.* **57:**78.

Csillery, G., Tobias, I., and Rusko, J., 1983, A new pepper strain of tomato mosaic virus, *Acta Phytopathol. Acad. Sci. Hung.* **18:**195.

Dawson, J. R. O., 1965, Contrasting effects of resistant and susceptible tomato plants on tomato mosaic virus multiplication, *Ann. Appl. Biol.* **56:**485.

Dawson, J. R. O., 1967, The adaptation of tomato mosaic virus to resistant tomato plants, *Ann. Appl. Biol.* **60:**209.

Dawson, J. R. O., Rees, M. W., and Short, M. N., 1975, Protein composition of unusual tobacco mosaic virus strains, *Ann. Appl. Biol.* **79:**189.

Dawson, J. R. O., Rees, M. W., and Short, M. N., 1979, Lack of correlation between the coat protein composition of tobacco mosaic virus isolates and their ability to infect resistant tomato plants, *Ann. Appl. Biol.* **91:**353.

Fenner, F., 1976, Classification and nomenclature of viruses, *Intervirology* **7:**4.

Fletcher, J. T., 1963, Tobacco mosaic virus of sweet peppers, *Plant Pathol.* **12:**113.

Fletcher, J. T., 1969, Studies on the overwintering of tomato mosaic in root debris, *Plant Pathol.* **18:**97.

Fletcher, J. T., 1973, Tomato mosaic, in: *The U.K. Tomato Manual* (H. G. Kingham, ed.), pp. 196–208, Grower Books, London.

Fletcher, J. T., 1978, The use of avirulent strains to protect plants against the effects of virulent strains, *Ann. Appl. Biol.* **89:**110.

Fletcher, J. T., and Butler, D., 1975, Strain changes in populations of tobacco mosaic virus from tomato crops, *Ann. Appl. Biol.* **81:**409.

Fletcher, J. T., and MacNeill, B. H., 1971, Influence of environment, cultivar and virus strain on the expression of tobacco mosaic virus symptoms in tomato, *Can. J. Plant Sci.* **51:**101.

Fletcher, J. T., and Rowe, J. M., 1975, Observations and experiments on the use of an avirulent mutant strain of tobacco mosaic virus as a means of controlling tomato mosaic, *Ann. Appl. Biol.* **81:**171.

Franklin, R. E., 1956, Location of ribonucleic acid in the tobacco mosaic virus particle,*Nature* (*London*) **177:**928.

Fraser, R. S. S., and Loughlin, S. A. R., 1980, Resistance to tobacco mosaic virus in tomato: Effects of the Tm-1 gene on virus multiplication, *J. Gen. Virol.* **48:**87.

Fraser, R. S. S., and Loughlin, S. A. R., 1982, Effects of temperature on the Tm-1 gene for resistance to tobacco mosaic virus in tomato, *Physiol. Plant Pathol.* **20:**109.

Fraser, R. S. S., Loughlin, S. A. R., and Conner, J. C., 1980, Resistance to tobacco mosaic virus in tomato: Effects of the Tm-1 gene on symptom formation and multiplication of strain 1, *J. Gen. Virol.* **50:**221.

Fukuda, M., Okada, Y., Otsuki, Y., and Takebe, I., 1980, The site of initiation of rod assembly on the RNA of a tomato and a cowpea strain of tobacco mosaic virus, *Virology* **101**:493.
Fukuda, M., Meshi, T., Okada, Y., Otsuki, Y., and Takebe, I., 1981, Correlation between particle multiplicity and location on virion RNA of the assembly initiation site for viruses of tobacco mosaic virus group, *Proc. Natl. Acad. Sci. USA* **78**:4231.
Gibbs, A. J., 1977, Tobamovirus group, *CMI/AAB Descriptions of Plant Viruses No. 184.*
Gibbs, A. J., 1980, How ancient are the tobamoviruses?, *Intervirology* **14**:101.
Gilmer, R. M., and Wilks, J. M., 1967a, Seed transmission of tobacco mosaic virus in apple and pear, *Phytopathology* **57**:214.
Gilmer, R. M., and Wilks, J. M., 1967b, Apple chlorotic leaf spot and tobacco mosaic viruses in cherry, *Plant Dis. Rep.* **51**:823.
Gilmer, R. M., and Wilks, J. M., 1968, Transmission of tobacco mosaic virus in grape seeds, *Phytopathology* **58**:277.
Hall, T. J., 1978, *Rep. Glasshouse Crops Res. Inst. 1977* **39**.
Hall, T. J., 1979, *Rep. Glasshouse Crops Res. Inst. 1978* **40**.
Hall, T. J., 1980, Resistance at the Tm-2 locus in the tomato to tomato mosaic virus, *Euphytica* **29**:189.
Hall, T. J., and Bowes, S. A., 1981, Screening for disease resistance in tomato, *Rep. Glasshouse Crops Res. Inst. 1980* **157**.
Hamilton, R. I., Edwardson, J. R., Francki, R. I. B., Hsu, H. T., Hull, R., Koenig, R., and Milne, R. G., 1981, Guidelines for the identification and characterisation of plant viruses, *J. Gen. Virol.* **54**:223.
Harris, K. F., and Bradley, R. H. E., 1973, Importance of leaf hairs in the transmission of tobacco mosaic virus by aphids, *Virology* **52**:295.
Harrison, B. D., Finch, J. T., Gibbs, A. J., Hollings, M., Shepherd, R. J., Valenta, V., and Wetter, C., 1971, Sixteen groups of plant viruses, *Virology* **45**:356.
Henderson Smith, J., 1928, Experiments with a mosaic disease of tomato, *Ann. Appl. Biol.* **15**:155.
Henderson Smith, J., 1930, Intracellular inclusions in mosaic of *Solanum nodiflorum, Ann. Appl. Biol.* **17**:213.
Hennig, B., and Wittmann, H. G., 1972, Tobacco mosaic virus: Mutants and strains, in: *Principles and Techniques in Plant Virology* (C. I. Kado and H. O. Agrawal, eds.), pp. 546–594, Van Nostrand–Reinhold, Princeton, N.J.
Herold, F., and Munz, K., 1967, An unusual kind of crystalline inclusion of tobacco mosaic virus, *J. Gen. Virol.* **1**:375.
Hollings, M., 1957, Reactions of some additional plant viruses in *Chenopodium amaranticolor, Plant Pathol.* **6**:133.
Hollings, M., and Huttinga, H., 1976, Tomato mosaic virus, *CMI/AAB Descriptions of Plant Viruses No. 156.*
Holmes, K. C., and Franklin, R. E., 1958, The radial density in some strains of tobacco mosaic virus, *Virology* **6**:328.
Horvath, J., Horvath, A., Lonhard, M., Mamula, D., and Besada, W. H., 1978, Natural occurrence of a strain of tomato mosaic virus in potato in Hungary, *Acta Phytopathol. Acad. Sci. Hung.* **13**:299.
Jaegle, M., and Van Regenmortel, M. H. V., 1985, Use of ELISA for measuring the extent of serological cross-reactivity between plant viruses, *J. Virol. Methods* **11**:189.
Jensen, M. H., 1969, Cross protection as a means of reducing severity of TMV in greenhouse tomatoes, *Diss. Abstr. B* **29**:3991.
Johnson, J., and Ogden, W. B., 1929, The overwintering of the tobacco mosaic virus, *Bull. Wis. Agric. Exp. Stn.* **95**.
Jones, L. K., and Burnett, G., 1935, Virous diseases of greenhouse-grown tomatoes, *Bull. Wash. State Agric. Exp. Stn.* **308**.
Juretić, N., Horvath, J., Besada, W. H., Horvath, A., and Lonhard, M., 1977, Serological relationship of tomato mosaic virus isolated from potato to two members of tobamovirus group, *JIPA.* **4**:64.

Kassanis, B., and Sheffield, F. M. L., 1941, Variations in the cytoplasmic inclusions induced by three strains of tobacco mosaic virus, *Ann. Appl. Biol.* **28:**360.
Kassanis, B., Woods, R. D., and White, R. F., 1972, Some properties of potato mop-top virus and its serological relationships to tobacco mosaic virus, *J. Gen. Virol.* **14:**123.
Kirkpatrick, H. C., and Lindner, R. C., 1964, Recovery of tobacco mosaic virus from apple, *Plant Dis. Rep.* **48:**855.
Knight, C. A., Silva, D. M., Dahl, D., and Tsugita, A., 1962, Two distinctive strains of tobacco mosaic virus, *Virology* **16:**236.
Komochi, S. T., Goto, T., and Oshima, N., 1966, Studies on the control of plant virus diseases by vaccination of attenuated virus. 3. Reduction in fruit setting as a shock reaction resulting from infection of tomato plants with TMV (in Japanese, English summary). *J. hort. Assoc. Japan* **35:**269.
Komuro, Y., Iwaki, M., and Nakahara, M., 1966, Viruses isolated from tomato plants showing mosaic and/or streak in Japan, with special reference to tomato strain of tobacco mosaic virus (in Japanese, English summary). *Ann. phytopath. Soc. Japan* **32:**130.
Kramer, E., and Wittmann, H. G., 1958, Elektrophoretische Untersuchungen der A-protein dreier Tobakmosaikvirus—Stamme, *Z. Naturforsch.* **136:**30.
Kukla, B. A., Guilley, H. A., Jonard, G. X., Richards, K. E., and Mundry, K. W., 1979, Characterisation of long guanosine-free RNA sequences from the dahlemense and U2 strains of tobacco mosaic virus, *Eur. J. Biochem.* **98:**61.
Kunkel, L. O., 1934, Studies on acquired immunity with tobacco and aucuba mosaics, *Phytopathology* **24:**437.
Kurachi, K., Funatsu, M., and Hidaka, S., 1972, Partial amino acid sequence of cucumber green mottle mosaic virus coat protein, *Agric. Biol. Chem.* **36:**1109.
Leonard, D. A., and Zaitlin, M., 1982, A temperature sensitive strain of tobacco mosaic virus defective in cell-to-cell movement generates an altered viral-coded protein, *Virology* **117:**416.
MacNeill, B. H., 1962, A specialized tomato form of the tobacco mosaic virus in Canada, *Can. J. Bot.* **40:**49.
MacNeill, B. H., and Fletcher, J. T., 1971, Changes in strain characteristics of tobacco mosaic virus after passage through grafted tomato hosts, *Phytopathology* **61:**130.
MacNeill, B. H., and Ismen, H., 1960, Studies on the virus-streak syndrome in tomatoes, *Can. J. Bot.* **38:**277.
McRitchie, J. J., and Alexander, L. J., 1963, Host-specific *Lycopersicon* strains of tobacco mosaic virus, *Phytopathology* **53:**394.
Mayhew, D. E., Hedin, P., and Thomas, D. L., 1984, Corky ringspot: A new strain of tomato mosaic virus in California, *Plant Dis.* **68:**623.
Melchers, G., Schramm, G., Trusnit, H., and Frederick-Freska, H., 1940, Biological, chemical and electron-microscopical investigation of a mosaic from tomatoes, *Biol. Zentralbl.* **60:**524.
Meshi, T., Ohno, T., and Okada, Y., 1982a, Nucleotide sequence and its character of cistron coding for the 30K protein of tobacco mosaic virus, *J. Biochem.* (*Tokyo*) **92:**1441.
Meshi, T., Ohno, T., and Okada, Y., 1982b, Nucleotide sequence of the 30K protein cistron of cowpea strain of tobacco mosaic virus, *Nucleic Acids Res.* **10:**6111.
Meshi, T., Kiyama, R., Ohno, T., and Okada, Y., 1983, Nucleotide sequence of the coat protein cistron and the 3′ non-coding region of cucumber green mottle mosaic virus (watermelon strain) RNA, *Virology* **127:**54.
Migliori, A., Ginoux, G., Peyrierc, J., Marrou, J., Musard, M., and Fauvrel, M., 1972, Application de la premunition contre le virus de la mosaique du tabac dans les cultures de tomate sous serre, *Rev. Horticole* **132:**15.
Miličić, D., and Juretić, N., 1971, Zelleinschluss der Viren der Tabakmosaik-Gruppe, *Tagungsber. Dtsch. Akad. Landwirtschaftswiss. Berlin* **115:**141.
Miller, P. M., and Thornberry, H. H., 1958, A new viral disease of tomato and pepper, *Phytopathology* **48:**665.

Mosch, W. H. M., Huttinga, H., and Rast, A. T. B., 1973, Some chemical and physical properties of 18 tobacco mosaic virus isolates from tomato, *Neth. J. Plant Pathol.* **79:**104.
Mossop, D. W., and Proctor, C. H., 1975, Cross-protection of glasshouse tomatoes against tobacco mosaic virus, *N.Z. J. Exp. Agric.* **3:**343.
Motoyoshi, F., and Oshima, N., 1975, Infection with tobacco mosaic virus of leaf mesophyll protoplasts from susceptible and resistant lines of tomato, *J. Gen. Virol.* **29:**81.
Motoyoshi, F., and Oshima, N., 1977, Expression of genetically controlled resistance to tobacco mosaic virus infection in isolated tomato leaf mesophyll protoplasts, *J. Gen. Virol.* **34:**499.
Mundry, K. W., 1965, A model of the coat protein cistron of tobacco mosaic virus and its biochemical investigation, *Z. Vererbungsl.* **97:**281.
Nicolaieff, A., and Van Regenmortel, M. H. V., 1980, Specificity of trapping of plant viruses on antibody-coated electron microscope grids, *Ann. Virol.* (*Inst. Pasteur*) **131E:**95.
Nishiguchi, M., Nozu, Y., and Oshima, N., 1984, Comparative study on coat proteins of an attenuated, a temperature sensitive and their parent strains of tobacco mosaic virus, *Phytopathol. Z.* **109:**104.
Nitzany, F. E., 1960, Transmission of tobacco mosaic virus through tomato seed and virus inactivation by methods of seed extraction and seed treatments, *Ktavim Rec. Agric. Res. Stn.* **10:**63.
Ohno, T., Takamatsu, N., Meshi, T., Okada, Y., Nishiguchi, M., and Kiho, Y., 1983, Single amino acid substitution in 30K protein of TMV defective in virus transport function, *Virology* **131:**255.
Ohno, T., Aoyagi, M., Yamanashi, Y., Saito, H., Ikawa, S., Neshi, T., and Okada, Y., 1984, Nucleotide sequence of the tobacco mosaic virus (tomato strain) genome and comparison with the common strain genome. *J. Biochem.* (*Tokyo*) **96:**1915.
Opel, H., Kegler, H., and Richter, J., 1969, Occurrence and characterization of TMV strains of pome fruit, *Acta Phytopathol. Acad. Sci. Hung.* **4:**1.
Oshima, N., 1975, The control of tomato mosaic disease with attenuated virus of tomato strain of TMV, *Rev. Plant Prot. Res.* **8:**126.
Oshima, N., 1981, Control of tomato mosaic disease by attenuated virus, *Jpn. Agric. Res. Q.* **14:**222.
Oshima, N., Komoti, S., and Goto, T., 1965, Study on control of plant virus diseases by vaccination of attenuated virus. (i) Control of tomato mosaic disease, *Bull. Hokkaido Agric. Exp. Stn.* **85:**23.
Oshima, N., Tanada, T., and Nozu, Y., 1976, Coat protein of an attenuated tomato strain ($L_{11}A$) of TMV, *Ann. Phytopathol. Soc. Jpn.* **42:**62.
Oster, G., 1951, The isoelectric points of some strains of tobacco mosaic virus, *J. Biol. Chem.* **190:**55.
Paludan, N., 1968, Tobacco mosaic virus (TMV): Investigation concerning TMV in different plant genera, the virulence of TMV strains, virus attenuation by heat treatment, cross protection and yield (Danish, English summ.), *Tidsskr. Planteavl* **72:**69.
Paludan, N., 1973, Tobak-mosaik-virus: Infektionsforsog, krydsbeskyttelse, smittetidspunkt og udbytte med tomatlinier af TMV hos tomat, *Tidsskr. Planteavl* **77:**495.
Palukaitis, P., and Symons, R. H., 1980, Nucleotide sequence homology of thirteen tobamovirus RNAs as determined by hybridization analysis with complementary DNA, *Virology* **107:**354.
Pecaut, P., 1964, Résistance a la mosaique du tabac, *Rep. Stn. d'Amel. Pl. Maraich.* (*INRA*) *1961* **17.**
Pecaut, P., 1964, Tomate: Selection pour la resistance aux maladies, *Rep. Stn. d'Amel. Pl. Maraich.* (*INRA*) *1963* **51.**
Pelham, J., 1966, Resistance in tomato to tobacco mosaic virus, *Euphytica* **15:**258.
Pelham, J., 1972, Strain-genotype interaction of tobacco mosaic virus in tomato, *Ann. Appl. Biol.* **71:**219.
Pelham, J., Fletcher, J. T., and Hawkins, J. H., 1970, The establishment of a new strain of

tobacco mosaic virus resulting from the use of resistant varieties of tomato, *Ann. Appl. Biol.* **65**:293.
Phillip, M. J., Honma, S., and Murakishi, H. H., 1965, Studies on the inheritance of resistance to tobacco virus in the tomato, *Euphytica* **14**:231.
Price, W. C., and Fenne, S. B., 1951, Tomato rosette, a severe disease caused by a strain of tobacco mosaic virus, *Phytopathology* **41**:1091.
Rast, A. T. B., 1965, A yellow ringspot strain of tobacco mosaic virus (TMV) from tomato, *Neth. J. Plant Pathol.* **71**:91.
Rast, A. T. B., 1967, Yield of glasshouse tomatoes as affected by strains of tobacco mosaic virus, *Neth. J. Plant Pathol.* **73**:147.
Rast, A. T. B., 1972, M11–16, an artificial symptomless mutant of tobacco mosaic virus for seedling inoculation of tomato crops, *Neth. J. Plant Pathol.* **78**:110.
Rast, A. T. B., 1973, Systemic infection of tomato plants with tobacco mosaic virus following inoculation of seedling roots, *Neth. J. Plant Pathol.* **79**:5.
Rast, A. T. B., 1975, Variability of tobacco mosaic virus in relation to control of tomato mosaic in glasshouse tomato crops by resistance breeding and cross protection, *Agric. Res. Rep.* Wageningen **834**.
Sheffield, F. M. L., 1931, The formation of intracellular inclusions in solanaceous hosts infected with aucuba mosaic of tomato, *Ann. Appl. Biol.* **18**:471.
Sugiyama, T., and Fraenkel-Conrat, H., 1963, The end-groups of tobacco mosaic virus RNA. II. Nature of the 3'-linked chain end in TMV and of both ends in four strains, *Biochemistry* **2**:332.
Takamatsu, N., Ohno, T., Meshi, T., and Okada, Y., 1983, Molecular cloning and nucleotide sequence of the 30K and the coat protein cistron of TMV (tomato strain) genome, *Nucleic Acids Res.* **11**:3767.
Taylor, R. H., Grogan, R. G., and Kimble, K. A., 1961, Transmission of tobacco mosaic virus in tomato seed, *Phytopathology* **51**:837.
Termohlen, G., and van Dorst, H. J. M., 1959, Verschillen tussen *Lycopersicum* virus 1 en *Nicotiana* virus 1, *Jversl. Proefstn. Groenten Fruitteelt Glas, Naaldwijk 1958* 125.
Thomas, B. J., 1973, Tomato mosaic virus, *Ann. Rep. Glasshouse Crops Res. Inst. 1972* **30**.
Thomas, B. J., 1974, Tobacco (tomato) mosaic virus, *Rep. Glasshouse Crops Res. Inst. 1973* **42**.
Tomaru, K., Suyama, K., and Kubo, S., 1970, Efficiency of local lesion formation and multiplication of tobacco mosaic virus strains in the inoculated leaves of some host plants [in Japanese; English summary], *Ann. Phytopathol. Soc. Jpn.* **36**:74.
Tsugita, A., 1962, The proteins of mutants of TMV: Classification of spontaneous and chemically evoked strains, *J. Mol. Biol.* **5**:293.
Upstone, M. E., 1974, Effects of inoculation with the Dutch mutant strain of tobacco mosaic virus on the cropping of commercial glasshouse tomatoes, *Rep. Agric. Dev. Adv. Serv. 1972* **162**.
Vanderveken, J., and Coutissi, S., 1975, Trial of the control of tobacco mosaic virus in tomato by cross protection, *Meded. Fac. Landbouwwet. Rijksuniv. Gent* **40**:791.
Van De Walle, M. J., and Siegel, A., 1976, A study of nucleotide sequence homology between strains of tobacco mosaic virus, *Virology* **73**:413.
Van De Walle, M. J., and Siegel, A., 1982, Relationships between strains of tobacco mosaic virus and other selected plant viruses, *Phytopathology* **72**:390.
van Dorst, H. J. M., 1970, Virus diseases of cucumbers, *Jversl. Proefstn. Groenten Fruitteelt Glas, Naaldwijk, 1969* 75.
Van Regenmortel, M. H. V., 1967, Serological studies on naturally occurring strains and chemically induced mutants of tobacco mosaic virus, *Virology* **31**:467.
Van Regenmortel, M. H. V., 1975, Antigenic relationships between strains of tobacco mosaic virus, *Virology* **64**:415.
Van Regenmortel, M. H. V., 1981, Tobamoviruses, in: *Handbook of Plant Virus Infections* (E. Kurstak, ed.), p. 541, Plenum Press, New York.

Van Regenmortel, M. H. V., and Burckard, J., 1980, Detection of a wide spectrum of tobacco mosaic virus strains by indirect enzyme-linked immunosorbent assays (ELISA). *Virology* **106**:327.

Van Regenmortel, M. H. V., and von Wechmar, M. V., 1974, Serological relationships of strains of tobacco mosaic virus, *S. Afr. J. Sci.* **70**:59.

van Winckel, A., 1965, Tobakmosaiekvirus op Tomatenzaad, *Agricultura* (*Louvain*) **13**:721.

van Winckel, A., 1967a, Bijdrage tot de studie van het tabakmozaiekvirus (TMV) bij tomaten, Thesis, Kath. Univ. Leuven.

van Winckel, A., 1967b, Invloed van het TMV op de vegetatieve ontwikkeling van tomatenplanten, *Agricultura* (*Louvain*) **15**:67.

van Winckel, A., 1968, Natuurlijke inactivering van het tabakmozaiekvirus (TMV) op tomatenzaad, *Parasitica* **24**:1.

Walter, J. M., 1956, Hereditary resistance to tobacco mosaic virus in tomato, *Phytopathology* **46**:513.

Walter, J. M., 1967, Hereditary resistance to disease in tomato, *Annu. Rev. Phytopathol.* **5**:131.

Wang, A. L., and Knight, C. A., 1967, Analysis of protein components of tomato strains of tobacco mosaic virus, *Virology* **31**:101.

Warmke, H. E., 1967, Aucuba strain of tobacco mosaic virus in an unusual aggregate, *Science* **156**:262.

Warmke, H. E., 1968, Fine structure of inclusions formed by the aucuba strain of tobacco mosaic virus, *Virology* **34**:149.

Warmke, H. E., 1969, A reinterpretation of amorphous inclusions in the aucuba strain of tobacco mosaic virus, *Virology* **39**:695.

Warmke, H. E., 1974, Direction of rotation of aucuba (TMV) angled layer aggregates, *Virology* **59**:591.

Westerdijk, J., 1910, Die Mosaikkrankheit der Tomaten, *Med. Phytopathol. Lab. 'Willie Commelin Scholten', Baarn* 1.

Wetter, C., 1984, Serological identification of four tobamoviruses infecting pepper, *Plant Dis.* **68**:597.

Wetter, C., Conti, M., Altschuh, D., Tabillion, R., and Van Regenmortel, M. H. V., 1984, Pepper mild mottle virus, a tobamovirus infecting pepper cultivars in Sicily, *Phytopathology* **74**:405.

Wittmann-Liebold, B., and Wittmann, H. G., 1963, Die primäre Proteinstruktur von Stämmen des Tabakmosaikvirus. Aminosäuresequenzen des Proteins des Tabakmosaikvirus stammes Dahlemense. Teil III. Diskussion der Ergebnisse. *Z. Vererbungsl.* **94**:427.

Wittmann-Liebold, B., and Wittmann, H. G., 1967, Coat proteins of strains of two RNA viruses: Comparison of their amino acid sequences, *Mol. Gen. Genet.* **100**:358.

Zaitlin, M., Beachy, R. N., and Bruening, G., 1977, Lack of molecular hybridisation between RNAs of two strains of TMV: Reconsideration of the criteria for strain relationships, *Virology* **82**:237.

Zimmerman-Gries, S., and Pelowsky, M., 1975, Experiments for protecting tomatoes from tobacco mosaic virus (TMV) by prior infection with the avirulent strain M11–16, *Phytoparasitica* **3**:75.

CHAPTER 10

Tobacco Mild Green Mosaic Virus

CARL WETTER

I. INTRODUCTION

In this chapter, the designation tobacco mild green mosaic virus (TMGMV) is used for an ensemble of serologically closely related viruses (Table I) containing among others the well-characterized U2 strain (Siegel and Wildman, 1954; Wetter, 1984a,b). TMGMV was detected in 1927 by McKinney (1929) in *Nicotiana glauca* plants on the Canary Islands and labelled mild dark green tobacco mosaic virus. It was regarded for a long time as a silent contaminant or a rare companion of type tobacco mosaic virus (TMV) but recent findings have indicated that TMGMV is of considerable economic importance. When the first groups of plant viruses were established, none of the strains of TMGMV were regarded as distinct viruses of the tobamovirus group (Brandes and Wetter, 1959; Harrison *et al.*, 1971). The TMGMV virus was so named by McKinney to describe the mild symptoms induced in tobacco leaves in contrast to the much more pronounced yellow mosaic symptoms caused by common TMV. Mild virus strains from tobacco, closely related to TMGMV, were detected on several independent occasions in West Germany and in the United States (Wetter, 1984b). The biophysical, biochemical, and biological properties of the virus are sufficiently unique to justify it being considered a distinct species (member) of the tobamovirus group. Some of the properties of TMGMV described in this chapter represent unpublished data of the author.

CARL WETTER • Department of Botany, University of Saarland, D-6600 Saarbrücken, West Germany.

TABLE I. Strains of Tobacco Mild Green Mosaic Virus and Their Original Hosts

Isolate/strain	Original host	Reference
Tobacco mild green mosaic virus (TMGMV), ATCC PV 226	*Nicotiana glauca*	McKinney (1929)
Para-tobacco mosaic virus (PTMV)	Burley tobacco	Köhler and Panjan (1943)
Mild mosaic strain of TMV	Field tobacco or *Eryngium aquaticum*(?)	Johnson (1947)
South Carolina mottling strain of TMV, ATCC PV 228	*Capsicum frutescens*	McKinney (1952)
Strain U2 of TMV	Field tobacco or *Eryngium aquaticum*(?)	Johnson (1947); Siegel and Wildman (1954)
Green-tomato atypical mosaic virus (G-TAMV)	*Capsicum annuum*(?)	Knight *et al.* (1962)

II. BIOLOGICAL PROPERTIES

A. Virus Strains and Their Economic Importance

As reference strain of TMGMV, the isolate of the American Type Culture Collection (1981) known as PV 226 was used for comparative studies (Table I) (Wetter, 1984b). Many TMGMV isolates obtained from cigarette tobacco have been described as para-tobacco mosaic virus (PTMV) (Wetter and Bernard, 1977). TMGMV is of economic importance for tobacco production in West Germany as it is the most common virus causing mosaic disease in this crop (Köhler and Panjan, 1943; Wetter, 1980). If no measures are taken to prevent mechanical transmission, a high percentage of field plants become infected. At the end of the growing season, it is not uncommon to find large fields of burley tobacco that do not contain a single healthy plant. The virus occurs also in field-grown pepper in the United States and Italy (McKinney, 1952; Miller and Thornberry, 1958; Wetter, 1984a; Conti and Marte, 1983) and is also widespread in genera of the family Gesneriaceae in which it is much more prevalent than common TMV (Zettler and Nagel, 1983). It is possible that cigarette tobacco containing TMGMV constitutes part of the infection cycle of the virus (Wetter, 1980; Zettler and Nagel, 1983).

B. Host Range and Symptomatology

Only a limited number of host plants naturally infected with TMGMV have been used in host range studies. These include plants of the families Solanaceae, Umbelliferae, and Gesneriaceae. The most prominent symptoms occur in wild plants of *N. glauca* (McKinney, 1929). This wooden bush is widespread on subtropical coasts. In the arid climate of the Canary Islands, *N. glauca* is planted along the roadways and disseminates from there (Fig. 1A). A very high percentage of plants are infected with TMGMV and develop a bright yellow mosaic (McKinney, 1929; Bald and Goodshild, 1960; Randles *et al.*, 1981) (Fig. 1B). In Samsun tobacco, the first signs of systemic infection are a yellowing of the veins (Fig. 1C). The subsequent leaves show symptoms of decreasing severity. A characteristic morphological symptom on older leaves is the appearance of teeth on the leaf margins (Fig. 1D). On burley tobacco, line pattern, oak leaf pattern, and ring symptoms are characteristic symptoms (Wetter, 1980). In pepper cultivars, TMGMVcauses the same type of necrotic infections followed by leaf drop as TMV and tomato mosaic virus (ToMV). However, in certain cultivars such as Sperling's Merit, in contrast to TMV and ToMV, TMGMV becomes systemic (Wetter, 1984b). A high percentage of pepper plants that become systemically infected die.

The following host plants are useful for identifying TMGMV. Local lesions are induced in *N. sylvestris, N. glutinosa,* Xanthi-nc, and Burley-nc tobacco as well as in *Datura stramonium* while *Lycopersicon esculentum* is immune to infection (McKinney, 1929; Johnson, 1947; Knight *et al.*, 1962; Wetter, 1984b). *Eryngium aquaticum* L. (Johnson, 1947) or *E. planum* L. (Wetter, 1984b) are useful for eliminating TMV from mixtures since TMV remains local in the inoculated leaves and TMGMV becomes systemic. The first signs of systemic infection with TMGMV are yellow flecks (Fig. 1E); in older plants, leaves showing no symptoms may contain a considerable amount of virus. Although pepper mild mottle virus does not infect *E. planum* systemically, in some hosts it causes the same symptoms as TMGMV (Wetter *et al.*, 1984). It is thus important to confirm any tentative diagnosis based on host reactions by serological tests.

C. Cytopathological Effects

A characteristic intracellular feature of infection with TMGMV strains is the formation of angled-layer aggregates. Such inclusions were reported for the U5 (~ U2) strain in mesophyll cells of tobacco (Shalla, 1968; Shalla *et al.*, 1975). In tobacco leaf tissue infected with TMGMV, only crystals with a multilayer structure were observed in a limited number of ultrathin sections. Abundant angled-layer aggregates were also induced in *E. planum* (Fig. 2A). Virus aggregates were observed in chloroplasts and nuclei (Fig. 2B,C) and were found also in proplastids. Since the virus inclusions are not surrounded by a membrane, they are probably synthesized within these organelles (Shalla, 1968). This view is supported

FIGURE 1. Systemic symptoms of tobacco mild green mosaic virus in different host plants. (A) Diseased bush of *Nicotiana glauca* at the roadside in Fuerteventura, Canary Islands. (B) Leaf of the plant in (A), showing strong yellow mottle symptoms. (C) Early systemic infection in Samsun tobacco with vein clearing. (D) Systemically infected leaf from the top of a Samsun plant showing teeth as a morphogenetic symptom. (E) Early systemic infection of *Eryngium planum* with yellow fleck symptoms.

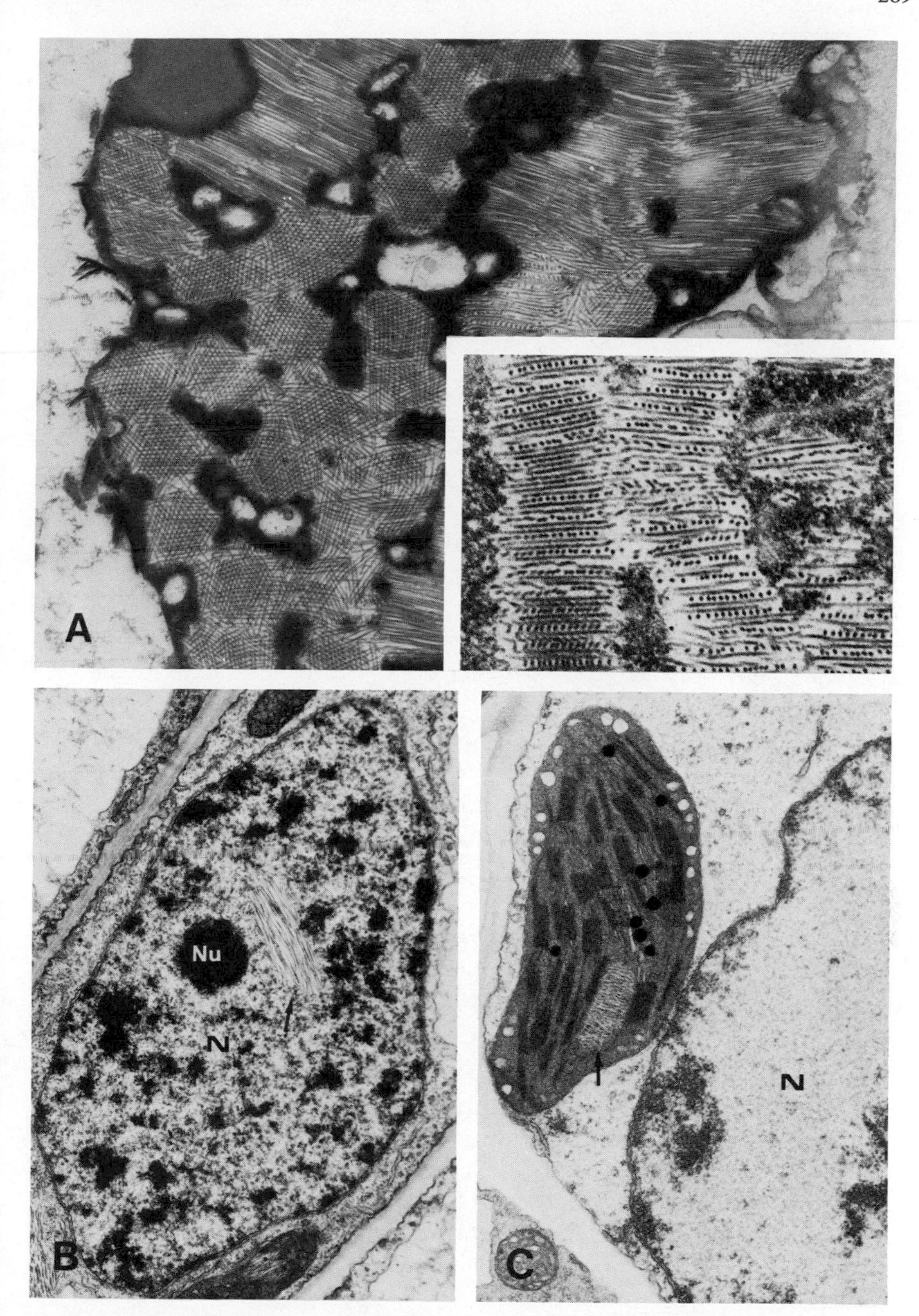

FIGURE 2. Ultrathin sections of *Eryngium planum* cells infected with tobacco mild green mosaic virus. (A) Layer aggregates and angled-layer aggregates in a mesophyll cell. The black regions are caused by degenerated cytoplasm. Inset: longitudinal section through an angled-layer aggregate. (B) Nucleus (N) with virus aggregate (arrow) and nucleolus (Nu). (C) Chloroplast containing virus aggregate (arrow). Ernwein and Wetter (unpublished).

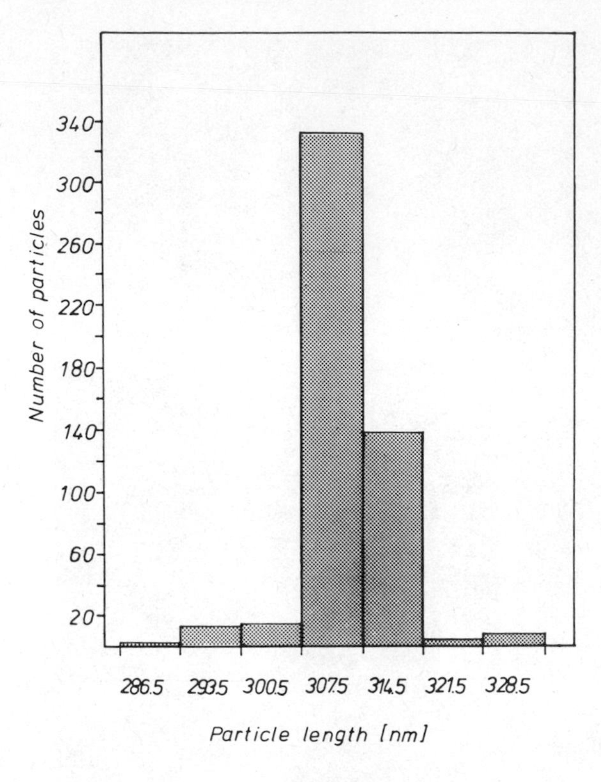

FIGURE 3. Histogram of particle length distribution of tobacco mild green mosaic virus from a preparation obtained by density gradient centrifugation.

by the finding that pseudovirions contain RNA that is complementary to the DNA of chloroplasts or nuclei (Siegel, 1971).

III. PHYSICOCHEMICAL PROPERTIES

A. Purification

TMGMV can be purified by any of the many procedures used for purifying TMV. To obtain a high proportion of monodisperse particles, the method of Boedtker and Simmons (1958) is recommended. Such preparations can be used to produce iridescent gels which are able to form two-dimensional micro- and macrocrystals (Kreibig and Wetter, 1980). Colloidal crystalline preparations are stable for several years.

B. Particle Morphology

Particles of TMGMV have the same morphology as TMV. The normal length (Brandes and Wetter, 1959) of the particles is 309.7 nm and the modal length is 308.3 nm (Fig. 3). These values were determined in comparative measurements on 627 particles taking the normal length of type TMV (300 nm) as a standard. Other tobamoviruses such as PMMV are also 10 to 20 nm longer than TMV (Wetter *et al.*, 1984). The diameter of TMGMV particles is 18 nm.

C. Virion Properties

Because of the similar morphology, many of the intrinsic properties of TMGMV may be taken to be similar to those of TMV (Zaitlin and Israel, 1975). The sedimentation coefficient ($S_{20,w}$) at infinite dilution is ~ 190 S; the particle molecular weight is ~ 39.4×10^6 and the diffusion coefficient ($D_{20,w}$) is ~ 4.4×10^{-8} cm²/sec. Other values, mostly determined for the U2 strain of TMGMV, differ from TMV because of the different virion composition. The isoelectric point is 4.17 for the U2 strain (Ginoza and Atkinson, 1955), and 3.96 for PTMV. The electrophoretic mobility is -4.9×10^{-5} cm²/sec per V (Siegel and Wildman, 1954; Ball, 1966; Wetter and Bernard, 1977), which is much lower than that of TMV. The extinction coefficient ($E^{0.1\%}_{1\,cm}$) at 260 nm, uncorrected for light scattering, is 3.16. The ratio of absorption at 260 nm/280 nm is 1.22; the ratio of minimum to maximum absorption is 1.1. The buoyant density in CsCl is 1.307 g/ml, compared to 1.310 for U1 (Siegel and Hudson, 1959; Skotnicki, *et al.*, 1976). The thermal inactivation point is lower than for TMV. Sap heated for 10 min at 85°C was still infective, but not so after being heated to 90°C.

D. Nucleotide Sequence Homology

The nucleotide sequence homology between strains of TMGMV and other members of the tobamovirus group has been investigated using different hybridization methods (Van De Walle and Siegel, 1976, 1982; Palukaitis and Symons, 1980). The different studies show that a high percentage of hybridization between U2 and G-TAMV and other isolates of TMGMV occurs. However, no relationship or a very weak one under less stringent conditions is observed between TMGMV and other tobamoviruses. These findings strongly support the view that TMGMV and TMV should be regarded as separate viruses.

E. Protein Composition

The amino acid sequence of the U2 protein was determined with the exception of a few residues by Wittmann (1965) and Rentschler (1967).

More recently, the sequence was completed and some positions were corrected (Altschuh *et al.*, 1981). According to this final determination, the protein subunit contains 158 amino acids. The sequence and the gross amino acid composition are presented in Chapter 8.

IV. SEROLOGY

The serological interrelationships between strains and isolates of TMGMV, TMV, ToMV, and other tobamoviruses have been analyzed mainly by immunodiffusion tests (Dudman, 1965; Van Regenmortel, 1967; Wetter and Bernard, 1977; Zettler and Nagel, 1983; Wetter, 1984a,b). The serological differentiation index (SDI) between different tobamoviruses which is used to express degrees of relationship was found to be correlated with the relative number of amino acid exchanges between the viruses (Van Regenmortel, 1975, 1982; Hennig and Wittmann, 1972; Wetter, 1984b). Serological relationships between different intact viruses and different protein subunits were also investigated using the enzyme-linked immunosorbent assay (ELISA) (Van Regenmortel and Burckard, 1980). In the highly strain-specific double-antibody sandwich technique, no cross-reaction between U2, TMV, ToMV, and other tobamoviruses was observed. In contrast, in the indirect ELISA method the presence of cross-reactions between these viruses could be easily demonstrated (Van Regenmortel and Burckard, 1980). In recent studies with TMGMV strains including the isolate from gesneriads, no serological differences could be detected with only one exception (Fig. 4A,B). The strain It(alian) III of TMGMV isolated from pepper in Italy gave a fusion of precipitin bands (reaction of identity) in immunodiffusion tests when antiserum to It III was reacted with It III antigen next to TMGMV antigen (Fig. 4C). Three antisera to It III were used, one of which was obtained after repeated booster immunizations over a period of 20 months. All antisera gave identical results. By contrast, antisera to type TMGMV, U2 strain, G-TAMV, and Johnson's strain gave reactions of "partial identity" with a strong spur when these antigens were compared to strain It III (Fig. 4D). Similar results were also obtained in double-antibody sandwich ELISA (Fig. 5A,B). To our knowledge, this is the first example of nonreciprocity of heterologous serological reactions in naturally occurring strains (Van Regenmortel, 1982). These results indicate that It III is lacking a specific determinant which the other strains of TMGMV possess. Although there are only three amino acid residue exchanges in the coat protein of It III compared to type TMGMV, they seem to have a decisive influence on the antigenic structure. A study of strain It III should be of considerable interest for elucidating the relationship between protein tertiary structure and the serological properties.

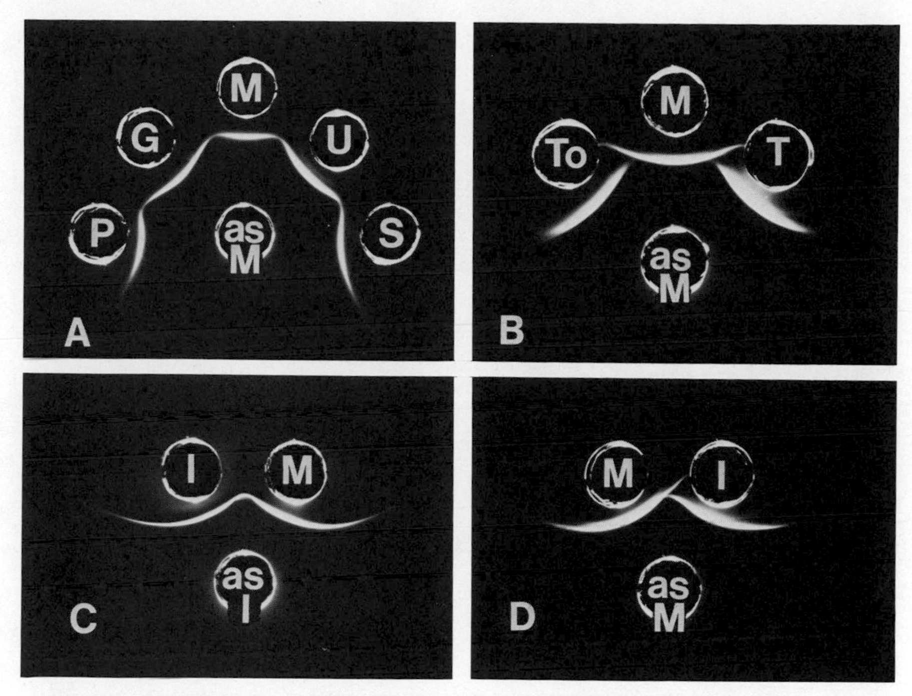

FIGURE 4. Reactions of tobacco mild green mosaic virus and related antigens in homologous and heterologous immunodiffusion tests. Antigens are abbreviated according to Table I. Antisera are labeled "as": (A) P = PTMV, G = G-TAMV, M = TMGMV, U = U2 strain, S = South Carolina mottling strain. (B) Comparison of TMGMV (M) with tomato mosaic virus (To), and TMV (T) as heterologous antigens. (C, D) Comparison of Italian III (I) with TMGMV (M) antigens in reciprocal tests.

V. INTERACTIONS OF TMGMV WITH OTHER TOBAMOVIRUSES

A. Cross-Protection and Interference

The discovery of the cross-protection phenomenon is generally attributed to McKinney (1929). He found that tobacco plants infected with a green mosaic strain of TMV developed no generalized yellow mosaic symptoms when subsequently inoculated with a challenging yellow mosaic strain of the same virus (U1). He also found that TMGMV did not protect tobacco plants against infection with the challenging yellow strain of U1. These tests showed that cross-protection occurred between closely related strains of one virus but not between more distantly related ones.

Interference between Johnson's (1947) severe (U1) and mild (U2) strains was investigated in detail by Fulton (1951) using *N. sylvestris* as host plant. It was found that cross-protection was partial rather than complete and that it depended on different factors. Leaves which were

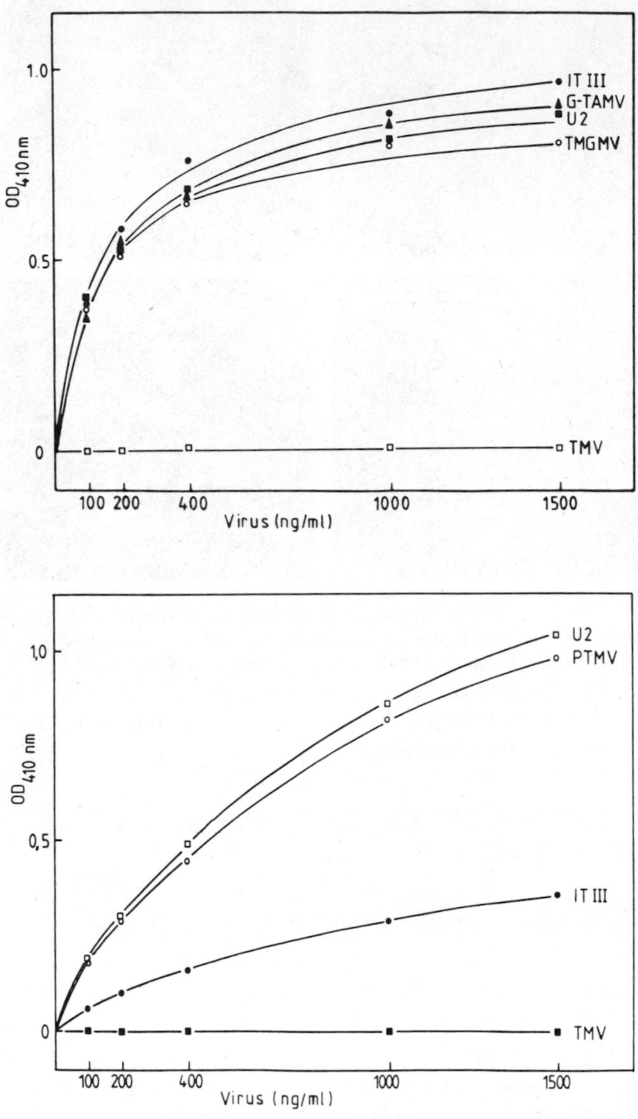

FIGURE 5. Double-antibody sandwich ELISA with tobacco mild green mosaic virus (TMGMV) strains. (A) Coating of wells was done with 5 μg/ml rabbit anti-It III globulins for 2 hr. The different antigens were incubated overnight. Rabbit anti-It III conjugate, diluted 1:1000, was incubated for 4 hr. Substrate hydrolysis time was 1 hr. (B) Coating was done with 5 μg/ml rabbit anti-PTMV globulins for 2 hr. The different antigens were incubated overnight. Rabbit anti-PTMV conjugate, diluted 1:1000, was incubated for 4 hr. Substrate hydrolysis time was 1 hr.

inoculated with U1 developed a nearly complete immunity to the challenging U2 within 4 days after inoculation, but no resistance was observed when older systemically infected leaves showing mosaic symptoms were inoculated with U2. In a later study it was found that specific and nonspecific factors were involved in cross-protection. Apparently, the interaction results from the prevention of uncoating of the related challenge virus in light green tissue of *N. sylvestris* (Sherwood and Fulton, 1982).

Numerous mixed infection experiments using strains U1 and U2 have been reported. Cohen *et al.* (1957) found that after a mixed infection there was a slight predominance of U1 depending, however, on the ratio of U1/U2 in the inoculum. Even with a high proportion of U1 in the mixture, at least some U2 was maintained in mixed systemic infections. Since U2 does not cause severe symptoms in tobacco and tends to become more or less latent in old leaves, its presence as a contaminant of U1 has often been overlooked. In order to keep U1 free of TMGMV, it is recommended that one cultivate U1 in *N. sylvestris* and use an inoculum of this plant for mass production in Samsun tobacco. To keep U2 free from U1, *E. planum* is recommended as host.

A strong interference has been observed in *N. sylvestris* upon inoculation with U1/U2 mixtures (Siegel, 1959). The U1 strain was able to inhibit the production of U2 lesions up to 90%. To explain this, it was suggested that if an infectible site of the cell is blocked by one particle of one strain, particles of the other strain are excluded from infection. This hypothesis means that one particle initiates infection and that the interaction is a competition for infectible sites. Similar results were obtained using *Phaseolus vulgaris* as indicator plant for U1/U2 interference (Wu and Rappaport, 1961). The interference between the two strains was explained quite differently by Helms (1965), who suggested that the interference is caused by a more rapid invasion of one virus (U2) inducing metabolic changes which inhibit the slower virus (U1). In a simliar vein, Benda (1959) and Shalla and Petersen (1978) attributed the interference of U1/U2 in the experiments of Siegel (1959) to a cellular immunization mechanism. In spite of the numerous studies that have been devoted to interference between U1 and U2 strains, it is clear that our understanding of this phenomenon is still fragmentary (Kassanis, 1963; Siegel and Zaitlin, 1964; Matthews, 1981).

It is easier to understand the survival of both viruses in mixed infections. Although U1 is more stable (see below) and predominates in certain hosts, U2 invades tissues more rapidly (Helms, 1965). Also, U2 is able to move closer to the apex than U1 (Solberg and Bald, 1963). Apparently, evolution produced a balanced system in which both viruses have equal survival chances.

B. Interference at an Early Stage of Infection

The early events of infection of tobacco with a mixture of U1 and U2 strains were investigated by using ultraviolet light for inactivation of the viruses, their nucleic acids, and the infectible sites (Siegel and Wildman, 1956; Siegel *et al.*, 1957).

It was found that U2 is 5½ times more sensitive to UV irradiation than U1. The U2 virus was only a little more sensitive *in vitro* than *in vivo*. The infective centers are probably very close to the leaf surface since in the first 2 hr the effect of UV irradiation is mainly on the virus and not on the host. The two viruses differ in their behavior during the early steps of the infection process. The lag phase is 2.5 hr for the U2 strain and 5 hr for the U1 strain. The coat protein protects the nucleic acid from damage by irradiation much better in the case of U1 than of U2. The U1 virus is also more resistant to chemical degradation and heat denaturation than U2 (Siegel *et al.*, 1956). These differences are due either to structural differences in the coat proteins or to a weaker binding between protein and nucleic acid in U2. This is reflected in the different densities of U1 and U2 virions (Siegel and Hudson, 1959; Skotnicki *et al.*, 1976). The U1 strain is more densely packed than U2 which may explain the better protecting ability of U1 protein.

C. Structural Interactions of Particles

If U1 and U2 replicate in the same host cell, interactions such as genomic masking or phenotypic mixing (Dodds and Hamilton, 1976) could be expected. However, no such structural interactions of the U1 and U2 virions were observed either *in vivo* or in reassembly experiments *in vitro* (Atabekova *et al.*, 1975; Taliansky *et al.*, 1977). This contrasts with the formation of "pseudovirions" containing chloroplast or nuclear RNA encapsidated in the viral coat protein. Such particles can make up 2.5% of a purified U2 preparation (Siegel, 1971).

Phenotypic mixing was observed when *N. glutinosa* plants were mixedly infected by U2 and the sunn-hemp mosaic virus. It was found that particles of an intermediate density contain both coat proteins and that these occur in the same proportion in all particles (Skotnicki *et al.*, 1976, 1977).

VI. VIRUS MUTATION VERSUS HOST-INDUCED MUTATION

The appearance of yellow spots on tobacco plants infected with TMV undoubtedly is due to spontaneous mutation (McKinney, 1929, 1935). TMGMV does not give rise to yellow mutants in samsun tobacco (McKinney, 1929, 1935; Köhler and Panjan, 1943). However, the amino acid exchange in the coat proteins of U2 and G-TAMV (Van Regenmortel, 1967) demonstrates that U2 is also subjected to point mutation. Tsugita (1962) provided evidence that mutation of G-TAMV could be chemically induced by *N*-bromosuccinimide leading to the amino acid exchange Ala→Gly. On the other hand, Bald *et al.* (1974) proposed that a host-induced mutation could lead to a "transition" of U1 to M5 (~ U2) virus, which would imply multiple base transitions leading to 16 amino acid substitutions. Such changes cannot be explained by simultaneous multiple point mutations. The results of a recent investigation make such a process of transition quite unlikely, since it was demonstrated that tomato, which is supposed to induce the transition at high temperatures, does not provide any replication of the virus (Wetter, 1984b). It seems safe to assume that TMV and TMGMV evolved over a long period of time from a common ancestor by the usual processes of mutation and selection.

REFERENCES

Altschuh, D., Reinbolt, J., and Van Regenmortel, M. H. V., 1981, Sequence and antigenic activity of the region 93–113 of the coat protein of strain U2 of tobacco mosaic virus, *J. Gen. Virol.* **52**:363.

American Type Culture Collection, 1981, *Catalogue of Strains II*, Rockville, Md.

Atabekova, T. I., Taliansky, M. E., and Atabekov, J. G., 1975, Specificity of protein–RNA and protein–protein interaction upon assembly of TMV *in vivo* and *in vitro*, *Virology* **67**:1.

Bald, J. G., and Goodshild, D. J., 1960, Tobacco mosaic virus in Nicotiana glauca, *Phytopathology* **50**:497.

Bald, J. G., Gumpf, D. J., and Heick, J., 1974, Transition from common tobacco mosaic virus to the *Nicotiana glauca* form, *Virology* **59**:467.

Ball, E. M., 1966, A technique for comparing the electrophoretic mobility rates of viruses or virus strains, *Arch. Biochem. Biophys.* **114**:547.

Benda, G. T. A., 1949, Concerning: Mutual exclusion of strains of tobacco mosaic virus, *Virology* **9**:712.

Boedtker, H., and Simmons, N. S., 1958, The preparation and characterization of essentially uniform tobacco mosaic virus particles, *J. Am. Chem. Soc.* **80**:2550.

Brandes, J., and Wetter, C., 1959, Classification of elongated plant viruses on the basis of particle morphology, *Virology* **8**:99.

Cohen, M., Siegel, A., Zaitlin, M., Hudson, W. R., and Wildman, S. G., 1957, A study of tobacco mosaic strain predominance and an hypothesis for the origin of systemic virus infection, *Phytopathology* **47**:694.

Conti, M., and Marte, M., 1983, Virus, virosi e micoplasmosi del peperone, *Ital. Agric.* **120**:132.

Dodds, J. A., and Hamilton, R. I., 1976, Structural interactions between viruses as a consequence of mixed infections, *Adv. Virus Res.* **20**:33.

Dudman, W. F., 1965, Differentiation of strains of tobacco mosaic virus by immune diffusion in agar plates, *Phytopathology* **55**:635.

Fulton, R. W., 1951, Superinfection by strains of tobacco mosaic virus, *Phytopathology* **41**:579.

Ginoza, W., and Atkinson, D. E., 1955, Comparison of some physical and chemical properties of eight strains of tobacco mosaic virus, *Virology* **1**:253.

Harrison, B. D., Finch, J. T, Gibbs, A. J., Hollings, M., Shepherd, R. J., Valenta, V., and Wetter, C., 1971, Sixteen groups of plant viruses, *Virology* **45**:356.

Helms, K., 1965, Interference between two strains of tobacco mosaic virus in leaves of pinto bean, *Virology* **27**:346.

Hennig, B., and Wittmann, H. G., 1972, Tobacco mosaic virus: Mutants and strains, in: *Principles and Techniques in Plant Virology* (C. J. Kado and H. O. Agrawal, eds.), pp. 546–672, Van Nostrand–Reinhold, Princeton, N.J.

Johnson, J., 1947, Virus attenuation and the separation of strains by specific hosts, *Phytopathology* **37:**822.

Kassanis, B., 1963, Interactions of viruses in plants, *Annu. Rev. Phytopathol.* **10:**219.

Knight, C. A., Silva, D. M., Dahl, D., and Tsugita, A., 1962, Two distinctive strains of tobacco mosaic virus, *Virology* **16:**236.

Köhler, E., and Panjan, M., 1943, Das Paramosaikvirus der Tabakpflanze, *Ber. Dtsch. Bot. Ges.* **61:**175.

Kreibig, U., and Wetter, C., 1980, Light diffraction of *in vitro* crystals of six tobacco mosaic viruses, *Z. Naturforsch.* **35c:**750.

McKinney, H. H., 1929, Mosaic diseases in the Canary Islands, West Africa, and Gibraltar, *J. Agric. Res.* **39:**557.

McKinney, H. H., 1935, Evidence of virus mutation in the common mosaic of tobacco, *J. Agric. Res.* **51:**951.

McKinney, H. H., 1952, Two strains of tobacco mosaic virus one of which is seed-borne in an etch-immune pungent pepper, *Plant Dis. Rep.* **36:**184.

Matthews, R. E. F., 1981, *Plant Virology*, 2nd ed., Academic Press, New York.

Miller, P. M., and Thornberry, H. H., 1958, A new viral disease of tomato and pepper, *Phytopathology* **48:**665.

Palukaitis, P., and Symons, R. H., 1980, Nucleotide sequence homology of thirteen tobamovirus RNAs as determined by hybridization analysis with complementary DNA, *Virology* **107:**354.

Randles, J. W., Palukaitis, P., and Davies, C., 1981, Natural distribution, spread and variation, in the tobacco mosaic virus infecting *Nicotiana glauca* in Australia. *Ann. appl. Biol.* **98:**109.

Rentschler, L., 1967, Aminosäuresequenzen und physikochemisches Verhalten des Hüllproteins eines Wildstammes des Tabakmosaikvirus. I. Analyse der Primärstruktur (Pos. 62–134) des Hüllproteins von Wildstamm U2, *Mol. Gen. Genet.* **100:**84.

Shalla, T. A., 1968, Virus particles in chloroplasts of plants infected with the U5 strain of tobacco mosaic virus, *Virology* **35:**194.

Shalla, T. A., and Petersen, L. J., 1978, Studies on the mechanism of viral cross protection, *Phytopathology* **68:**1681.

Shalla, T. A., Petersen, L. J., and Giunchedi, L., 1975, Partial characterization of virus-like particles in chloroplasts of plants infected with the U5 strain of TMV, *Virology* **66:**94.

Sherwood, J. L., and Fulton, R. W., 1982, The specific involvement of coat protein in tobacco mosaic virus cross protection, *Virology* **119:**150.

Siegel, A., 1959, Mutual exclusion of strains of tobacco mosaic virus, *Virology* **8:**470.

Siegel, A., 1971, Pseudovirions of tobacco mosaic virus, *Virology* **46:**50.

Siegel, A., and Hudson, W., 1959, Equilibrium centrifugation of two strains of tobacco mosaic virus in density gradients, *Biochim. Biophys. Acta* **34:**254.

Siegel, A., and Wildman, S. G., 1954, Some natural relationships among strains of tobacco mosaic virus, *Phytopathology* **44:**277.

Siegel, A., and Wildman, S. G., 1956, The inactivation of the infectious centers of tobacco mosaic virus by ultraviolet light, *Virology* **2:**69.

Siegel, A., and Zaitlin, M., 1964, Infection process in plant virus diseases, *Annu. Rev. Phytopathol.* **2:**179.

Siegel, A., Wildman, S. G., and Ginoza, W., 1956, Sensitivity to ultra-violet light of infectious tobacco mosaic virus nucleic acid, *Nature* (*London*) **178:**1117.

Siegel, A., Ginoza, W., and Wildman, S. G., 1957, The early events of infection with tobacco mosaic virus nucleic acid, *Virology* **3:**554.

Skotnicki, A., Scotti, P. D., and Gibbs, A., 1976, On the nature of the difference in the densities of the particles of two tobamoviruses, *Intervirology* **7:**292.

Skotnicki, A., Scotti, P. D., and Gibbs, A., 1977, Particles produced during a mixed infection by two tobamoviruses contain coat proteins of both viruses, *Intervirology* **8**:60.

Solberg, R. A., and Bald, J. G., 1963, Distribution of a natural and an alien form of tobacco mosaic virus in the shoot apex of *Nicotiana glauca* Grah., *Virology* **21**:300.

Taliansky, M. E., Atabekova, T. I., and Atabekov, J. G., 1977, The formation of phenotypically mixed particles upon mixed assembly of some tobacco mosaic virus (TMV) strains, *Virology* **76**:701.

Tsugita, A., 1962, The proteins of mutants of TMV: Classification of spontaneous and chemically evoked strains, *J. Mol. Biol.* **5**:293.

Van De Walle, M. J., and Siegel, A., 1976, A study of nucleotide sequence homology between strains of tobacco mosaic virus, *Virology* **73**:413.

Van De Walle, M. J., and Siegel, A., 1982, Relationships between strains of tobacco mosaic virus and other selected plant viruses, *Phytopathology* **72**:390.

Van Regenmortel, M. H. V., 1967, Serological studies on naturally occurring strains and chemically induced mutants of tobacco mosaic virus, *Virology* **31**:467.

Van Regenmortel, M. H. V., 1975, Antigenic relationships between strains of tobacco mosaic virus, *Virology* **64**:415.

Van Regenmortel, M. H. V., 1982, *Serology and Immunochemistry of Plant Viruses*, Academic Press, New York.

Van Regenmortel, M. H. V., and Burckard, J., 1980, Detection of a wide spectrum of tobacco mosaic virus strains by indirect enzyme-linked immunosorbent assays (ELISA), *Virology* **106**:327.

Wetter, C., 1980, Occurrence of para-tobacco mosaic virus in field tobacco in South-West Germany, *Z. Pflanzenkr. Pflanzenschutz.* **87**:150.

Wetter, C., 1984a, Serological identification of four tobamoviruses infecting pepper, *Plant Dis.* **68**:597.

Wetter, C., 1984b, Antigenic relationships between isolates of mild dark-green tobacco mosaic virus and the problem of host-induced mutation, *Phytopathology* **74**:1308.

Wetter, C., and Bernard, M., 1977, Identifizierung, Reinigung und serologischer Nachweis von Tabakmosaikvirus und Para-Tabakmosaikvirus aus Zigaretten, *Phytopathol. Z.* **90**:257.

Wetter, C., Conti, M., Altschuh, D., Tabillion, R., and Van Regenmortel, M. H. V., 1984, Pepper mild mottle virus, a tobamovirus infecting pepper cultivars in Sicily, *Phytopathology* **74**:405.

Wittmann, H. G., 1965, Die primäre Proteinstruktur von Stämmen des Tabakmosaikvirus. Teil IV. Aminosäuresequenzen (Pos. 1–61 und 135–158) des Proteins des Tabakmosaikvirus-Stammes U2, *Z. Naturforsch.* **20b**:1213.

Wu, J. H., and Rappaport, I., 1961, An analysis of the interference between two strains of tobacco mosaic virus on *Phaseolus vulgaris* L., *Virology* **14**:259.

Zaitlin, M., and Israel, H. W., 1975, Tobacco mosaic virus (type strain), *CMI/AAB Descriptions of Plant Viruses No. 151*.

Zettler, F. W., and Nagel, J., 1983, Infection of cultivated gesneriads by two strains of tobacco mosaic virus, *Plant Dis.* **67**:1123.

CHAPTER 11

Ribgrass Mosaic Virus

CARL WETTER

I. INTRODUCTION

Holmes (1941) was the first to recognize that a virus isolated from ribgrass and broad-leaf plantain was distinct from tobacco mosaic virus (TMV). But the history of the virus began earlier because Holmes called the attention of Valleau and Johnson (1943) to a similar virus that these authors had found in *Plantago major* and which they had transferred to tobacco in 1930. It was found subsequently that this virus was the inciting pathogen of a necrotic virus disease in burley tobacco in Kentucky (Valleau and Johnson, 1943). In the following decades, ribgrass mosaic virus (RMV) was found to be widely distributed in North America, Europe, and Asia, but it has apparently not produced serious epidemic diseases. This virus is often called Holmes ribgrass virus (HRV).

II. BIOLOGICAL PROPERTIES

A. Isolates and Strains of RMV and Their Economic Importance

Many variants of RMV have been isolated. Records of isolation are listed in Table I. Although most isolates could be identified by the fact that the symptoms they induced were similar to those caused by Holmes's type strain, some isolates showed a remarkable variability in chemical and serological properties. There have been isolated reports of local disease outbreaks in crops such as tobacco (Valleau and Johnson, 1943; Kovachevsky, 1963) and tomato in which an internal browning of the fruits was observed (Holmes, 1950). A mosaic disease of *Digitalis lanata* and *D. purpurea* cultivated as drug plants in East Germany was

CARL WETTER • Department of Botany, University of Saarland, D-6600 Saarbrücken, West Germany.

TABLE I. Host Plants of Isolates/Strains of Ribgrass Mosaic Virus and Location of Occurrence

Host plant	Location	Reference
Plantago lanceolata L., *P. major* L.	Princeton, N.J.	Holmes (1941)
P. major	Arlington, Va.	McKinney (1943)
P. major, burley tobacco	Lexington, Ky.	Valleau and Johnson (1943)
P. rugelii Decne.	Beltsville, Md.; Arlington, Va.	McKinney and Fulton (1949)
Lycopersicon esculentum L.	New Jersey	Holmes (1950)
P. major, *Silene alba* Mill., tobacco, tomato	Bulgaria	Kovachevsky (1953, 1963)
P. major	Manhattan, Kans.	McKinney (1954) (cited in ATCC, 1981)
P. lanceolata, *P. major*	Scotland	Harrison (1956)
Rorippa sylvestris L. Bess. (syn. *Radicula*)	Northern Japan	Goto and Oshima (1962)
Brassica rapa L.	China	Pei (1962)
Digitalis lanata Ehrh., *D. purpurea* L.	East Germany	Schumann (1963)
Cardaria draba (L.) Desv., *Sisymbrium loeselii* L.	Czechoslovakia	Polák (1964)
Eutrema wasabi Maxim.	Central Japan	Tochihara *et al.* (1964)
Anchusa officinalis L.	Czechoslovakia	Polák (1966)
Silene alba (syn. *Lychnis*)	Montana	Chessin *et al.* (1967)
Primula obconica Hance	Bulgaria	Kovachevsky (1969)
P. media L.	Yugoslavia	Juretić *et al.* (1969)
R. amphibia L.	Yugoslavia	Juretić *et al.* (1973)
D. ciliata Trautv.	Yugoslavia	Juretić and Miličić (1977)
P. media	Missoula, Mont.	Chessin *et al.* (1980).

found to be caused by RMV. The proportion of diseased plants that were heavily reduced in growth ranged from 5 to 100% (Schumann, 1963). Although RMV has been found to infect many cruciferous crops in Asia, the economic importance of these diseases has not been assessed.

B. Symptomatology and Host Range

In ribgrass and in broad-leaved plantain, RMV causes a chlorotic mottling (Fig. 1A). Inoculation of Samsun tobacco initiates numerous primary lesions (Fig. 1D). Usually, the lesions appear as white necrotic flecks and rings. The secondary lesions on uninoculated leaves consist of rings and lines of necrotic tissues that follow the veins (Fig. 1E,F). The next leaves that become systemically infected show chlorotic mottling with less necrosis. The necrotic reaction of Samsun tobacco is characteristic of RMV infection. The virus can be clearly distinguished from TMV because the latter is not able to infect ribgrass systemically (Holmes, 1941).

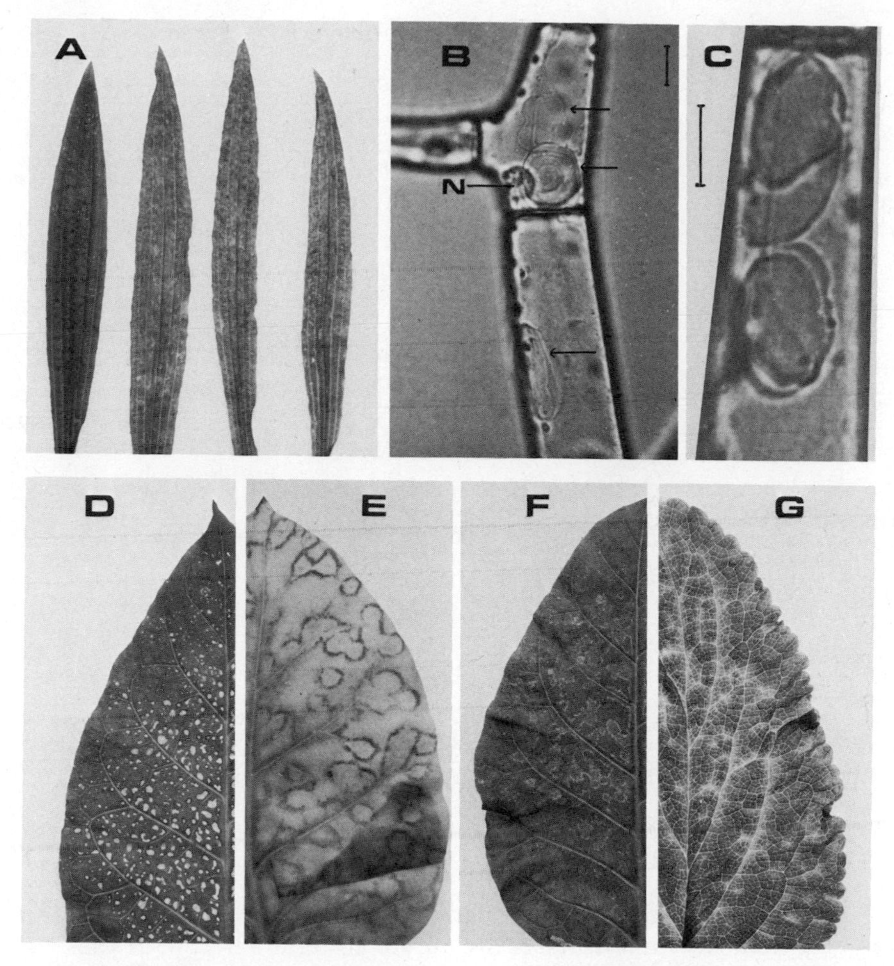

FIGURE 1. Symptoms induced by ribgrass mosaic virus (RMV) in different host plants. (A) Three leaves of ribgrass infected with RMV (right); one healthy leaf is shown at the left. (B, C) Hair cells of Samsun tobacco with rounded plates (arrows). N, nucleus; bar = 10 μm. (D) Primary lesions on a leaf of Samsun tobacco. (E, F) Secondary necrotic ring symptoms on Samsun leaves. (G) Systemic flecks induced on a leaf of *Digitalis purpurea*.

About 50 species in 10 plant families can be infected by inoculation (Oshima and Harrison, 1975). The virus is found in nature to infect plants of the families Caryophyllaceae, Cruciferae, Primulaceae, Plantaginaceae, and Solanaceae. In spite of the fact that RMV has been found to cause a necrotic fruit disease in tomato (Holmes, 1950), most investigations agree that tobacco, tomato, and pepper are relatively resistant to infection (Kovachevsky, 1963). Holmes's RMV, the *Lychnis* strain, and strains C and W (Table II) could infect many cruciferous species inciting mosaic and mottle symptoms. Many species of *Brassica* may be infected symptomlessly (Oshima *et al.*, 1974). Possibly because the natural hosts

TABLE II. Amino Acid Composition of RMV Strains

Amino acid	Strain[a]					
	HRV	C	W	M	Y	Ly[b]
Asp	17	15	13	16	14	16
Thr	13	12	11	11	11	13
Ser	13	14	14	15	14	16
Glu	22	23	22	23	25	21
Pro	8	8	7	10	11	9
Gly	4	3	3	4	3	3
Ala	18	17	17	18	19	17
Val	10	9	10	10	10	10
Met	3	4	4	3	2	4
Ile	8	7	8	7	7	7
Leu	11	15	13	12	12	12
Tyr	7	6	6	6	6	7
Phe	6	6	6	5	5	5
His	1	1	1	1	1	1
Lys	2	2	3	2	2	2
Arg	10	11	11	9	9	11
Cys	1	1	1	1	3	1
Trp	2	2	2	3	2	3
Total	156	156	152	156	156	158
Exchanges	—	8	8	7	11	7

[a] Abbrevation of strains and composition is adapted from: HRV = Holmes's ribgrass virus (Wittmann *et al.*, 1969); C = virus from *Rorippa sylvestris*, and W = virus from *Eutrema wasabi* (Oshima *et al.*, 1971); M = Missoula isolate, and Y = Yugoslavian isolate (Chessin *et al.*, 1980); Ly = *Lychnis* isolate (Chessin *et al.*, 1967).

[b] The composition of Ly probably has to be corrected to 156 residues. This does not alter the number of exchanges in relation to HRV.

of RMV are rather resistant to infection with other tobamoviruses (Oshima *et al.*, 1974), RMV isolates are rarely contaminated with these viruses. RMV seems to be the most common virus in *Plantago* species. Of these, *P. lanceolata* and *P. major* belong to the most successful colonizing perennial weeds of the world (Hammond, 1982). Especially *P. lanceolata* is widely dispersed in crops, pastures, and arable fields. Although one would expect plantains to be an important reservoir of RMV from which it infects cultivated plant species (Hammond, 1982), this spread apparently does not occur. This may be due to the tendency of RMV to cause a necrotic reaction in many hosts. In tobacco, RMV reaches much lower concentration levels than TMV and some of its strains (Veldee and Fraenkel-Conrat, 1962; Rappaport and Wildman, 1957). It seems that RMV is not able to compete successfully with other tobamoviruses but that it is especially adapted to certain hosts such as plantains.

C. Cytopathology

In hair cells of tobacco plants infected with RMV, characteristic crystalline inclusions can be observed in the light microscope (Fig. 1B,C).

Such inclusions were called rounded plates by Miličić *et al.* (1968) and were first described by Goldin (1953). Ultrathin sections of infected tissue showed that the rounded plates consisted of layers of virions. But in contrast to TMV, the virions within the layers were mostly not tilted from layer to layer but were oriented parallel to the long axis of the plate (Juretić *et al.*, 1969). Recent investigations on the ultrastructure of crystalline inclusions of several RMV strains in different hosts revealed that a variety of intracellular structures can be formed (Ernwein and Wetter, unpublished).

In Samsun tobacco, the virus isolate PV 229 (ATCC, 1981) induced rounded plates which consisted of four to seven layers of parallel-oriented virions. The regular arrangement is mostly disturbed by cytoplasmic inclusions and particles that are irregularly packed. The same virus induced, in young xylem cells of *Capsicum annuum* cv. Liebesapfel, angled-layer aggregates similar to those found in tissues infected with pepper mild mottle virus (Wetter *et al.*, 1984) and mild dark-green tobacco mosaic virus (Wetter, this volume). Another type of crystalline structure was found in a phloem cell of pepper. This inclusion consisted of virion layers wound around a rotation axis parallel to the long axis of the particles. Abundant virus aggregates of this structure were found in cells of *D. purpurea* infected with the isolate of Schumann (1963) (Fig. 1G). As can be seen, the layers take the form of a spiral and the virions become circularly bent around a center (Fig. 2A–C). The orientation of the particles changes in such a way that particles that are 300 nm apart are cut in the longitudinal and in the transverse direction (Fig. 2B). Similar viral inclusions were observed by Purcifull *et al.* (1966) in pea infected with clover yellow mosaic virus, a potexvirus. In view of the similarity between the two structures, we also call this type of inclusion, which is new for RMV, "spiral aggregate." The resemblance of these structures to fingerprints is striking. It is possible that the particles of RMV are more elastic than those of other tobamoviruses.

III. PHYSICOCHEMICAL PROPERTIES

A. Purification

RMV can be purified by any of the many procedures used in the case of TMV. To obtain a high proportion of monodisperse particles, the method of Boedtker and Simmons (1958) is recommended. Such preparations can be used to produce iridescent gels (Oster, 1950) which are able to form two-dimensional micro- and macrocrystals (Kreibig and Wetter, 1980). Colloidal crystalline preparations are stable for several years.

B. Particle Morphology

Particles of RMV have a morphology similar to that of TMV. They exhibit a unimodal length distribution; the normal length (Brandes and

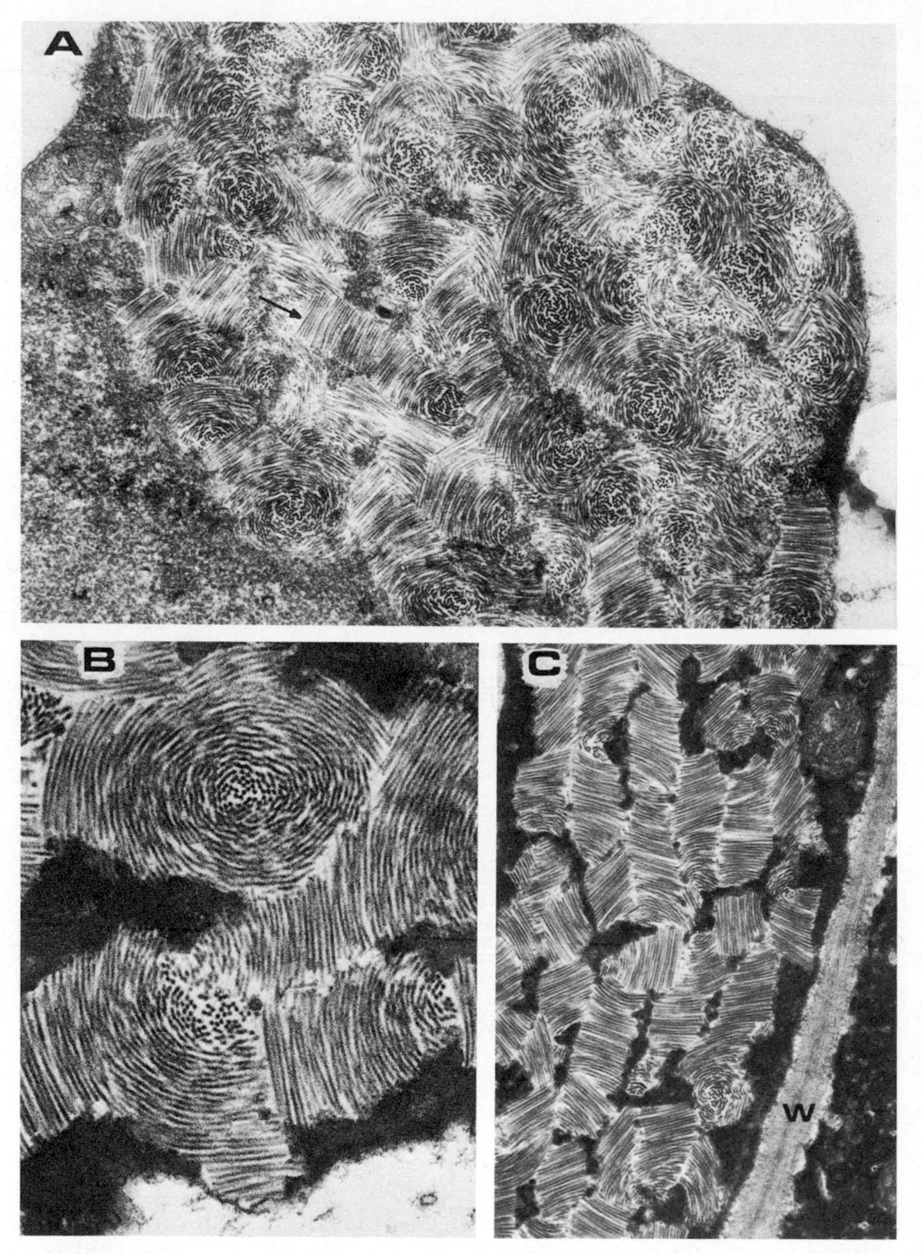

FIGURE 2. Ultrathin sections of *Digitalis purpurea* cells infected with ribgrass mosaic virus *Digitalis* strain. (A) Numerous spiral aggregates in a mesophyll cell. Arrow indicates a layer of virions. (B) Layers of virions bent and turned in a spiral fashion. (C) Layers with a few spiral aggregates. w, cell wall.

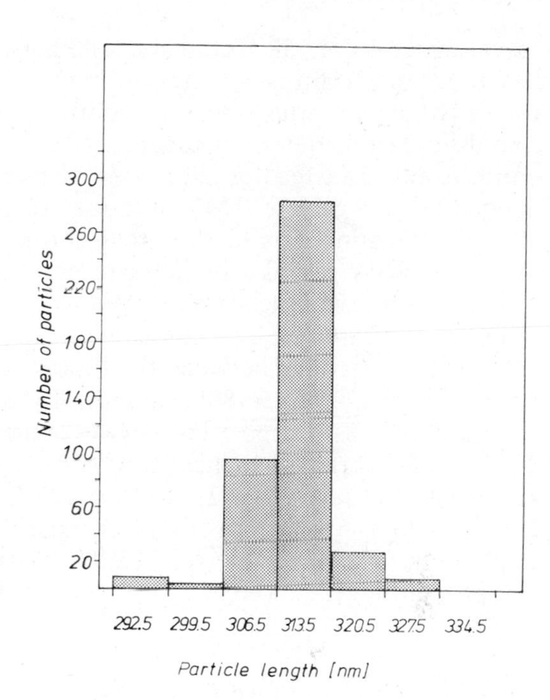

FIGURE 3. Histogram of particle length distribution of ribgrass mosaic virus (RMV) from a preparation obtained by density gradient centrifugation.

Wetter, 1959) of the particles is 311.4 nm and the modal length is 312.9 nm (Fig. 3). These values were determined in comparative measurements on 476 particles using the length of TMV as a standard. About 90% of the particles belonged to the main maximum. The diameter of the particles is 18 nm.

C. Virion Properties

Owing to similar morphology, some of the intrinsic properties of RMV such as the sedimentation coefficient must be very similar to those of TMV (Zaitlin and Israel, 1975). Other values differ from TMV because of the different virion composition. Several values have been reported for the isoelectric point of RMV: pH 4.49 (Oster, 1951), pH 4.02 (Gordon and Price, 1953), and pH 4.18 (Ginoza and Atkinson, 1955). The electrophoretic mobility is -7.7×10^{-5} cm^2/sec per V at pH 7.1 (Knight and Lauffer, 1942). The extinction coefficient ($E^{0.1\%}_{1\,cm}$) at 260 nm, uncorrected for light scattering, is 3.2. The ratio of absorbance at 260 nm/280 nm is 1.2. The thermal inactivation point is ~ 93°C; the dilution endpoint is 10^{-6} to 10^{-7}. RMV is about three times as sensitive to ultraviolet irradiation as U1 (Oshima *et al.*, 1971).

D. The Coat Protein

The amino acid sequence of the coat protein of RMV has been determined by Wittmann *et al.* (1969). In contrast to all other tobamoviruses, which have 158 amino acid residues in the protein subunit, RMV has 156. One strain is reported to have 152 residues (Table II). RMV is the only tobamovirus whose protein contains one histidine residue. It also differs from the other tobamoviruses in having two to four methionine residues instead of zero to two. The difference in amino acid sequence with regard to TMV amounts to 56% (Hennig and Wittmann, 1972). Determinations of the gross composition of naturally occurring strains of RMV reveal a higher rate of variation than in other tobamoviruses. Holmes's type strain differs from the Montana isolate, also obtained from *Plantago*, by 7 exchanges (Table II). The highest number, 11 exchanges, was found between the type strain of RMV and a Yugoslavian strain (Chessin *et al.*, 1980) (Table II). In comparison, the number of exchanges in gross composition between the two virus species TMV and ToMV is 8, while the highest number of exchanges found among TMV strains is 3 (Tsugita, 1962), and among ToMV strains is 2 (Wang and Knight, 1967). Apart from the great variability in coat protein composition, all strains of RMV are similar with respect to host range and symptomatology.

E. Structural Interactions

Because of the great differences in amino acid sequence, it is not surprising that mixtures of RMV and TMV coat proteins neither exhibited any interaction under weakly alkaline conditions nor formed mixed aggregates at low pH values (Sarkar, 1960). Reconstitution experiments (Fraenkel-Conrat and Singer, 1957) using RNA of RMV and protein of TMV and the reverse combination revealed that the hybrid particles were only slightly less infectious than homologous reconstituted virus. However, the most important aspect of this work was the incontrovertible proof for the genetic activity of RNA, in that RMV RNA, even when reconstituted with TMV protein, yielded progeny with the histidine- and methionine-containing coat protein characteristic for RMV and not TMV, as well as showing the symptomatology of RMV. Reconstituted viruses had the same stability under mechanical stress and in weak alkaline solutions, regardless of the origin of the components (Holoubek, 1962).

IV. SEROLOGY

RMV strains seem to be rather distantly related to TMV as judged from the sequence of the coat proteins. In spite of this, the serological differentiation index (SDI) between RMV and TMV is only 2.1 (Van Regenmortel, 1982) which indicates a closer serological relationship than between the pair TMV–U2 (SDI = 2.7) differing in sequence by only 26%. In other words, RMV seems to be more closely related to TMV serolog-

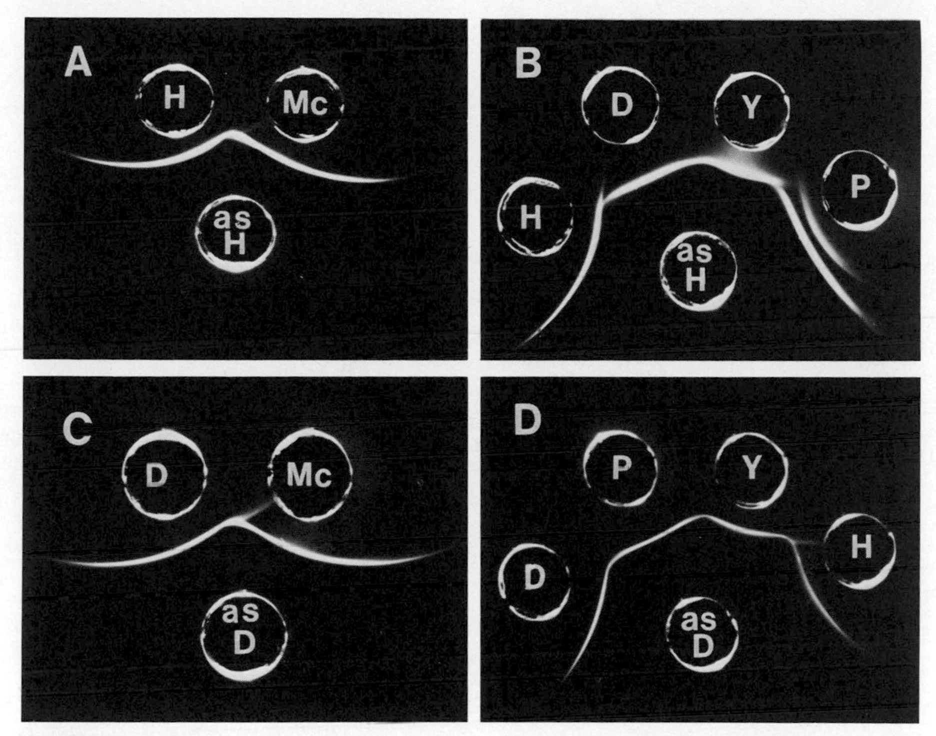

FIGURE 4. Reactions of serotypes of ribgrass mosaic virus (RMV) with their antisera (as) in immunodiffusion tests. (A) Antigens are: H = Holmes's original strain of RMV; Mc = McKinney's strain. Note the very weak spur. (B) Comparison between H = Holmes's strain, D = *Digitalis* strain from East Germany, Y = Yugoslavian strain, and P = *Primula* strain from Bulgaria. (C) *Digitalis* strain (D) is compared with McKinney's strain (Mc). (D) *Digitalis* strain (D) compared with the *Primula* strain (P), the Yugoslavian strain (Y), and Holmes's original strain (H).

ically than would be expected from the difference in coat protein sequence (see Chapter 4 for an explanation of this discrepancy).

The serological relationships between five isolates from different host plants and geographic locations were investigated by immunodiffusion tests (Juretić and Wetter, 1973). It was found that four European isolates could be distinguished from Holmes's type strain, while two isolates, the East German isolate from *Digitalis lanata* and the Yugoslavian isolate from *Plantago media*, could not be distinguished in this type of test (Fig. 4B). The serological differences were visualized by the formation of spurs between the precipitin lines. The European isolates, regardless from which hosts they came, were more closely related to the *Primula* isolate from Bulgaria (Kovachevsky, 1963) than the American isolates of Holmes (Fig. 4D) and McKinney (Fig. 4C). The serological interrelationships among RMV strains seem to reflect the variability in amino acid composition (Table II).

A small spur was also seen at the intersection of the precipitin lines

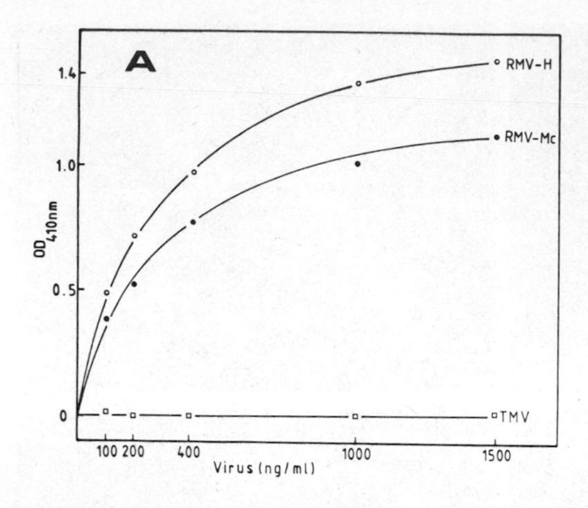

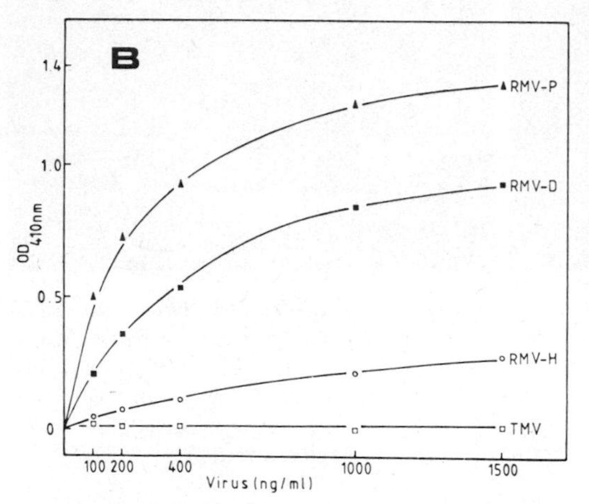

FIGURE 5. Double-antibody sandwich ELISA with ribgrass mosaic virus (RMV) serotypes. (A) Coating of wells was done with 5 μg/ml rabbit anti-RMV-H (H = Holmes's original strain) globulins for 2 hr. The antigens RMV-H and RMV-Mc (Mc = McKinney's strain) were incubated overnight. Rabbit anti-RMV-H conjugate, diluted 1:1000, was incubated for 4 hr. Substrate hydrolysis time was 1 hr. (B) Coating was done with 5 μg/ml rabbit anti-RMV-P (P = *Primula* strain) globulins for 2 hr. The antigens RMV-P, RMV-D (D = *Digitalis* isolate), and RMV-H (H = Holmes's strain) were incubated overnight. Rabbit anti-RMV-P conjugate, diluted 1:1000, was incubated for 4 hr. Substrate hydrolysis time was 1 hr.

formed by Holmes's RMV isolate and the isolate of McKinney (ATCC, 1981) (Fig. 4A). Intragel cross-absorption tests also revealed a minor antigenic difference between the two isolates. Small differences between different isolates were also clearly seen in double-antibody sandwich (DAS)-ELISA (Fig. 5A,B). These examples demonstrate the suitability of DAS-ELISA for the investigation of close relationships between virus strains (Van Regenmortel and Burckard, 1980). Another example of the variability of RMV strains was observed by Juretić (1974). An isolate of RMV from *Rorippa amphibia* could not be distinguished in immunodiffusion tests from an isolate obtained from *P. media*, both originating from Yugoslavia. However, both isolates were clearly distinguishable by a large difference in electrophoretic mobility.

REFERENCES

American Type Culture Collection, 1981, *Catalogue of Strains II*, Rockville, Md.

Boedtker, H., and Simmons, N. S., 1958, The preparation and characterization of essentially uniform tobacco mosaic virus particles, *J. Am. Chem. Soc.* **80**:2550.

Brandes, J., and Wetter, C., 1959, Classification of elongated plant viruses on the basis of particle morphology, *Virology* **8**:99.
Chessin, M., Zaitlin, M., and Solberg, R. A. 1967, A new strain of tobacco mosaic virus from *Lychnis alba, Phytopathology* **57**:452.
Chessin, M., Juretić, N., Miličić, D., Perryman, J., and Giri, L., 1980, A distinctive strain of ribgrass mosaic virus from Montana, USA, *Phytopathol. Z.* **97**:295.
Fraenkel-Conrat, H., and Singer, B., 1957, Virus reconstitution. II. Combination of protein and nucleic acid from different strains, *Biochim. Biophys. Acta* **24**:540.
Ginoza, W., and Atkinson, D. E., 1955, Comparison of some physical and chemical properties of eight strains of tobacco mosaic virus, *Virology* **1**:253.
Goldin, M., 1953, A mosaic of ribgrass [in Russian], *C.R. Acad. Sci. URSS N.S.* **88**:933.
Gordon, R. B., and Price, W. C., 1953, Differentiation of tobacco mosaic virus strains by differences in pH of maximum optical density, *Arch. Biochem. Biophys.* **45**:117.
Goto, T., and Oshima, N., 1962, A strain of tobacco mosaic virus isolated from a wild crucifer plant, *Radicula sylvestris* Druce, *Ann. Phytopathol. Soc. Jpn.* **27**:109.
Hammond, J., 1982, *Plantago* as a host of economically important viruses, *Adv. Virus Res.* **27**:103.
Harrison, B. D., 1956, A strain of tobacco mosaic virus infecting *Plantago* spp. in Scotland, *Plant Pathol.* **5**:147.
Hennig, B., and Wittmann, H. G., 1972, Tobacco mosaic virus: Mutants and strains, in: *Principles and Techniques in Plant Virology* (C. J. Kado and H. O. Agrawal, eds.), pp. 546–672, Van Nostrand–Reinhold, Princeton, N.J.
Holmes, F. O., 1941, A distinctive strain of tobacco-mosaic virus from *Plantago, Phytopathology* **31**:1089.
Holmes, F. O., 1950, Internal-browning disease of tomato caused by strains of tobacco-mosaic virus from *Plantago, Phytopathology* **40**:487.
Holoubek, V., 1962, Mixed reconstitution between protein from common tobacco mosaic virus and ribonucleic acid from other strains, *Virology* **18**:401.
Juretić, N., 1974, Serological properties and inclusion bodies of a tobamovirus isolated from *Roripa amphibia, Phytopathol. Z.* **79**:16.
Juretić, N., and Miličić, D., 1977, Natural infection of *Digitalis ciliata* with ribgrass mosaic virus, *Acta Bot. Croat.* **36**:23.
Juretić, N., and Wetter, C., 1973, Serological relationship among some necrotic strains of the Holmes' ribgrass mosaic virus, in: *Plant Virology*, Proc. 7th Conf. Czech. Plant Virologists, High Tatras, 1971, pp. 369–375, Publishing House Slovak Acad. Sci., Bratislava.
Juretić, N., Wrischer, M., and Polák, Z., 1969, A strain of Holmes' ribgrass virus occurring in Yugoslavia, *Biol. Plant.* **11**:284.
Juretić, N., Miličić, D., and Mamula, D., 1973, Identification of a tobamovirus isolated from *Roripa amphibia, Acta Bot. Croat.* **32**:29.
Knight, C. A., and Lauffer, M. A., 1942, A comparison of the alkaline cleavage products of two strains of tobacco mosaic virus, *J. Biol. Chem.* **144**:411.
Kovachevsky, I. C., 1953, Zhivovlechnata mosaika po tyutyuna (Das Wegerichmosaik des Tabaks), *Izv. Mikrobiol. Inst. Bulg. Acad. Nauk.* **4**:109.
Kovachevsky, I. C., 1963, Untersuchungen über das Wegerichmosaik in Bulgarien, *Phytopathol. Z.* **49**:127.
Kovachevsky, I. C., 1969, Some new diseases in crops, in: *Plant Virology*, Proc. 6th Conf. Czech. Plant Virologists, Olomuc, 1967, pp. 250–252, Publishing House Slovak Acad. Sci., Bratislava.
Kreibig, U., and Wetter, C., 1980, Light diffraction of *in vitro* crystals of six tobacco mosaic viruses, *Z. Naturforsch.* **35c**:750.
McKinney, H. H., 1943, Reaction of resistant tobaccos to certain strains of Nicotiana 1 and other viruses, *Phytopathology* **33**:551.
McKinney, H. H., and Fulton, R. W., 1949, Local susceptibility of cotyledons and leaves of cucumber to tobacco mosaic virus, *Phytopathology* **39**:806.
Miličić, D., Štefanac, Z., Juretić, N., and Wrischer, M., 1968, Cell inclusions of Holmes' ribgrass virus, *Virology* **35**:356.

Oshima, N., and Harrison, B. D., 1975, Ribgrass mosaic virus, *CMI/AAB Descriptions of Plant Viruses No. 152.*

Oshima, N., Ohashi, Y., and Umekawa, M., 1971, Studies on some strains of tobacco mosaic virus pathogenic to crucifer plants. I. Physical and chemical studies, *Ann. Phytopathol. Soc. Jpn.* **37**:319.

Oshima, N., Ohashi, Y., and Umekawa, M., 1974, Studies on some strains of tobacco mosaic virus pathogenic to crucifer plants. 2. Host range, *Ann. Phytopathol. Soc. Jpn.* **40**:243.

Oster, G., 1950, Two-phase formation in solutions of tobacco mosaic virus and the problem of long-range forces, *J. Gen. Physiol.* **33**:445.

Oster, G., 1951, The isoelectric points of some strains of tobacco mosaic virus, *J. Biol. Chem.* **190**:55.

Pei, M. Y., 1962, Preliminary studies on several isolates of TMV from different plants [Chinese with Engl. summary], *Acta Microbiol. Sin.* **8**:420.

Polák, Z., 1964, In nature occurring distinctive necrotic strain of tobacco mosaic virus, in: *Plant Virology,* Proc. 5th Conf. Czech. Plant Virologists, Prague, 1962, pp. 168–169, Publishing House Slovak Acad. Sci., Bratislava.

Polák, Z., 1966, *Anchusa*-mosaic caused by TMV-infection in nature, *Biol. Plant.* **8**:431.

Purcifull, D. E., Edwardson, J. R., and Christie, R. G., 1966, Electron microscopy of intracellular aggregates in pea (*Pisum sativum*) infected with clover yellow mosaic virus, *Virology* **29**:276.

Rappaport, I., and Wildman, S. G., 1957, A kinetic study of local lesion growth on *Nicotiana glutinosa* resulting from tobacco mosaic virus infection, *Virology* **4**:265.

Sarkar, S., 1960, Interaction and mixed aggregation of proteins from tobacco mosaic virus strains, *Z. Naturforsch.* **15b**:778.

Schumann, K., 1963, Untersuchungen zur Charakterisierung und Identifizierung der Erreger des "Digitalis-Mosaik." I. Das Tabakmosaikvirus (*Marmor tabaci* H.), *Phytopathol. Z.* **48**:1.

Tochihara, H., Komuro, Y., and Kobari, Y., 1964, *Proc. Kanto Plant Prot. Soc.* **11**:46 (quoted according to Oshima *et al.*, 1974, Studies on some strains of tobacco mosaic virus pathogenic to crucifer plants. 2. Host range, *Ann. Phytopathol. Soc. Jpn.* **40**:243).

Tsugita, A., 1962, The proteins of mutants of TMV: Classification of spontaneous and chemically evoked strains, *J. Mol. Biol.* **5**:293.

Valleau, W. D., and Johnson, E. M., 1943, An outbreak of plantago virus in burley tobacco, *Phytopathology* **33**:210.

Van Regenmortel, M. H. V., 1982, *Serology and Immunochemistry of Plant Viruses,* Academic Press, New York.

Van Regenmortel, M. H. V., and Burckard, J., 1980, Detection of a wide spectrum of tobacco mosaic virus strains by indirect enzyme-linked immunosorbent assays (ELISA), *Virology* **106**:327.

Veldee, S., and Fraenkel-Conrat, H., 1962, The characterization of tobacco mosaic virus strains by their productivity, *Virology* **18**:56.

Wang, A. L., and Knight, C. A., 1967, Analysis of protein components of tomato strains of tobacco mosaic virus, *Virology* **31**:101.

Wetter, C., Conti, M., Altschuh, D., Tabillion, R., and Van Regenmortel, M. H. V., 1984, Pepper mild mottle virus, a tobamovirus infecting pepper cultivars in Sicily, *Phytopathology* **74**:405.

Wittmann, H. G., Hindennach, I., and Wittmann-Liebold, B., 1969, Die primäre Proteinstruktur von Stämmen des Tabakmosaikvirus. Teil VI: Aminosäuresequenz (Positionen 62–156) des Proteins des Tabakmosaikvirusstammes Holmes rib grass, *Z. Naturforsch.* **24b**:877.

Zaitlin, M., and Israel, H. W., 1975, Tobacco mosaic virus (type strain), *CMI/AAB Descriptions of Plant Viruses No. 151.*

CHAPTER 12

Odontoglossum Ringspot Virus

J. R. EDWARDSON AND F. W. ZETTLER

I. HISTORICAL REVIEW

Odontoglossum ringspot virus (ORSV) was first described by Jensen and Gold (1951) as a pathogen of *Odontoglossum grande*. This pathogen, sometimes referred to as the orchid strain of tobacco mosaic virus (Kado *et al.*, 1968), infects numerous commercially important orchid genera and may be found wherever orchids are cultivated, including Taiwan (Ko, 1982; Inouye and Leu, 1983), Hawaii (Ishii and Martînez, 1973), and Florida (Zettler *et al.*, 1978). Diseases reportedly caused by ORSV include Cymbidium diamond mottle (Jensen, 1953), Cattleya infectious blossom necrosis (Izadpanah *et al.*, 1968), Cattleya mild flower break (Jensen, 1955), and Cattleya color break (Murakishi, 1958). ORSV and the potexvirus, Cymbidium mosaic virus, are considered to be the most prevalent and troublesome of the 22 orchid viruses listed by Lawson and Horst (1984).

ORSV was classified on the basis of particle length and serological affinities by Brandes and Wetter (1959) as a "group 3" member, which also included the tobacco mosaic (TMV), Bawden's cowpea mosaic, and cucumber green mottle viruses. Later, this group was renamed the tobamovirus group by the Plant Virus Subcommittee of the International Committee on Nomenclature of Viruses (Harrison *et al.*, 1971).

II. STRUCTURE

Normal length measurements of ORSV particles in purified and leaf dip preparations range from 280 to 325 nm (Table I). Reported diameters

J. R. EDWARDSON • Plant Virus Laboratory, Agronomy Department, University of Florida, Gainesville, Florida 32611. F. W. ZETTLER • Plant Pathology Department, University of Florida, Gainesville, Florida 32611.

TABLE I. Particle Length Measurements Reported for Odontoglossum Ringspot Virus

Length (nm)	Sample preparation[a]	Host	Reference
275–300	P, D/Sh	*Tetragonia expansa*	Inouye (1966)
280	P/Sh	*Odontoglossum*	Jensen and Gold (1951)
291	—/Sh	*Cattleya–Brassivola* hybrid	Murakishi (1958)
296	D/Sh	*Nicotiana tabacum*	Brandes and Chessin (1965)
296	P/Sh	*Cymbidium*	Francki (1966)
300	P/Sh	—	Newton and Rosberg (1952)
300	—	—	Shvedchikova and Protsenko (1977)
300	D/Sh	—	Brandes (1964)
300	P/Sh	*N. tabacum*	Paul *et al.* (1965)
300	P/Sh	*Cattleya*	Lawson (1970b)
300	P/Ns	—	Inouye (1983)
304	D/Ns	*Cochleanthes*	Clifford (1977)
308	D/Ns	*Gomphrena globosa*	Chagas et al. (1977)
314	D/Sh	*Cymbidium*	Van Regenmortel *et al.* (1964)
317	D/Sh	*Chenopodium quinoa*	Van Regenmortel *et al.* (1964)
320	P/Sh	*G. globosa*	Corbett (1967)
324	P/Sh, Ns	*C. quinoa*	Izadpanah *et al.* (1968)
325	P/Sh	*Cattleya*	Thornberry and Phillippe (1964)

[a] P = purified preparations; D = leaf dips; Sh = shadow cast; Ns = negative staining; — = information not provided.

range from 18 to 27 nm, and measurements derived from shadowed preparations probably represent overestimations. In negatively stained preparations, the particles exhibit distinct channels (Izadpanah *et al.*, 1968; Paul, 1975; Inouye, 1983). At higher magnifications, the coat appears to be constructed of stacked disks (Corbett, 1974). These aspects of ORSV particles and their rigid outline are typical of tobamovirus particles. No discernible morphological changes of ORSV particles were noted when inactivated by exposure to γ and/or neutron rays (Mori and Inouye, 1977).

Analytical centrifugation studies of purified ORSV preparations revealed two peaks ($S_{20,w}$), a broad one at 119 S and a sharper one at 211.6 S. However, only single peaks were noted in similarly treated preparations of TMV (Paul *et al.*, 1965). During sucrose density-gradient centrifugation, ORSV particles were separated into distinct top and middle zones. These zones corresponded to the two peaks noted during analytical centrifugation. A third, fainter bottom zone was also noted. The top zone consisted of fragmented virions which were not infectious, and the middle and bottom zones contained intact, infectious particles about 300 nm in length. In the bottom zone, aggregated particles with lengths greater than

300 nm were also noted (Paul *et al.*, 1965). Chemical and spectrophotometric analyses show that ORSV particles in the middle and bottom zones contain 5–6% RNA; little or no RNA is present in the top zones. $A_{260/280}$ ratios of 1.1 (Clifford, 1977) and 0.93 (Wisler, 1981) have been reported for purified preparations of ORSV.

ORSV capsid molecular weights of 17,598 and 17,300 have been determined from sequence analysis (Hennig, 1972) and equilibrium centrifugation studies (Paul and Buchta, 1971), respectively. Paul (1975) reported the capsid subunit of ORSV to contain 157 amino acid residues, one less than the 158 reported for TMV (Gibbs, 1977).

III. ORSV MUTANTS OR STRAINS

ORSV is considered to be a definitive member of the tobamovirus group (Brandes and Wetter, 1959; Brandes, 1964; Brandes and Bercks, 1965; Harrison *et al.*, 1971; Fenner, 1976; Gibbs, 1977; Matthews, 1979, 1982), although some have referred to it as being a serotype (Francki, 1966) or strain (Van Regenmortel *et al.*, 1964; Corbett, 1967) of TMV. In molecular hybridization analyses, a strain of TMV which induces bright yellow mosaic symptoms on *Nicotiana glauca* exhibited little or no relationship to ORSV (Randles *et al.*, 1981). Kado *et al.* (1968) described seven isolates of TMV from orchids, of which five were distinctive with respect to their serological and chemical characteristics. Isolates 01, 03, 04, 05, and 06 appeared similar to type TMV, while isolates 02 and 07 were more closely related to an ORSV isolate analyzed by Paul *et al.* (1965). Kado *et al.* (1968) also described biological differences between five of the seven isolates based on symptoms induced in various cultivars and species of *Nicotiana*. Other studies do not indicate that TMV is a serious pathogen of orchids. Mechanical inoculation of *Cattleya* seedling leaves has been reported to result in infections with TMV, but the virus did not move systemically (Corbett, 1967). Izadpanah *et al.* (1968) reported TMV and a similar orchid-infecting tobamovirus to have distinct host ranges: whereas TMV could not infect *Cattleya*, ORSV could not infect "Samsum-type" tobaccos susceptible to TMV. Likewise, Lawson and Brannigan (1986) inoculated *Cattleya* seedlings with TMV isolated from commercial cigarettes, but were unable to reisolate the virus from any of the inoculated plants.

Paul (1975) has cautioned that some of the distinctions reported between ORSV and other possible orchid-infecting tobamoviruses should be regarded as tentative since conclusive comparative studies are lacking. Styer and Corbett (1975) found neither biological nor serological differences between an isolate of "Odontoglossum ringspot virus" and two isolates of the "orchid strain of tobacco mosaic virus." However, an isolate of ORSV from California (Jensen and Gold, 1951) differed from a Dutch isolate of this virus in not being able to infect either Xanthi tobacco or *Nicotiana glutinosa* (Paul *et al.*, 1965). Differences between ORSV strains were also noted in direct ELISA studies conducted by Adams and Allison (1981).

The available information indicates that ORSV exists in the form of different strains and that at least some of the so-called isolates of "TMV-O" could be considered to be synonymous with ORSV.

IV. ANTIGENIC STRUCTURE

ORSV has been reported to be serologically related to but distinct from TMV (Corbett, 1967), distantly related to TMV (Paul *et al.*, 1965; Inouye, 1966, 1983; Thomson and Smirk, 1967), and identical to two isolates of "the orchid strain of tobacco mosaic virus" (Styer and Corbett, 1975). The following tobamoviruses have been reported to be serologically related to ORSV: *Chara corallina* virus (Gibbs *et al.*, 1975), cucumber virus-4 (Kado *et al.*, 1968), cucumber green mottle (Paul, 1975), ribgrass mosaic (Kado *et al.*, 1968; Paul, 1975), U2-TMV (Paul, 1975), and tomato mosaic (Kado *et al.*, 1968; Paul, 1975). These and other serological results are summarized in Table II.

Since antisera with similar homologous titers may vary according to the cross-reacting antibodies they may contain, it may be difficult to distinguish viral genetic factors from those of immunization modes and/or schedules (Van Regenmortel, 1966). Use of high-titer antisera has often shown a continuum of serological relationships between different virus strains, and thus decisions regarding the degree of relationship between viruses may be somewhat arbitrary. Specific serological criteria for differentiating virus strains from distinct members of a group are not yet unequivocally established (Bercks, 1966; Van Regenmortel, 1966; Van Regenmortel and von Wechmar, 1970), although recently the serological differentiation index (SDI) has been used to indicate the degree of serological cross-reactivity between two viruses (Van Regenmortel, 1982). This is expressed by the number of twofold dilution steps which separate homologous from heterologous precipitin titers. Viruses are usually considered different viruses rather than strains of one another when their SDI exceed 3–5. However, the reliability of the SDI decreases where it represents a single measurement or an average of very few measurements (Van Regenmortel, 1975). In general, based on SDI differences between ORSV and other tobamoviruses, it appears that ORSV should be considered a virus distinct from other tobamoviruses, including U2-TMV, tomato mosaic virus, and TMV.

V. PATHOGENICITY

ORSV has been reported to infect numerous orchid genera, and species in the families Aizoaceae, Amaranthaceae, Chenopodiaceae, Com-

TABLE II. Serological Relationships Reported for Odontoglossum Ringspot Virus

Technique	Reference	Relationship
Gel diffusion	Corbett (1976)	ORSV is related but not identical to TMV
Tube precipitation	Francki (1966)	ORSV, TMV, and U2-TMV are serologically related to but can be distinguished as serotypes of TMV
Microagglutination	Inouye (1966)	ORSV is distantly related to TMV
Gel diffusion	Inouye (1983)	ORSV is distantly related to TMV
Gel diffusion	Izadpanah *et al.* (1968)	"Cattleya straight rod" virus is related but not identical to TMV or ORSV
Gel diffusion	Kado *et al.* (1968)	"01, 04, 05, 06" are indistinguishable from TMV but distinct from each other. "02" and "07" are very similar to ORSV
Liquid precipitation, gel diffusion	Paul *et al.* (1965)	ORSV is distantly related to TMV
—[a]	Styer and Corbett (1975)	ORSV is identical to two isolates of "TMV-O"
Gel diffusion	Paul (1975)	ORSV is related to Opuntia ringspot, sunn-hemp mosaic, cucumber green mottle, ribgrass mosaic, and tomato mosaic viruses and to TMV and U2-TMV
—	Murakishi (1958)	"Cattleya color break" virus reacted strongly with ORSV
—	Gibbs *et al.* (1975)	ORSV is related to *Chara corallina* virus
—	Thomson and Smirk (1967)	ORSV and TMV are distinguishable; at least 50% of their antigenic groups are dissimilar
Liquid precipitin	Perez and Cortez-Monllor (1960)	"Orchid virus" reacts with TMV antiserum
—	Brunt *et al.* (1982)	Tobamovirus from *Ullucus tuberosus* is unrelated to ORSV
ELISA	Adams and Allison (1981)	Differences exist between ORSV strains when direct, but not indirect, methods are used

[a] — = information not provided.

positae, Leguminosae, and Solanaceae. However, this virus apparently is of agricultural significance only to orchidaceous plants. Studies of inoculating dicotyledonous indicator species have yielded conflicting results. For example, *Cassia occidentalis*, reported to be susceptible by Corbett (1967), was described as insusceptible to the "ORSV-Cy-1" isolate

(Inouye, 1966). Also, Jensen (1970a) has observed that *Gomphrena globosa* is not a host of Cymbidium mosaic virus (CyMV) but is susceptible to ORSV. He states that it is probable that inoculum from *G. globosa*, purported to be CyMV by Hollings (1959), was actually ORSV. If this supposition is correct, then additional hosts of ORSV occur in the families Caryophyllaceae, Lythraceae, Phytolaccaceae, Polygonaceae, Portulacaceae, and Primulaceae.

VI. TRANSMISSION, EPIDEMIOLOGY, CONTROL

No natural vectors of tobamoviruses are known. Namba and Ishii (1971) tested the ability of the aphids *Myzus persicae* and *Cerataphis orchidearum* to transmit ORSV from infected *Cattleya* plants to virus-free *Cattleya* plantlets. The aphids were observed to make brief probes and to engage in prolonged feeding on the healthy plantlets. The plantlets did not develop symptoms during a 5-week period, and attempts to infect indicator plants of *G. globosa* from triturated plantlets failed. Inouye (1968) likewise was unable to demonstrate aphid transmission of ORSV. An early report (Jensen, 1949) of transmission of a virus inducing color break in *Cattleya* by *M. persicae* has not been correlated with a known virus.

Michon (1982) reported ORSV particles (and those of the potexvirus, CyMV) in pollinia of singly infected *Phalaenopsis* plants. Hamilton and Valentine (1984) described ORSV particles in stacked plates in 10% of the *Cymbidium* pollen examined. However, there is no evidence of pollen or seed transmission of ORSV (Inouye, 1968; Paul, 1975; Wisler, 1981). Thus, while seed transmission of ORSV may be possible, rates would appear to be quite low.

The distribution of ORSV in commercial orchids is worldwide. Burnett (1974) suggested that ORSV might spread in native forests because the virus was detected in some *Cattleya* plants from South America upon their arrival in the United States. However, this was not confirmed in studies comparing native and cultivated orchid genera for ORSV infections using serological and electron microscopic techniques (Zettler *et al.*, 1978; Wisler *et al.*, 1979, 1982, 1983; Wisler, 1981). ORSV was not detected in 84 native *Cattleya* accessions or in 354 native orchids in 64 other genera. However, ORSV was detected in 4 of 160 cultivated *Cattleya* accessions and in 43 of 1152 cultivated orchids in 69 other genera. Wisler *et al.* (1979) and Wisler (1981) compared the incidence of ORSV in orchids in cultivation less than 1 year with orchids in cultivation 1 to 4 years and for more than 4 years. No ORSV infections were detected in the accessions cultivated less than 1 year or in those cultivated for 1 to 4 years. However, in the plants cultivated more than 4 years, 43 ORSV infections were detected in 1233 plants of 9 genera. Thus, cultivated,

TABLE III. Physical Properties Reported for Odontoglossum Ringspot Virus

Physical properties[a]				
TIP	DEP	LIV	Host	Reference
70–75	10^{-6}	>2 yrs	*Tetragonia expansa*	Inouye (1966)
82–84	10^{-6}	>30 days	*Chenopodium quinoa*	Izadpanah *et al.* (1968)
85–90	10^{-6}	59 days	*Gomphrena globosa*	Chagas *et al.* (1977)
86–88	—[b]	>8 months	—	Styer and Corbett (1975)
>90	—	—	—	Perez *et al.* (1956)
90–95	10^{-7}	>10 years	Orchid	Inouye (1983)
93	—	—	—	Paul *et al.* (1965)
80–90	10^{-6}	40 days	—	Pokaew and Sutabutra (1974)

[a] TIP = thermal inactivation point (°C); DEP = dilution endpoint; LIV = longevity *in vitro*.
[b] — = information not provided.

rather than wild, orchids seem to be of major significance in the spread of ORSV.

The cultivators of orchids appear to be the most important agents in disseminating ORSV within restricted growing areas and over long distances. ORSV is spread from one orchid to another by using contaminated cutting tools, in dividing plants and harvesting flowers, as well as through use of contaminated pots, sphagnum, and water (Inouye, 1968). This mode of transmission is made possible by the extreme stability of ORSV (Table III). Kado (1964a), Lawson and Ali (1975), Lawson and Brannigan (1986) and Zettler *et al.* (1984), emphasize the importance of hygienic practices in growing orchids. They recommend burning of virus-infected plants to prevent infection of healthy orchids. In cases where the grower retains infected plants, they should be removed from greenhouses containing healthy orchids. Lawson and Ali (1975) recommended washing faucet handles, sinks, towels, doorknobs, benches, tools, and pots with bleach. Lawson (1967) reports reduction of ORSV infectivity by 86 to 96% through treating cutting tools with a 2% NaOH solution, but recommends inactivating viruses on cutting tools by heat. Na_3PO_4 has also been shown to be effective for sterilizing tools (Inouye, 1968). Lawson and Ali (1975) suggest that the extra work and time involved in retaining a diseased plant be considered before saving it from destruction.

While successful for CyMV, initial attempts to obtain ORSV-free orchids through meristem cutting were unsuccessful (Lawson, 1970b; Ishii, 1974). However, ORSV-free *Cymbidium* plants were obtained when apical meristems or protocorms were immersed for 1 hr in ORSV antiserum prior to being transferred to culture medium (Inouye, 1984).

Kado (1964b) states that symptoms induced by thrips, scales, and mites may be confused with virus symptoms. He also states that in winter, virus symptoms may be masked or disappear. In the latter case, growers may conclude that their plants have "outgrown" the virus (Kado, 1964b). High light intensity, air pollution, excessive salts in irrigation water, and pesticide sprays may induce symptoms resembling virus-induced symptoms in orchids. Symptoms induced by different orchid viruses may differ from cultivar to cultivar in the same genus and among different genera, while some virus infections induce no foliage symptoms. Although CyMV by itself caused flower necrosis in *Cattleya* orchids, ORSV appeared to have no causal relation to the disease. ORSV was implicated, however, as causing color-break symptoms in lavender-colored *Cattleya* blooms (Lawson, 1970a,b).

It is obvious from the examples in the preceding paragraph that the appearance of symptoms is unreliable in the detection of ORSV and other orchid viruses. Several methods for diagnosing virus infections in orchids are available, however (Lawson and Ali, 1975; Christie and Edwardson, 1977; Ko *et al.*, 1985a,b; Wisler and Zettler, 1985).

ORSV inoculated on *Chenopodium quinoa, C. amaranticolor* (Jensen, 1970b), *Gomphrena globosa*, and *Tetragonia expansa* (Inouye, 1963) produced local lesions. While other orchid viruses may also induce local lesions on these hosts, the investigator can use such results to eliminate virus-symptom-mimicking phenomena in orchids as causal agents of the symptoms.

Serological tests which are specific and rapid may be more practical than bioassays for indexing large numbers of plants. ORSV antiserum is now commercially available and as such is being used to index growers' plants. The chloroplast agglutination test (Bradley, 1953), while not highly sensitive, is useful where the virus to be tested is in high concentration in plant sap, as ORSV is. Inouye (1977) described a simple, reliable microagglutination test for ORSV involving centrifugation and filtration of plant samples to eliminate precipitin-inducing impurities that make the recording of antibody–antigen reactions difficult. Agar diffusion techniques utilizing sodium dodecyl sulfate to degrade the virus particles and enhance the diffusion of viral protein into the agar (Purcifull and Batchelor, 1977; Clifford and Zettler, 1977) have been modified (Wisler *et al.*, 1982) to detect ORSV in infected leaf disks.

ELISA techniques (Voller *et al.*, 1976) are sensitive in detecting virus in low concentrations and require small quantities of antiserum. Several ELISA techniques have been used to detect ORSV in orchids. ELISA tests were as sensitive as serologically specific electron microscopy and negative staining (Allison *et al.*, 1981), and marked serological differences between ORSV strains were evident when direct, but not indirect, ELISA methods were used (Adams and Allison, 1981). Indirect and direct methods proved essentially equal, however, in their ability to detect ORSV antigen (Chee and Allison, 1981). ELISA was also used successfully to detect ORSV infections in both *Cymbidium* mother plants and explants in various stages of their *in vitro* culture (Toussaint and Albouy, 1982).

Talens *et al.* (1979) described a "counterimmunoelectrophoretic" technique for detecting nanogram quantities of ORSV in orchid leaf extracts. In this procedure, viral and antibody proteins subjected to electrophoresis through an agar gel medium migrate in opposite directions according to their net surface charge, until they meet to produce a precipitin reaction. Results are available in 30–60 min, and faint precipitin reactions can be visualized by staining in Coomassie blue examined under darkfield illumination. According to Talens *et al.*, 200–400 orchid samples can be processed daily by this method.

Electron microscopy of ORSV particles in dip preparations is a sensitive method of detecting the particles in low concentrations (Corbett, 1974; Lawson and Ali, 1975). The morphology of ORSV particles is diagnostic for infection in orchids since viruses in other groups have different particle morphologies. Derrick (1973) and Derrick and Brlansky (1976) have described serologically specific electron microscopic techniques for coating grids with antiserum, which attaches virus particles to the grids. These techniques could be used for ORSV detection. Decoration techniques may also be applied to identification of ORSV particles. Here, leaf dips (Brandes, 1964) are modified by using drops of antiserum diluted with a volatile buffer on the grids. Cut edges of leaves are then dipped into the antiserum drop. After drying, the negatively stained particles appear coated wth antibodies, while unrelated particles are not coated (Ball and Brakke, 1968; Milne and Luisoni, 1977).

Light and elecron microscopy of inclusions induced by ORSV have been used to identify ORSV infections (Christie and Edwardson, 1977; Ko, 1982; Ko *et al.*, 1985b). The light microscopic techniques are rapid, specific for ORSV infections, and inexpensive (see Section VII).

Disc electrophoresis in acrylamide gels and benzidine staining revealed no qualitative changes in peroxidase isozymes of ORSV-infected plants, although quantitative differences were noted (Suchonlimakul and Ishii, 1974).

VII. CYTOPATHOLOGY

The induction of cytoplasmic crystalline inclusions consisting of virions aligned parallel to each other and to the lateral faces of the inclusion is a main characteristic of the tobamoviruses (Matthews, 1982). These inclusions may consist of a single plate of virions, termed a "monolayer," or they may be composed of two or more plates, in which case they are termed "stacked-plate inclusions." While the crystalline inclusions induced by TMV (Sheffield, 1934) and tomato mosaic virus (Warmke, 1968) are usually angular and often hexagonal in polar views, the crystalline inclusions induced by cucumber green mottle mosaic (Hatta and Ushiyama, 1973), ORSV (Miličić and Juretić, 1971; Miličić and Štefanac, 1971; Corbett, 1974; Christie and Edwardson, 1977; Ko, 1982; Ko *et al.*, 1985b), ribgrass mosaic (Miličić *et al.*, 1968), and sunn-hemp mosaic (Resonich, 1961) viruses are rounded or elongate in polar views. Unlike the crystalline inclusions induced by TMV and tomato mosaic, where the diameters of the plates forming the stacked plate inclusions are the same, ORSV, ribgrass mosaic, and sunn-hemp mosaic stacked plate inclusions are composed of plates of decreasing diameters. The plate with the smallest diameter is the farthest removed from the basal or central plate. This stacking of consistently smaller plates gives a "stair-step" appearance to the inclusions in lateral views.

The crystalline monolayer and stacked plate inclusions induced by ORSV and other tobamoviruses stain green in the calcomine orange–Luxol brilliant green (OG) combination. They stain magenta in the Azure A stain after heating (Christie and Edwardson, 1977; Ko *et al.*, 1985b). The staining reaction of the crystalline inclusions indicates that they contain large amounts of ribonucleoprotein. Tobamovirus crystalline inclusions may be disrupted into masses of virus particles by certain fixatives or mechanical stresses. The masses of particles react to the OG and Azure A stains in the same manner as when they are arrayed in crystals. Masses of ORSV particles have been depicted in thin-sectioned cells (Ko, 1982; Inouye, 1983; Ko *et al.*, 1985b).

Cytoplasmic paracrystalline inclusions composed of ORSV particles have been described in light microscopic studies of *Warscewiczella discolor* cells (Christie and Edwardson, 1977) and in thin-sectioned cells of ORSV-infected *Cymbidium* (Ko, 1982; Ko *et al.*, 1985b). Similar paracrystalline inclusions have also been reported in tissues infected with other tobamoviruses.

Cytoplasmic "angled-layer aggregate" inclusions as described by Warmke (1968) have been reported in ORSV-infected *Calanthe* (Corbett, 1974) and *Cymbidium* (Inouye, 1983). Angled-layer aggregate inclusions have also been reported in tissues infected with tomato mosaic (Warmke, 1968), beet necrotic yellow vein (Putz and Vuittenez, 1980), peanut clump (Thouvenel and Fauquet, 1981; Reddy *et al.*, 1983), and pepper mild mottle (Wetter *et al.*, 1984) viruses. They also occur in tissues infected with a pepper strain (Herold and Munz, 1967) and a necrotic strain of TMV (Vela and Rubio-Huertos, 1972), as well as U5 (Shalla, 1968). These inclusions are composed of many layers of aligned and parallel particles oriented across the long axes of the aggregates, one layer flat against the adjacent one and rotated at an angle of slightly less than 60° (or slightly over its supplement, 120°). In cross sections, angled-layer aggregates appear as three-way cross-hatched figures, and in longitudinal sections as short, parallel lines separated by rows of dots (Warmke, 1968). The angled-layer aggregates and paracrystalline inclusions exhibit the same reactions in the OG and Azure A stains as do the crystalline inclusions and the masses of virus particles. Cytoplasmic X-body inclusions induced by TMV have not been reported in ORSV-infected tissues.

The morphology and staining reactions of ORSV-induced crystalline, paracrystalline, and angled-layer aggregate inclusions are diagnostic for infection by this virus in comparisons with inclusions induced in orchids by viruses in other groups (Ko *et al.*, 1985b).

ACKNOWLEDGMENT. The authors are grateful to the American Orchid Society for their part in supporting our virus research program.

REFERENCES

Adams, E. B., and Allison, A. V., 1981, The specificity of ELISA for strains of TMV, *Phytopathology* **71**:103.

Allison, A. V., Adams, E. B., and Steinagel, L., 1981, Enzyme linked immunosorbent assay (ELISA) as a rapid method for screening for the presence of Cymbidium mosaic virus and Odontoglossum ringspot virus, *Phytopathology* **71**:103.

Ball, E. M., and Brakke, M. K., 1968, Leaf-dip serology for electron microscopic identification of plant viruses, *Virology* **36**:152.

Bercks, R., 1966, The significance of weak cross-reactions with high titer antisera, *Proc. Int. Conf. Plant Viruses, Wageningen* **1965**:205.

Bradley, R. H. E., 1953, A rapid method of testing plants in the field for potato virus X, *Am. Potato J.* **29**:289.

Brandes, J., 1964, Identifizierung von gestreckten pflanzenpathogenen Viren auf morphologische Grundlage, *Mitt. Biol. Bundesanst. Land Forstwirtsch. Berlin-Dahlem* **110**:1.

Brandes, J., and Bercks, R., 1965, Gross morphology and serology as a basis for classification of elongated plant viruses, *Adv. Virus Res.* **11**:1.

Brandes, J., and Chessin, M., 1965, An electron microscope study on the size of Sammons' Opuntia virus, *Virology* **25**:673.

Brandes, J., and Wetter, C., 1959, Classification of elongated plant viruses on the basis of particle morphology, *Virology* **8**:99.

Brunt, A. A., Phillips, S., Jones, R. A. C., and Kenten, R. H., 1982, Viruses detected in *Ullucus-tuberosus* Basellaceae from Peru and Bolivia, *Ann. Appl. Biol.* **101**:65.

Burnett, H. C., 1974, Orchid diseases, *Fla. Dep. Agric. Consumer Serv. Bull.* **10**:1.

Chagas, C. M., Noronha, A., and July, J. R., 1977, Ocorrencia de um complexo viral em *Cymbidium* no Brasil, *Biologico* **43**:72.

Chee, W. W., and Allison, A. V., 1981, A comparison of three ELISA techniques for sensitivity in detecting virus antigen, *Phytopathology* **71**:103.

Christie, R. G., and Edwardson, J. R., 1977, Light and electron microscopy of plant virus inclusions, *Fla. Agric. Exp. Stn. Monogr.* **9**:1.

Clifford, H. T., 1977, Immunodiffusion techniques for the detection of the Cymbidium and Odontoglossum ringspot viruses of orchids, M.S. thesis, University of Florida.

Clifford, H. T., and Zettler, F. W., 1977, Application of immunodiffusion tests for detecting Cymbidium mosaic and Odontoglossum ringspot viruses in orchids, *Proc. Am. Phytopathol. Soc.* **4**:121.

Corbett, M. K., 1967, Some distinguishing characteristics of the orchid strain of tobacco mosaic virus, *Phytopathology* **57**:164.

Corbett, M. K., 1974, Detection of viruses and diagnosis of plant viral diseases by electron microscopy, *Tech. Commun. Int. Soc. Hortic. Sci.* **36**:141.

Derrick, K. S., 1973, Quantitative assay for plant viruses using serologically specific electron microscopy, *Virology* **56**:652.

Derrick, K. S., and Brlansky, R. H., 1976, Assay for viruses and mycoplasmas using serologically specific electron microscopy, *Phytopathology* **66**:815.

Fenner, F., 1976, Classification and nomenclature of viruses, *Intervirology* **7**:1.

Francki, R. I. B., 1966, Isolation, purification and some properties of two viruses from cultivated *Cymbidium* orchids, *Aust. J. Biol. Sci.* **19**:555.

Gibbs, A. J., 1977, Tobamovirus group, *CMI/AAB Descriptions of Plant Viruses No. 184.*

Gibbs, A., Skotnicki, A. H., Gardiner, J. E., Walker, E. S., and Hollings, M., 1975, A tobamovirus of a green alga, *Virology* **64**:571.

Hamilton, R. I., and Valentine, B., 1984, Infection of orchid pollen by plant viruses, *Can. J. Plant Pathol.* **6**:185.
Harrison, B. D., Finch, J. T., Gibbs, A. J., Hollings, M., Shepherd, R. J., Valenta, V., and Wetter, C., 1971, Sixteen groups of plant viruses, *Virology* **45**:356.
Hatta, T., and Ushiyama, R., 1973, Mitochondrial vesiculation associated with cucumber green mottle mosaic virus-infected plants, *J. Gen. Virol.* **21**:9.
Hennig, B., 1972, D-287, in: *Atlas of Protein Sequence and Structure*, Volume 5 (M. O. Dayhoff, ed.), National Biomedical Research Foundation, Washington, D.C.
Herold, F., and Munz, K., 1967, An unusual kind of crystalline inclusion of tobacco mosaic virus, *J. Gen. Virol.* **1**:375.
Hollings, M., 1959, Host range studies with fifty-two plant viruses, *Ann. Appl. Biol.* **47**:98.
Inouye, N., 1963, Virus diseases of orchids. II. Symptoms and properties of viruses in Cymbidium, *Jpn. Orchid Soc. Bull.* **11**:1.
Inouye, N., 1966, A virus disease of Cymbidium caused by Odontoglossum ringspot virus, *Ber. Ohara Inst. Landwirtsch. Biol. Okayama Univ.* **13**:149.
Inouye, N., 1968, Some experiments on the transmission of Cymbidium mosaic virus and Odontoglossum ringspot virus, *Ber. Ohara Inst. Landwirtsch. Biol. Okayama Univ.* **14**:89.
Inouye, N., 1977, Serological diagnosis method for Cymbidium mosaic virus and Odontoglossum ringspot virus in orchids, *Nogaku Kenkyu* **56**:1.
Inouye, N., 1983, Host range and properties of a strain of Odontoglossum ringspot virus in Japan, *Nogaku Kenkyu* **60**:53.
Inouye, N., 1984, Effect of antiserum treatment on the production of virus-free *Cymbidium* by means of meristem culture, *Nogaku Kenkyu* **60**:123.
Inouye, N., and Leu, L. S., 1983, Survey on the viruses in orchids in Taiwan, *Nogaku Kenkyu* **60**:91.
Ishii, M., 1974, Partial elimination of virus from doubly infected orchids by meristem explant culture, *Acta Hortic.* **36**:229.
Ishii, M., and Martinez, R., 1973, Significant orchid diseases in Hawaii, *Hawaii Orchid J.* **2**:6.
Izadpanah, K., Thompson, M. R., and Thornberry, H. H., 1968, Orchid virus: Characterization of a type of tobacco-mosaic virus isolated from Cattleya plants with infectious blossom necrosis, *Phytopathol. Z.* **63**:272.
Jensen, D. D., 1949, Breaking of Cattleya or orchid flowers by orchid mosaic virus and its transmission by aphids, *Phytopathology* **39**:1056.
Jensen, D. D., 1953, Virus diseases of Cymbidiums, *Am. Orchid Soc. Bull.* **22**:800.
Jensen, D. D., 1955, Orchid disorders, with special reference to virus diseases, *Am. Orchid Soc. Bull.* **24**:756.
Jensen, D. D., 1970a, Virus diseases of orchids in the Netherlands, *Neth. J. Plant Pathol.* **76**:135.
Jensen, D. D., 1970b, Virus diseases of orchids, *Calif. Agric. Exp. Stn. Ext. Serv. Circular* No. 552.
Jensen, D. D., and Gold, A. H., 1951, A virus ring spot of Odontoglossum orchid: Symptoms, transmission, and electron microscopy, *Phytopathology* **41**:648.
Kado, C. I., 1964a, Orchid virus diseases, their identity and control, *Orchid Dig.* **28**:6.
Kado, C. I., 1964b, Viruses, villains of orchid disorders, *Am. Orchid Soc. Bull.* **33**:943.
Kado, C. I., Van Regenmortel, M. H. V., and Knight, C. A., 1968, Studies on some strains of tobacco mosaic virus in orchids. I. Biological, chemical and serological studies, *Virology* **34**:17.
Ko, N. -J., 1982, Occurrence and diagnosis of Odontoglossum ringspot virus in Taiwan, *Bull. Orchid Soc. Rep. China* **5**:319.
Ko, N. -J., Zettler, F. W., and Wisler, G. C., 1985a, A simplified bioassay technique for Cymbidium mosaic and Odontoglossum ringspot viruses, *Am. Orchid Soc. Bull.* **54**:1080.

Ko, N. -J., Zettler, F. W., Edwardson, J. R., and Christie, R. G., 1985b, Light microscopic techniques for detecting orchid viruses, *Acta Hortic.* **164**:241.

Lawson, R. H., 1967, Chemical inactivation of Cymbidium mosaic and Odontoglossum ringspot viruses, *Am. Orchid. Soc. Bull.* **36**:998.

Lawson, R. H., 1970a, Flower necrosis in Cattleya orchid, *Am. Orchid Soc. Bull.* **39**:306.

Lawson, R., 1970b, Virus induced color-breaking in Cattleya orchid flowers, *Am. Orchid Soc. Bull.* **39**:395.

Lawson, R. H., and Ali, S. 1975, Orchid viruses and their detection by bioassay, serology, and electron microscopy, in: *The Handbook On Orchid Pests and Diseases*, pp. 62–103, American Orchid Society, Cambridge, Mass.

Lawson, R. H., and Brannigan, M., 1986, Virus diseases of orchids, in: *Handbook on Orchid Pests and Diseases*, pp. 2–49, American Orchid Society, Cambridge, Mass.

Lawson, R. H., and Horst, R. K., 1984, Tests help prevent viruses in orchids, *Greenhouse Manager* **July**:102.

Matthews, R. E. F., 1979, Classification and nomenclature of viruses, *Intervirology* **12**:132.

Matthews, R. E. F., 1982, Classification and nomenclature of viruses, *Intervirology* **17**:4.

Michon, E., 1982, Virus, pollens, pollinies, *Rev. Cytol. Biol. Veg. Bot.* **5**:31.

Miličić, D., and Juretić, N., 1971, Zelleinschluss der Viren der Tabakmosaik-Gruppe, *Tagungsber. Dtsch. Akad. Landwirtschaftswiss. Berlin*, **115**:141.

Miličić, D., and Štefanac, Z., 1971, Cell inclusions of the cucumber green mottle mosaic virus and the Odontoglossum ringspot virus, *Acta Bot. Croat.* **30**:33.

Miličić, D., Štefanac, Z., Juretić, N., and Wrischer, M., 1968, Cell inclusions of Holmes' ribgrass virus, *Virology* **35**:356.

Milne, R. G., and Luisoni, E., 1977, Rapid immune electron microscopy of virus preparations, in: *Methods in Virology* (D. Maramorosch and H. Koprowski, eds.), Volume 6, pp. 265–281, Academic Press, New York.

Mori, I., and Inouye, N., 1977, Interaction effect of gamma rays and thermal neutrons on the inactivation of Odontoglossum ringspot virus isolated from orchid, *Annu. Rep. Res. React. Inst. Kyoto Univ.* **10**:54.

Murakishi, H. H., 1958, Serological and morphological relationships among orchid viruses, *Phytopathology* **48**:137.

Namba, R., and Ishii, M., 1971, Failure of aphids to transmit Odontoglossum ringspot and Cymbidium mosaic viruses to orchid plantlets derived from meristem cultures, *Phytopathology* **61**:582.

Newton, N., and Rosberg, J., 1952, Electron-microscope studies of a new orchid virus complex, *Phytopathology* **42**:79.

Paul, H. L., 1975, Odontoglossum ringspot virus, *CMI/AAB Descriptions of Plant Viruses No. 155*.

Paul, H. L., and Buchta, U., 1971, Molecular weights of plant virus protein subunits determined by the meniscus depletion method using solvents containing urea, *J. Gen. Virol.* **11**:11.

Paul, H. L., Wetter, C., Wittmann, H. G., and Brandes, J., 1965, Untersuchungen am Odontoglossum Ringspot Virus, einem erwandten des Tabakmosaik-Virus. I. Physikalische, Chemische, Serologische und Symptomatologische Befunde, *Z. Vererbungsl.* **97**:186.

Perez, J. E., and Cortez-Monllor, A., 1960, Survey for the presence of tobacco mosaic, Cymbidium mosaic, and other viruses in orchids, *J. Agric. Univ. P.R.* **44**:138.

Perez, J. E., Adsuar, J., and Sala, O., 1956, Tobacco mosaic virus in orchids in Puerto Rico, *Phytopathology* **46**:650.

Pokaew, S., and Sutabutra, T., 1974, Morphological and physical properties of tobacco mosaic virus orchid strain in Thailand, *Kasetsart J.* **8**:28.

Purcifull, D. E., and Batchelor, D. L., 1977, Immunodiffusion tests with sodium dodecyl sulfate (SDS)-treated plant viruses and plant viral inclusions, *Fla. Agric. Exp. Stn. Bull.* **788**.

Putz, C., and Vuittenez, A., 1980, The intracellular location of beet necrotic yellow vein virus, *J. Gen. Virol.* **50**:201.

Randles, J. W., Palukaitis, P., and Davies, C., 1981, Mosaic virus infecting *Nicotiana glauca* in Australia, *Ann. Appl. Biol.* **98:**109.

Reddy, D. V. R., Rajeshwari, R., Iizuka, N., Lesemann, D. E., Nolt, B. L., and Goto, T., 1983, The occurrence of Indian peanut clump, a soilborne virus disease of groundnuts (*Arachis hypogea*) in India, *Ann. Appl. Biol.* **102:**305.

Resonich, E. C., 1961, Interpretation of the forms of inclusions in bean systemically infected with tobacco mosaic virus, *Virology* **15:**16.

Shalla, T. A., 1968, Virus particles in chloroplasts of plants infected with the U5 strain of tobacco mosaic virus, *Virology* **35:**194.

Sheffield, F. M. L., 1934, Experiments bearing on the nature of intracellular inclusions in plant virus diseases, *Ann. Appl. Biol.* **21:**430.

Shvedchikova, N. G., and Protsenko, A. E., 1977, Intracellular inclusions in the virus disease of the orchid Cymbidium, *Bull. Glav. Bot. Sada* **104:**105.

Styer, E. L., and Corbett, M. K., 1975, Comparison of the orchid strain of tobacco mosaic virus and Odontoglossum ringspot virus, *Proc. Am. Phytopathol. Soc.* **2:**128.

Suchonlimakul, C., and Ishii, M., 1974, Peroxidase isozymes in some virus-infected orchids, *Acta Hortic.* **36:**273.

Talens, L. T., Chansilpa, N., and Dolores-Talens, A. C., 1979, Rapid detection of tobacco mosaic virus in orchids, *Philipp. Phytopathol.* **15:**137.

Thomson, A. A., and Smirk, B. A., 1967, An unusual strain of tobacco mosaic virus from orchids, *N. Z. J. Bot.* **5:**197.

Thornberry, H. H., and Phillippe, M. R., 1964, Orchid disease: Cattleya blossom brown necrotic streak, *Plant Dis. Rep.* **48:**936.

Thouvenel, J. C., and Fauquet, C., 1981, Peanut clump virus, *CMI/AAB Descriptions of Plant Viruses No. 235.*

Toussaint, A., and Albouy, J., 1982, Application de la methode immunoenzymatique a la detection de deux virus dans des plantes de *Cymbidium* d'ages differents, *Agronomie* **2:**901.

Van Regenmortel, M. H. V., 1966, Plant virus serology, *Adv. Virus Res.* **12:**207.

Van Regenmortel, M. H. V., 1975, Antigenic relationships between strains of tobacco mosaic virus, *Virology* **64:**415.

Van Regenmortel, M. H. V., 1982, *Serology and Immunochemistry of Plant Viruses*, pp. 174–177, Academic Press, New York.

Van Regenmortel, M. H. V., and von Wechmar, M. B., 1970, A re-examination of the serological relationship between tobacco mosaic virus and cucumber virus 4, *Virology* **41:**330.

Van Regenmortel, M. H. V., Hahn, J. S., and Fowle, L. G., 1964, Internal calibration of electron micrographs with an orchid virus, *S. Afr. J. Agric. Sci.* **7:**159.

Vela, A., and Rubio-Huertos, M., 1972, Ultraestructura de hojas de *Digitalis thlaspi* infectadas espontaneamente con dos virus, *Microbiol. Esp.* **25:**1.

Voller, A., Bartlett, D. E., Clark, M. F., and Adams, A. N., 1976, The detection of viruses by enzyme linked immunosorbent assay (ELISA), *J. Gen. Virol.* **32:**165.

Warmke, H. E., 1968, Fine structure of inclusions formed by the aucuba strain of tobacco mosaic virus, *Virology* **34:**149.

Wetter, C., Conti, M., Altschuh, D., Tabillion, R., and Van Regenmortel, M. H. V., 1984, Pepper mild mottle virus, a tobamovirus infecting pepper cultivars in Sicily, *Phytopathology* **74:**405.

Wisler, G. C., 1981, Development of serological techniques designed to index Cymbidium mosaic and Odontoglossum ringspot viruses in commercial orchids, M.S. thesis, University of Florida.

Wisler, G. C., and Zettler, F. W., 1985, Principles and applications of diagnostic techniques used for orchid virus detection, in: *Proceedings, Eleventh World Orchid Conference*, p. 402, American Orchid Society, Florida.

Wisler, G. C., Zettler, F. W., and Sheehan, T. J., 1979, Relative incidence of Cymbidium mosaic and Odontoglossum ringspot viruses in several genera of wild and cultivated orchids, *Proc. Fla. State Hortic. Soc.* **92:**339.

Wisler, G. C., Zettler, F. W., and Purcifull, D. E., 1982, A serodiagnostic technique for detecting Cymbidium mosaic and Odontoglossum ringspot viruses, *Phytopathology* **72:**835.

Wisler, G. C., Zettler, F. W., and Sheehan, T. J., 1983, A diagnostic technique developed for orchid growers to detect Cymbidium mosaic and Odontoglossum ringspot viruses, *Am. Orchid Soc. Bull.* **52:**255.

Zettler, F. W., Hennen, G. R., Bodnaruk, W. H., Clifford, H. T., and Sheehan, T. J., 1978, Wild and cultivated orchids surveyed in Florida for the Cymbidium mosaic and Odontoglossum ringspot viruses, *Plant Dis. Rep.* **62:**949.

Zettler, F. W., Wisler, G. C., Sheehan, T. J., Ko, N. -J., and Logan, A. E., 1984, Viruses—Some basics and a perspective for orchid growers, *Am. Orchid Soc. Bull.* **53:**505.

CHAPTER 13

Sunn-Hemp Mosaic Virus

ANUPAM VARMA

I. INTRODUCTION

Sunn-hemp mosaic virus (SHMV) occurs naturally in leguminous plants in Australia, India, Nigeria, and the United States (Boswell and Gibbs, 1983; Kassanis and Varma, 1975). It has been variously described as catjang (chavali cowpea) mosaic virus (Capoor *et al.*, 1947), Dolichos enation mosaic virus (Capoor and Varma, 1948), southern sunn-hemp mosaic virus (Capoor, 1950, 1962), *Crotalaria mucronata* mosaic virus (Raychaudhuri and Pathanian, 1950), cowpea mosaic virus (Lister and Thresh, 1955), among others, depending on the host from which the virus was isolated. As the most common natural host of the virus is sunn hemp, it is generally referred to as sunn-hemp mosaic virus* and other names are accepted as synonyms (Kassanis and Varma, 1975). Some variants causing chlorotic spot disease in cowpea (Sharma and Varma, 1975a) and severe leaf distortion disease in sunn-hemp (Verma and Awasthi, 1976) have also been described.

II. IMPORTANT DISEASES AND THEIR GEOGRAPHICAL DISTRIBUTION

In India, SHMV naturally infects beans (*Phaseolus vulgaris*; Nagaich and Vashisth, 1963), chavali cowpea (*Vigna catjang*; Capoor, *et al.*, 1947;

* Some other viruses described as sunn-hemp mosaic (Raychaudhuri, 1947; Dasgupta *et al.*, 1951) and Centrosema mosaic (van Velsen and Crowley, 1961) infect sunn hemp under natural conditions but are easily distinguished from SHMV on the basis of particle morphology and stability in extracts.

ANUPAM VARMA • Division of Mycology and Plant Pathology, Indian Agricultural Research Institute, New Delhi 110012, India.

Capoor and Varma, 1956), cowpea (*Vigna unguiculata*; Sharma and Varma, 1975b), *Crotalaria mucronata* (Raychaudhuri and Pathanian, 1950), *Dolichos lablab* (Capoor and Varma, 1948), and sunn hemp (*Crotalaria juncea*; Capoor, 1950, 1962), and is fairly widely distributed. In Nigeria, the virus naturally infects Bengal beans (*Mucuna aterrima*), *Centrosema* sp., cowpeas, sunn hemp, and sword beans (*Canavalia ensiformis*), but is not widely distributed (Lister and Thresh, 1955; Chant, 1959). In Australia, the virus has been isolated from sunn hemp (Gibbs and Varma, 1977), and in the United States, from cowpea (Toler, 1964). The extent of distribution in these countries is not known.

In sunn hemp, SHMV causes severe mosaic, puckering, blistering, malformation, and occasional enations. The plants are stunted and produce seeds of poor quality (Capoor, 1950, 1962). Other natural hosts are similarly affected.

III. HOST RANGE AND SYMPTOMS

More than 40 species belonging to 7 families have been found susceptible to various isolates of SHMV. Reaction of hosts to different isolates varies considerably. Some hosts are usefully employed for diagnosis.

A. Diagnostic Species

1. *Crotalaria* spp.

Crotalaria juncea is the main host. The virus also infects *C. laburnifolia*, *C. lanceolata*, *C. mucronata*, *C. mycorensis*, *C. retusa*, *C. sericea*, and *C. striata*, causing vein clearing, severe mosaic, blistering, deformation, and reduction in leaf size (Capoor, 1962; Niazi *et al.*, 1973).

2. *Cyamopsis tetragonoloba*

Distinct dark-brown lesions develop on inoculated cotyledonary leaves. The number of local lesions produced is greater in plants kept in the dark prior to inoculation (Capoor, 1962).

3. *Nicotiana glutinosa*

Necrotic local lesions less than 1 mm in diameter appear 4 days after inoculation (Bawden, 1958) at 22°C; if the plants are constantly maintained at 35°C, chlorotic local lesions of spreading type develop and some isolates even induce systemic chlorosis and mottle (Gibbs and Varma, 1977).

4. *Nicotiana tabacum* cv. White Burley

Primary lesions develop as thin white rings of necrotic tissue enclosing green areas followed by systemic mosaic mottle in about 65 days after inoculation (Capoor and Varma, 1948). Under normal glasshouse conditions, less than 10% of the plants develop systemic symptoms, and these plants tend to recover. However, if the plants are kept in the dark for about 5 days, symptoms quickly develop even in recovered leaves after the plants are switched back to normal light conditions. Preinoculation darkening of seedlings also increases susceptibility of the plants (Capoor, 1962).

5. *Phaseolus vulgaris*

French bean cv. The Prince develops white or yellow local lesions on inoculated primary leaves, vein clearing in young systemically infected trifoliate leaves, followed by severe mosaic, blistering, leaf deformities and enations (Bawden, 1958). Other cultivars like Black Turtle Soup and Pinto, however, react with necrotic spreading ringspots and milder systemic mosaic (Gibbs and Varma, 1977). At lower temperatures (10–15°C), the virus usually does not spread beyond inoculated leaves (Nagaich and Vashisth, 1963).

B. Inclusion Bodies

Characteristic inclusion bodies develop in *Dolichos lablab* within 4–12 hr after inoculation. The amorphous inclusions, which occasionally even engulf nuclei of epidermal cells, later break to form needlelike components. The needles gradually become indistinct (Sulochana and Solomon, 1970).

C. Physiological Changes

SHMV infection of *D. lablab* results in reduction in total carbohydrates and reducing sugars in shoots, roots, and nodules except at the early stages of infection when an increase is found (Raju, 1974). In contrast, SHMV-infected plants contain more nitrogen than healthy plants (Sadasivan, 1963). The amount of soluble proteins and total soluble and insoluble nitrogen in nodules also increases. The increase in nitrogen-fixing activity of the symbiotic rhizobia in nodules starts about 24 to 31 days after virus infection (Raju, 1974). The increase in nitrogen content of diseased plants coincides with the decrease in carbohydrate accumulation indicating rapid utilization of carbohydrates in nitrogen utilization by nodules.

SHMV-infected *D. lablab* plants develop a smaller number of rhizobial nodules than usual when grown in ordinary soil, and a larger number when grown in nitrogen-free sand culture conditions (Rajagopalan and Raju, 1972).

D. Interaction with Viruses and Fungi

SHMV not only infects Pinto beans readily but also helps in the transport and replication of brome mosaic virus, which normally does not infect Pinto beans. SHMV appears to complement transport function in the transport-deficient brome mosaic virus in this host (Taliansky *et al.*, 1982). SHMV also increases the severity of infection by type TMV in White Burley tobacco (Capoor, 1962).

Infection of cowpea with SHMV results in significant reduction in fresh weight, height, and leaf area of cowpea plants. The effect is more pronounced when plants are also infected with *Fusarium oxysporum* f. sp. *tracheiphilum* or f. sp. *phaseoli*. In such plants, virus concentration is also greater than in plants infected with the virus alone. Coinfection with *F. oxysporum* f. sp. *tracheiphilum* and SHMV causes death of most of the cowpea plants within 28 days of inoculation (Gbaja and Chant, 1983; Chant *et al.*, 1984). SHMV seems to predispose cowpeas to infection by fungi, resulting in extensive colonization of the xylem vessels by *Fusarium* spp. and disease severity (Gbaja and Chant, 1984).

IV. STRAINS

No distinctive strains of SHMV have been identified. The Nigerian isolate is considered as type strain. Other recognized isolates, such as those from sunn hemp and *Dolichos*, are minor variants. However, two other isolates—one causing severe mosaic in sunn hemp (SHMV-s; Verma and Awasthi, 1976) and the other causing chlorotic spots in cowpea (SHMV-cs; Sharma and Varma, 1975a) appear to be major variants of SHMV.

A. Probable Strains

SHMV strain cs causes severe mosaic, mottling, rolling of leaves, filiformy and rosetting in lateral shoots, reduced flowering, and seed formation in naturally infected plants. This results in considerable reduction in yield as 40–80% of the plants are naturally infected with the virus. SHMV-s has a rather limited host range, systemically infecting only sunn hemp and producing local lesions in *Chenopodium amaranticolor*, *Cyamopsis tetragonoloba*, and *Vigna unguiculata*. Modal length of the particles is reported to be 400 nm, which is more than that expected and requires confirmation. Serological affinities of the virus are not known. However, it does not react with antiserum to TMV type strain (Verma and Awasthi, 1976, 1978).

SHMV strain cs causes systemic chlorotic spots, fine vein clearing of leaves, severe reduction and distortion of leaf laminae of artificially or naturally infected cowpea plants. Diseased plants are greatly stunted and bushy. Host range is restricted to the Leguminosae and Chenopodiaceae, but it does not infect *Crotalaria juncea* and *Phaseolus vulgaris*, which are readily infected by other variants. It is also less stable than type SHMV as it loses infectivity in 5 days of storage at 28 ± 2°C. It resembles type SHMV in particle morphology and transmission. Little is known concerning the serological affinities of the strain cs (Sharma and Varma, 1975a).

B. Thermophilic Strain

A minor variant [N(ecrotic)-SHMV] causing discrete necrotic local lesions in the primary leaves of French bean has been reported (Kassanis and McCarthy, 1967). N-SHMV produces many defective particles when grown in plants at 20°C but not at 32°C. This is shown by the relatively greater reduction in infectivity as compared to serological titer of N-SHMV at 20°C. Preparations from such plants mainly contain disklike particles (Fig. 1), viral protein, and some free RNA, and have greater optical density at 280 nm than at 260 nm (Kassanis and McCarthy, 1967). N-SHMV resembles the thermophilic strain of type TMV (Lebeurier and Hirth, 1966), which multiplies better at temperatures above 24°C, and it seems to produce more protein at 16–20°C.

C. Host-Induced Changes

Bawden (1956, 1958) observed interesting host-induced reversible changes in two SHMV isolates from Nigeria and India. Both isolates when propagated in beans produced the "bean form" of the virus and in tobacco the "tobacco form." The latter closely resembled type TMV in host reactions and serological affinity. The "bean form," on the other hand, was serologically distantly related to type TMV and also differed in host reactions. The "tobacco form" reverted to the "bean form" when propagated in beans, and back to the "tobacco form" when propagated in tobacco. Both forms may occur in the same host but not in sufficient quantity to cause infection. Bawden (1958) explained these changes on the basis of hosts selecting variants that originate in them. However, others (Rees and Short, 1975; Kassanis, 1975) did not observe any reversible change upon extensive inoculation of SHMV from cowpea to tobacco and back to leguminous hosts. Furthermore, no difference in

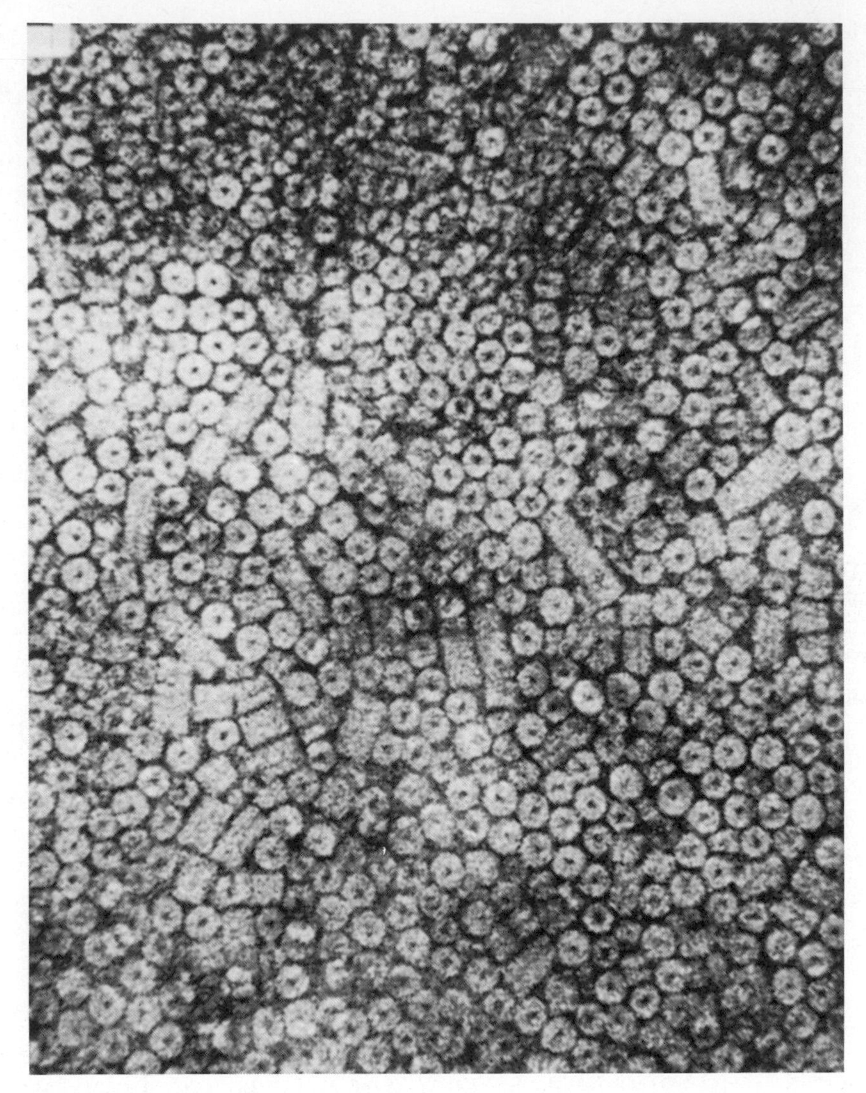

FIGURE 1. Disklike particles of N-SHMV produced in plants grown at 20°C. From Kassanis and McCarthy (1967).

amino acid composition of the virus was observed in virus preparations purified from solanaceous or leguminous hosts (Rees and Short, 1975).

It seems very likely that Bawden worked with a mixed culture of type TMV and SHMV, one growing well in tobacco and the other in beans but not completely excluding each other in either host.

V. TRANSMISSION

Like most of the tobamoviruses, SHMV spreads through contact and is easily sap-transmissible. It is also carried through seeds of some species.

A. Vector Transmission

Vector transmission has been reported for some of the isolates of SHMV. The chavali cowpea isolate from Poona, India is reported to be efficiently transmitted, in nonpersistent manner, by *Aphis craccivora*, *A. euonymi*, *A. gossypii*, *A. medicaginis*, and *Myzus persicae*, but not by *A. nerii* and thrip *Taeniothrips* spp. (Capoor *et al.*, 1947; Capoor and Varma, 1956; Nariani and Kandaswamy, 1961; Haque and Chenulu, 1967). Efficient transmission of a tobamovirus by aphids is rather unusual. It is probable that the culture of SHMV used for aphid transmission by various workers was mixed with some aphid-transmitted virus, especially as several aphid-transmitted viruses are known to naturally occur in cowpea in India, and later workers have failed to isolate aphid-transmitted SHMV (Chenulu and Varma, 1986). SHMV-cs is not transmitted by any aphid species tested even from plants also infected with aphid-transmitted viruses, or through direct membrane feeding (Sharma and Varma, 1982).

The galerucid beetle (*Ootheca mutabilis*) has been reported to transmit SHMV (Chant, 1959), but this claim probably refers to another virus and not SHMV (Kassanis and Varma, 1975).

The *Dolichos* isolate is known to be transmitted by nymphs and adults of the red spider mite *Tetranychus ludeni* (Rajagopalan, 1974). The mite could acquire the virus from infected plants in 7 days and transmit to healthy test plants in access periods of 4 days. Extracts of mites were found to contain the virus.

B. Seed Transmission

Some isolates are transmitted through seeds. The extent of seed transmission varies in different hosts and their cultivars.

The chavali isolate is carried through up to 23% of the seeds of cowpea (Capoor *et al.*, 1947; Capoor and Varma, 1956). Although the sunn hemp isolate from Poona, India is not carried through the seeds of sunn hemp (Capoor, 1962), the bean isolate from India (Nagaich and Vashisth, 1963) and the sunn hemp isolate from Queensland, Australia (Gibbs and Varma, 1977) are transmitted through the seeds of this host. The bean isolate, however, is not transmitted through the seeds of beans or peas (Nagaich and Vashisth, 1963).

SHMV-s is carried through up to 20% of the seeds of sunn hemp when parent plants are infected at the seedling stage and up to 10% when parent plants are infected later but before flowering (Verma and Awasthi, 1976). Similarly, seed transmission of SHMV-cs also depends on the time of infection and the growing conditions of the parent cowpea plants. The virus is present in all parts (seed coat, cotyledons, and plumule plus radicle) of immature and mature seeds. Presence of the virus in the seed embryo is unusual for a tobamovirus, but this may be a special adaptation

TABLE I. Serological affinity between TMV and SHMV[a]

Antiserum	Antigen used for absorption	Antigen used in precipitin tests	
		TMV	SHMV
TMV	SHMV	1024[b]	0
TMV	—	1024	16
SHMV	TMV	0	512
SHMV	—	16	512
N-SHMV	TMV	0	640
N-SHMV	—	40	640

[a] Data from Bawden and Kassanis (1968) and Kassanis and McCarthy (1967).
[b] Reciprocals of end titer.

to ensure its transmission, especially as the virus has a limited host range (Sharma and Varma, 1986).

VI. STABILITY IN SAP

SHMV is a highly stable virus. Most of the isolates have a dilution endpoint in extracts of over 10^{-6} and a thermal inactivation point between 85 and 95°C. They remain infective in extracts for more than 8 years (at 20–32°C), and they can also withstand desiccation in leaves (Capoor and Varma, 1948; Capoor, 1962). Some isolates are, however, less stable. For example, the chavali isolate loses infectivity after 19 days of storage and does not withstand desiccation (Capoor and Varma, 1956).

VII. SEROLOGICAL RELATIONSHIPS

SHMV is strongly immunogenic. In tube precipitin tests it forms flocculent precipitates, and two bands of precipitate in gel diffusion tests (Kassanis and McCarthy, 1967). Isolates from sunn hemp, cowpea, *Dolichos*, and beans are serologically very similar (Badami, 1963; Nagaich and Vashisth, 1963). The virus is serologically distantly related to TMV (Bawden and Kassanis, 1968), and the removal (from antisera to SHMV or TMV) of the antibodies that cross-react with the heterologous virus does not alter the homologous titer (Table I). In Ouchterlony double gel diffusion tests, precipitation bands of the two viruses cross each other, forming double spurs (Kassanis and McCarthy, 1967).

There is very little published information about the serological relationships between SHMV and other tobamoviruses. The affinities, however, seem to vary with different isolates; for example, frangipani mosaic virus is distantly related to the SHMV isolate from Queensland, but not to the isolate from Nigeria (Varma and Gibbs, 1978).

VIII. PROPERTIES OF PARTICLES

A. Structure and Sedimentation

Infective particles of SHMV are rods about 300 nm long and 17 nm wide (Kassanis and Varma, 1975). However, preparations of SHMV always contain many shorter particles, the majority of which are about 40 nm long (Dunn and Hitchborn, 1965; Kassanis and McCarthy, 1967; Morris, 1974; Whitfeld and Higgins, 1976). In addition to these two particle sizes, there also are a small number of particles about 80 nm long (Whitfeld and Higgins, 1976). In 0.06 M phosphate buffer (pH 8), these three classes of particles sediment as three components with coefficients of 20–50 S, 70–80 S, and 187 S (Kassanis and McCarthy, 1967). The proportion of short and full-length particles varies according to growing conditions of the plants and extraction procedures. For example, when plants are grown at 32°C a larger proportion of full-length particles are produced and in extracts at pH 8, more shorter particles are observed than at pH 5.2 (Kassanis and McCarthy, 1967). Earlier workers did not attribute any biological significance to short particles and considered these to be formed by breakage of full-length particles. Subsequent studies have, however, shown that short particles contain RNA (M_r ca. 0.35×10^6) which acts as messenger for the viral coat protein.

B. Ultraviolet Absorption

The UV absorption is typical of viruses containing about 5% nucleic acid, the $A_{260/280}$ ratio being 1.2 (Kassanis and Varma, 1975). Some minor differences in the UV absorption spectra of short and full-length particles have been noted. The absorption spectrum of short particles has a more marked trough at 247 nm, which may be partly due to differences in light scattering by the particles of different sizes. The $A_{260/280}$ ratio is 1.09 for short particles under conditions when the ratio for full-length particles is 1.15 (Whitfeld and Higgins, 1976).

IX. PARTICLE COMPOSITION

A. Protein

The molecular weight of SHMV protein is 18,062. Although earlier findings indicated that there were only small differences between the

```
      1
   Ac Ser-Tyr-Ser-Ile-Thr-Thr-Pro-Ser-Gln-Phe-Val-Phe-Leu-Ser-Ser-Ala-
   Ac Ala-Tyr-Ser-Ile-Pro-Thr-Pro-Ser-Gln-Leu-Val-Tyr-Phe-Thr-Glu-Asn-

   17
   Trp-Ala-Asp-Pro-Ile-Glu-Leu-Ile-Asn-Leu-Cys-Thr-Asn-Ala-Leu-Gly-
   Tyr-Ala-Asp-Tyr-Ile-Pro-Phe-Val-Asn-Arg-Leu-Ile-Asn-Ala-Arg-Ser-

   33
   Asn-Gln-Phe-Gln-Thr-Gln-Gln-Ala-Arg-Thr-Val-Val-Gln-Arg-Gln-Phe-
   Asn-Ser-Phe-Gln-Thr-Gln-Ser-Gly-Arg-Asp-Glu-Leu-Arg-Glu-Ile-Leu-

   49
   Ser-Gln-Val-Trp-Lys-Pro-Ser-Pro-Gln-Val-Thr-Val-Arg-Phe-Pro-Asp-
   Ile-Lys-Ser-Gln-Val-Ser-Val-Val-Ser-Pro-Ile-Ser-Arg-Phe-Pro-Ala-

   65
   Ser-Asp-Phe-Lys-Val-Tyr-Arg-Tyr-Asn-Ala-Val-Leu-Asp-Pro-Leu-Val-
Glu-Pro-Ala-Tyr-Tyr-Ile-Tyr-Leu-Arg-Asp-Pro-Ser-Ile-Ser-Thr-Val-Tyr-

   81
   Thr-Ala-Leu-Leu-Gly-Ala-Phe-Asp-Thr-Arg-Asn-Arg-Ile-Ile-Glu-Val-
   Thr-Ala-Leu-Leu-Gln-Ser-Thr-Asp-Thr-Arg-Asn-Arg-Val-Ile-Glu-Val-

   97
   Glu-Asn-Gln-Ala-Asn-Pro-Thr-Thr-Ala-Glu-Thr-Leu-Asp-Ala-Thr-Arg-
   Glu-Asn-Ser-Thr-Asp-Val-Thr-Thr-Ala-Glu-Gln-Leu-Asn-Ala-Val-Arg-

   113
   Arg-Val-Asp-Asp-Ala-Thr-Val-Ala-Ile-Arg-Ser-Ala-Ile-Asn-Asn-Leu-
   Arg-Thr-Asp-Asp-Ala-Ser-Thr-Ala-Ile-His-Asn-Asn-Leu-Glu-Gln-Leu-

   129
   Ile-Val-Glu-Leu-Ile-Arg-Gly-Thr-Gly-Ser-Tyr-Asn-Arg-Ser-Ser-Phe-
   Leu-Ser-Leu-Leu-Thr-Asn-Gly-Thr-Gly-Val-Phe-Asn-Arg-Thr-Ser-Phe-

   145
   Glu-Ser-Ser-Ser-Gly-Leu-Val-Trp-Thr-Ser-Gly-Pro-Ala-Thr.
   Glu-Ser-Ala-Ser-Gly-Leu-Trp-Leu-Val-Thr-Thr-Pro-Thr-Arg-Thr-Ala.
```

FIGURE 2. The amino acid sequences of SHMV (bottom line) and TMV (top line) proteins. The numbers indicate amino acid residues from the N-terminus of the proteins. Residue 65 of SHMV is shifted one step back to obtain best alignment (shading). There are 161 residues in SHMV and 158 in type TMV. Adapted from Rees and Short (1975).

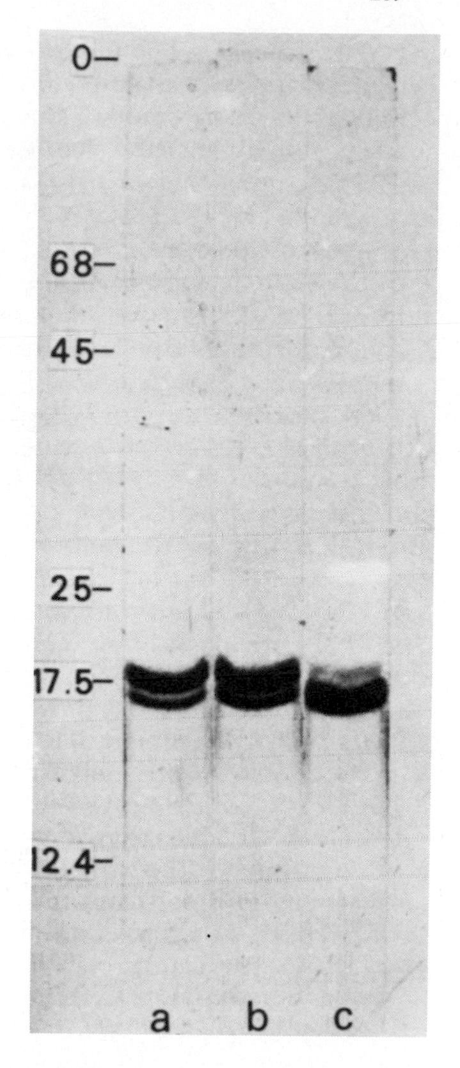

FIGURE 3. SDS–polyacrylamide gel electrophoresis of coat proteins from (a) unfractionated particles, (b) full-length particles, and (c) short particles of SHMV. Figures indicate molecular weights of markers. From Whitfeld and Higgins (1976).

coat proteins of TMV and SHMV (Rees and Short, 1965, 1972), later studies (Rees and Short, 1975) revealed considerable differences in the primary structure of the coat protein of the two viruses.

SHMV has three additional amino acid residues and 96 amino acid changes from TMV (Fig. 2). Two of the insertions have been found near the C-terminus of the chain. A third insertion appears to be in the region of residues 64–80, possibly at the end of the helical region so that the general shape of the folded molecule is least affected. In spite of the differences in amino acid sequence of SHMV and TMV coat proteins, they show similar pH-dependent reversible helical aggregation properties (Rees and Short, 1975).

Rees and Short (1975) showed that in SHMV, as in TMV, the residues 115 and 116, as well as 88 and 145 provide the carboxyl–carboxylate pairs that control the formation of protein disks.

Unfractionated particles of SHMV contain mainly a 18K dalton polypeptide, with a smaller proportion of 16.5K polypeptide, whereas protein from short particles consists mainly of a 16.5K component (Fig. 3). No serological differences have been observed between the two proteins. Tryptic peptide mapping also appears similiar except for the absence of one unidentified peptide of about 10 amino acids in the short particle protein (Whitfeld and Higgins, 1976). No conclusive evidence is available to explain this anomalous situation of two size classes of protein in a tobamovirus. It is, however, possible that the 16.5K protein arises by the action of naturally occurring carboxypeptidase on the 18K viral protein, as this enzyme has been found to attack SHMV protein (Rees and Short, 1965). If this is so, the occurrence of a greater proportion of the 16.5K polypeptide in the short particles indicates a higher susceptibility of the protein in short particles than in full-length particles.

B. Nucleic Acid

Two main sizes of RNA are found in SHMV. The full-length particles contain RNA of molecular weight 2×10^6, similar to TMV RNA. The size of the other major RNA species, isolated from short particles, is $0.3 \times 10^6 \pm 20{,}000$. In 2.2 and 5% polyacrylamide gels, some minor bands with molecular weights less than 0.3×10^6 are also observed. Whether these are normal constituents of SHMV particles or products of partial degradation of large RNA species has not been determined. Some RNAs with molecular weights 0.6×10^6 and greater than 1.3×10^6 have also been observed (Whitfeld and Higgins, 1976).

1. Base Ratio

The base composition of the two major RNAs varies slightly. The short RNAs have more uridylic acid and less adenylic acid than the full-length RNAs. The purine/pyrimidine ratio is also smaller in short RNAs than in full-length RNAs (Whitfeld and Higgins, 1976).

2. Nucleotide Sequence Homology

Using one isolate of SHMV, Palukaitis and Symons (1980) did not find any nucleotide sequence homology between SHMV RNA and the RNAs of cucumber 4, cucumber green mottle mosaic, frangipani mosaic, Odontoglossum ringspot, tobacco mosaic, and tomato mosaic viruses.

No information is available concerning nucleic acid homology between different strains of SHMV.

X. BIOLOGICAL SIGNIFICANCE OF THE TWO TYPES OF PARTICLES

A. Relative Production of Short and Full-Length Particles in Plants

In bean leaves, for every full-length particle 1.5 short particles are produced, whereas in tobacco leaves the ratio approaches approximately 2 short particles for every full-length particle (Whitfeld and Higgins, 1976). In extracts from infected plants, the ratio of genomic RNA molecules to polyribosome-associated coat protein mRNA is different from that found for virus particles (Skotnicki *et al.*, 1976).

B. Separation of Short and Full-Length Particles

It has been shown that both types of particles are normally produced in infected leaves and can be separated by differential centrifugation and judicious suspension of pellets (Whitfeld and Higgins, 1976). After high-speed centrifugation (30,000 rpm for 1–3 hr), the virus forms a translucent pellet (main pellet) covered by a loosely packed layer of yellow-green material. Nearly 60% of the short particles sediment along with the full-length particles to form the main pellet, and the remaining 40% remain in the yellow-green layer. Differential resuspension of different layers of pellet helps in separating the two types of particles. The short particles, however, have a tendency to aggregate end to end and sediment with full-length particles; for this reason, preparations of full-length particles always contain some amount of short particles.

C. Infectivity

Short particles or their RNA (M_r 0.3 $\times$ 10^6) are neither infectious nor have any detectable effect on the infectivity of full-length particles (Whitfeld and Higgins, 1976). Long particles of SHMV as well as 2 $\times$ 10^6 M_r RNA cause infection in susceptible plants, and in such plants short particles are also produced. This indicates that the 2 $\times$ 10^6 M_r RNA contains information for the production of both types of particles (Whitfeld and Higgins, 1976).

D. Translation of RNAs from Short and Full-Length Particles in Cell-Free Systems

Although the full-length RNA contains full genomic information, it is not able to program the synthesis of viral coat protein in an *in vitro*

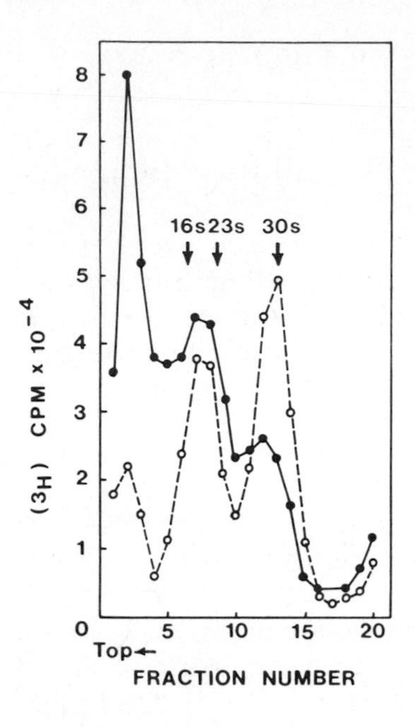

FIGURE 4. Distribution of radioactivity in polyribosome-associated RNAs from SHMV-infected (●) and healthy (○) leaves of beans fractionated in sucrose gradients. From Skotnicki *et al.* (1976).

protein-synthesizing system, as is the case also with TMV RNA. The short RNA of SHMV, however, efficiently programs the synthesis of viral coat protein under the same *in vitro* conditions (Higgins *et al.*, 1976).

The coat protein produced in the cell-free system is slightly larger than the normal 18K protein, and also incorporates methionine, which is normally not found in the SHMV coat protein (Higgins *et al.*, 1976). It is possible that this is also the case in plants infected with SHMV, but that the "extra" amino acid is cleaved off before assembly of particles *in vivo*.

The short RNA also leads to the production of some proteins smaller than the coat protein. Some of these may be incomplete coat proteins as they react with coat protein antiserum. The full-length RNA is also an efficient messenger in a cell-free system and results in the synthesis of a wide range of proteins of molecular weights ranging from 5000 to 130,000, none of which react with coat protein antiserum. One of these proteins may be the viral replicase (Higgins *et al.*, 1976).

These findings indicate that the cistron for coat protein occurs in full-length as well as short particles although the latter are not required for infection, a situation analogous to that found in many other plant viruses, e.g., alfalfa mosaic virus (Bol *et al.*, 1971).

E. Polyribosome-Associated Coat Protein mRNA in Plants

SHMV-infected bean plants contain larger amounts of "4–8 S polyribosome-associated (PA)-RNAs" and significantly smaller amounts of

"30 S PA-RNA" than virus-free plants (Fig. 4; Skotnicki *et al.*, 1976). The molecular weight of "4–8 S PA-RNAs" is mostly in the range from 0.25 $\times 10^6$ to 0.4 $\times 10^6$, and in cell-free systems these program the synthesis of polypeptide(s) very similar in size, antigenicity, and peptide mapping to the coat protein, clearly indicating that "4–8 S PA-RNAs" from infected plants mainly contain mRNA for coat protein. These RNAs assemble with coat protein as readily as RNAs from short particles or unfractionated particles, and form RNase-resistant particles (Skotnicki *et al.*, 1976). Similar "4–8 S PA-RNA" messengers for viral coat protein are also observed in plants infected with TMV (Higgins *et al.*, 1976) and T_2MV (Skotnicki *et al.*, 1976), indicating that short mRNAs are formed intracellularly in all tobamoviruses irrespective of their encapsidation as short particles in certain viruses (see Chapter 14). It is now understood that the formation of short rods is a consequence of the relative location of the rod assembly initiation site and the coat protein cistron in different tobamoviruses. In SHMV, these respective sites are 369–461 and 711 residues from the 3′-end of the genomic RNA (Meshi *et al.*, 1981) (see Chapter 14).

XI. ECOLOGY AND CONTROL

The incidence of SHMV-cs in cowpeas in northern India varies from 6.0 to 25% resulting in about 20% reduction in yield. In addition to direct loss in seed yield, nodulation and nitrogen fixation by cowpeas is also adversely affected (Sharma and Varma, 1975b, 1981). Final incidence of the virus in a crop depends on the extent of seed transmission. When seed transmission is less than 1%, final incidence is up to 5%; a higher incidence is found with a greater rate of seed transmission (Sharma and Varma, 1983). That secondary spread of SHMV-cs is mainly by contact is also supported by the almost similar incidence in cowpea crops grown with or without gramineous barrier crops which do help in reducing the spread of vectorborne viruses of cowpea (Sharma and Varma, 1984).

No definite control measures for reducing the seed infection and further spread of the virus are known, but losses due to the diseases caused by SHMV can be easily avoided by the use of healthy seeds.

Some promising inhibitors have also been found. For example, extracts of *Boerhaavia diffusa, Cuscuta reflexa, Datura metel,* and *Euphorbia hirta* have been shown to inhibit infection by SHMV-s up to 95% in hypersensitive and systemic hosts (Mukerjee *et al.*, 1982; Verma *et al.*, 1979). Leaf extracts of *Chenopodium ambrosoides* also prevent local lesion production by SHMV-s in *Cyamopsis tetragonoloba* and *Vigna sinensis,* but not in *C. amaranticolor* (Verma and Baranwal, 1983). Whether these or other inhibitors can be used for controlling field spread of the virus remains to be seen.

XII. GENERAL REMARKS

SHMV is mainly a virus of the tropics where it causes some economically important diseases in legumes. The extent of damage caused by the virus in the tropics is not known as it has been detected only in India and west Africa. Efficient transmission through seeds of different legumes, however, must have resulted in a much wider distribution especially in recent times of large-scale exchange of germ plasm. SHMV also occurs in some temperate areas where it must have been inadvertently introduced through seeds, but in these areas it is unlikely to cause serious disease problems due to its inability to multiply efficiently at lower temperatures.

SHMV is the first tobamovirus to have been shown to have at least two definitive sizes of particles, i.e., 300-nm full-length and 40-nm short particles. The short particles contain coat protein mRNA. Coat protein mRNAs are produced intracellularly by all tobamoviruses, but they are encapsidated only if the assembly initiation site is located within them (see Chapter 14).

REFERENCES

Badami, R. S., 1963, Some aspects in plant virus serology, *Bull. Natl. Inst. Sci. India* **24:**166.

Bawden, F. C., 1956, Reversible, host induced, changes in a strain of tobacco mosaic virus, *Nature (London)* **177:**302.

Bawden, F. C., 1958, Reversible changes in strains of tobacco mosaic virus from leguminous plants, *J. Gen. Microbiol.* **18:**751.

Bawden, F. C., and Kassanis, B., 1968, The serological relationship between tobacco mosaic virus and cucumber viruses 3 and 4, *Virology* **34:**174.

Bol, J. F., van Vloten-Doting, L., and Jaspars, E. M. J., 1971, A functional equivalence of top component *a* RNA and coat protein in the initiation of infection by alfalfa mosaic virus, *Virology* **46:**73.

Boswell, K. F., and Gibbs, A. J., 1983, Viruses of legumes 1983: Descriptions and keys from VIDE, The Australian National University, Research School of Biological Sciences, Canberra.

Capoor, S. P., 1950, A mosaic disease of sunn-hemp in Bombay, *Curr. Sci.* **19:**22.

Capoor, S. P., 1962, Southern sunn-hemp mosaic virus: A strain of tobacco mosaic virus, *Phytopathology* **52:**393.

Capoor, S. P., and Varma, P. M., 1948, Enation mosaic of *Dolichos lablab* Linn, a new virus disease, *Curr. Sci.* **17:**57.

Capoor, S. P., and Varma, P. M., 1956, Studies on a mosaic disease of *Vigna cylindrica* Skeels, *Indian J. Agric. Sci.* **26:**95.

Capoor, S. P., Varma, P. M., and Uppal, B. N., 1947, A mosaic disease of *Vigna catjang* Walp., *Curr. Sci.* **16:**151.

Chant, S. R., 1959, Viruses of cowpea, *Vigna unguiculata* L. (Walp.), in Nigeria, *Ann. Appl. Biol.* **47:**565.

Chant, S. R., Gbaja, I. S., and Kang, A. S., 1984, Effect of nutrition on the interaction of *Fusarium oxysporum* and sunn-hemp mosaic virus in cowpea seedlings, *Trop. Agric. (Trinidad)* **61:**87.

Chenulu, V. V., and Varma, A., 1986, Virus and virus-like diseases of pulse crops commonly grown in India, in: *Recent Research on Pulse Crops in India*, Indian Council Agricultural Research, New Delhi (in press).

Dasgupta, N. N., De, M. L., and Raychaudhuri, S. P., 1951, Structure of sunn-hemp (*Crotalaria juncea* Linn) mosaic virus with the electron microscope, *Nature* (*London*) **168:**114.

Dunn, D. B., and Hitchborn, J. H., 1965, The use of bentonite in the purification of plant viruses, *Virology* **25:**171.

Gbaja, I. S., and Chant, S. R., 1983, Effects of co-infection by *Fusarium oxysporum* and sunn-hemp mosaic virus on the growth of cowpea (*Vigna unguiculata* (L.) Walp), *Trop. Agric.* (*Trinidad*) **60:**272.

Gbaja, I. S., and Chant, S. R., 1984, Scanning electron-microscopy of the colonization of cowpea (*Vigna unguiculata* (L.) Walp) by host and non-host *Fusarium oxysporum, Trop. Agric.* (*Trinidad*) **61:**92.

Gibbs, A. J., and Varma, A., 1977, Queensland isolate of sunn-hemp mosaic virus, Unpublished findings.

Haque, S. Q., and Chenulu, V. V., 1967, Studies on the relationship of cowpea mosaic virus and its vector *Aphis craccivora* Koch, *Phytopathol. Z.* **59:**277.

Higgins, T. J. V., Goodwin, P. B., and Whitfeld, P. R., 1976, Occurrence of short particles in beans infected with cowpea strain of TMV. 2. Evidence that short particles contain the cistron for coat-protein, *Virology* **71:**486.

Kassanis, B., 1975, in Kassanis and Varma (1975).

Kassanis, B., and McCarthy, D., 1967, The quality of virus as affected by the ambient temperature, *J. Gen. Virol.* **1:**425.

Kassanis, B., and Varma, A., 1975, Sunn-hemp mosaic virus, *CMI/AAB Descriptions of Plant Viruses No. 153.*

Lebeurier, G., and Hirth, L., 1966, Effect of elevated temperatures on the development of two strains of tobacco mosaic virus, *Virology* **29:**385.

Lister, R. M., and Thresh, J. M., 1955, A mosaic disease of leguminous plants caused by a strain of tobacco mosaic virus, *Nature* (*London*) **175:**1047.

Meshi, T., Ohno, T., Iba, H., and Okada, Y., 1981, Nucleotide sequence of a cloned cDNA copy of TMV (cowpea strain) RNA, including the assembly origin, the coat protein cistron, and the 3′ non-coding region, *Mol. Gen. Genet.* **184:**20.

Morris, T. J., 1974, Two nucleoprotein components associated with the cowpea strain of TMV, *Am. Phytopathol. Soc. Proc.* **1:**83.

Mukerjee, K., Awasthi, L. P., and Verma, H. N., 1982, Further studies on the antiviral resistance induced in host plants treated with the leaf extracts of *Datura metel* L., *Indian J. Bot.* **5:**161.

Nagaich, B. B., and Vashisth, K. S., 1963, A virus causing atypical mosaic disease of beans, *Indian J. Microbiol.* **3:**113.

Nariani, T. K., and Kandaswamy, T. K., 1961, Studies on a mosaic disease of cowpea (*Vigna sinensis* Savi), *Indian Phytopathol.* **14:**77.

Niazi, F. R., Chandra, K. J., and Prakash, N., 1973, A new strain of tobacco mosaic virus infecting sunn-hemp, *Indian Phytopathol.* **26:**115.

Palukaitis, P., and Symons, R. H., 1980, Nucleotide sequence homology of thirteen tobamovirus RNAs as determined by hybridization analysis with complementary DNA, *Virology* **107:**354.

Rajagopalan, K., 1974, First record of spidermite *Tetranychus ludeni* Zacher transmitting Dolichos enation mosaic virus, *Curr. Sci.* **43:**488.

Rajagopalan, N., and Raju, P. N., 1972, The influence of infection by Dolichos enation mosiac virus on nodulation and nitrogen fixation by field bean (*Dolichos lablab* L.), *Phytopathol. Z.* **73:**285.

Raju, P.N., 1974, Dolichos enation mosaic virus (DEMV) and symbiotic nitrogen fixation in field bean (*Dolichos lablab* Linn.), *Proc. Indian Natl. Sci. Acad.* **40B:**629.

Raychaudhuri, S. P., 1947, A note on mosaic virus of sunn-hemp (*Crotalaria juncea* Linn.) and its crystallisation, *Curr. Sci.* **16:**26.

Raychaudhuri, S. P., and Pathanian, P. S., 1950, A mosaic disease of *Crotalaria mucronata* Desv. (*C. striata* DC), *Curr. Sci.* **19:**213.

Rees, M. W., and Short, M. N., 1965, Variation in the composition of two strains of tobacco mosaic virus in relation to their host, *Virology* **26:**596.

Rees, M. W., and Short, M. N., 1972, The tryptic peptides and terminal sequences of the protein from the cowpea strain of tobacco mosaic virus, *Virology* **50:**772.

Rees, M. W., and Short, M. N., 1975, The amino acid sequence of the cowpea strain of tobacco mosaic virus protein, *Biochim. Biophys. Acta* **393:**15.

Sadasivan, T. S., 1963, Physiology of virus infected plants, *J. Indian Bot. Soc.* **42:**339.

Sharma, S. R., and Varma, A., 1975a, Three sap transmissible viruses from cowpea in India, *Indian Phytopathol.* **28:**292.

Sharma, S. R., and Varma, A., 1975b, Natural incidence of cowpea viruses and their effect on yield of cowpea, *Indian Phytopathol.* **28:**330.

Sharma, S. R., and Varma, A., 1981, Assessment of losses caused by cowpea banding mosaic (CpBMV) and cowpea chlorotic spot (CpCSV) viruses in cowpea, *Proc. 3rd Int. Symp. Plant Pathol.* p. 95.

Sharma, S. R., and Varma, A., 1982, Aphid transmission of two cucumoviruses from plants also infected with tobamovirus, *Zibl. Mikrobiol.* **137:**415.

Sharma, S. R., and Varma, A., 1983, Seed transmission in cowpea viruses, *Vegetable Sci.* **10:**55.

Sharma, S. R., and Varma, A., 1984, Effect of cultural practices on virus infection in cowpea, *Z. Acker Pflanzenbau* **153:**23.

Sharma, S. R., and Varma, A., 1986, Transmission of cowpea banding mosaic and cowpea chlorotic spot viruses through the seeds of cowpea, *Seed Sci. Technol.* **14:**1.

Skotnicki, A., Gibbs, A., and Shaw, D. C., 1976, *In vitro* translation of polyribosome-associated RNAs from tobamivirus infected plants, *Intervirology* **7:**256.

Sulochana, C. B., and Solomon, J. J., 1970, Local lesion formation in *Cyamopsis tetragonoloba* Taub. with DEMV and other plant viruses, *Proc. Indian Acad. Sci. Sect. B* **71:**56.

Taliansky, M. E., Malyshenko, S. I., Pshennikova, E. S., and Atabekov, J. G., 1982, Plant virus-specific transport function. II. A factor controlling virus host range, *Virology* **122:**327.

Toler, R. W., 1964, Identity of a mosaic virus of cowpea, *Phytopathology* **54:**910.

van Velsen, R. J., and Crowley, N. C., 1961, Centrosema mosaic: A plant virus disease transmitted by both aphids and plant bugs, *Nature (London)* **189:**4767.

Varma, A., and Gibbs, A. J., 1978, Frangipani mosaic virus, *CMI/AAB Descriptions of Plant Viruses No. 196.*

Verma, H. N., and Awasthi, L. P., 1976, Sunn-hemp rosette: A new virus disease of sunn-hemp, *Curr. Sci.* **45:**642.

Verma, H. N., and Awasthi, L. P., 1978, Further studies on a rosette virus of *Crotalaria juncea, Phytopathol. Z.* **92:**83.

Verma, H. N., and Baranwal, V. K., 1983, Antiviral activity and the physical properties of the leaf extract of *Chenopodium ambrosoides* L., *Proc. Indian Acad. Sci. Plant Sci.* **92:**461.

Verma, H. N., Awasthi, L. P., and Mukerjee, K., 1979, Prevention of virus infection and multiplication by extracts from medicinal plants, *Phytopathol. Z.* **96:**71.

Whitfeld, P. R., and Higgins, T. J. V., 1976, Occurrence of short particles in beans infected with the cowpea strain of TMV. 1. Purification and characterization of short particles, *Virology* **71:**471.

CHAPTER 14

Cucumber Green Mottle Mosaic Virus

Yoshimi Okada

I. INTRODUCTION

Cucumber green mottle mosaic virus (CGMMV) was first described by Ainsworth (1935). Since then, several strains have been isolated in Europe, India, and Japan.

The viruses are sap-transmissible, but most strains have a restricted host range. They are found in cucumber, watermelon, and melon crops. The virus is transmitted by foliage contact, handling of plants during cultivation, soil contamination, or through the cucurbit seeds. No biological vector is known.

The viruses are easily purified from systemically infected cucumber plants (Nozu *et al.*, 1971; Tung and Knight, 1972). Several strains are distinguished serologically and by the differential responses in *chenopodium amaranticolor* and *Datura stramonium* (Tochihara and Komuro, 1974).

The CGMMV are rod-shaped viruses, 300 nm in length and 18 nm in diameter, and are morphologically indistinguishable from tobacco mosaic virus (TMV). The biophysical and biochemical properties of these viruses are also similar to those of TMV (Knight, 1952). The CGMMV are characterized by the production of short rods, with a modal length of about 50 nm RNA containing the coat protein mRNA, in addition to the normal-sized particle (Fukuda *et al.*, 1981) as in the case of the cowpea strain of TMV now termed sunn-hemp virus (Bruening *et al.*, 1976; Higgins *et al.*, 1976) (see Chapter 13).

YOSHIMI OKADA • Department of Biophysics and Biochemistry, Faculty of Science, University of Tokyo, Hongo Bunkyo-ku, Tokyo, 113, Japan.

II. STRAINS

A. Type Strain (Cucumber Virus 3)

The type strain was isolated in Europe (Ainsworth, 1935). In cucumbers the type strain causes leaf mottling, blistering, and distortion with stunted growth. This strain was maintained at the University of California as Berk CV3.

The amino acid composition and C-terminal threonine of the coat protein were reported (Tung and Knight, 1972).

B. Cucumber Aucuba Mosaic Strain (Cucumber Virus 4)

Cucumber virus 4 (CV4) was found in Europe (Ainsworth, 1935). The aucuba mosaic strain causes bright yellow leaf mottling, with only slight leaf distortion and stunting. The fruit may show yellow- or silver-colored streaks and flecks. This strain was maintained at the University of California as Berk CV4.

The amino acid composition (Tung and Knight, 1972), N-terminal acetyl-alanine (Narita, 1959), and C-terminal alanine (Niu *et al.*, 1958; Tung and Knight, 1972) of the coat protein were reported.

C. Watermelon Strain

The watermelon strain (CGMMV-W) was isolated in Japan (Komuro *et al.*, 1971). In watermelons, this strain produces slight leaf mottling and dwarfing, but infection at fruit-set can induce serious internal discoloration and decomposition of the fruit.

The coat protein, also having N-terminal acetyl-alanine and C-terminal alanine (Nozu *et al.*, 1971), is composed of 160 amino acid residues (Meshi *et al.*, 1983). The amino acid compositions of the CGMMV-W and CV3 coat proteins are identical except for the number of serine and isoleucine residues, which are 22 and 8 in the CGMMV-W protein and 23 and 7 in the CV3 protein, respectively. This suggests that CGMMV-W is closely related to CV3.

The sequence of 1071 nucleotides from the 3′-end of CGMMV-W RNA was determined (Meshi *et al.*, 1983). The amino acid sequence of the coat protein was deduced from the nucleotide sequence. In terms of biochemistry, CGMMV-W has been the most extensively studied among the strains of CGMMV.

Recently, an attenuated strain of CGMMV-W, 3H33b, was isolated (Motoyoshi and Nishiguchi, 1984) and used for controlling diseases of melon.

D. Japanese Cucumber Strain

A cucumber strain (CGMMV-C) was found in Japan (Inouye *et al.*, 1967). It causes severe fruit distortion in cucumbers.

The amino acid composition and the partial amino acid sequence of the coat protein of CGMMV-C were reported by Kurachi *et al.* (1972). The homology of the amino acid sequences of CGMMV-W and CGMMV-C is at most 48% in this region. Therefore, CGMMV-C would be distantly related to CGMMV-W. No further studies of this strain have been reported and the rest of this chapter will thus deal only with the watermelon strain.

III. VIRUS PARTICLE

CV3, CV4, and CGMMV-W were prepared from systemically infected cucumber leaves by differential centrifugation procedures commonly employed for purifying TMV (Nozu *et al.*, 1971; Tung and Knight, 1972). CGMMV resembles TMV in the size and shape of particles, and in the content of RNA (Knight, 1952). The concentration of virus particles can be determined spectrophotometrically by using $E_{260}^{0.1\%} = 3.0$ (Ohno *et al.*, 1977).

Tobamoviruses can be divided into two subgroups according to the location of the assembly origin (Fig. 1) (Okada *et al.*, 1980; Fukuda *et al.*, 1981). Common TMV (vulgare, OM) and tomato mosaic virus belong to subgroup 1, in which the assembly origin is outside the coat protein cistron (Lebeurier *et al.*, 1977; Otsuki *et al.*, 1977; Fukuda *et al.*, 1980; Takamatsu *et al.*, 1983).

In the case of subgroup 2 viruses, the assembly origin is within the coat protein cistron. CGMMV as well as SHV (formerly the cowpea strain of TMV; see Chapter 13) belong to subgroup 2 (Fukuda *et al.*, 1981; Meshi *et al.*, 1981) and, therefore, a virus particle preparation of CGMMV contains, in addition to full-length 300-nm viral rods, short subviral particles containing subgenomic coat protein mRNA (Whitfeld and Higgins, 1976; Fukuda *et al.*, 1981). The length distribution of CGMMV-W particles from a cucumber leaf extract was reported (Fukuda *et al.*, 1981). The length of CGMMV-W short rods (about 50 nm) is significantly greater than that of SHV (about 35 nm). This was confirmed by comparing the sizes of the RNA extracted from these rods.

IV. GENOMIC AND SUBGENOMIC RNA

A. Nucleotide Composition and Sequence of Genomic RNA

The base compositions of CV3 RNA and CV4 RNA are 18.3% C, 25.8% A, 30.8% U, and 25.5% G, and 19.3% C, 25.8% A, 29.5% U, and

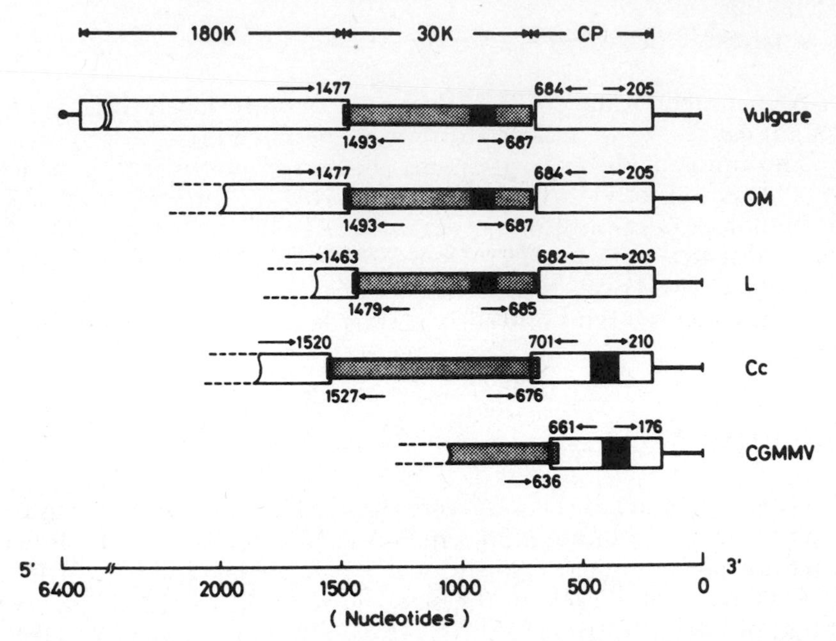

FIGURE 1. Diagrammatic summary of genome structure at the 3′-end portion of several tobamoviruses. Coding regions are indicated by boxes. Numbers refer to the first letter of the initiation codon or to the third letter of the termination codon from the 3′-end of the genomic RNA. Darkened areas show the assembly origin (see Fig. 3).

25.8% G in molar percent, respectively (Knight, 1952). The base composition of CGMMV-W RNA is 20.6%, 24.6% A, 31.6% U, and 23.2% G (Kurisu *et al.*, 1976). CGMMV RNA contains significantly less A and more U than TMV RNA.

The 5′-terminus of CGMMV RNA is blocked, as is the case in TMV RNA (Kurisu *et al.*, 1976).

The sequence of 1071 nucleotides from the 3′-end of the CGMMV-W genomic RNA, which contains the coat protein cistron, the assembly origin, and the noncoding region, was determined from cloned cDNA copies of CGMMV-W RNA (Fig. 2) (Meshi *et al.*, 1983).

B. Coat Protein and 30K Protein Cistron

The coat protein cistron is located at residues 176–661 from the 3′-end of the genomic RNA. It starts with AUG at residues 659–661 and terminates with UAG at residues 176–178 (Fig. 2).

In the 5′ flanking region of the coat protein cistron, a long open reading frame was found. This is likely to be the 30K protein cistron because the amino acid sequence deduced from the genetic code shows obvious homology with the C-terminal halves of the 30K proteins of

```
----CTCTGTTAGGTTCATACCTAATTATTCTGTCGTGGCTGCGGATGCCCTTCGCGATCCTTGGTCTTTATTTGTGAGACTCTCTAATG  986

TAGGTATTAAAGATGGTTTCCATCCTTTGACCTTAGAGGTCGCTTGTTTAGTCGCTACAACTAACTCTATTATCAAAAAGGGTCTTAGAG  896

CTTCTGTAGTCGAGTCTGTCGTCTCTTCCGATCAGTCCATTGTCCTAGATTCTTTATCCGAGAAAGTTGAACCTTTCTTTGATAAAGTTC  806

CTATTTCGGCGGCTGTAATGGCAAGAGACCCCAGTTATAGGTCTAGGTCGCAGTCTGTCAGTGGTCGTGGTAAGCGGCATTCTAAACCTC  716

                                                          30K Protein cistron
CAAATCGGAGGTTGGACTCTGCTTCTGAAGAGTCCAGTTCTGTTTCTTTCGACGATGGCTTACAATCCGATCACACCTAGCAAACTTATT  626
                                                        AlaTyrAsnProIleThrProSerLysLeuIle  (11)

GCGTTTAGTGCTTCTTATGTTCCCGTCAGGACTTTACTTAATTTTCTAGTTGCTTCACAAGGTACCGCCTTCCAGACTCAAGCGGGAAGA  536
AlaPheSerAlaSerTyrValProValArgThrLeuLeuAsnPheLeuValAlaSerGlnGlyThrAlaPheGlnThrGlnAlaGlyArg  (41)

GATTCTTTCCGCGAGTCCCTGTCTGCGTTACCCTCGTCTGTCGTAGATATTAATTCTAGGTTCCCAGATGCGGGTTTTTACGCTTTCCTC  446
AspSerPheArgGluSerLeuSerAlaLeuProSerSerValValAspIleAsnSerArgPheProAspAlaGlyPheTyrAlaPheLeu  (71)

AACGGTCCTGTGTTGAGGCCTATCTTCGTTTCGCTTCTCAGCTCCACGGATACGCGTAATAGGGTCATTGAGGTTGTAGATCCTAGCAAT  356
AsnGlyProValLeuArgProIlePheValSerLeuLeuSerSerThrAspThrArgAsnArgValIleGluValValAspProSerAsn  (101)

CCTACGACTGCTGAGTCGCTTAACGCTGTAAAGCGTACTGATGACGCATCTACGGCCGCTAGGGCTGAAATAGATAATTTAATAGAGTCT  266
ProThrThrAlaGluSerLeuAsnAlaValLysArgThrAspAspAlaSerThrAlaAlaArgAlaGluIleAspAsnLeuIleGluSer  (131)

ATTTCTAAGGGTTTTGATGTTTATGATAGGGCTTCATTTGAAGCCGCGTTTTCGGTAGTCTGGTCAGAGGCTACCACCTCGAAAGCTTAG  176
IleSerLysGlyPheAspValTyrAspArgAlaSerPheGluAlaAlaPheSerValValTrpSerGluAlaThrThrSerLysAla     (160)

TTTCGAGGGTCTTCTGATGGTGGTGCACACCAAAGTGCATAGTGCTTTCCCGTTCACTTAAATCGAACGGTTTGCTCATTGGTTTGCGGA  86

AACCTCTCACGTGTGACGTTGAAGTTTCTATGGGCAGTAATTCTGCAAGGGGTTCGAATCCCCCCTTTCCCCGGGTAGGGGCCCA(OH)  1
```

FIGURE 2. The sequence of 1071 nucleotides from the 3′-end of CGMMV-W RNA. The numbering of nucleotides begins with the 3′-terminal A of the genomic RNA and proceeds in the 5′-direction. The amino acid sequence of the coat protein deduced from the nucleotide sequence is also shown under the corresponding region, numbering of which (in parenthesis) begins with the N-terminal Ala. The postulated termination codon of the 30K protein cistron is also indicated.

common TMV (Guilley *et al.*, 1979; Goelet *et al.*, 1982; Meshi *et al.*, 1982a), tomato mosaic virus (Takamatsu *et al.*, 1983; Ohno *et al.*, 1984), and SHV (Meshi *et al.*, 1982b).

The frame of the 30K protein terminates with UAG at residues 636–638 in the coat protein cistron. Thus, the coat and 30K protein cistrons of CGMMV overlap each other (Fig. 1).

C. Nucleotide Sequence of the Assembly Origin and Comparison with That of TMV

The assembly origin of CGMMV-W RNA was located at about 320 nucleotides away from the 3′-end by electron microscopic serology (Fukuda *et al.*, 1981). In the predicted region, a possible assembly origin was found. In this region, the target sequence, GAXGUUG, and the triplet-

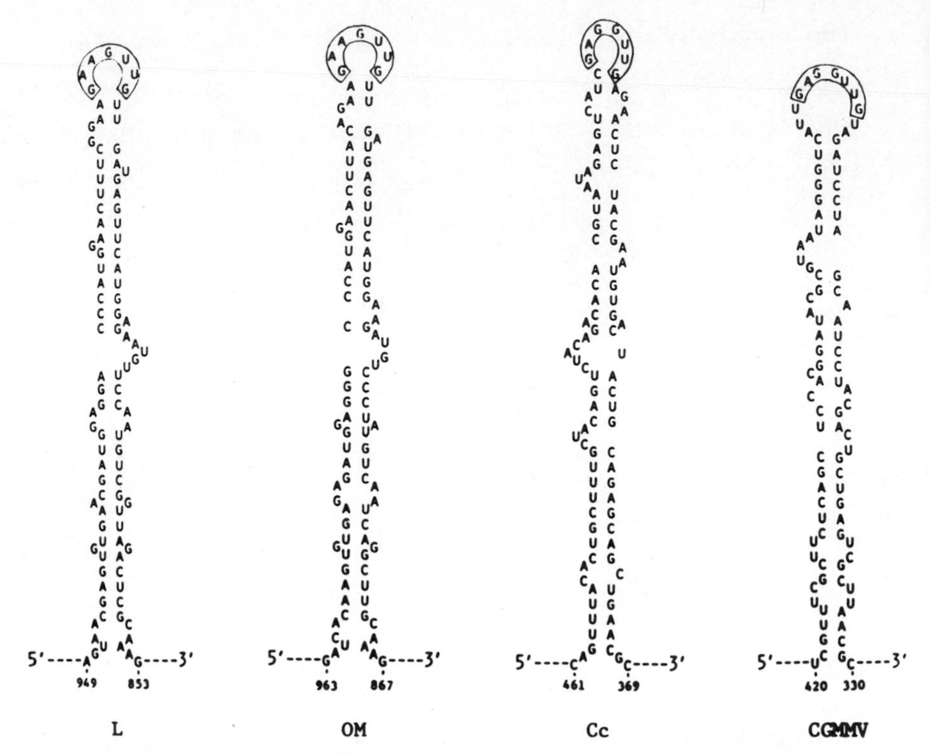

FIGURE 3. Possible secondary structures folded in the assembly origin of tobamoviruses. The numbering of nucleotides begins with the 3′-terminal end of the genomic RNA. The circumscribed nucleotides indicate the target sequences found in all tobamoviruses sequenced so far.

repeated purine base tract was found, and this sequence can be folded into a highly base-paired hairpin loop structure similar to those found in the other sequenced tobamoviruses (Fig. 3) (Zimmern, 1977; Meshi *et al.*, 1981, Takamatsu *et al.*, 1983). According to the rules of Tinoco *et al.* (1973), the thermodynamic stability was calculated to be -32.6, -35.8, -30.3, and -25.2 kcal/mole for CGMMV-W, SHV, common strain OM, and tomato mosaic virus, respectively.

D. The 3′ Noncoding Region

The 3′ noncoding region of CGMMV-W shows extensive homology to that of the common strain. The conserved region (156 nucleotides from the 3′-end) can be further divided into three subregions by the extent of conservation: residues 1–53 of CGMMV (85% homology), residues 54–87 of CGMMV (36% homology), and residues 88–155 of CGMMV (76% homology). The homology in the region upstream from residue 156 and in the coat protein cistron decreases drastically.

The 3′-end of the RNA of tobamoviruses can be aminoacylated with histidine (Oberg and Philipson, 1972; Carriquiry and Litvak, 1974) except for SHV RNA, which accepts only valine (Beachy *et al.*, 1976). The 3′-terminal region of CGMMV is extensively homologous to that of common TMV and, as expected, can be aminoacylated with histidine efficiently (Joshi, unpublished result).

A three-dimensional model for the tRNA-like structure of CGMMV RNA has been presented (Rietvelt *et al.*, 1984).

E. Subgenomic RNA

Subgenomic coat protein mRNA was isolated from the short rods in a CGMMV preparation (Fukuda *et al.*, 1981). The RNA induced in the wheat-germ cell-free protein-synthesizing system a protein which comigrated with the CGMMV coat protein and was precipitated with antibody against the latter.

The subgenomic mRNA for the 30K protein has not yet been isolated.

V. COAT PROTEIN

A. Amino Acid Sequence

The amino acid sequence of the CGMMV-W coat protein was determined from the nucleotide sequence (Fig. 2) (Meshi *et al.*, 1983). The coat protein is composed of 160 amino acid residues (i.e., two residues longer than that of common TMV) and lacks cysteine, methionine, and histidine. The molecular weight was calculated to be 17,261, considering *N*-acetylation (Nozu *et al.*, 1971).

The partial amino acid sequence determined by Nozu *et al.* (1971), corresponding to residues 91–122 from the N-terminus, coincides with that deduced from the nucleotide sequence.

The partial amino acid sequence of the CGMMV-C coat protein (residues 42–129) was also reported (Kurachi *et al.*, 1972). The homology between CGMMV-W and CGMMV-C is at most 48% in this region.

The amino acid sequence of the coat protein of CGMMV-W was compared with those of TMV vulgare and SHV (Meshi *et al.*, 1983). The homologies between CGMMV-W and TMV vulgare, CGMMV-W and SHV, and vulgare and SHV are 36, 44, and 46%, respectively. There is a highly conserved region common to both SHV and CGMMV-W composed of residues 82–120 from the N-terminus, in which 30 residues out of 38 are identical. The results indicated that the CGMMV-W protein is more homologous to SHV protein than to the vulgare protein. This supports the grouping of tobamoviruses according to the location of the assembly origin (Fig. 1).

B. Anomalous Mobility of the CGMMV Coat Protein on SDS–Polyacrylamide Gel Electrophoresis

On SDS–polyacrylamide gel electrophoresis, the CGMMV-W protein was clearly separated from the TMV (common strain) protein despite the fact that they have similar molecular weights (Ohno *et al.*, 1977; Sano *et al.*, 1978). That is, the CGMMV protein migrated faster than the TMV protein at various gel concentrations, and the molecular weight was calculated to be 16,200 from the Ferguson plot (Sano *et al.*, 1978). When the gel electrophoresis was carried out in the presence of 8 M urea, both coat proteins migrated at the same rate.

The anomalous mobility of the CGMMV protein may be attributable to the higher intrinsic negative charge density, smaller hydrodynamic volume, and slightly higher α-helix content as compared with the TMV protein.

C. Aggregation State of the Coat Protein

In 0.1 M phosphate buffer (pH 7.2, 3.6 mg/ml) at 25°C, the coat protein of CGMMV-W exists as 21 S aggregates and as 2.8 S subunits (Fukuda and Okada, 1982). However, in 0.25 M phosphate (pH 7.2) at 25°C, the protein exists mainly as 21 S aggregates with a small amount of 14 S aggregates. The equilibrium of the CGMMV protein lies highly to the side of the 21 S aggregates, and the 2.8 S subunit is not observed on analytical ultracentrifugation. This was the case in 0.1 M pyrophosphate buffer (pH 7.2) at 25°C (Ohno *et al.*, 1975). In some CGMMV preparations, the coat protein formed 14 S aggregates predominantly (Ohno *et al.*, 1972a, 1975).

Electron microscopic observations (Ohno *et al.*, 1975; Nonomura and Ohno, 1974) indicated that 21 S aggregates of the CGMMV protein correspond to the disk aggregates of common TMV. The 14 S aggregate is concluded to be a single-layer disk from hydrodynamic data (Ohno *et al.*, 1972a). The CD spectrum in the near-UV region of the CGMMV protein aggregate was measured (Inoue *et al.*, 1974).

The tertiary structure of the vulgare strain protein has been extensively studied by x-ray analysis, and the functional groups involved in disk formation have been elucidated (Bloomer *et al.*, 1978). The lateral interaction between subunits of the disk aggregate has been shown to comprise a polar one (Arg 122, Thr 89 and 118, Asp 88) and a hydrophobic one. The axial contacts have been proposed to be composed of a polar interaction (Thr 59, Ser 147 and 148), a hydrophobic interaction (Pro 54, Ala 74, Val 75), and a salt bridge system (Asp 19, Glu 22, 50, and 131, Lys 53, Asn 127, Arg 134).

The secondary structure and the distribution of the hydrophilic region over the CGMMV coat protein sequence were shown to be similar to those of the vulgare protein (Meshi *et al.*, 1983). Arginine residues 90, 92, 113, and 41, which are functional residues in the binding to RNA (Stubbs *et al.*, 1977), are all conserved. Thus, the tertiary structure of the CGMMV protein is assumed to be similar to that of the vulgare protein. Comparing the above-described functional groups of the two coat proteins, the amino acid residues involved in lateral interactions are well conserved. However, most of those involved in axial interactions are replaced in the CGMMV protein. This suggests that there are some different interactions among subunits in the axial direction in the disk aggregate of the CGMMV protein. This will influence the formation of single-layered disks (14 S) other than that of double-layered ones (21 S).

Acidification of the CGMMV protein solution causes the disks or subunits to polymerize into long helical rods as in the case of the TMV protein (Nonomura and Ohno, 1974). The CD spectra of the CGMMV protein under different conditions, i.e., in 0.01 M pyrophosphate buffer (pH 7.2), 0.1 M pyrophosphate buffer (pH 7.2), and 0.1 M phosphate buffer (pH 5.5), corresponding to 2.8 S subunits, 14 S or 21 S aggregates, and helical rods, were reported (Inoue *et al.*, 1974).

D. RNA Location in the CGMMV Protein Disk

The location of the RNA in CGMMV protein disk aggregates was determined by a negative staining method (Nonomura and Ohno, 1974). In negatively stained images of protein disks of CGMMV, a dense narrow ring around the central hole was clearly visible as shown in Fig. 4. Helical rods of the CGMMV protein without RNA also showed a ringlike groove in a position similar to that in the disks. On the other hand, no natural CGMMV particles showed any ringlike dense area. This indicates that the RNA occupies a position corresponding to that of the ringlike area in the samples. In some TMV disks, a clear ringlike groove was also visible in the position expected from the X-ray model of TMV.

Figure 4 shows selected views of TMV and CGMMV protein disks and the density distribution along the diameter of a representative disk of each type. The density in both cases shows an almost symmetrical distribution from the deep groove of the central hole toward the edges. For comparison, the radial density distributions of TMV rods with and without RNA determined by X-ray analysis by Franklin and Holmes (1958) are also shown. The peak at 4 nm from the center in Fig. 4c corresponds to the presence of RNA, and this position corresponds to the location of the decrease in density distribution in the electron micrograph of the TMV protein disk.

In the CGMMV protein disk, the diameter is slightly larger, 19 to 23 nm. Some flattening may have occurred in this case. The diameter of the central hole is smaller than that for TMV, 2 to 3 nm, but the distance of the RNA from the center is the same as that in TMV, nearly 4 nm.

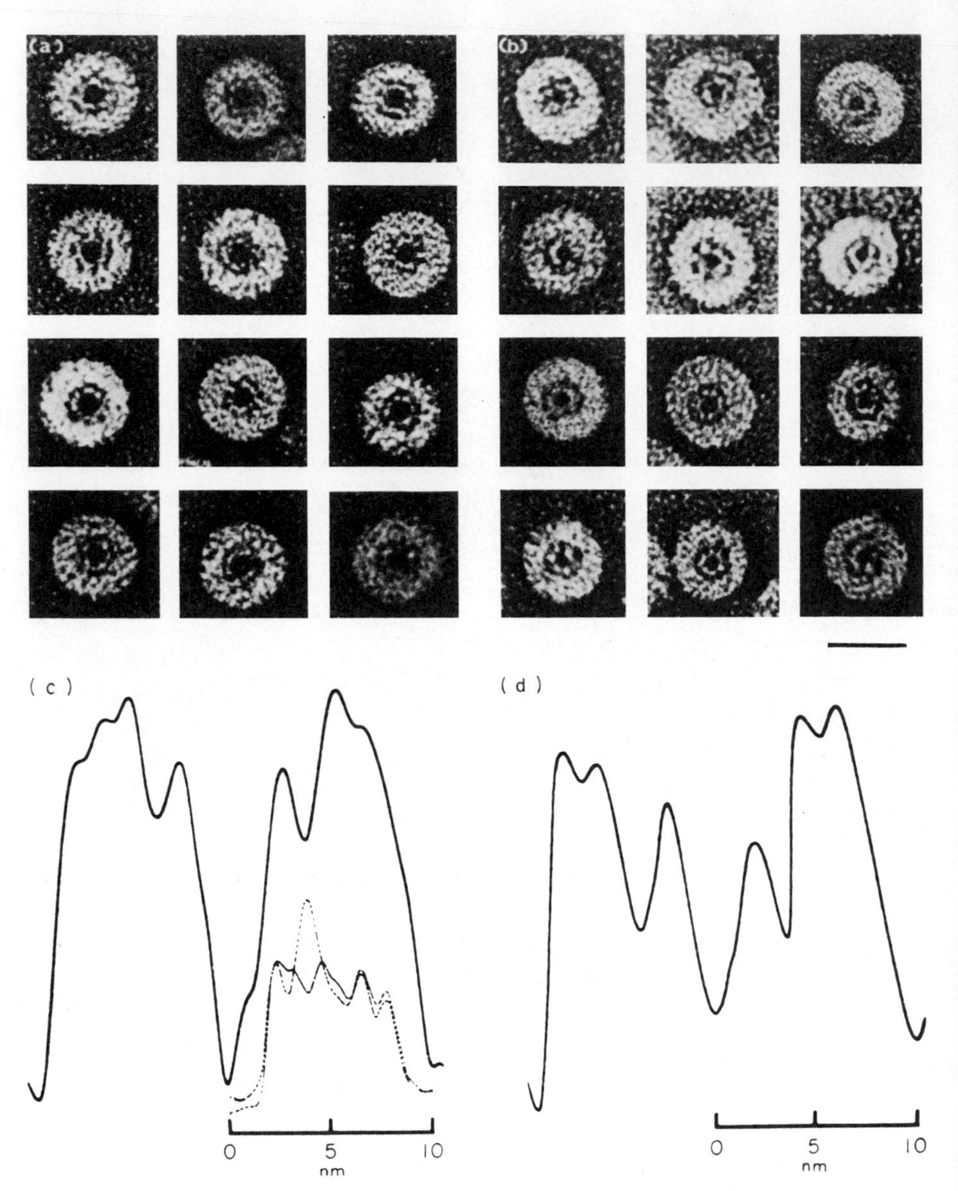

FIGURE 4. Selected views of TMV and CGMMV protein disks with clear ring-shaped staining and radial density distribution of selected disks. Bar = 20 nm. (a) TMV protein disks. (b) CGMMV protein disks. (c) Radial density distribution in one of the electron micrographs of the TMV protein disk. Inserted plots at the lower right were obtained by adapting the radial density distributions of TMV rods with and without RNA obtained from X-ray analysis by Franklin and Holmes (1958). (d) Radial density distribution in one of the electron micrographs of the CGMMV protein disk.

VI. VIRUS PARTICLE ASSEMBLY IN VITRO

A. Specificity of the Assembly Reaction *in Vitro*

CGMMV particles can be reconstitued efficiently *in vitro* from CGMMV RNA and CGMMV protein (Kurisu *et al.*, 1976), as in the case of TMV RNA and TMV protein (Fraenkel-Conrat and Williams, 1955). CGMMV RNA can also form virus particles with TMV protein (Kurisu *et al.*, 1976).

TMV RNA can be reconstituted into infectious particles with the CGMMV protein as efficiently as with the TMV protein and the hybrid virus was able to infect the same hosts as TMV (Ohashi *et al.*, 1969; Ohno *et al.*, 1972a). With a lower protein content in the reconstitution mixture of TMV RNA with TMV or CGMMV protein, partially reconstituted RNA was formed by coating at or near the assembly origin with one or the other of the proteins (Ohno *et al.*, 1971). The partially reconstituted RNA containing either TMV or CGMMV protein could be assembled into complete infectious particles only with the same protein, even when both proteins were present (Okada *et al.*, 1970).

B. Studies on TMV Assembly Using the CGMMV Protein

The first step in TMV assembly is the interaction between the assembly origin on RNA and the 20 S disk aggregates of the coat protein (Butler and Klug, 1971; Okada and Ohno, 1972). The second step is rod elongation which proceeds in two directions (Lebeurier *et al.*, 1977; Otsuki *et al.*, 1977). Contrary to the wide agreement as to the requirement of disks in assembly initiation, the role of the disks in elongation has been the subject of considerable controversy.

Butler and Klug (1971) showed that disks must be active in the elongation reaction, and that elongation can occur more rapidly with disks than with A-protein or subunits (Butler and Finch, 1973; Butler, 1974a,b). In contrast, Richards and Williams (1972, 1973) and the Tokyo group (Okada and Ohno, 1972; Ohno *et al.*, 1972a, 1972b, 1977) demonstrated that elongation could occur with A-protein more rapidly and that disks are not suitable. The matter is complicated by the fact that the coat protein of the TMV common strain (vulgare or OM) exists as an equilibrium mixture of disks and A-protein under the conditions of the reconstitution reaction (Durham and Klug, 1971). To eliminate this complication, Okada and co-workers used the CGMMV protein (Ohno *et al.*, 1972a, 1975; Fukuda and Okada, 1982, 1985).

In the assembly reaction of TMV RNA and CGMMV protein, the disk aggregates are needed for formation of the initial assembly complex. However, the disks cannot effect growth of the initial complex. Rod elongation takes place only after the addition of CGMMV protein subunits.

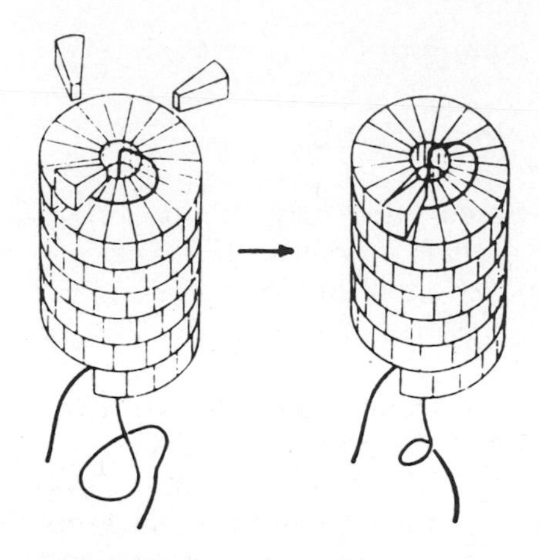

FIGURE 5. A possible mechanism of elongation of tobamoviruses in the major direction.

The 260-nm rod, which is the intermediate product of 3′-to-5′ elongation from the internal assembly origin during TMV assembly (Fukuda *et al.*, 1978), appeared after 5 min as in the case of the assembly reaction with TMV RNA and TMV protein. The rate and mode of the elongation by the CGMMV protein subunit, therefore, were the same as those observed in the system with the usual equilibrium disk preparation of TMV (Fukuda and Okada, 1982). The results proved that the rapid elongation of TMV assembly following the assembly initiation is the outcome of preferential incorporation of TMV subunits.

Elongation from the 260-nm intermediate to the 3′-end of the RNA was also investigated. The results showed that the 21 S aggregate is kinetically favored as the protein source for the 5′-to-3′ elongation (Fukuda and Okada, 1982).

Butler and Lomonossoff (1978) followed the TMV assembly reaction by protecting RNA from nuclease digestion and concluded that the protected RNA was quantitized into discrete lengths of about 50 or 100 nucleotides which correspond to one or two turns of the virus. However, similar protected RNA can also be observed in the assembly reaction between TMV RNA and CGMMV protein, where the protein was shown to be incorporated into a growing rod only as the subunit protein (Fukuda and Okada, 1985). This indicates that the banding pattern of the protected RNA observed during the assembly reaction is not due to the stepwise addition of the 20 S disk protein. The latter workers further showed that the banding pattern of the protected RNA differs depending on what kind of RNA is used. From these results, as shown in Fig. 5, Fukuda and Okada (1985) proposed for TMV assembly a model of elongation by protein subunits toward the 5′-direction.

REFERENCES

Ainsworth, G. C., 1935, Mosaic disease of cucumber, *Ann. Appl. Biol.* **22:**55.

Beachy, R. N., Zaitlin, M., Bruening, G., and Isnael, H. W., 1976, A genetic map for the cowpea strain of TMV, *Virology* **73:**498.

Bloomer, A. C., Champness, J. N., Bricogne, G., Staden, R., and Klug, A., 1978, Protein disk of tobacco mosaic virus at 2.8Å resolution showing the interactions within and between subunits, *Nature (London)* **276**:362.

Bruening, G., Beachy, R. N., Scalla, R., and Zaitlin, M., 1976, In vitro and in vivo translation of the ribonucleic acids of a cowpea strain of tobacco mosaic virus, *Virology* **73**:363.

Butler, P. J. G., 1974a, Structures and roles of the polymorphic forms of tobacco mosaic virus protein. IX. Initial stages of assembly of nucleoprotein rods from virus RNA and the protein disk, *J. Mol. Biol.* **82**:343.

Butler, P. J. G., 1974b, Structure and roles of the polymorphic forms of tobacco mosaic virus protein. VIII. Elongation of nucleoprotein rods of the virus RNA and protein, *J. Mol. Biol.* **82**:333.

Butler, P. J. G., and Finch, J. T., 1973, Structures and roles of the polymorphic forms of tobacco mosaic virus protein. VII. Lengths of the growing rods during assembly into nucleoprotein with the viral RNA, *J. Mol. Biol.* **78**:637.

Butler, P. J. G., and Klug, A., 1971, Assembly of the particle of tobacco mosaic virus from RNA and disks of protein, *Nature New Biol.* **229**:47.

Butler, P. J. G., and Lomonossoff, G. P., 1978, Quantitized incorporation of RNA during assembly of tobacco mosaic virus from protein disks, *J. Mol. Biol.* **126**:877.

Carriquiry, E., and Litvak, S., 1974, Further studies on the enzymatic aminoacylation of TMV-RNA by histidine, *FEBS Lett.* **38**:287.

Durham, A. C. H., and Klug, A., 1971, Polymerization of tobacco mosaic virus protein and its control, *Nature New Biol.* **229**:42.

Fraenkel-Conrat, H., and Williams, R. C., 1955, Reconstitution of active TMV from its inactive protein and nucleic acid components, *Proc. Natl. Acad. Sci. USA* **41**:690.

Franklin, R. E., and Holmes, K. C., 1958, Tobacco mosaic virus; Application of the method of isomorphic replacement to the determination of the helical parameters and radical density distribution, *Acta Crystallogr.* **11**:213.

Fukuda, M., Ohno, T., Okada, Y., Otsuki, Y., and Takebe, I., 1978, Kinetics of biphasic reconstitution of tobacco mosaic virus in vitro, *Proc. Natl. Acad. Sci. USA* **75**:1727.

Fukuda, M., and Okada, Y., 1982, Mechanism of tobacco mosaic virus assembly: Role of subunit and larger aggregate protein, *Proc. Natl. Acad. Sci. USA* **79**:5833.

Fukuda, M., and Okada, Y., 1985, Elongation in the major direction of tobacco mosaic virus assembly, *Proc. Natl. Acad. Sci. USA* **82**:3631.

Fukuda, M., Okada, Y., Otsuki, Y., and Takebe, I., 1980, The site of initiation of rod assembly on the RNA of a tomato and a cowpea strain of tobacco mosaic virus, *Virology* **101**:493.

Fukuda, M., Meshi, T., Okada, Y., Otsuki, Y., and Takebe, I., 1981, Correlation between particle multiplicity and location on virion RNA of the assembly initiation site for viruses of the tobacco mosaic virus group, *Proc. Natl. Acad. Sci. USA* **78**:4231.

Goelet, P., Lomonossoff, G. P., Butler, P. J. G., Akam, M. E., Gait, M. J., and Karn, J., 1982, Nucleotide sequence of tobacco mosaic virus RNA, *Proc. Natl. Acad. Sci. USA* **79**:3961.

Guilley, H., Jonard, G., Kukla, B., and Richards, K. E., 1979, Sequence of 1000 nucleotides at the 3′ end of tobacco mosaic virus RNA, *Nucleic Acids Res.* **6**:1287.

Higgins, T. J. V., Goodwin, P. B., and Whitfield, P. R., 1976, Occurrence of short particles in beans infected with the cowpea strain of TMV. II. Evidence that short particles contain the cistron for coat protein, *Virology* **71**:286.

Inoue, H., Kuriyama, K., Ohno, T., and Okada, Y., 1974, Circular dichroism and sedimentation studies on the reconstitution of tobacco mosaic virus, *Arch. Biochem. Biophys.* **165**:34.

Inouye, T., Inouye, N., Asatani, M., and Mitsuhata, K., 1967, Studies on cucumber green mottle mosaic virus in Japan [in Japanese], *Nogaku Kenkyu* **51**:175.

Knight, C. A., 1952, The nucleic acids of some strains of tobacco mosaic virus, *J. Biol. Chem.* **197**:241.

Komuro, Y., Tochihara, H., Funatsu, R., Nagi, Y., and Yomeyama, S., 1971, Cucumber green mottle mosaic virus in watermelon and its bearing on deterioration of watermelon fruit known as Konnyaku disease [in Japanese], *Ann. Phytopathol. Soc. Jpn.* **37**:34.

Kurachi, K., Funatsu, G., Funatsu, M., and Hidaka, S., 1972, Partial amino acid sequence of cucumber green mottle mosaic virus protein, *Agric. Biol. Chem.* **36:**1109.

Kurisu, M., Ohno, T., Okada, Y., and Nozu, Y., 1976, Biochemical characterization of cucumber green mottle mosaic virus ribonucleic acid, *Virology* **70:**214.

Lebeurier, G., Nicolaieff, A., and Richards, K. E., 1977, Inside-out model for self-assembly of tobacco mosaic virus, *Proc. Natl. Acad. Sci. USA* **74:**149.

Meshi, T., Ohno, T., Iba, H., and Okada, Y., 1981, Nucleotide sequence of a cloned cDNA copy of TMV (cowpea strain) RNA, including the assembly origin, the coat protein cistron, and the 3′ non-coding region, *Mol. Gen. Genet.* **184:**20.

Meshi, T., Ohno, T., and Okada, Y., 1982a, Nucleotide sequence and its character of cistron coding for the 30K protein of tobacco mosaic virus, *J. Biochem.* (*Tokyo*) **91:**1441.

Meshi, T., Ohno, T., and Okada, Y., 1982b, Nucleotide sequence of the 30K protein cistron of the cowpea strain of tobacco mosaic virus, *Nucleic Acids Res.* **10:**6111.

Meshi, T., Kiyama, R., Ohno, T., and Okada, Y., 1983, Nucleotide sequence of the coat protein cistron and the 3′ noncoding region of cucumber green mottle mosaic virus (watermelon strain) RNA, *Virology* **127:**54.

Motoyoshi, F., and Nishiguchi, M., 1984, Isolation and utilization of attenuated viruses for controlling cucurbit virus diseases [in Japanese], *Plant Prot.* **38:**353.

Narita, K., 1959, The isolation of acetylpeptide from the protein of cucumber virus 4, *Biochim. Biophys. Acta* **31:**372.

Niu, C., Shore, V., and Knight, C. A., 1958, The peptide chain of some plant viruses, *Virology* **6:**226.

Nonomura, Y., and Ohno, T., 1974, Direct visualization of the RNA location in cucumber green mottle mosaic virus and tobacco mosaic virus protein disk, *J. Mol. Biol.* **90:**523.

Nozu, Y., Tochihara, H., Komuro, Y., and Okada, Y., 1971, Chemical and immunological characterization of cucumber green mottle mosaic virus (watermelon strain) protein, *Virology* **45:**577.

Oberg, B., and Philipson, L., 1972, Binding of histidine to tobacco mosaic virus RNA, *Biochem. Biophys. Res. Commun.* **48:**927.

Ohashi, Y., Ohno, T., Nozu, Y., and Okada, Y., 1969, Reconstitution of tobacco mosaic virus in vitro: Inhibition by coat protein of other strains, *Proc. Jpn. Acad.* **45:**919.

Ohno, T., Nozu, Y., and Okada, Y., 1971, Polar reconstitution of tobacco mosaic virus, *Virology* **44:**510.

Ohno, T., Inoue, H., and Okada, Y., 1972a, Assembly of rod-shaped virus in vitro: Reconstitution with cucumber green mottle mosaic virus protein and tobacco mosaic virus RNA, *Proc. Natl. Acad. Sci. USA* **69:**3680.

Ohno, T., Yamaura, R., Kuriyama, K., Inoue, H., and Okada, Y., 1972b, Structure of N-bromosuccinimide-modified tobacco mosaic virus protein and its function in the reconstitution process, *Virology* **50:**76.

Ohno, T., Okada, Y., Nonomura, Y., and Inoue, H., 1975, Assembly of a rod-shaped virus: Disk aggregate of cucumber green mottle mosaic virus protein and its function, *J. Biochem.* (*Tokyo*) **77:**313.

Ohno, T., Takahashi, M., and Okada, Y., 1977, Assembly of tobacco mosaic virus in vitro: Elongation of partially reconstituted RNA, *Proc. Natl. Acad. Sci. USA* **74:**552.

Ohno, T., Aoyagi, M., Yamanashi, Y., Saito, H., Ikawa, S., Meshi, T., and Okada, Y., 1984, Nucleotide sequence of the tobacco mosaic virus (tomato strain) genome and comparison with the common strain genome, *J. Biochem.* (*Tokyo*) **96:**1915.

Okada, Y., and Ohno, T., 1972, Assembly mechanism of tobacco mosaic virus particle from its ribonucleic acid and protein, *Mol. Gen. Genet.* **114:**205.

Okada, Y., Ohashi, Y., Ohno, T., and Nozu, Y., 1970, Sequential reconstitution of tobacco mosaic virus, *Virology* **42:**243.

Okada, Y., Fukuda, M., Takebe, I., and Otsuki, Y., 1980, Initiation site for assembly of several strains of TMV and its relation to occurrence of the short particles in infected plants, *Biosystems* **12:**257.

Otsuki, Y., Takebe, I., Ohno, T., Fukuda, M., and Okada, Y., 1977, Reconstitution of tobacco mosaic virus rods occurs bidirectionally from an internal initiation region: Demonstration by electron microscopic serology, *Proc. Natl. Acad. Sci. USA* **74:**1913.

Richards, K. E., and Williams, R. C., 1972, Assembly of tobacco mosaic virus in vitro: Effect of state of polymerization of the protein component, *Proc. Natl. Acad. Sci. USA* **69:**1121.

Richards, K. E., and Williams, R. C., 1973, Assembly of tobacco mosaic virus in vitro: Elongation of partially assembled rods, *Biochemistry* **12:**4574.

Rietvelt, K., Linschooten, K., Pleij, C. W. A., and Bosch, L., 1984, The three-dimensional folding of the tRNA-like structure of tobacco mosaic virus RNA, a new building principle applied twice, *EMBO J.* **3:**2613.

Sano, Y., Nozu, Y., and Inoue, H., 1978, Anomalous mobility of cucumber green mottle mosaic virus protein in sodium dodecyl sulfate–polyacrylamide gel electrophoresis, *J. Biochem. (Tokyo)* **84:**1041.

Stubbs, G., Warren, S., and Holmes, K., 1977, Structure of RNA and RNA binding site in tobacco mosaic virus from 4-Å map calculated from X-ray fibre diagrams, *Nature (London)* **267:**216.

Takamatsu, N., Ohno, T., Meshi, T., and Okada, Y., 1983, Molecular cloning and nucleotide sequence of the 30K and the coat protein cistron of TMV (tomato strain) genome, *Nucleic Acids Res.* **11:**3767.

Tinoco, J. I., Borer, P. N., Dengler, B., Levine, M. D., Uhlenbeck, O. C., Crothers, D. M., and Gralla, S., 1973, Improved estimation of secondary structure in ribonucleic acids, *Nature New Biol.* **246:**40.

Tochihara, H., and Komuro, Y., 1974, Infectivity test and serological relationships among various isolates of cucumber green mottle mosaic virus [in Japanese], *Ann. Phytopathol. Soc. Jpn.* **40:**52.

Tung, J.-S., and Knight, C. A., 1972, The coat protein subunits of cucumber viruses 3 and 4 and a comparison of methods for determining their molecular weights, *Virology* **48:**574.

Whitfield, P. R., and Higgins, T. J. V., 1976, Occurrence of short particles in beans infected with the cowpea strain of TMV, *Virology* **71:**471.

Zimmern, D., 1977, The nucleotide sequence at the origin for assembly on tobacco mosaic virus RNA, *Cell* **11:**463.

CHAPTER 15

Miscellaneous Tobamoviruses

Alan A. Brunt

I. INTRODUCTION

Because some tobamoviruses naturally infect few species, are newly recognized, or have a limited geographical distribution, they have not been characterized as fully as tobacco mosaic (TMV), tomato mosaic (ToMV), mild dark-green tobacco mosaic (TMGMV), and similar viruses which have a very wide geographical distribution, debilitate economically important crops, and/or are of great intrinsic interest. Some of these miscellaneous tobamoviruses were previously described as strains of TMV, but are now considered by many to be sufficiently different to be recognized as distinct viruses. The natural occurrence, properties, and affinities of the better known of these viruses are reviewed here (Table I).

II. THE VIRUSES AND THEIR NATURAL OCCURRENCE

A. The Viruses and Their Hosts

1. *Chara corallina* Virus (CCV)

CCV was first found infecting the alga *Chara corallina* var. *corallina*, a large ecorticate dioecious charaphyte (Chlorophyceae) collected from the Murrumbidgee River near Canberra, Australia (Gibbs *et al.*, 1975). It was subsequently found in *C. corallina* collected from two other upland rivers (the Shoalhaven and Bombala) to the south and east of Can-

ALAN A. BRUNT • Glasshouse Crops Research Institute, Littlehampton, West Sussex BN17 6LP, England.

TABLE I. Miscellaneous Tobamoviruses

Virus	Host(s)	Geographical distribution	Reference
Chara corallina (CCV)	*Chara corallina*[a]	Australia	Gibbs *et al.* (1975)
Frangipani mosaic (FMV)	*Plumeria acutifolia*	Australia	Francki *et al.* (1971)
	P. alba	India	Varma and Gibbs (1978)
Maracuja mosaic (MMV)	*Passiflora* spp.	Argentina	C. E. Fribourg, R. Koenig, and D. E. Lesemann (unpublished information)
Pepper mild mottle (PMMV)	*Capsicum annuum*	Australia	Pares (1985)
		England	B. J. Thomas and A. A. Brunt (unpublished)
		Hungary	Tobias *et al.* (1982)
		Italy	Wetter *et al.* (1984)
		Netherlands	Rast (1979)
		Spain	C. Wetter (unpublished)
		United States	McKinney (1952, 1968)
Peziza ostracoderma (POV)	*Peziza ostracoderma*	Netherlands	Dieleman-van Zaayen (1967)
Potato 14R (PV14R)	*Solanum tuberosum*	Peru	Salazar (1977)
Rose tobamovirus (RTV)	*Rosa* sp.	England	Hicks and Frost (1984)
Sammons's Opuntia (SOV)	*Opuntia* spp.	United States	Sammons and Chessin (1961)
Ullucus mild mottle (UMMV)	*Ullucus tuberosus*	Peru	Brunt *et al.* (1982)

[a] This alga, although initially identified as *C. corallina* (Gibbs *et al.*, 1975) from information given by Wood (1972), is now considered to be *C. australis* (Vernon W. Proctor, personal communication to A. J. Gibbs).

berra, but not in other *Chara* species, nor in *Nitella* or *Tolypella* species (Skotnicki *et al.*, 1976).

CCV can be injected into virus-free *C. corallina* cells which, within 8 days, become chlorotic and flaccid. The virus, however, failed to infect any of 13 plant species from five dicotyledonous families.

2. Frangipani Mosaic Virus (FMV)

FMV, also sometimes known as temple tree mosaic virus, is reported to occur naturally in frangipani (*Plumeria acutifolia* and *P. alba*) in eastern Australia and northern India (Francki *et al.*, 1971; Varma and Gibbs, 1978). It induces chlorosis, ringspots, vein-banding, and bronzing of *P. acutifolia* leaves, and ringspots, distortion, and necrosis in those of *P. alba* (Fig. 1A).

The experimental host range of FMV, although not investigated extensively, is probably fairly restricted (Francki *et al.*, 1971). However, more species are infected, and symptoms appear more quickly, in plants

FIGURE 1. Symptoms induced by frangipani mosaic virus in (A) frangipani (*Plumeria acutifolia*), (B) *Datura stramonium*, and (C) *Nicotiana glutinosa*. Courtesy of R. I. B. Francki.

grown at higher temperatures (>25°C). FMV infected six of eight species from two families (Solanaceae and Chenopodiaceae), inducing systemic infection only in *Nicotiana tabacum* seedlings and, with one strain, in *Datura stramonium* plants grown at 35°C. The virus otherwise infected the inoculated leaves only of other hosts, of which *D. stramonium* is convenient for local lesion assays and *N. glutinosa* for virus propagation (Fig. 1B,C).

Three strains of FMV (from Adelaide, Allahabad, and Delhi) have been distinguished, principally by the symptoms they induce in *D. stramonium* and *N. tabacum* cv. Virginia Gold (Varma and Gibbs, 1978).

3. Maracuja Mosaic Virus (MMV)

This virus, first isolated in Peru from *Passiflora edulis* with mosaic leaf symptoms, was subsequently shown to be a hitherto undescribed tobamovirus (C. E. Fribourg, R. Koenig, and D.-E. Lesemann, personal communication). MMV is readily sap-transmissible to 27 species in 9 families, inducing systemic infection in *P. edulis* and *Nicotiana clevelandii,* but local lesions only in other susceptible species. The virus is best propagated in *N. clevelandii* and assayed in *Chenopodium quinoa.* Although the virus has yet to be further characterized, it has been shown to be serologically distinct from other tobamoviruses.

4. Pepper Mild Mottle Virus (PMMV)

At least four tobamoviruses are known to occur naturally in pepper (*Capsicum annuum*). Strains of TMV, ToMV, and TMGMV infect pepper in the United States, Europe, and possibly elsewhere (e.g., Johnson, 1930; Holmes, 1934; McKinney, 1952; Miller and Thornberry, 1958; Conti and Masenga, 1977; Conti and Marte, 1983; Tobias *et al.*, 1982; Demski, 1981; Wetter, 1984). The fourth, now designated PMMV, was isolated in Sicily from naturally infected plants of *C. annuum* cv. Lamuyo, a genotype known to be resistant to common strains of TMV and ToMV. PMMV was subsequently shown to be serologically related to, but different from, other tobamoviruses and thus to justify its recognition as a distinct virus (Wetter *et al.*, 1984). The virus is now known to occur in *C. annuum* in the Netherlands, Italy, Spain, United States, England, Hungary, and Australia (Wetter *et al.*, 1984; Rast, 1979, 1982; Marte and Wetter, personal communication; McKinney, 1952; Murakishi, 1960; B. J. Thomas and A. A. Brunt, unpublished information; Pares, 1985).

Infected plants of *C. annuum* in Sicily were slightly stunted and had fruits that were abnormally small, deformed, mottled, and occasionally necrotic (Fig. 2A); leaves on plants sown during October usually developed no conspicuous symptoms, whereas those of plants sown later were often mottled and slightly deformed (Fig. 2B).

A virus infecting pepper in the United States, and long known as the

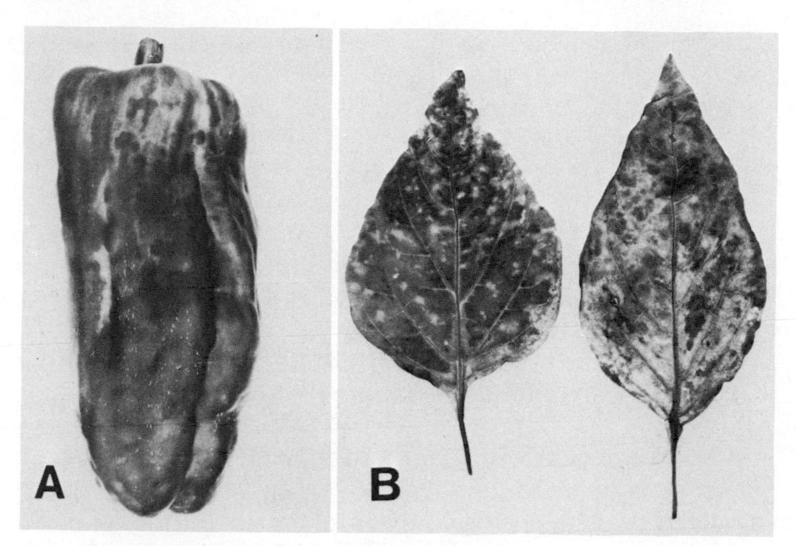

FIGURE 2. Symptoms induced by pepper mild mottle virus in (A) fruit of *Capsicum annuum* cv. Lamuyo and (B) leaves of cv. Quadrato d'Asti. Courtesy of C. Wetter.

Samsun latent strain of TMV (SLTMV) (McKinney, 1952, 1968; Greenleaf *et al.*, 1964), was subsequently shown to be serologically indistinguishable from PMMV. SLTMV is thus best considered to be a synonym of PMMV (Wetter, 1984).

It is uncertain whether or not the tobamovirus from pepper described as an unusual strain of TMV in Argentina (Feldman and Oremianer, 1972) is also a strain of PMMV. Although some host range and serological tests indicated its similarity to PMMV (Tobias *et al.*, 1982), subsequent serological tests and amino acid analyses suggest that it is a distinct virus (Wetter *et al.*, 1984; C. Wetter, personal communication).

PMMV was initially differentiated from TMV, ToMV, and TMGMV by differences in its host range and its symptomatology in some common solanaceous hosts. Thus, unlike TMV, PMMV induces symptomless infection only in inoculated leaves of *N. tabacum* cv. Samsun, and, unlike ToMV but like TMGMV, it fails to infect tomato. TMGMV, however, differs from PMMV in inducing conspicuous systemic infection in *N. tabacum* cv. Samsun and *N. glauca* (Wetter *et al.*, 1984).

5. *Peziza ostracoderma* Virus (POV)

A tobamo-like virus has been found in the Netherlands in *Peziza ostracoderma*, a saprophytic ascomycete commonly found growing on straw bales and compost in mushroom farms (Dieleman-van Zaayen, 1967; Dieleman-van Zaayen *et al.*, 1970). POV, which apparently causes symptomless infection in *P. ostracoderma*, was detected in only some of the samples tested. It was not transmitted to the edible mushroom (*Agar-*

icus bisporus), from which similar viruslike particles have sometimes been recovered (Dieleman-van Zaayen, 1967), or to commonly used herbaceous hosts of tobamoviruses.

6. Potato 14R Virus (PV14R)

This virus, tentatively designated PV14R, was found infecting potato (*Solanum tuberosum* ssp. *andigena* x ssp. *tuberosum* cv. Aleli) in Huancayo, Junin, Peru (Salazar, 1977). Infected plants, which also contained potato virus X, had leaves with yellow mosaic symptoms. Although readily sap-transmissible to other *Solanum* species and other potato cultivars, PV14R induced symptomless infection in the inoculated leaves only of most hosts.

PV14R has a moderately wide experimental host range. In addition to *Solanum* species, it infected 45 of 57 species from 7 of 12 families, inducing necrotic spots or rings only in inoculated leaves of most hosts but chlorosis or symptomless infection also in systemically infected leaves of *Chenopodium ambrosioides, C. amaranticolor, C. quinoa, Lycopersicon esculentum*, and *Petunia hybrida*. The virus thus has the unusual property of failing to induce systemic infection in most hosts, even in potato and other solanaceous species in which it attains a high concentration in inoculated leaves.

Although critical comparisons have yet to be made, PV14R is possibly similar to a tobamovirus reported to infect potato in Chile (Accatino, 1966). It is, however, distinct from the strain of TMV reported to infect potato in India (Phatak and Verma, 1967).

7. Rose Tobamovirus (RTV)

A tobamovirus has been isolated with difficulty in England from a climbing rose (probably cv. Danse de Feu) with broken flowers, although not from plants of eight other cultivars with similar symptoms (Hicks and Frost, 1984). The virus was isolated originally only by partially purifying extracts from the infected rose, and subculturing the virus from the lesions they induced in inoculated leaves of *Chenopodium amaranticolor*. However, as no flower symptoms were induced in sap- or graft-inoculated roses, the etiology of flower-breaking in rose is unproven.

8. Sammons's Opuntia Virus (SOV)

This virus, subsequently designated SOV (Brandes, 1964), was first found together with a potexvirus (cactus virus I) in *Opuntia monocantha* f. *variegata* and four other *Opuntia* species in California and Montana (Sammons and Chessin, 1961). SOV and cactus virus I were subsequently found in many cactus species in Arizona, Nevada, and Utah (Chessin, 1965, 1969, Chessin and Lesemann, 1972).

SOV is sap-transmissible to healthy *Opuntia* species, but plants are infected symptomlessly. Although in initial tests it failed to infect any species from six dicotyledonous families (Sammons and Chessin, 1961), it was subsequently shown to induce lesions in inoculated leaves of *Chenopodium amaranticolar, C. album, C. murale,* and *C. quinoa*. The virus can be conveniently propagated in *C. quinoa* (Brandes and Chessin, 1965; Wetter and Paul, 1967).

9. Ullucus Mild Mottle Virus (UMMV)

This virus, now designated ullucus mild mottle virus, was detected in all *Ullucus tuberosus* ("ulluco") plants from seven locations in Peru but in none of those from five areas of Bolivia (Brunt *et al.*, 1982). The infected plants also contained ullucus mosaic, ullucus C, and usually also papaya mosaic viruses; UMMV, however, was readily separated from naturally occurring mixtures because it was the only virus of the four to induce local lesions in *Nicotiana glutinosa*. UMMV was readily sap-transmissible to healthy ulluco plants obtained by meristem-tip culture, but alone it induced only inconspicuous chlorosis in inoculated leaves and virtually symptomless infection in systemically infected leaves.

In addition to *U. tuberosus*, UMMV infected 21 of 30 species from

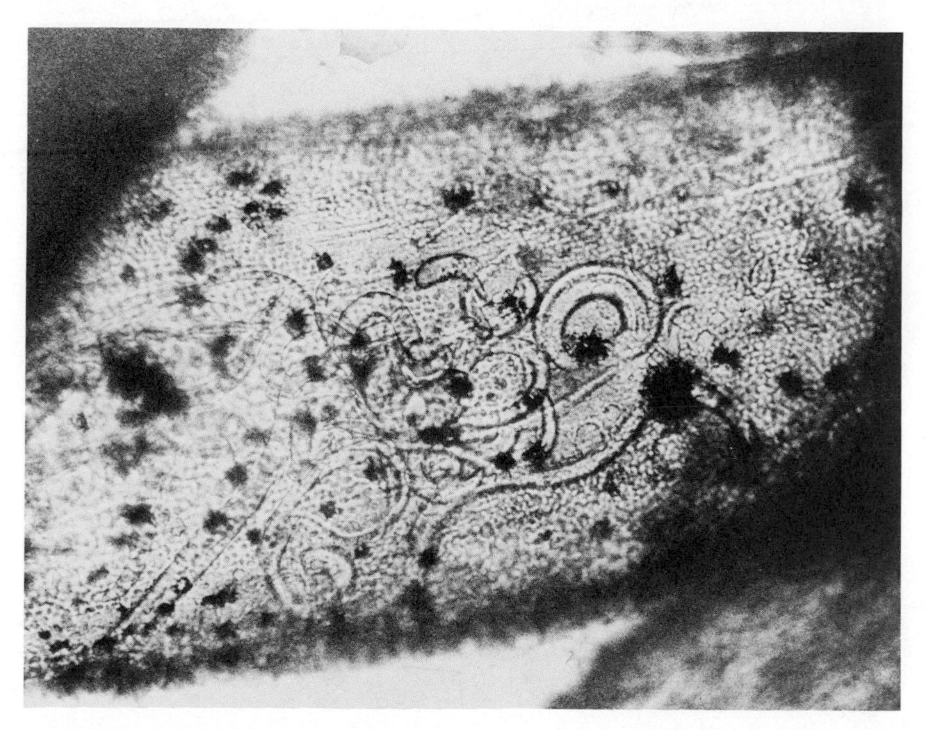

FIGURE 3. Elongated hyaline inclusions induced by *Chara corallina* virus in *C. australis* (syn. *C. corallina*). Courtesy of A. J. Gibbs.

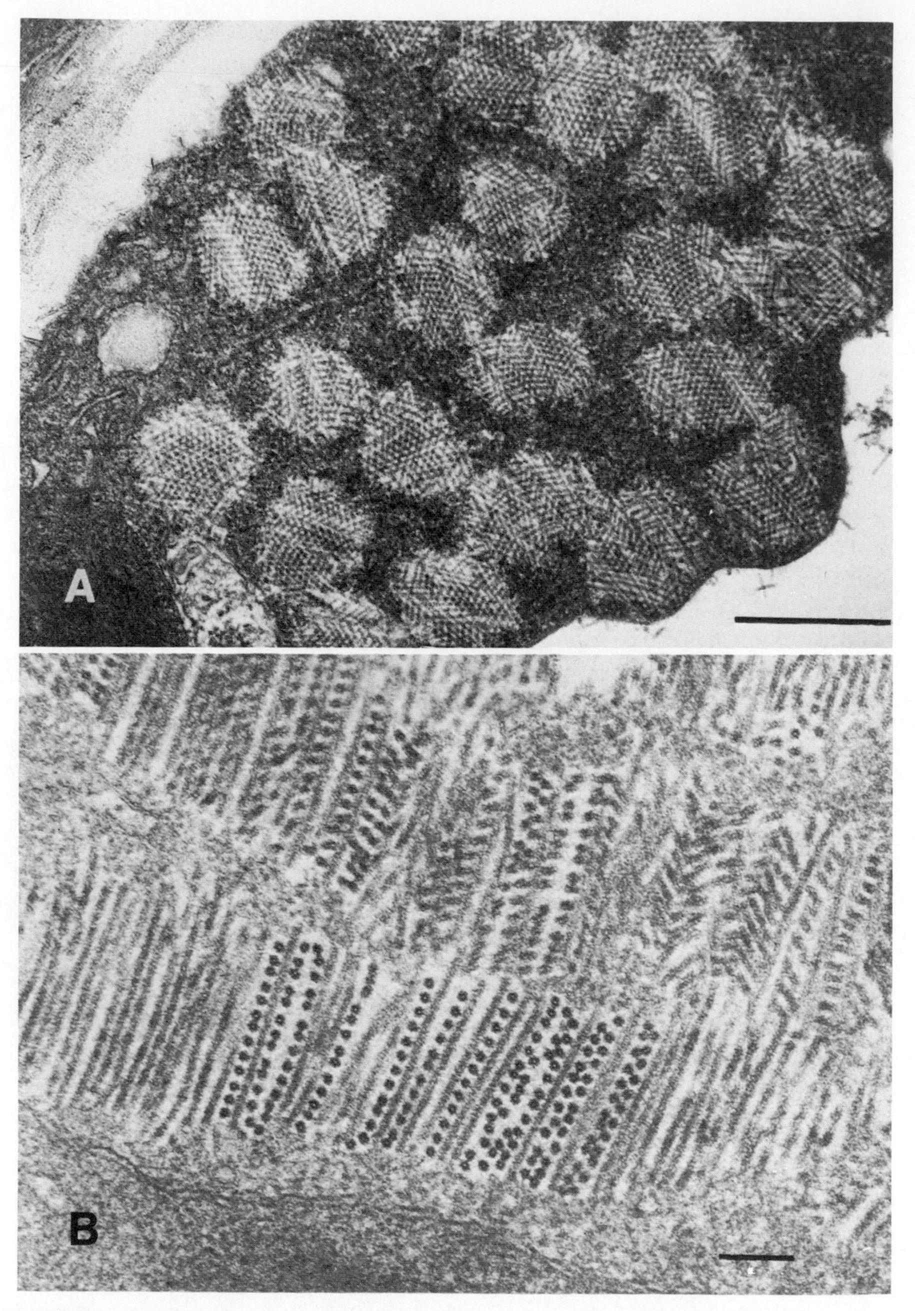

FIGURE 4. Angled-layer aggregates induced by pepper mild mottle virus in (A) *C. annuum* (bar = 500 nm) and (B) at higher magnification (bar = 100 nm). Courtesy of Drs. E. Ernwein and C. Wetter.

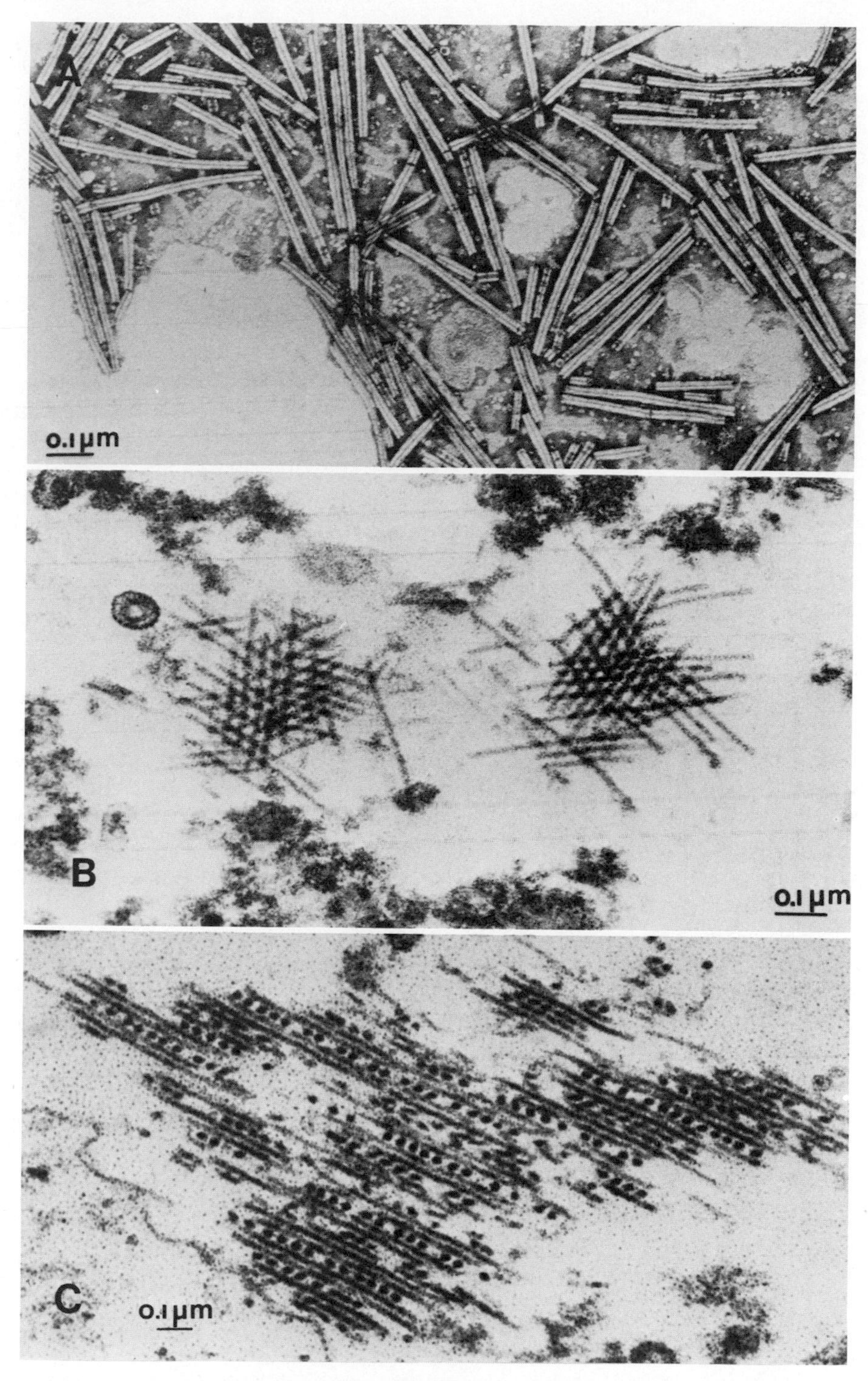

FIGURE 5. *Peziza ostracoderma* virus. (A) Viruslike particles negatively stained with phosphotungstic acid (pH 5.5). (B) Aggregates of viruslike particles in vacuole of apothecial cell. (C) Angled-layer aggregate. Courtesy of Dr. A. van Zaayen.

6 of 8 families. It infected 12 of 14 species of the Solanaceae, infecting the inoculated leaves only of most but inducing conspicuous mosaic symptoms in systemically infected leaves of *N. clevelandii, N. benthamiana*, and *Datura metel* in which the virus can be conveniently cultured; the virus is best assayed in *N. glutinosa*.

B. Intracellular Occurrence of the Viruses

The intracellular occurrence of only five (CCV, FMV, MMV, PMMV, and POV) of the nine miscellaneous tobamoviruses has been investigated.

CCV induces hyaline, birefringent, elongated or banana-shaped inclusions (Fig. 3) which, in infected living cells, can be seen by light microscopy to move during cyclosis of cytoplasm (Gibbs *et al.*, 1975). The inclusions in ultrathin sections, like those of TMV, were found by electron microscopy to consist of numerous rod-shaped particles in parallel array. Similar spindle-shaped inclusions ("Eiweissspindeln") occurring also in *Opuntia* spp. infected with SOV and cactus virus I (Sammons and Chessin, 1961) are probably induced by the latter virus (Miličić and Udjbinac, 1961), although tobamo-like X-bodies were found previously in virus-infected *Opuntia* and *Epiphyllum* species (Amelunxen, 1958).

FMV causes no obvious damage to organelles of *Datura stramonium*, but in parenchymatous cells its particles aggregate into microcrystals of various shapes and sizes (Varma and Gibbs, 1978). The particles of MMV sometimes aggregate to form platelike crystals ca. 300 nm thick. The virus also consistently induces the formation of granular cytoplasmic inclusions (C. E. Fribourg, R. Koenig, and D.-E. Lesemann, personal communication).

In pepper, PMMV induces true two-dimensional crystals typical of TMV and angled-layer aggregates (Fig. 4) consisting of monolayers of particles alternating at angles of 60–80° (Wetter *et al.*, 1984).

Ultrastructural studies on infected apothecia of *P. ostracoderma* have shown that the particles mainly occur within the cytoplasm and vacuoles of cells immediately below asci (Fig. 5). The particles are usually present in crystalline arrays which, with alternate layers at right angles, are sometimes seen in transverse or longitudinal section; in oblique section, however, they are seen as cross-hatched structures (Dieleman-van Zaayen *et al.*, 1970) similar to those of PMMV (see above) and the aucuba strain of TMV (Warmke, 1968).

III. PROPERTIES

A. General Properties of Particles

1. Stability in Sap

The better studied of the viruses, like most other tobamoviruses (Van Regenmortel, 1981), occur within infected plants in high concentration and are very stable *in vitro*. Thus, five of the viruses (FMV, PMMV, RTV, PV14R, and UMMV) are infective in sap heated for 10 min at 90 but not 95°C, and in sap diluted from 10^{-5} to 10^{-11} (Varma and Gibbs, 1978; Wetter *et al.*, 1984; Salazar, 1977; Hicks and Frost 1984; Brunt, unpublished information).

The viruses are also readily purified. Saps can be clarified with little loss of virus by treatment with organic solvents. Precipitation of virus from such clarified extracts with polyethylene glycol (Hebert, 1963) yields from 80 to 1300 mg of PV14R and PMMV per kg of leaf tissue (Salazar, 1977; Wetter *et al.*, 1984); similar yields of UMMV are obtainable by differential centrifugation of infective sap followed by chromatography on controlled-pore glass beads (Brunt *et al.*, 1982).

2. Particle Sizes

FMV, MMV, UMMV, and RTV have rod-shaped particles mostly measuring 18 × 300 nm (Francki *et al.*, 1971; C. E. Fribourg, R. Koenig, and D.-E. Lesemann, personal communication; Brunt *et al.*, 1982; Hicks and Frost, 1984). PV14R, PMMV, SOV, and POV, however, have morphologically similar particles with normal lengths reported to be 310, 312, 317, and 350 nm, respectively (Salazar, 1977; Wetter *et al.*, 1984; Brandes and Chessin, 1965; Dieleman-van Zaayen *et al.*, 1970). CCV is unusual in having rod-shaped particles mostly measuring 18 × 532 nm (Gibbs *et al.*, 1975; Skotnicki *et al.*, 1976). The helical pitch of most of the viruses, easily measured on particles mounted for electron microscopy in uranyl formate or uranyl acetate, is 2.4 nm although that of the algal and fungal tobamoviruses is ca. 2.7 nm.

3. Other Properties

The UV absorption spectra of the purified viruses are typical of tobamoviruses, with maximum and minimum absorption at 260–262 nm and 248–250 nm, respectively, and $A_{max/min}$ and $A_{260/280}$ ratios of ca. 1.11 and 1.21, respectively (Francki *et al.*, 1971; Wetter *et al.*, 1984; Salazar, 1977; Skotnicki *et al.*, 1976). The extinction coefficient (E_{260} mg/cm^3) of PMMV, the only virus of the nine for which a value has been determined, is 3.18 (Wetter *et al.*, 1984).

Homogeneous preparations of UMMV, PV14R, FMV, and SOV contain a major component with a sedimentation coefficient of 176–196 S (Brunt *et al.*, 1982; Salazar, 1977; Varma and Gibbs, 1978; Wetter and Paul, 1967), but the longer particles of CCV sediment at 230 S (Gibbs *et al.*, 1975; Skotnicki *et al.*, 1976). UMMV particles have a buoyant density in cesium chloride of 1.32 g/cm^3 (Brunt *et al.*, 1982). PMMV and CCV have isoelectric precipitation points of pH 3.7–3.8 and 3.4–3.7, respectively (Wetter *et al.*, 1984; Skotnicki *et al.*, 1976).

TABLE II. Amino Acid Compositions of Tobacco Mosaic Virus (TMV) and Five Miscellaneous Tobamoviruses

	No. of residues of					
	TMV[a]	CCV[b]	FMV[c]	PMMV[d]	SOV[e]	UMMV[f]
Ala	14	14	14	17	12	18
Arg	11	8	11	9	8	11
Asp	18	25	17	18	12	11
Cys	1	ND[g]	1	1	1	1
Glu	16	15	16	18	12	24
Gly	7	12	9	8	5	5
His	0	1	1	0	0	2
Ile	9	6	11	5	7	7
Leu	12	12	13	16	12	14
Lys	2	10	4	2	4	3
Met	0	3	0	1	1	1
Phe	8	14	7	7	6	7
Pro	8	9	4	6	6	10
Ser	16	15	14	10	9	15
Thr	16	14	13	20	11	11
Trp	3	ND	5	2	ND	2
Tyr	4	4	5	4	5	5
Val	14	8	13	14	9	11
Total	158	170	158	158	ND	158

[a] Wittmann-Liebold and Wittmann (1967).
[b] Skotnicki *et al.* (1976).
[c] Francki *et al.* (1971).
[d] Wetter *et al.* (1984).
[e] Gibbs (1977).
[f] M. H. V. Van Regenmortel and A. A. Brunt (unpublished data).
[g] ND, not determined.

B. Composition

1. Protein

Spectrophotometric analyses of purified preparations suggest that the viruses contain ca. 95% protein. Polyacrylamide gel electrophoresis (PAGE) of PV14R and UMMV proteins has shown that each contains a single polypeptide with estimated molecular weights of 19.4×10^3 and 21.0×10^3, respectively (Salazar, 1977; Brunt *et al.*, 1982), values close to those obtained by PAGE for other tobamoviruses and for labile fungus-transmitted rod-shaped viruses (Brunt and Shikata, this volume). CCV protein also contains a single polypeptide, but with a molecular weight of $16.5–17.0 \times 10^3$ or at least 18,716 when estimated by PAGE and amino acid analyses, respectively (Gibbs *et al.*, 1975; Skotnicki *et al.*, 1976). MMV protein produces two bands in PAGE (M_r 17.4 and 15.9×10^3), the smaller of which is apparently not a degradation product of the larger (C. E. Fribourg, R. Koenig, and D.-E. Lesemann, personal communication).

The reported amino acid compositions of CCV, FMV, PMMV, SOV,

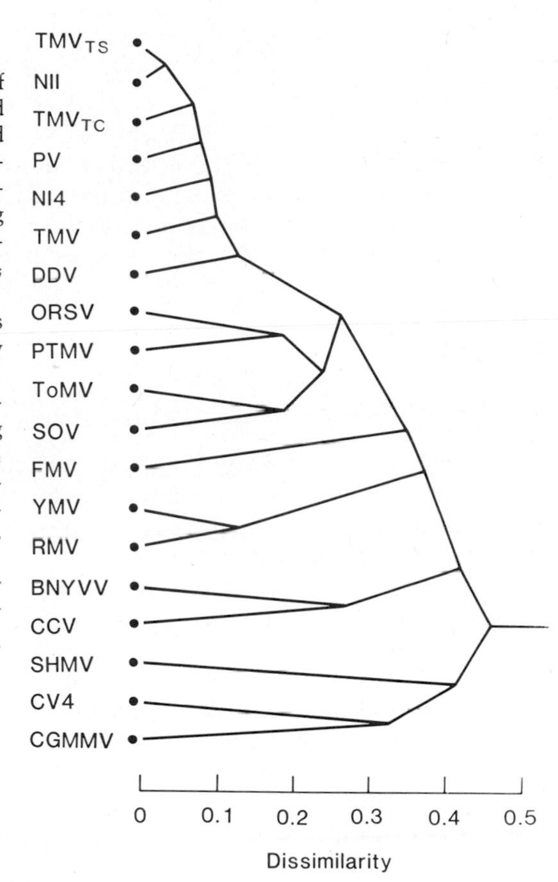

FIGURE 6. Classification of some tobamoviruses computed from their molar amino acid compositions; agglomerative hierarchical classification, Canberra metric and median sorting strategy. TMV_{TS} = common tomato mosaic virus (Chinese); N11 = TMV_{N11}; TMV_{TC} = common tobacco mosaic virus (Chinese); PV = Populus bushy top virus (Chinese); N14 = TMV_{N14}; TMV = tobacco mosaic virus (type) DDV = Dihuang degeneration virus (Chinese); ORSV = Odontoglossum ringspot virus; PTMV = para-tobacco mosaic virus; ToMV = tomato mosaic virus; SOV = Sammons's Opuntia virus; FMV = frangipani mosaic virus; YMV = Youcai mosaic virus (Chinese); RMV = ribgrass mosaic virus; BNYVV = beet necrotic yellow vein virus; CCV = *Chara corallina* virus; SHMV = sunn-hemp mosaic virus; CV4 = cucumber virus 4; CGMMV = cucumber green mottle mosaic virus. From Gibbs *et al.* (1982).

and UMMV are compared with that of TMV in Table II. The polypeptide of FMV, PMMV, and UMMV contains 158 residues (Francki *et al.*, 1971; Wetter *et al.*, 1984; Van Regenmortel and Brunt, unpublished data); that of CCV contains 170 residues when calculated from a molecular weight of 18,716 (Skotnicki *et al.*, 1976) but 158 when calculated from a molecular weight of 17,500. The compositions of these polypeptides generally resemble, but differ specifically from, that of TMV. For example, the FMV polypeptide has a low proline and a high tryptophan content, and that of CCV has unusually high contents of aspartic acid, glycine, lysine, and phenylalanine; in comparisons with other tobamoviruses, the polypeptide composition of PMMV differs from that of TMV, ToMV, TMGMV, sunn-hemp mosaic (SHMV), ribgrass mosaic (RMV), and cucumber green mottle mosaic (CGMMV) viruses by 16, 15, 15, 21, 22, and 26 exchanges respectively (Wetter *et al.*, 1984), results that correlate well with the degree of serological relationship between the viruses (Van Regenmortel, 1975).

A principal coordinates analysis of tobamoviruses computed from their amino acid compositions clearly illustrates the degree of relationship between CCV, FMV, SOV, and other tobamoviruses (Gibbs, 1977). The classification of tobamoviruses computed from their amino acid compositions (Gibbs *et al.*, 1982) can also be represented in a dendrogram (Fig. 6) which shows that FMV and SOV cluster with, but differ from, TMV and other better-characterized tobamoviruses and that CCV shows the greatest dissimilarity to other tobamoviruses. Similar computer analyses have shown that, although distinct from other tobamoviruses, UMMV is most like RMV, and FMV like TMV and ToMV (A. J. Gibbs, unpublished data).

2. Nucleic Acid

The nucleic acids of only two of the nine viruses have been partially characterized. Those of PV14R and CCV were each shown by PAGE to be a single species of ssRNA with a molecular weight of 2.4×10^6 and 3.6×10^6, respectively (Salazar, 1977; Skotnicki *et al.*, 1976). The base composition of CCV was calculated to be 24.5% G, 28.0% A, 20.0% C, and 27.5% U (Skotnicki *et al.*, 1976).

Molecular hybridization analyses using ^{32}P-labeled cDNA to FMV RNA showed that, under stringent conditions in 0.12 M NaCl, it did not hybridize with RNA from TMV, ToMV, TMGMV, or odontoglossum ringspot virus (ORSV); similar analyses under less stringent conditions in 0.56 M NaCl, however, indicated that FMV is distantly related to ToMV and TMGMV (Palukaitis and Symons, 1980). Under nonstringent conditions, cDNA to a strain of ToMV was later shown to have 15% nucleotide sequence homology with FMV RNA (Palukaitis *et al.*, 1981; Gibbs *et al.*, 1982).

IV. SEROLOGICAL AFFINITIES

Some of the nine less well-known viruses are either unrelated, or only distantly so, to the well-characterized tobamoviruses; others are closely

TABLE III. The Serological Relationship between Frangipani Mosaic (FMV), Tobacco Mosaic (TMV), and Mild Dark-Green Tobacco Mosaic Virus (TMGMV)[a]

Antiserum	Test antigen	Titer: Homologous	Titer: Heterologous	Serological differentiation index
FMV	TMV	1024[b]	16	6
TMV	FMV	2048	256	3
FMV	TMGMV	1024	0	>10
TMGMV	FMV	8192	64	7
TMV	TMGMV	2048	128	4
TMGMV	TMV	8192	256	5

[a] Data from Francki et al. (1971).
[b] Reciprocals of maximum reacting antiserum dilutions in precipitin tube tests.

TABLE IV. Serological Relationships, Determined in Microprecipitin Tests, between Pepper Mild Mottle Virus (PMMV), Mild Dark-Green Tobacco Mosaic Virus (TMGMV), Tobacco Mosaic Virus (TMV), and Tomato Mosaic Virus (ToMV)[a]

	Antigen			
Antiserum to	PMMV	TMV	TMGMV	ToMV
PMMV	4096[b] (0)	32 (7)	128 (5)	256 (4)
TMV	128 (5)	4096 (0)	512 (3)	1024 (2)
TMGMV	16 (6)	256 (2)	1024 (0)	128 (3)
ToMV	16 (7)	512 (2)	256 (3)	2048 (0)

[a] Data from Wetter *et al.* (1984).
[b] Reciprocals of antiserum dilutions with, in parentheses, the serological differentiation indices.

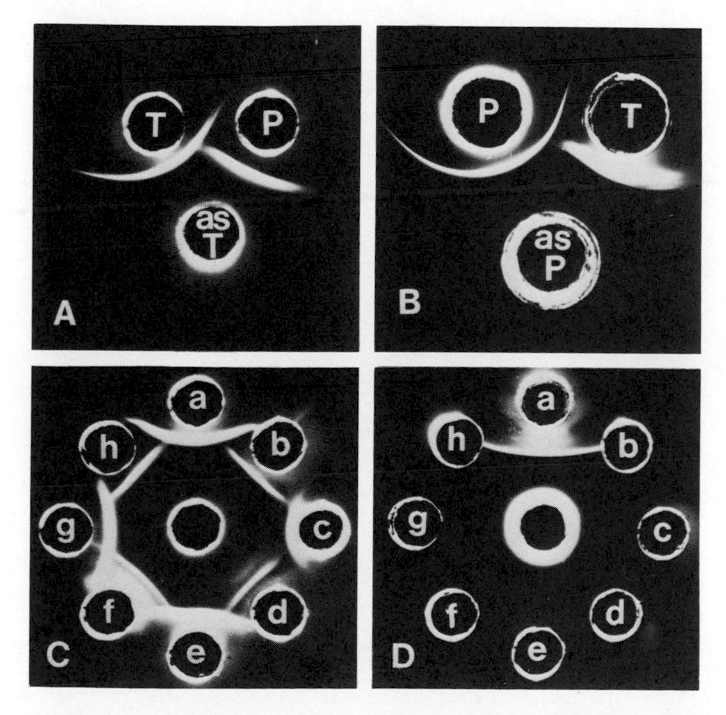

FIGURE 7. Reactions in gel diffusion tests between pepper mild mottle virus (P) and tobacco mosaic virus (T) with antisera (as) to (*A*) TMV and (*B*) PMMV. Reactions between (*C*) PMMV antiserum (central well) and PMMV (a), ToMV (b), PTMV (c), RMV (d), TMV, strain Ohio III from tomato (e), SHMV (f), CV4 (g), and TMV (h), and (*D*) after intragel absorption with viruses b to h. From Wetter *et al.* (1984).

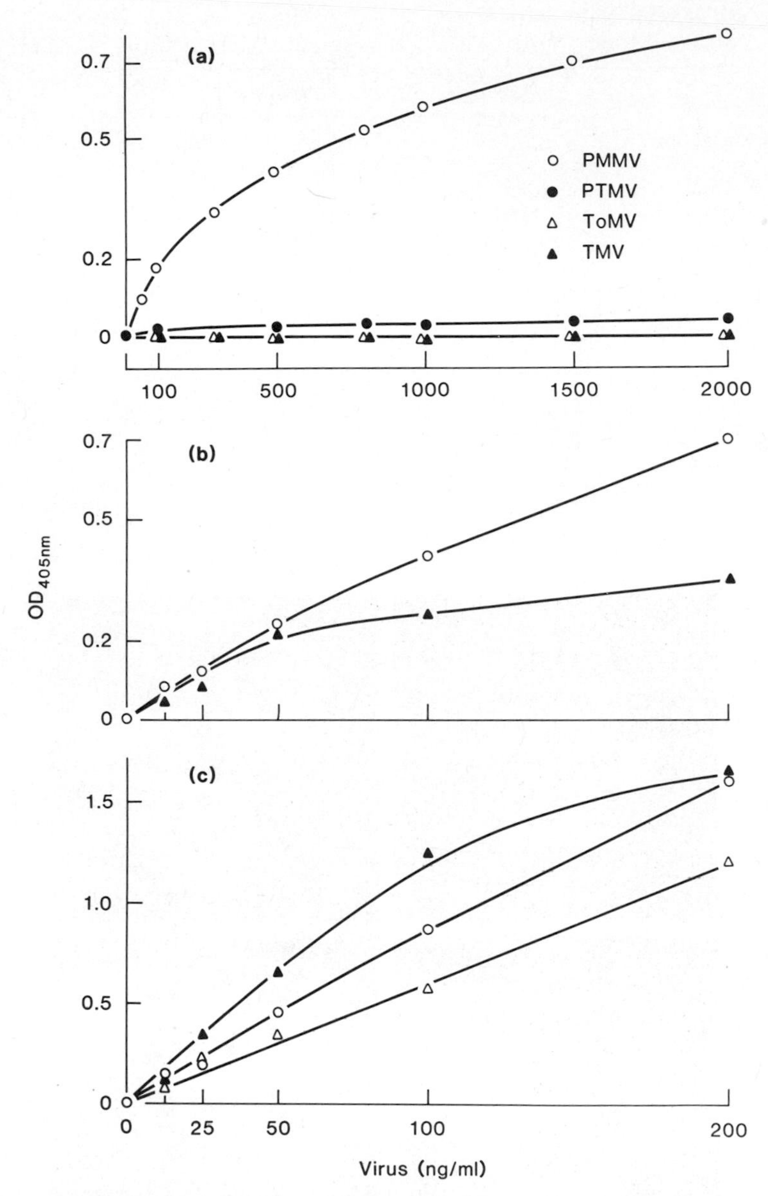

FIGURE 8. Reactions in different ELISA procedures of pepper mild mottle virus (PMMV), tobacco mosaic virus (TMV), para-tobacco mosaic virus (TMGMV), and tomato mosaic virus (ToMV) with TMGMV antiserum. (a) Double-antibody sandwich ELISA with PMMV and other tobamoviruses. (b) Indirect ELISA with PMMV and TMV. (c) Indirect ELISA with PMMV, TMV, and ToMV. From Wetter *et al.* (1984).

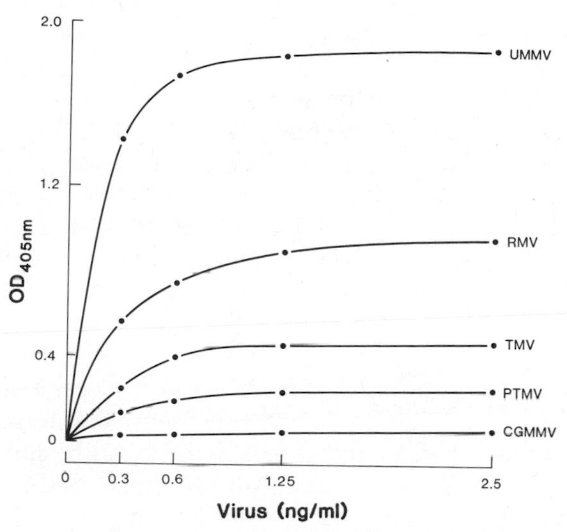

FIGURE 9. Reactions in indirect ELISA tests between ullucus mild mottle virus (UMMV) antiserum and five tobamoviruses. M. H. V. Van Regenmortel and A. A. Brunt (unpublished data).

related to TMV, but are sufficiently different in other respects (especially in pathology) to be recognized as distinct viruses. Thus, POV is unrelated to TMV (Dieleman-van Zaayen *et al.*, 1970), and PV14R to TMV, SHMV, SOV and to fungus-transmitted labile rod-shaped viruses such as potato mop-top and *Nicotiana velutina* mosaic viruses (Salazar, 1977). Although FMV is unrelated to SOV, it is distantly related to CGMMV and an isolate of SHMV from Australia (but not one from west Africa), and very distantly to TMV, TMGMV, ToMV, and RMV (Francki *et al.*, 1971; Varma and Gibbs, 1978). The degree of interrelationship between FMV, TMV, and TMGMV (Francki *et al.*, 1971) is shown in Table III.

CCV is distantly related to TMV and ORSV (serological differentiation indices of 5–9 and 7, respectively, with high-titered and broadly specific antisera) but unrelated to CGMMV, RMV, SHMV, and SOV (Gibbs *et al.*, 1975). In gel diffusion and intragel absorption tests, SOV was shown to be related to TMV and SHMV, but unrelated to ORSV (Wetter and Paul, 1967).

In precipitin tube tests, UMMV showed a distant relationship to RMV (SDI 5), to four of eight strains of ToMV (SDI 6–9), and to two of six strains of TMV (SDI 6–8); the virus, however, is unrelated to CGMMV, FMV, PV14R, ORSV, and SHMV (Brunt *et al.*, 1982). In a comprehensive study, PMMV was shown by microprecipitin tests to be distantly related to TMV, ToMV, and TMGMV (Table IV) and, in double-diffusion gel tests (Fig. 7), to be related also to CGMMV, RMV, and SHMV (Wetter *et al.*, 1984). MMV showed only distant serological relationships to other tobamoviruses in microprecipitin, indirect ELISA, and ISEM tests, although such relationships were more readily detected in electroblot immunoassays (C. E. Fribourg, R. Koenig, and D.-E. Lesemann, personal communication).

In double-antibody sandwich ELISA, antiserum to PMMV failed to react with TMV, ToMV, and TMGMV (Wetter *et al.*, 1984); in less specific indirect forms of the test, however, PMMV was shown to be related to TMV and ToMV (Fig. 8). Indirect ELISA tests confirmed that UMMV is distantly related to RMV, very distantly related to TMV and TMGMV (Fig. 9), but unrelated to CGMMV (Van Regenmortel and Brunt, unpublished data).

ACKNOWLEDGMENTS. I am most grateful to Drs. Cesar Fribourg, Adrian Gibbs, Renate Koenig, Dietrich Lesemann, Marc Van Regenmortel, and Carl Wetter for permission to quote unpublished results, and to Drs. Annamarie van Zaayen, Richard Francki, Adrian Gibbs, and Carl Wetter for supplying illustrations. I also thank Mrs. Sue Bewsey for preparing the manuscript and Mr. Maurice Bone for redrawing the illustrations.

REFERENCES

Accatino, L. P., 1966, Virus desconocido determinado en dos variedades de papa (*Solanum tuberosum* L.) autoctonas de Chile, *Agric. Tec.* (*Santiago*) **26:**85.

Amelunxen, F., 1958, Die Virus-Eiweissspindeln der Kakteen: Darstellung, electronenmikroskopische und biochemische Analyse des Virus, *Protoplasma* **49:**140.

Brandes, J., 1964, Identifizierung von gestreckken pflanzenpathogenen Viren auf morphologische Grundlage, *Mitt. Biol. Bundesanst. Land Forstwirtsch. Berlin-Dahlem* **110**.

Brandes, J., and Chessin, M., 1965, An electron microscope study on the size of Sammon's Opuntia virus, *Virology* **25:**673.

Brunt, A. A., Phillips, S., Jones, R. A. C., and Kenten, R. H., 1982, Viruses detected in *Ullucus tuberosus* (Basellaceae) from Peru and Bolivia, *Ann. Appl. Biol.* **101:**65.

Chessin, M., 1965, Wild plant hosts of a cactus virus, *Phytopathology* **55:**933.

Chessin, M., 1969, Viruses of wild and cultivated cacti, *Phytopathology* **59:**10.

Chessin, M., and Lesemann, D., 1972, Distribution of cactus viruses in wild plants, *Phytopathology* **62:**97.

Conti, M., and Marte, M., 1983, Virus, virosi e micoplasmosi del peperone, *Ital. Agric.* **120:**132.

Conti, M., and Masenga, V., 1977, Identification and prevalence of pepper viruses in northwest Italy, *Phytopathol. Z.* **90:**212.

Demski, J. W., 1981, Tobacco mosaic virus is seedborne in pimiento peppers, *Plant Dis.* **65:**723.

Dieleman-van Zaayen, A., 1967, Virus-like particles in a weed mould growing on mushroom trays, *Nature* (*London*) **216:**595.

Dieleman-van Zaayen, A., Igesz, O., and Finch, J. T., 1970, Intracellular appearance and some morphological features of viruslike particles in an ascomycete fungus, *Virology* **42:**534.

Feldman, J. M., and Oremianer, S., 1972, An unusual strain of tobacco mosaic virus from pepper, *Phytopathol. Z.* **75:**250.

Francki, R. I. B., Zaitlin, M., and Grivell, C. J., 1971, An unusual strain of tobacco mosaic virus from *Plumeria acutifolia, Aust. J. Biol. Sci.* **24:**815.

Gibbs, A. J., 1977, Tobamovirus group, *CMI/AAB Descriptions of Plant Viruses No. 184.*

Gibbs, A. J., Skotnicki, A. H., Gardiner, J. E., Walker, E. S., and Hollings, M., 1975, A tobamovirus of a green alga, *Virology* **64**:571.

Gibbs, A. J., Tien, P., Kang, L.-Y., Tian, Y.-C., and Randles, J., 1982, Classification of several tobamoviruses isolated in China on the basis of the amino acid composition of their virion proteins, *Intervirology* **18**:160.

Greenleaf, W. H., Cook, A. A., and Heyn, A. N. J., 1964, Resistance to tobacco mosaic virus in *Capsicum*, with reference to the Samsun latent strain, *Phytopathology* **54**:1367.

Hebert, T. T., 1963, Precipitation of plant viruses by polyethylene glycol, *Phytopathology* **53**:362.

Hicks, R. G. T., and Frost, R. R., 1984, Isolation of a tobamovirus from rose, *Plant Pathol.* **33**:581.

Holmes, F. O., 1934, A masked strain of tobacco mosaic virus, *Phytopathology* **24**:845.

Johnson, E. M., 1930, Virus diseases of tobacco in Kentucky, *Ky. Agric. Exp. Stn. Bull.* **306**:285.

McKinney, H. H., 1952, Two strains of tobacco mosaic virus, one of which is seed-borne in an etch-immune pungent pepper, *Plant Dis. Rep.* **36**:184.

McKinney, H. H., 1968, Further studies of the latent strain of the tobacco mosaic virus, *Plant Dis. Rep.* **52**:919.

Miličić, D., and Udjbinac, Z., 1961, Virus-Eiweissspindeln der Kakteen in Lokalläsionen von *Chenopodium, Protoplasma* **53**:584.

Miller, P. M., and Thornberry, H. H., 1958, A new viral disease of tomato and pepper, *Phytopathology* **48**:665.

Murakishi, H. H., 1960, A necrotic pod streak of pepper caused by tobacco mosaic virus, *Phytopathology* **50**:464.

Palukaitis, P., and Symons, R. H., 1980, Nucleotide sequence homology of thirteen tobamoviruses as determined by hybridisation analysis with complementary DNA, *Virology* **107**:354.

Palukaitis, P., Randles, J. W., Tian, Y.-C., Kang, L.-Y., and Tien, P., 1981, Taxonomy of several tobamoviruses from China as determined by molecular hybridisation analysis with complementary DNA, *Intervirology* **16**:136.

Pares, R. D., 1985, A tobamovirus infecting capsicum in Australia, *Ann. Appl. Biol.* **106**:469.

Phatak, H. C., and Verma, V. S., 1967, A strain of tobacco mosaic virus from potato, *Phytopathol. Z.* **59**:141.

Rast, A. T. B., 1979, Pepper strains of TMV in the Netherlands, *Meded. Fac. Landbouw. wet. Rijksuniv. Gent* **44**(2):617.

Rast, A. T. B., 1982, Resistance of *Capsicum* species to tobacco, tomato and pepper strains of tobacco mosaic virus, *Neth. J. Plant Pathol.* **88**:163.

Salazar, L. F., 1977, Studies on three viruses from South American potatoes, Ph.D. thesis, University of Dundee.

Sammons, I. M.., and Chessin, M., 1961, Cactus virus in the United States, *Nature* (*London*) **191**:517.

Skotnicki, A., Gibbs, A. J.., and Wrigley, N. G., 1976, Further studies on *Chara corallina* virus, *Virology* **75**:457.

Tobias, I., Rast, A. T. B., and Maat, D. Z., 1982, Tobamoviruses of pepper, eggplant and tobacco: Comparative host reactions and serological relationships, *Neth. J. Plant Pathol.* **88**:257.

Van Regenmortel, M. H. V., 1975, Antigenic relationships between strains of tobacco mosaic virus, *Virology* **64**:415.

Van Regenmortel, M. H. V., 1981, Tobamoviruses, in: *Handbook of Plant Virus Infections and Comparative Diagnosis* (E. Kurstak, ed.), pp. 541–564, Elsevier/North-Holland, Amsterdam.

Varma, A., and Gibbs, A. J., 1978, Frangipani mosaic virus, *CMI/AAB Descriptions of Plant Viruses No. 196*.

Warmke, H. E., 1968, Fine structure of inclusions formed by the aucuba strain of tobacco mosaic virus, *Virology* **34**:149.

Wetter, C., 1984, Serological identification of four tobamoviruses infecting pepper, *Plant Dis.* **68:**597.

Wetter, C., and Paul, H. L., 1967, Serologische und physikalische Eigenschaften von Sammons' Opuntia Virus, *Phytopathol. Z.* **60:**92.

Wetter, C., Conti, M., Altschuh, D., Tabillion, R., and Van Regenmortel, M. H. V., 1984, Pepper mild mottle virus, a tobamovirus infecting pepper cultivars in Sicily, *Phytopathology* **74:**405.

Wittmann-Liebold, B., and Wittmann, H. G., 1967, Coat proteins of strains of two RNA viruses: Comparison of their amino acid sequences, *Mol. Gen. Genet.* **100:**358.

Wood, R. D., 1972, Characeae of Australia. *Nova Hedwigia Z. Kryptogamenkd.* **22.**

PART II

FUNGUS-TRANSMITTED AND SIMILAR LABILE ROD-SHAPED VIRUSES

CHAPTER 16

Fungus-Transmitted and Similar Labile Rod-Shaped Viruses

ALAN A. BRUNT AND EISHIRO SHIKATA

I. INTRODUCTION

At least 12 viruses (Table I) are known to have rod-shaped particles which, although superficially similar to those of authentic tobamoviruses (Matthews, 1982; Gibbs, 1977), differ in that they occur within plants in low concentration, are relatively unstable *in vitro*, and have particles of two or more modal lengths which are purified and separated with difficulty; moreover, unlike tobamoviruses, most are transmitted by plasmodiophoromycete fungi (Table II), and three are reported to have bipartite genomes (Shirako and Brakke, 1984a; Richards *et al.*, 1985; Mayo and Reddy, 1985). Three of these labile viruses [broad bean necrosis (BBNV), potato mop-top (PMTV), and soilborne wheat mosaic viruses (SBWMV)] are reported to be serologically related to one or more tobamoviruses (Kassanis *et al.*, 1972; Powell, 1976; Nakasone and Inouye, 1978) and, since 1976, have been recognized as possible members of the tobamovirus group (Fenner, 1976). It has been suggested, however, that SBWMV and similar viruses should now be included in a new group for which the name furovirus (*fu*ngus-transmitted *ro*d-shaped viruses) group has been proposed (Shirako and Brakke, 1984a). Although evidence is accumulating to support the formation of a new group, the taxonomic status of the fragile tobamo-like viruses is still uncertain and will be considered further (see Section VI).

ALAN A. BRUNT • Glasshouse Crops Research Institute, Littlehampton, West Sussex BN17 6LP, England. EISHIRO SHIKATA • Department of Botany, Faculty of Agriculture, Hokkaido University, Sapporo, Japan.

TABLE I. Natural Occurrence of Fungus-Transmitted and Similar Rod-Shaped Viruses

Virus	Hosts	Geographical occurrence	References
Beet necrotic yellow vein (BNYVV)	*Beta vulgaris* (sugar beet), *Spinacia oleracea* (spinach)	Austria	Koch (1982)
		China	Gao *et al.* (1983)
		Czechoslovakia	Novak and Lanzova (1984)
		France	Putz and Vuittenez (1974)
		West Germany	Hamdorf *et al.* (1977)
		Greece	Ioanides (1978)
		Italy	Canova (1959), Faccioli and Giunchedi (1974), Rana *et al.* (1978)
		Japan	Tamada and Baba (1973), Tamada *et al.* (1971)
		Yugoslavia	Sutic and Milovanovic (1978)
		United States	Al Musa and Mink (1981)
Beet soil-borne (BSBV)	*Beta vulgaris* (sugar beet)	England	Ivanovic *et al.* (1983)
Broad bean necrosis (BBNV)	*Vicia faba* (broad bean)	Japan	Fukano and Yokoyama (1952), Inouye and Asatani (1968)
Fern mottle (FMV)	*Phyllitis scolopendrium* (Hart's tongue fern)	England	Hull (1968)

Hypochoeris mosaic (HMV)	*Hypochoeris radicata* (cat's ear), *Leontodon autumnalis* (autumnal hawkbit)	Australia	Greber and Finlay (1981)
		Canada	Brunt and Stace-Smith (1978a), Singh and McDonald (1980)
		England	Brunt (unpublished data)
Indian peanut clump (IPCV)	*Arachis hypogaea* (groundnut, peanut)	India	Reddy *et al.* (1983)
Nicotiana velutina mosaic (NVMV)	*Nicotiana velutina*	Australia	Randles *et al.* (1976)
Oat golden strip virus (OGSV)	*Avena sativa* (oat)	England	Plumb and MacFarlane (1977)
Peanut clump (PCV)	*Arachis hypogaea* (groundnut, peanut), *Sorghum arundinacearum* (great millet)	Ivory Coast	Thouvenel *et al.* (1976)
		Senegal	Thouvenel *et al.* (1974), Bouhot (1967)
		Upper Volta	Germani and Dhery (1973)
Potato mop-top (PMTV)	*Solanum tuberosum* (potato)	Peru	Salazar and Jones (1975)
		Western Europe	Calvert and Harrison (1966), Harrison and Jones (1970)
		Japan	Imoto *et al.* (1981)
Rice stripe necrosis (RSNV)	*Oryza sativa* (rice)	Ivory Coast	Fauquet and Thouvenel (1983)
Soilborne wheat mosaic (SBWMV)	*Triticum aestivum* (winter wheat), *Hordeum vulgare* (barley)	Italy	Canova (1966b)
		Japan	Inouye (1969), Saito *et al.* (1964)
		United States	McKinney (1923), Brakke *et al.* (1965), Brandes *et al.* (1964)
		China	Cai *et al.* (1983)

TABLE II. Natural Modes of Spread of Labile Rod-Shaped Viruses

Virus	Transmission through Seed	Transmission through Soil	Vector	References
BNYVV	NR[a]	+	*Polymyxa betae*	Tamada *et al.* (1975), Vuittenez *et al.* (1977)
BSBV	NR	+	*Polymyxa betae*	Ivanovic *et al.* (1983)
BBNV	–	+	? *Polymyxa* sp.	Fujikawa (1963)
FMV	–	+	Unknown	Hull (1968)
HMV	–	NR	Unknown	Brunt and Stace-Smith (1978a), Greber and Finlay (1981)
IPCV	+	+	*Polymyxa graminis*	Reddy, in Mayo and Reddy (1985)
NVMV	+	–	Unknown	Randles *et al.* (1976)
OGSV	–	+	*Polymyxa graminis*	Plumb and MacFarlane (1977)
PCV	+	+	*Polymyxa graminis*	Thouvenel and Fauquet (1980, 1981), Thouvenel *et al.* (1976)
PMTV	NR	+	*Spongospora subterranea*	Jones and Harrison (1969)
RSNV	–	+	*Polymyxa graminis*	Fauquet and Thouvenel (1983)
SBWMV	–	+	*Polymyxa graminis*	Estes and Brakke (1966), Canova (1966b)

[a] NR, no records; +, transmission; –, no transmission.

Tobacco stunt (TS) and lettuce big vein (LBV) diseases are also reported to be induced by fungus-transmitted viruses with labile rod-shaped particles mostly measuring 18–22 × 330–375 nm (Kuwata and Kubo, 1981, 1984; Kuwata *et al.*, 1983; Masri and Hiruki, 1983). Unlike fungus-transmitted tobamo-like viruses, however, TSV and LBVV are transmitted by zoospores of the chytrid fungus *Olpidium brassicae*, they have a capsid protein with a molecular weight (48,000–52,000) that is almost threefold greater, and contain two molecules (each 4.5×10^6) of dsRNA; infected plants also contain four species of virus-specific dsRNA with molecular weights of 3.4, 2.3, 2.1, and 0.72×10^6 (Masri and Hiruki, 1983; Dodds and Mirkov, 1983). Because both viruses have properties that differ markedly from those of other labile rod-shaped viruses, Masri and Hiruki (1983) have suggested that they should be placed in a separate group. Neither virus will be considered further here.

The labile tobamo-like viruses also differ from cereal viruses such as barley yellow mosaic, rice necrosis mosaic, wheat yellow mosaic, oat mosaic, and wheat spindle streak viruses which, although also transmitted by *Polymyxa graminis*, have longer filamentous particles and induce in infected plants cylindrical cytoplasmic inclusions (Inouye and Fujii, 1977; Inouye and Saito, 1975; Saito *et al.*, 1968; Hebert and Panizo, 1975; Hooper and Wiese, 1972).

Here, we will discuss the pathology, properties, and possible affinities of labile rod-shaped viruses that are transmitted by, and can persist in, plasmodiophoromycete fungi and viruses with very similar properties.

II. THE VIRUSES AND THEIR NATURAL HOSTS

A. The Viruses and the Diseases They Cause

The labile rod-shaped viruses each infect crop, weed, or ornamental species grown in temperate or tropical regions in one or more of all five continents. Most of the viruses infect major food crops such as cereals, potatoes, legumes, and sugar beet, but two have been detected only in weeds and another is known to occur only in ornamental ferns. Each is briefly considered below.

1. Beet Necrotic Yellow Vein Virus (BNYVV)

This virus induces the "rhizomania" ("root madness") disease of sugar beet (*Beta vulgaris* var. *saccarifera*) in several countries and occurs also in Swiss chard (*B. vulgaris* var. *cycla oleracea*) in southern Italy (Di Franco and Gallitelli, 1980). Infected sugar beet plants were first found over 30 years ago in Italy, and 9 years later the virus was known to be present in 17 Italian provinces (Canova, 1959, 1966a). It has since been found in Austria (Koch, 1982), China (Gao *et al.*, 1983), France (Putz and Vuittenez, 1974), the German Democratic Republic (Hamdorf *et al.*, 1977), Greece (Ioanides, 1978), Japan (Tamada *et al.*, 1971), Yugoslavia (Sutic and Milovanovic, 1978), Czechoslovakia (Novak and Lanzova, 1984), and the United States (Al Musa and Mink, 1981).

Infected sugar beet plants are usually stunted and have leaves that show chlorosis, yellowing, necrotic vein-yellowing, crinkling, and wilting of variable intensity; the tap root is severely stunted with profuse proliferation of rootlets to produce "rhizomania" symptoms. The virus is often restricted to the roots of naturally infected plants.

2. Broad Bean Necrosis Virus (BBNV)

This virus induces a severe disease of broad bean (*Vicia faba*), its only known natural host, in the Kyushu district of Japan (Fukano and Yokoyama, 1952). Infected plants in early spring often have young leaves with chlorotic vein-spotting, and older leaves with reddish-brown spots, rings, veinal necrosis, and streaks. Severely affected plants are usually also stunted and have leaves that senesce prematurely. Plants infected later, however, have less conspicuous symptoms.

3. Hypochoeris Mosaic Virus (HMV)

HMV, a virus first found infecting *Hypochoeris radicata* ("cat's ear") in western Canada (Brunt and Stace-Smith, 1978a), has since been found in this species in Australia (Greber and Finlay, 1981) and Britain (A. A.

Brunt, unpublished information) and to occur in *Leontodon autumnalis* ("autumnal hawkbit") in eastern Canada (Singh and McDonald, 1980). Naturally infected plants of both hosts usually have some leaves that are chlorotic and others that are either symptomless or are inconspicuously chlorotic.

4. Indian Peanut Clump Virus (IPCV)

A disease of groundnut (*Arachis hypogaea*), now presumed to be induced by IPCV, was first observed in India over 80 years ago (Sundararaman, 1927). It was later shown to be widespread in crops grown in sandy or sandy loam soils in the Sangru district of Punjab state and to occur also in the states of Andhra Pradesh, Gujerat, and Rajasthan (Reddy *et al.*, 1979, 1983).

Naturally infected plants characteristically occur in isolated groups; they are usually very stunted, produce small pods, and have leaves which, although initially chlorotic, later turn dark green. Roots are abnormally dark colored and have cortical tissues which are easily sloughed off (Reddy *et al.*, 1979).

IPCV and West African peanut clump virus are similar, but are serologically distinct (Reddy *et al.*, 1979). More recently, however, their RNAs have been shown to have 23–41% sequence homology (B. D. Harrison, personal communication).

5. *Nicotiana velutina* Mosaic Virus (NVMV)

This virus induces bright yellow leaf chlorosis in *Nicotiana velutina*, its only known natural host in southern Australia (Randles *et al.*, 1976).

6. Oat Golden Stripe Virus (OGSV)

A virus now known as OGSV (Plumb and MacFarlane, 1978) was first found infecting oat (*Avena sativa*) in southwest England (MacFarlane *et al.*, 1968). It was later found in Kent (Plumb and MacFarlane, 1977) and in west Wales, where it was tentatively designated "oat tubular virus" (Catherall *et al.*, 1977). OGSV often occurs in complex with oat mosaic virus, although alone it can induce bright yellow stripes in flag and second leaves of mature plants.

7. Peanut Clump Virus (PCV)

A disease of groundnut (*A. hypogaea*), first described as "peanut clump" in Senegal (Bouhot, 1967), was later found in Upper Volta (Germani and Dhery, 1973; Germani *et al.*, 1975) and the Ivory Coast (Thouvenel *et al.*, 1976). Natural infection was also found in Upper Volta in *Sorghum arundinacearum*, a species which is also a host of the vector *Polymyxa graminis* (Dollet *et al.*, 1976; Thouvenel and Fauquet, 1980).

Infected groundnut plants are usually severely stunted with poorly developed roots and abnormally small and dark-green leaves. One virus strain, however, induces severe leaf necrosis but no stunting.

8. Potato Mop-Top Virus (PMTV)

In potato (*Solanum tuberosum*), its only known natural host, the severity of symptoms induced by PMTV in leaves and tubers is dependent on environmental conditions and the tolerance of the cultivar (Calvert, 1968; Harrison and Jones, 1971). The tubers of some cultivars, such as Arran Pilot, develop in the year of natural infection so-called "spraing" symptoms; these consist of internal areas of brown necrotic tissue which are sometimes evident as brown rings on tuber surfaces. The necrosis, which occurs at the boundary of healthy and infected tissue as a result of a change in temperature, does not prevent systemic spread of the virus (Harrison and Jones, 1971). Plants of other cultivars when infected produce tubers with only raised superficial rings (Calvert, 1968). Tubers produced during the second year of infection may be either symptomless or cracked and distorted; stolons may also develop internal brown necrotic arcs.

Symptoms on the leaves of intolerant cultivars usually develop only in cool conditions (5–15°C). Some leaves may then develop yellow blotches or chlorotic chevron patterns on leaflets. Shoots on intolerant cultivars are often severely stunted to give "mop-top" symptoms.

PMTV occurs in Peru, western Europe, Japan, and possibly other countries where potatoes are grown (Harrison, 1974; Imoto *et al.*, 1981).

9. Rice Stripe Necrosis Virus (RSNV)

A disease of upland rice (*Oryza sativa*), first observed in the Ivory Coast in 1977, was later shown to be induced by a newly recognized virus with labile rod-shaped particles and designated RSNV (Fauquet and Thouvenel, 1983). Affected plants are severely stunted, have a reduced number of tillers, and leaves with long chlorotic stripes which later sometimes become necrotic. The virus is reported to occur only in the Ivory Coast.

10. Soilborne Wheat Mosaic Virus (SBWMV)

This virus, first described over 60 years ago in the United States (McKinney, 1923), is now known to occur also in Italy (Canova, 1966b), Japan (Saito *et al.*, 1964) and China (Cai *et al.*, 1983). In winter wheat (*Triticum aestivum*) and barley (*Hordeum vulgare*), it induces mosaic diseases the severity of which is dependent on virus strain, crop cultivars, and environmental conditions (Brakke, 1971). Some infected wheat cultivars, such as Harvest Queen and its derivatives, are severely stunted and have excessive tillering. Light-green to yellow mosaic leaf symptoms in wheat usually first appear in spring, but occasionally develop in late autumn. Affected areas of fields, most frequently those in low-lying areas, can then be seen from a distance because the leaves of the crop are noticeably yellow or light green. Symptoms, however, become much less conspicuous as plants mature and mean ambient temperatures increase (Koehler *et al.*, 1952).

11. Other Viruses

Two other inadequately characterized viruses occurring in England resemble labile, fungus-transmitted rod-shaped viruses. One, here designated fern mottle virus (FMV), was initially considered to resemble tobraviruses (Hull, 1968), although it has properties more like labile tobamoviruses. It was first detected in hart's tongue fern (*Phyllitis scolopendrium*) with mottled and deformed fronds collected from Cornwall and Cambridge, England (Hull, 1968). It was subsequently found also in other ferns (*Pellaea falcata, Polystichum falcatum, Pteris cretica* var. *albo-lineatum* and *P. "Čhildsii"*) in the collection of the Cambridge University Botanical Garden (Hull, 1968).

The second virus, now (R. A. C. Jones, personal communication) designated beet soil-borne virus (BSBV), was isolated from sugar beet in Norfolk, England. Like BNYVV, it has particles ca. 19 nm in diameter with modal lengths of 65, 150, and 300 nm and is transmitted by *Polymyxa betae.* The two viruses, however, are serologically distinct (Ivanovic *et al.*, 1983).

B. Intracellular Occurrence and Cytopathology

The intracellular occurrence of only five labile rod-shaped viruses (BBNV, BNYVV, HMV, PMTV, and SBWMV) has been investigated in detail. The particles of all occur in cytoplasm and vacuoles of parenchyma cells, but not within vascular tissues or organelles. Although the viruses do not obviously damage organelles, some such as BBNV and SBWMV induce the formation of inclusions which are readily detectable by light microscopy (Inouye, 1971; McKinney *et al.*, 1923). The viruses usually occur *in vivo* in aggregates of several types, and some induce the formation of inclusions which in ultrathin section appear to consist of interwoven masses of tubules, ribosomes, and virus particles (Inouye, 1971; White *et al.*, 1972; Hibino *et al.*, 1974a,b; Fraser, 1976; Stocky *et al.*, 1977; Brunt and Stace-Smith, 1978b; Putz and Vuittenez, 1980; Giunchedi *et al.*, 1981; Russo *et al.*, 1981).

III. TRANSMISSION AND DISEASE CONTROL

A. Fungal Vectors and Virus–Vector Relationships

Ten of the better known labile rod-shaped viruses are soilborne and nine of these have, or are thought to have, plasmodiophoromycete fungi as vectors (Table II); neither of the viruses affecting weeds (HMV and NVMV) is known to be soilborne, but adequate transmission tests have not been made.

The plasmodiophoromycete vectors, species of *Polymyxa* and *Spongospora*, are obligate parasites that naturally colonize relatively few species but will colonize many more under experimental conditions (Jones and Harrison, 1969, 1972). The fungi, although initially thought to be widespread only in temperate regions, are now known to occur also in tropical Africa in association with *Arachis hypogaea, Oryza sativa*, and *Sorghum arundinacearum* (Thouvenel and Fauquet, 1980, 1981; Fauquet and Thouvenel, 1983), and also in India in association with *A. hypogaea* (Reddy *et al.*, 1983; Mayo and Reddy, 1985). *Polymyxa betae* occurs also in California (Falk and Duffs, 1977) and Ontario (Barr, 1979), but is not known to be a virus vector in either area.

Polymyxa graminis was the first plasmodiophoromycete to be identified as a virus vector. It was initially found to transmit SBWMV (Estes and Brakke, 1966; Rao and Brakke, 1969; Brakke *et al.*, 1965), but it is also now known to be the vector of OGSV, RSNV, PCV, and probably also IPCV and BBNV. Although a vector of PCV, *P. graminis* does not usually parasitize groundnut roots in West Africa; it is, however, a common parasite of *S. arundinacearum*, a natural host of PCV from which the virus is probably transmitted to groundnut (Thouvenel and Fauquet, 1980, 1981).

Another species, *P. betae*, was first found to be associated with the spread of BNYVV in Japan (Kanzawa, 1970; Tamada *et al.*, 1970, 1971; Kanzawa and Ui, 1972) and later shown to be the vector of the virus (Tamada *et al.*, 1975; Fujisawa and Sugimoto, 1977). Thus, sugar beet seedlings remained healthy after the addition of virus to sterilized soil, but became infected after the addition of resting spores or zoospores from infected plants; moreover, a virus-free isolate of *P. betae* was shown to acquire BNYVV from infected plants. The rod-shaped virus infecting sugar beet in England (BSBV), although apparently distinct from BNYVV, is transmitted similarly by *P. betae* (Ivanovic *et al.*, 1983).

Spongospora subterranea, the vector of PMTV, can, like *Polymyxa* spp., acquire virus from infected plants but not from virus-containing suspensions; viruliferous zoospores released from resting spores transmit virus from infected to healthy plants (Jones and Harrison, 1969). Viruliferous resting spores, which can retain virus internally for at least 2 years, can be detected in soils by air-drying samples at 20°C for 2 weeks, then adding water to induce the release of zoospores, and subsequently testing "bait" *Nicotiana debneyi* seedlings for infection (Jones and Harrison, 1969).

B. Seed Transmission

Of the 12 viruses, only PCV and NVMV are known to be seedborne. PCV is seedborne in *A. hypogaea* but not in *Phaseolus mungo, Sorghum arundinacearum*, or *Nicotiana benthamiana*. From 6 to 14% of seeds collected from naturally infected groundnut plants and 19–24% of those from artificially infected plants can be infected (Thouvenel *et al.*, 1978; Thouvenel and Fauquet, 1981). It is interesting that IPCV, a similar virus infecting *A. hypogaea* in India, is not seedborne in groundnut (Reddy *et al.*, 1983).

NVMV is seedborne to a high level (up to 72%) in *Nicotiana glutinosa, N. clevelandii, N. debneyi*, and *N. rustica* but not in *Beta macrocarpa* (Randles *et al.*, 1976). This high level of seed transmission possibly permits NVMV to survive in a semiarid environment in which vegetative growth and activity of any soilborne vector is brief and intermittent (Randles *et al.*, 1976).

IV. VIRUS PURIFICATION

Labile rod-shaped viruses are purified with difficulty mainly because they are unstable *in vitro*, tend to aggregate and/or fragment during purification, occur within plants in low concentration, and/or are dissociated only slowly from the intracellular inclusions in which they occur *in vivo*. Thus, early attempts to purify SBWMV were very laborious and then often only partially successful (Thornberry *et al.*, 1953; Hebert and Coleman, 1955; Hebert, 1963; Brandes *et al.*, 1964; Saito *et al.*, 1964; Rao and Brakke, 1969). The first reliable procedure (Gumpf, 1971), in which up to 150 mg/kg leaf was obtained, involved extraction of SBWMV-infected winter wheat leaves (2 g/ml) in 0.5 M sodium orthoborate at pH 9.0 containing 0.1% potassium cyanide, clarification of extracts with an equal volume of a mixture (1:2 by volume) of ethylene dichloride and chloroform, and differential centrifugation; sedimented virus was further purified by rate zonal centrifugation in sucrose columns containing a detergent. Gumpf (1971) noted, however, that longer particles fragmented throughout the procedure so that shorter particles and/or particle fragments were always present with longer particles in the faster-sedimenting component, even in preparations obtained by two cycles of differential centrifugation. A similar procedure, in which Triton X-100 was used for clarification, consistently yields 20–30 mg BWMV/kg leaf tissue (Powell, 1976; Shirako and Brakke, 1984b).

Similar procedures, variously reported to yield 0.75 to 25 mg virus/kg leaf tissue, have been used for the purification of BBNV (Nakasone and Inouye, 1978), IPCV (Reddy *et al.*, 1983), RSNV (Fauquet and Thouvenel, 1983), PCV (Thouvenel *et al.*, 1977; Thouvenel and Fauquet, 1981), BNYVV (Putz and Kuszala, 1978; Koenig *et al.*, 1984), PMTV (Harrison and Jones, 1970; Kassanis *et al.*, 1972), and NVMV (Randles *et al.*, 1976).

It is difficult to assess the relative efficiency of the various methods of purifying different labile rod-shaped viruses. Although most of the viruses are difficult to purify, the methods described have yielded sufficient virus for the physicochemical characteristics of some to be determined.

V. PROPERTIES OF VIRUS PARTICLES

A. Morphology, Size, Structure, and Infectivity

1. Morphology

The labile fungus-transmitted rod-shaped viruses have particles which, although superficially similar to those of tobamoviruses, differ in some important features. Thus, unlike those of authentic tobamoviruses, the particles characteristically occur in low concentration *in vivo*, are labile *in vitro*, frequently have obvious defects in their protein helices, and have two or more modal lengths less than 300 nm (e.g., Harrison and Jones, 1970; Kassanis *et al.*, 1972; Brunt and Stace-Smith, 1978a; Reddy *et al.*, 1983). The particles of some such as PMTV and HMV have helices which frequently uncoil at one end, have fissures throughout their lengths or have parts missing, expanded, or partially disrupted (Figs. 1, 2). HMV particles frequently fragment where helical disks separate from

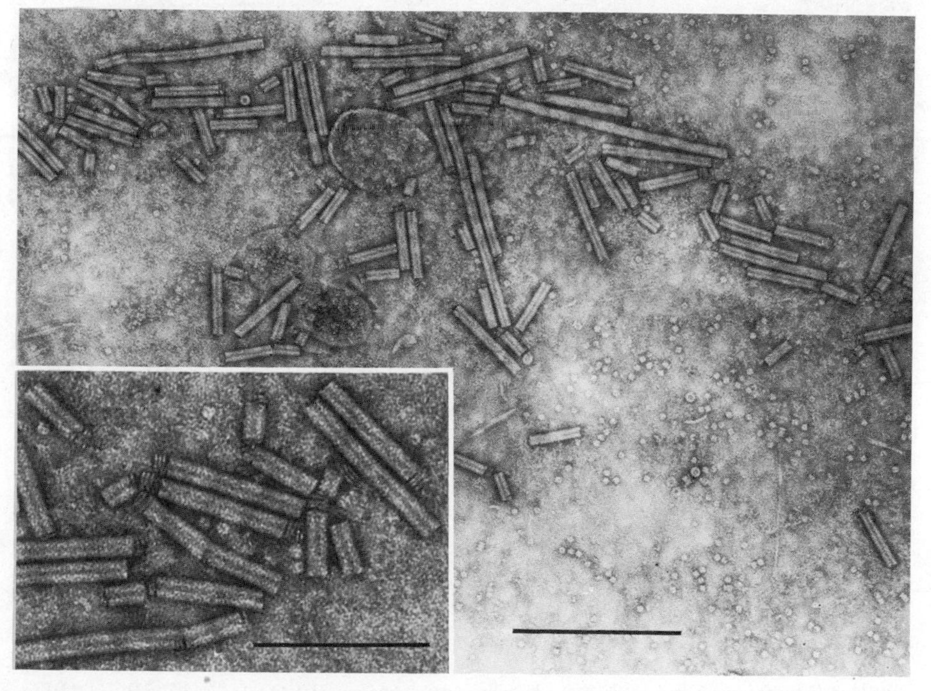

FIGURE 1. Potato mop-top virus particles in partially purified preparation (bar = 300 nm); inset, particles at higher magnification (bar = 150 nm) showing unraveling of protein helix. Courtesy of R. D. Woods.

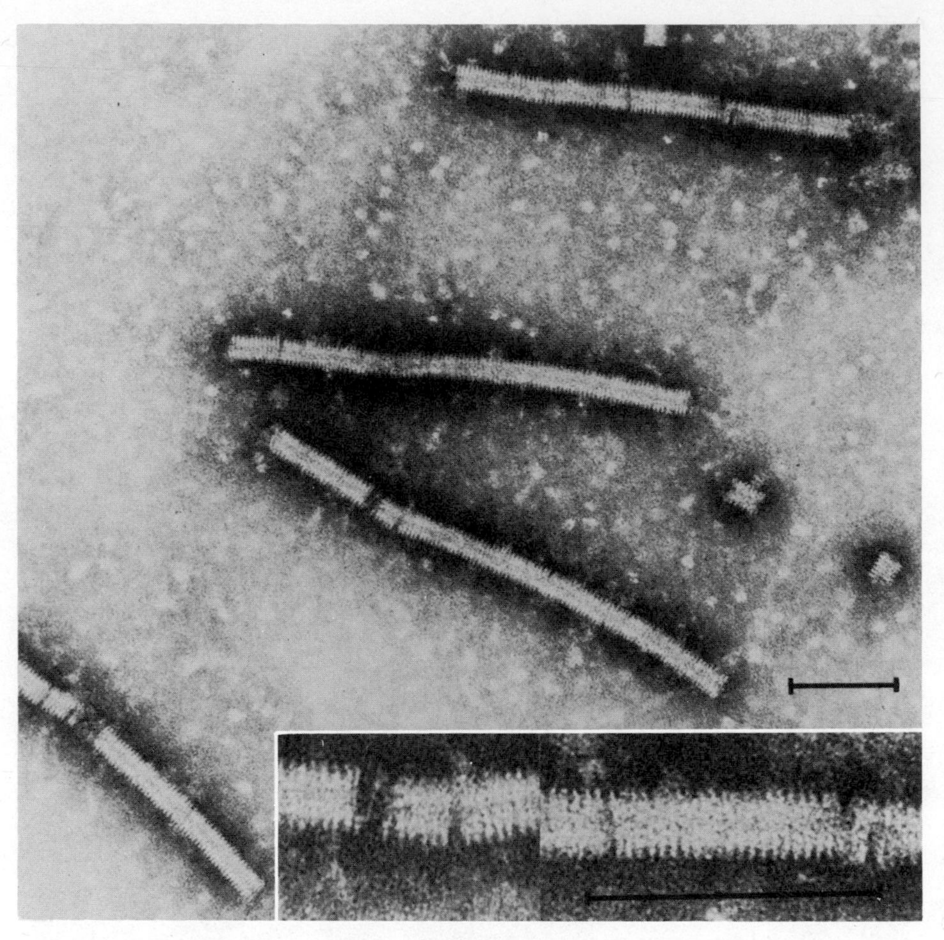

FIGURE 2. Hypochoeris mosaic virus particles in sap from *Nicotiana clevelandii*; inset, at higher magnification. Bar = 100 nm.

the particles; such disks can sometimes be seen either attached to particle fragments or, after separation, in end-on view. Fragmentation of HMV particles is not prevented by first fixing leaves in 3% glutaraldehyde or osmium tetroxide (Brunt and Stace-Smith, 1978a).

2. Dimensions

Negatively stained particles of the viruses have diameters which, although reported to range from 17.0 to 22.5 nm, are mostly 18–20 nm wide (Table III). There is no uniformity, however, in the particle length distributions of different viruses or in the number of their predominant modal lengths. Thus, the particles of some (BBNV, HMV, IPCV, PCV, PMTV, and FMV) are usually of two modal lengths, and those of two others (RNSV and SBWMV) are of three lengths. Particles of Japanese and

TABLE III. Reported Particle Dimensions of Labile Rod-Shaped Viruses

Virus	Preparation	Negative stain[a]	Width (nm)	Lengths (nm)				References
BBNV	Sap	UF	25	—	150	250	—	Inouye and Asatani (1968)
BNYVV								
Japan	—	—	—	65–105	—	270	390	Tamada (1975)
France	Sap	UA (2%)	20	85	100	265	390	Putz (1977)
HMV	Sap	PTA	21–22.5	—	120–140	220–240	—	Brunt and Stace-Smith (1978a)
IPCV	Sap (fixed)	UA	24	60–100	184	249	—	Reddy *et al.* (1983)
		PTA	20		169	239	—	Reddy *et al.* (1983)
NVMV	Purified	AM/PTA	18–19	—	125–175	—	—	Randles *et al.* (1976)
OGSV	—	—	20	—	152	305	—	Plumb and MacFarlane (1977)
PCV	Sap	UA (1%)	21	—	190	245	—	Thouvenel *et al.* (1976)
PMTV	Sap (clarified)	UF	17	—	100–150	250–300	—	Kassanis *et al.* (1972)
RSNV	Purified	UA (0.5%)	20	—	110–160	270	380	Fauquet and Thouvenel (1983)
SBWMV	—	—	20	92–110	138–160	281–300	—	Gumpf (1971), Brakke *et al.* (1965), Tsuchizaki *et al.* (1973), Brakke (1977)

[a] UF, uranyl formate; UA, uranyl acetate; PTA, potassium phosphotungstate; AM, ammonium molybdate.

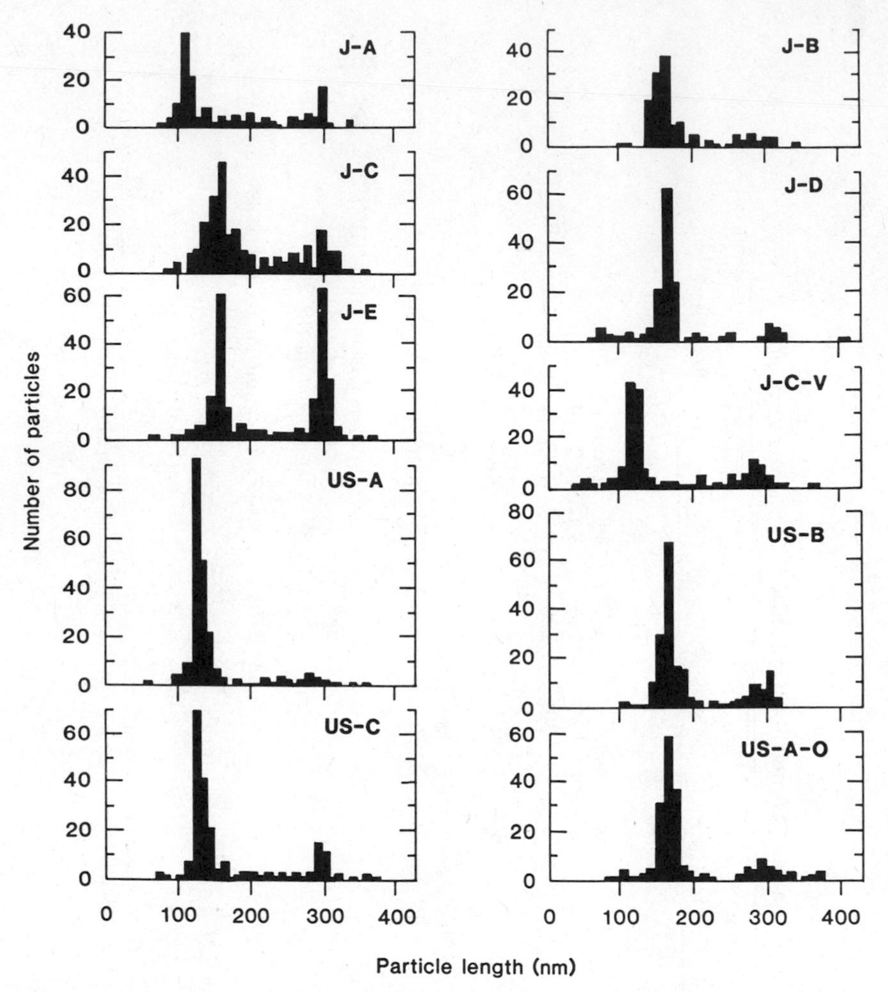

FIGURE 3. Particle length distribution of some isolates of soilborne wheat mosaic virus. Data from Tsuchizaki *et al.* (1975).

French isolates of BNYVV are of three and four predominant lengths, respectively (Table III).

Most of the viruses have particles which are either about 150 or 250 nm long. A few such as BNYVV and RSNV also have some particles 380–390 nm long, while SBWMV and BNYVV (French isolate) in addition have some particles 85–92 nm long. Japanese isolates of SBWMV have particles of different but characteristic lengths; thus, Tsuchizaki *et al.* (1975) differentiated three groups of strains which, in addition to particles 280–300 nm long, had others with lengths of 110, 120, or 160 nm (Fig. 3).

The particle length distribution of these viruses is often difficult to establish. For example, although particles of NVMV in extracts during early stages of purification are predominantly 250–300 nm long, those in

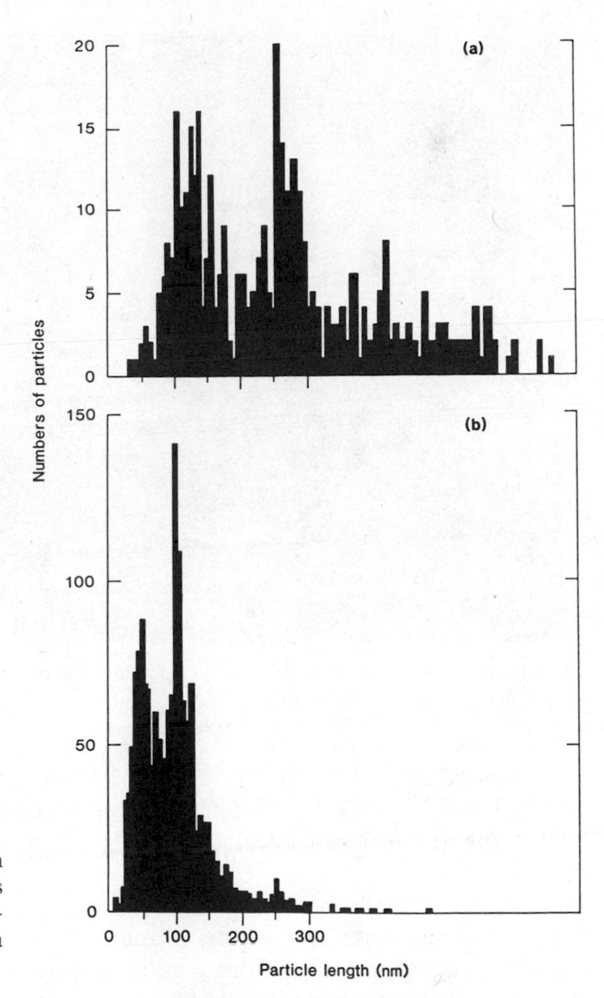

FIGURE 4. Length distribution of potato mop-top virus particles in (a) clarified sap and (b) partially purified preparation. Data from Kassanis *et al.* (1972).

purified preparations are mostly 100–175 nm long; the increase in the number of short particles was accompanied by a concomitant decrease in the number of longer particles (Randles *et al.*, 1976). Similarly, whereas PMTV particles mostly measure 100–150 and 250–300 nm in clarified sap (Fig. 4), those in purified preparations are mostly about 40 and 80 nm long (Kassanis *et al.*, 1972). With HMV (Fig. 2) it was also noted that during purification the gradual increase in the number of short particles was correlated with a decrease in the number of long particles (Brunt and Stace-Smith, 1978a). It has been noted with other viruses such as IPCV that purified virus preparations, unlike clarified sap, contain an unusually large number of particle fragments 60–100 nm long (Reddy *et al.*, 1983).

These difficulties may help to explain why SBWMV, the most intensively studied of the viruses, has been variously reported to have particles with a single modal length of either 128 nm (Gold *et al.*, 1957) or 170 nm (Saito *et al.*, 1964), a bimodal distribution of 160 and 300 nm

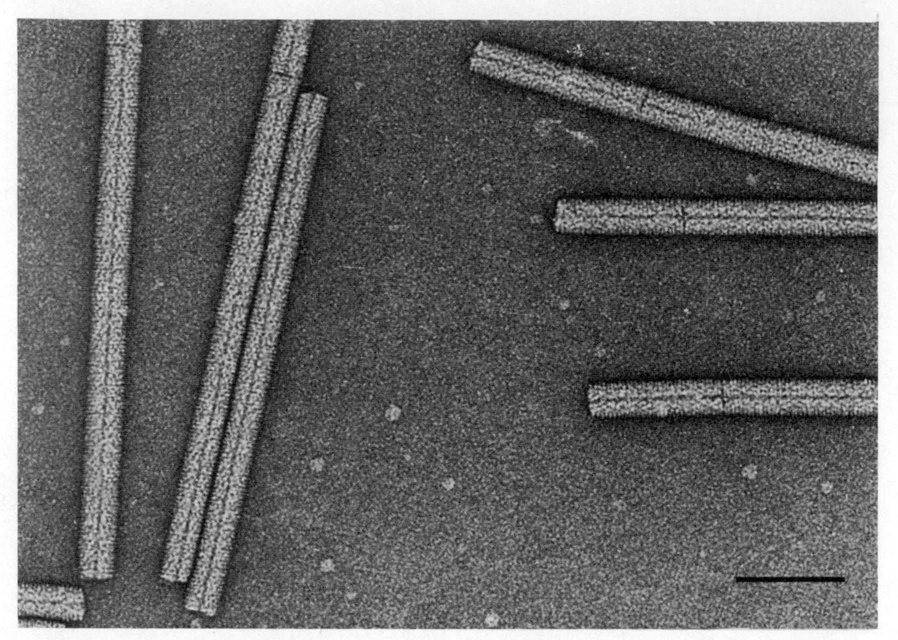

FIGURE 5. Beet necrotic yellow vein virus particles in purified preparation showing signs of fragmentation. Courtesy of Dr. A. Steven.

(Brandes *et al.*, 1964; Brakke *et al.*, 1965) or 148 and 295 nm (Gumpf, 1971; Tsuchizaki *et al.*, 1975) and, more recently, a trimodal distribution with particles mostly 82, 138, and 281 nm long (Shirako and Brakke, 1984a).

The occurrence of a very wide range of particle lengths, apparent fragility of the particles, and an abundance of fragments in purified preparations suggest that the shorter particles, or at least a large proportion of them, and even perhaps those of both predominant lengths, are fragments of longer fragile particles. However, the occurrence of abnormally short particles in extracts of plants containing SBWMV (Tsuchizaki *et al.*, 1973; Brakke, 1977; Hibino *et al.*, 1981) has more recently been attributed to the spontaneous occurrence of deletion mutations in which deletion of 700 or 1050 bases from RNA in particles 138 nm long will result in the production of particles 100 and 82 nm long, respectively (Shirako and Brakke, 1984b).

3. Structure

The detailed structure of only BNYVV particles (Fig. 5) has been reported. Optical diffraction and image processing techniques have shown that the particles of the French isolate, although of four different lengths, are identical in structure (Steven *et al.*, 1981). The protein helix of each particle type is, like that of authentic tobamoviruses, a single right-

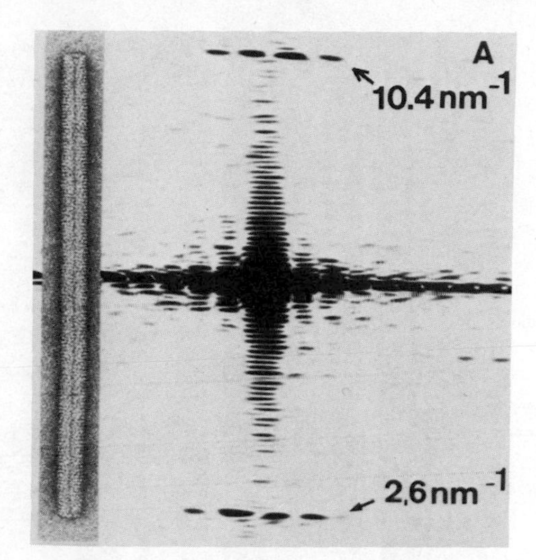

FIGURE 6. Electron micrograph (× 155,000) of a long (390 nm) BNYVV particle with its optical diffraction pattern, showing a strong layer line at a meridional spacing of 2.6 nm^{-1} and an intense layer line at 10.4 nm^{-1}. Courtesy of Dr. A. Steven.

handed strand; its pitch was calculated to be 2.6 nm (Fig. 6) which, with 49 structural subunits in four complete turns, has $12\frac{1}{4}$ subunits per helical turn, probably packed as shown in the computer-generated model of Fig. 7; this model is in accordance with filtered electron micrographs (Fig. 8).

4. Infectivity of Particles

There are conflicting opinions on the possible contribution of particles of different lengths to infectivity. Thus, in early studies on PMTV particles separated by rate-zonal sucrose density gradient centrifugation, fractions containing relatively few long particles (250-–300 nm) were more infective than those containing many more shorter particles (Kassanis *et al.*, 1972). Dilution infectivity curves of PMTV also resembled a one-hit, rather than a two-hit curve, results indicating that it has a mon-

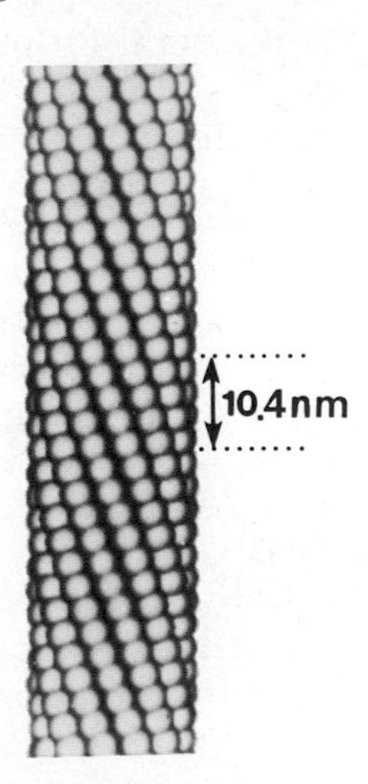

FIGURE 7. Computer-generated model showing helical packing of subunits in BNYVV particles (× 950,000). Courtesy of Dr. A. Steven.

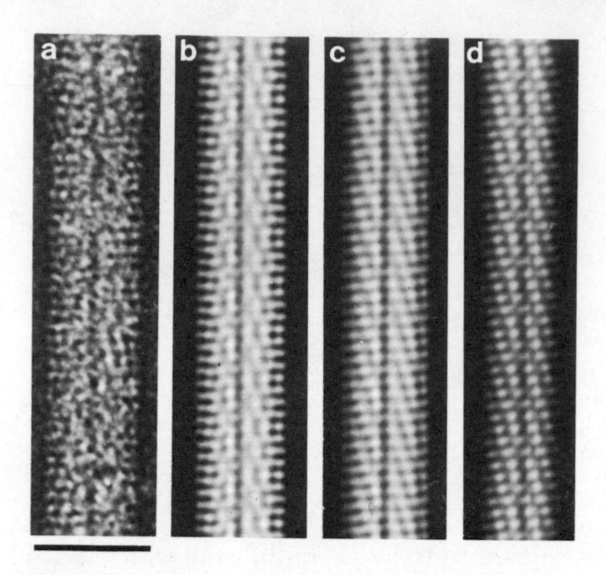

FIGURE 8. Computer-filtered micrographs (× 620,000) of BNYVV particles. Unprocessed image (a), original image containing a "signal" component (b), and single-sided images of projected helical particles (c, d). Courtesy of Dr. A. Steven.

opartite genome (Kassanis *et al.*, 1972). However, Tsuchizaki *et al.* (1975) later found that longer particles of SBWMV alone were infective, but very much more so when present with shorter particles. In contrast to PMTV, the infectivity curve of SBWMV inocular containing both long and short particles was of a double-hit type.

In infectivity tests with RNAs from SBWMV particles of three lengths (92, 138, and 281 nm), infectivity and subsequent virus replication occurred only in plants inoculated with mixtures of RNA from the longest particles plus RNA from either of the shorter particles (Shirako and Brakke, 1981, 1984a).

B. General Properties

1. Stability and Concentration in Sap

Most of the viruses are very labile, occur within infected plants in low concentration, and/or have a low specific infectivity. Thus, unlike typical tobamoviruses, the infectivity of the viruses is usually abolished when sap is heated for 10 min at 65–70°C or stored for more than a few days at room temperature (Inouye and Asatani, 1968; Tamada and Baba, 1973; Brunt and Stace-Smith, 1978a; Reddy *et al.*, 1983; Randles *et al.*, 1976; Thouvenel *et al.*, 1976; Kassanis *et al.*, 1972; McKinney *et al.*, 1965). Although some viruses seem to be a little more stable than others, infectivity tests need to be interpreted cautiously. Thus, PMTV is totally inactivated in sap heated for 10 min at 80°C, yet its infectivity is halved after 10 min at only 50°C and is almost abolished at 70°C (Kassanis *et al.*, 1972); similarly, although some infectivity of NVMV survives after 10 min at 65–70°C, much is lost after 10 min at 50°C (Randles *et al.*, 1976).

2. Properties in Purified Preparations

a. *Ultraviolet Absorption*

Because of difficulties in obtaining pure preparations, the UV absorption properties of only a few of the viruses have been determined. The better estimates show that the viruses have maximum and minimum absorption at 260–263 and 249–250 nm, respectively. The $A_{260/280}$ absorption ratios of 1.19 and 1.21 for NVMV and SBWMV (Randles *et al.*, 1976; Gumpf, 1971) are typical of a nucleoprotein containing about 5% nucleic acid. The extinction coefficients (E_{260} mg/cm^3) of PCV, NVMV, and SBWMV, the only three viruses for which values have been established experimentally (Fauquet and Thouvenel, 1983; Randles *et al.*, 1976; Gumpf, 1971), are 3.0 2.86, and 3.1, respectively.

b. *Sedimentation Coefficients*

Four of the viruses that have been studied have two or more sedimenting components. Preparations of HMV, PCV, PMTV, and SBWMV each have a component sedimenting at 171–183 S (Brunt and Stace-Smith, 1978a; Thouvenel and Fauquet, 1981; Kassanis *et al.*, 1972; Gumpf, 1971) which, by analogy with the sedimentation coefficient of 187 S for TMV (e.g., Kassanis *et al.*, 1972), is probably attributable to particles 250–300 nm long. PMTV components of 126 and 236 S are probably dimers of short (80–100 nm) and long (250–300 nm) particles, respectively (Kassanis *et al.*, 1972). Sedimenting components of SBWMV of 173 and 212 S have been attributed to short (148 nm) and long (295 nm) particles (Gumpf, 1971), and PCV components of 183, 224, and 330 S to the two particle types and to aggregated particles, respectively (Thouvenel and Fauquet, 1981). HMV is unusually unstable in having up to five sedimenting components, with those of 173 and 143 S probably containing particles 220–240 and 120–140 nm long, respectively, and smaller components of 114, 45, and 15 S containing particle fragments of different lengths (Brunt and Stace-Smith, 1978a).

c. *Buoyant Density in Cesium Chloride*

The buoyant density of PCV, the only virus for which a value has been reported, is 1.32 g/cm^3 (Thouvenel and Fauquet, 1981).

TABLE IV. Molecular Weights of Capsid Polypeptides and Nucleic Acids Determined by Polyacrylamide Gel Electrophoresis

Virus	Polypeptide $M_r(10^{-3})$	RNA $M_r(10^{-6})$ and (in parentheses) No. of bases	References
BNYVV	21.0	2.3 (6553), 1.8 (5128), 0.7 (1994), 0.6 (1709)	Putz (1977), Putz *et al.* (1979)
		2.49 (7100), 1.68 (4800), 0.63 (1800), 0.535 (1500)	Richards *et al.* (1985)
HMV	24.5	—	Brunt and Stace-Smith (1978a)
IPCV	—	1.9 (5412), 1.6 (4558)	Mayo and Reddy (1985)
NVMV	21.5	2.3 (6553)	Randles *et al.* (1976), Randles (1978)
PMTV	18.5–20.0	—	Kassanis *et al.* (1972), Randles *et al.* (1976)
PCV	23.0	2.1 (5983), 1.7 (4857)	Fauquet and Thouvenel (1980)
SBWMV	19.7	2.28 (6500), 1.23 (3500), 0.9 (2800), 0.86 (2450)	Shirako and Brakke (1984a)
	17.5[a]	1.84 (5242), 0.95 (2706)	Powell (1976), Gumpf (1971)

[a] Estimated by coelectrophoresis with TMV coat protein.

C. Composition

1. Capsid Protein

a. Molecular Weights

Spectrophotometry of purified preparations indicates that labile tobamo-like viruses contain about 95% protein (Brunt and Stace-Smith, 1978a; Randles *et al.*, 1976; Thouvenel and Fauquet, 1981; Fauquet and Thouvenel, 1983). The protein of the six viruses that have been investigated each contains a single polypeptide with a molecular weight, determined by polyacrylamide gel electrophoresis (PAGE), of 18,500–24,500 (Table IV).

There is, however, some confusion about their actual molecular weights. Because the polypeptides of SBWMV and tobacco mosaic virus (TMV) comigrated, that of SBWMV was assumed to be 17,500 (Powell, 1976). The SBWMV polypeptide was later found to be 19,700 by comparison with marker proteins (Shirako and Brakke, 1984a). Although the TMV polypeptide has a molecular weight of 17,500 when determined by amino acid analyses, values of 18,500–21,000 are usually obtained in estimates made by PAGE (Kassanis *et al.*, 1972; Salazar, 1977; R. J. Barton, personal communication). A molecular weight of 19,700 for the SBWMV polypeptide, and comparable values for similar viruses, indicates that labile tobamo-like viruses are similar to, rather than dissimilar from (Shirako and Brakke, 1984a), TMV.

TABLE V. Amino Acid Compositions of BNYVV and PCV

	BNYVV		PCV	
	Reported[a]	Recalculated[b]	Reported[c]	Recalculated[b]
Asp	24	19	27	20
Thr	15	12	8	6
Ser	18	14	13	9
Glu	14	11	25	17
Pro	10	8	12	8
Gly	14	11	26	18
Ala	19	15	21	15
Cys	0	0	0	0
Val	14	11	20	14
Met	7	6	0	0
Ile	5	4	11	8
Leu	19	15	18	12
Tyr	4	3	5	4
Phe	6	5	7	5
Lys	12	10	7	5
His	2	2	5	4
Arg	10	8	19	13
Try	4	3	0	0
Total	197	157	224	158

[a] Putz (1977) (calculated from molecular weight of 21,000).
[b] Calculated from molecular weight of 17,500 (see text).
[c] Fauquet and Thouvenel (1980) (calculated from molecular weight of 23,000).

b. *Amino Acid Composition*

The compositions of only BNYVV (Putz, 1977) and PCV polypeptides (Fauquet and Thouvenel, 1980) have been reported. Based on molecular weights of 21,000 and 23,000 estimated by PAGE, the BNYVV and PCV polypeptides were reported to contain 197 and 224 residues, respectively (Table V). Recalculation of their compositions on the basis of a molecular weight of 17,500 indicates that the polypeptides, like that of TMV, each contain 157 or 158 residues. Computer analyses suggest that the composition of BNYVV protein is similar to that of some tobamoviruses (Gibbs *et al.*, 1982).

2. Nucleic Acid

a. *Type and Size*

Although the nucleic acid of some labile rod-shaped viruses has yet to be isolated and characterized, particles of five (HMV, NVMV, PCV, SBWMV, and BNYVV) are known to contain about 5% ssRNA (Brunt and Stace-Smith, 1978a; Randles *et al.*, 1976; Thouvenel and Fauquet, 1981; Gumpf, 1971; Putz, 1977; Putz and Fritsch, 1978; Shirako and Brakke,

1984a). Preparations of NVMV were found by PAGE to contain a major RNA species of 2.3×10^6 molecular weight and eight minor components (Randles, 1978). Unfractionated preparations of PCV and IPCV were each found to contain two RNAs with molecular weights of 2.1 and 1.7×10^6 and 1.9 and 1.6×10^6, respectively (Fauquet and Thouvenel, 1980; Mayo and Reddy, 1985). Initial studies indicated that SBWMV contained two RNAs whose molecular weights, when calculated, from their sedimentation coefficients of 24 and 35 S, were 0.95 and 1.84×10^6, respectively (Gumpf, 1971). More recently, two major and two minor RNA species have been detected in each of two isolates (designated WTW and Lab.1) of SBWMV (Shirako and Brakke, 1984a). The molecular weights of the RNAs after separation by three cycles of sucrose density gradient centrifugation, determined by PAGE after denaturation with formaldehyde or glyoxal, were 2.28×10^6 (6500 bases), 1.23×10^6 (3500 bases), 0.97×10^6 (2800 bases), and 0.86×10^6 (2450 bases). The two smaller RNAs are considered to be deletion mutants of that with a molecular weight of 1.23×10^6, and were always tenfold more abundant than the two larger species (Shirako and Brakke, 1984b).

In initial analyses by PAGE, the French isolate of BNYVV produced four bands with estimated molecular weights in nondenaturing gels of 2.3, 1.8, 0.7, and 0.6×10^6 (Putz, 1977). The RNAs, as determined by agarose gel electrophoresis using methylmercury hydroxide as a denaturant, were later found to be 2.49, 1.68, 0.63, and 0.53×10^6 (Richards *et al.*, 1985).

The 3′-terminus of BNYVV RNA is polyadenylated (Putz *et al.*, 1983), although those of SBWMV and IPCV RNAs are apparently not (Hsu and Brakke, 1985; Mayo and Reddy, 1985). Furthermore, the 5′-end of SBWMV RNA 2 shows heterogeneity and apparently has no genome-linked protein or other cap structure (Hsu and Brakke, 1985).

cDNA has been prepared by reverse transcription of BNYVV RNA using as primers oligodeoxynucleotides derived from calf thymus DNA (Richards *et al.*, 1985). Recombinant plasmids containing BNYVV cDNAs were then prepared from nonfractionated BNYVV RNAs, using as primer *Pst*I-linearized pUC-9 possessing oligo(dT) tails at the 3′-termini. Twelve of the clones so obtained were found by restriction enzyme analyses to contain cDNA inserts of at least 1000 base pairs; of these, five were found to hybridize with RNA 3, one with RNA 4, five with RNA 1, and one with RNA 2. Although some sequence homology was detected between RNAs 3 and 4, clones which hybridized with RNAs 3 and 4 showed no homology with RNAs 1 and 2, indicating that they are not derived from the larger species and thus that BNYVV has a bipartite genome.

Although no relationship could be detected between three isolates of IPCV by immunosorbent electron microscopy, their RNAs showed 51–86% sequence homology in tests with unfractionated cDNA to one isolate. Moreover, the three isolates had 23–41% sequence homology with the serologically unrelated PCV (B. D. Harrison, personal communication).

More recently, cDNA to wild-type SBWMV RNA 1 was shown to hybridize with RNA 1, but not RNA 2, of three SBWMV isolates (Hsu and Brakke, 1985). Similarly, cDNA to its RNA 2 hybridized with RNA 2 of all three isolates, results confirming that two laboratory virus isolates (Lab.1 and Lab.2) arose by deletion of part of wild-type RNA 2.

b. In Vitro Translation

The ability of SBWMV, BNYVV, and IPCV to direct protein synthesis in cell-free rabbit reticulocyte translation systems has been reported recently (Hsu and Brakke, 1983; Richards *et al.*, 1985; Mayo and Reddy, 1985). SBWMV RNA 1 (from particles 281 nm long) directed the synthesis of a polypeptide of 200,000 molecular weight, whereas the coat protein polypeptide of 19,700 was the main product of RNAs 2, 3, and 4 (from particles 138, 110, and 92 nm, respectively). The results indicated that the coat protein gene is located at the 5′-terminus of RNA 2, and that deletion may occur at its 3′-terminus to produce the two shorter RNAs (Hsu and Brakke, 1983). Unfractionated RNAs of BNYVV directed the synthesis of three translation products (M_r 22,000, 27,000, and 31,000). The smallest product, shown to be the coat protein, was consistently present in the highest concentration; the two larger proteins varied in concentration in different preparations.

The four RNAs present in BNYVV (French isolate) were later separated and their individual coding capacity investigated. RNA 2 (from particles about 265 nm long) was found to direct the synthesis of the 22,000 viral coat protein, RNA 3 (from particles about 100 nm long) the 27,000 peptide, and RNA 4 (from particles 85 nm long) the 31,000 peptide. RNA 1 (from particles 390 nm long) induced no detectable translation products, although it is not yet known whether this is attributable to a fundamental property of the molecule, to the presence of inhibitory contaminants in the reaction mixture, or to an insufficient concentration of RNA to initiate translation (Richards *et al.*, 1985).

Similar studies with IPCV have shown that its RNA 1 (from particles 249 nm long) predominantly directs the synthesis of polypeptides of 195,000 and 143,000 molecular weight, whereas RNA 2 (from particles 184 nm long) initiated the production of a polypeptide, shown to be coat protein of 24,500 molecular weight. In addition, low concentrations of RNA of $0.5–0.9 \times 10^6$ molecular weight in nucleic acid preparations coded for a polypeptide of 50,000 molecular weight. It is not known, however, whether the smallest nucleic acid is a genomic, subgenomic, or satellite RNA (Mayo and Reddy, 1985).

D. Serological Affinities

Although SBWMV shows a distant serological relationship to BBNV (Nakasone and Inouye, 1978) and PMTV (Cooper and Harrison, 1972;

Randles *et al.*, 1976), other labile tobamo-like viruses seem to be distinct from each other; thus, NVMV is apparently unrelated to PMTV, SBWMV, or BNYVV; HMV to BBNV, BNYVV, PCV, PMTV, NVMV, or SBWMV; PCV to BNYVV, IPCV, or NVMV; and RSNV to BNYVV, NVMV, PCV, or SBWMV (Brunt and Stace-Smith, 1978a; Fauquet and Thouvenel, 1983; Nakasone and Inouye, 1978; Randles *et al.*, 1976; Reddy *et al.*, 1983; Thouvenel and Fauquet, 1981).

The taxonomic status of the labile rod-shaped virus infecting sugar beet in England is uncertain because it is serologically unrelated to three isolates of BNYVV and to PMTV, SBWMV, HMV, and TMV (Ivanovic *et al.*, 1983).

Although the serological affinities of these viruses have yet to be thoroughly studied by sensitive assay procedures, three (BBNV, PMTV, and SBWMV) are reported to be serologically related to one or more authentic tobamoviruses. Using rabbit antisera produced by prolonged immunization procedures to enhance cross-reactivity with related viruses, a very distant relationship was detected between PMTV and TMV (Kassanis *et al.*, 1972). Thus, antiserum to TMV with a homologous titer of 1/1600 reacted with PMTV in tube precipitin tests to a dilution of only 1/5 [serological differentiation index (SDI) of 8.3]. In reciprocal tests, antiserum to PMTV with a homologous titer of 1/2048 reacted with TMV to only 1/8 (SDI of 8), but only when serum and antigen were diluted in 0.1 M EDTA instead of saline solution. The apparent relationship between the two viruses seemed to be supported by the results of cross-protection and infectivity interference tests in tobacco; prior inoculation of tobacco cv. White Burley leaves with TMV reduced by more than 60% the number of lesions subsequently induced by PMTV, and adding TMV to PMTV caused the latter to induce fewer, and different, lesions in *N. tabacum* cv. Xanthi-nc (Kassanis *et al.*, 1972).

From comprehensive microprecipitin, gel-diffusion, sucrose-density gradient centrifugation, and competitive infectivity tests, Powell (1976) concluded that SBWMV is fairly closely related to TMV. Thus, in conventional serological tests, the viruses reacted with their heterologous antisera to SDIs of 0–4 (Table VI). In further tests, the addition of SBWMV antiserum, diluted to 1/4, 1/8, and 1/16, to 200-μg aliquots of TMV before sucrose-density gradient centrifugation of the mixtures removed 100, 68, and 18% of the virus; by contrast, homologous antiserum removed all the virus, and normal serum none. The possible relationship between the two viruses was apparently confirmed by the ability of SBWMV RNA 2 (M_r 1.0 × 10^6) to suppress the infectivity of TMV; in infectivity assays in *Phaseolus vulgaris* cv. Pinto, the addition of 200, 100, and 20 μg of SBWMV to 10, 5, and 1 μg of TMV, respectively, reduced the infectivity of TMV by 80, 68, and 20%, whereas infectivity was not reduced in comparable controls. A similar reduction in infectivity attributable to competition occurred when inocula contained mixtures of SBWMV RNA and TMV RNA (Powell, 1976). Using immunosorbent electron microscopy

TABLE VI. Results of Microprecipitin Serological Tests between SBWMV and TMV[a]

Buffer	Virus	Titer of antiserum to SBWMV	Titer of antiserum to TMV	Serological differentiation index
0.14 M NaCl and 0.01 M KH_2PO_4 (pH 7.0)	SBWMV	1024	1024	0
	TMV	32	512	4
0.014 M NaCl and 0.001 M KH_2PO_4 (pH 7.0)	SBWMV	256	128	1
	TMV	128	1024	3
0.15 M NaCl and 0.015 M Na citrate (pH 7.0)	SBWMV	1024	64	4
	TMV	32	512	4
0.2 M KH_2PO_4–Na_2HPO_4 (pH 7.0)	SBWMV	1024	128	3
	TMV	32	512	4
0.6 M Tris, 0.06 M NaH_2PO_4, and 0.003 M EDTA (pH 7.3)	SBWMV	1024	64	4
	TMV	32	512	4

[a] Data from Powell (1976).

(ISEM), Randles *et al.* (1976) also detected a distant serological relationship between SBWMV and TMV.

BBNV has since been reported to be distantly related to two tobamoviruses (cucumber green mottle mosaic and odontoglossum ringspot viruses) as well as SBWMV (Nakasone and Inouye, 1978). No relationship, however, has been detected between TMV and BBNV, HMV, and PCV (Brunt and Stace-Smith, 1978a; Nakasone and Inouye, 1978; Randles *et al.*, 1976; Reddy *et al.*, 1983; Thouvenel and Fauquet, 1981; Fauquet and Thouvenel, 1983).

VI. TAXONOMY

Although there is evidence that SBWMV and BBNV are distantly related to PMTV (Nakasone and Inouye, 1978; Cooper and Harrison, 1972; Randles *et al.* 1976), other labile fungus-transmitted and similar rod-shaped viruses are apparently serologically distinct. Nevertheless, because the viruses are so similar in many other respects, there is little doubt that they should be grouped together.

The wider affinities of the viruses, however, are uncertain. Their particles are morphologically similar to those of tobraviruses, hordeiviruses, and tobamoviruses but the viruses differ in other properties; moreover, two (NVMV and RSNV) show no serological relationship to tobacco rattle virus and four (NVMV, HMV, RSNV, and PCV) are unrelated to barley stripe mosaic virus (Brunt and Stace-Smith, 1978a; Randles *et al.*, 1976; Reddy *et al.*, 1983; Thouvenel and Fauquet, 1981; Fauquet and Thouvenel, 1983). The viruses have particles which, although they fragment easily, are morphologically similar, and have protein molecular weights and amino acid compositions comparable with those of tobamoviruses. Moreover, PMTV and SBWMV are reported to be serologically related to TMV, and BBNV to CGMMV and ORSV; the relationship of PMTV and BBNV to tobamoviruses is reported to be very distant (Nakasone and Inouye, 1978; Kassanis *et al.*, 1972) although that between SBWMV and TMV, with SDIs of 0–4 in gel-diffusion and microprecipitin tests, is much closer (Powell, 1976). PMTV and SBWMV have, therefore, been included as possible members of the tobamovirus group since 1976 (Fenner, 1976). As previously noted, fungus-transmitted labile rod-shaped viruses, although superficially similar to authentic tobamoviruses, differ from most in their relative instability *in vitro*, their occurrence within infected plants in low concentration, and the transmission of most by plasmodiophoromycetes. Moreover, evidence is accumulating that SBWMV, IPCV, and BNYVV, unlike authentic tobamoviruses, have bipartite genomes (Shirako and Brakke, 1984a,b; Mayo and Reddy, 1985; Richards *et al.*, 1985).

Available evidence, therefore, indicates that the labile rod-shaped viruses, although mostly serologically distinct, are sufficiently similar to be grouped together. It is less certain, however, whether they should form a subgroup of the tobamovirus group or be included in the newly proposed furovirus group (Shirako and Brakke, 1984a). Unfortunately, the labile rod-shaped viruses have been differentiated from tobamoviruses mainly by comparative studies with stable strains of TMV; no comparisons have been made with a range of tobamoviruses nor with labile or defective strains of TMV which, like fungus-transmitted rod-shaped viruses, occur within infected plants in low concentration, are unstable *in vitro*, have particles that easily fragment, and are purified with great difficulty (Kassanis and Woods, 1969). Although the viruses resemble defective strains of TMV in many respects, they differ in that some have bipartite genomes and most have fungal vectors. It is probably best, therefore, for fungus-transmitted rod-shaped viruses to be included in the furovirus group, the wider affinities of which need to be determined by further comparative investigations.

REFERENCES

Al Musa, A. M., and Mink, G. I., 1981, Beet necrotic yellow vein in North America, *Phytopathology* **71**:773.

Barr, D. J. S., 1979, Morphology and host range of *Polymyxa graminis*, *Polymyxa betae*, and *Ligniera pilorum* from Ontario and some other areas, *Can. J. Plant Pathol.* **1**:85.

Bouhot, D., 1967, Observations sur quelques affections des plantes cultivées au Sénégal, *Agron. Trop. (Maracay, Venez.)* **22**:888.

Brakke, M. K., 1971, Soil-borne wheat mosaic virus, *CMI/AAB Descriptions of Plant Viruses No. 70*.

Brakke, M. K., 1977, Sedimentation coefficients of the virions of soil-borne wheat mosaic virus, *Phytopathology* **67**:1433.

Brakke, M. K., Estes, A. P., and Schuster, M. L., 1965, Transmission of soil-borne wheat mosaic virus, *Phytopathology* **55**:79.

Brandes, J., Phillippe, M. R., and Thornberry, H. H., 1964, Electron microscopy of particles associated with soil-borne wheat mosaic virus, *Phytopathol. Z.* **50**:181.

Brunt, A. A., and Stace-Smith, R., 1978a, Electron microscopy of a virus from *Hypochoeris radicata* in sap and infected plant tissue, *J. Gen. Virol.* **41**:207.

Brunt, A. A., and Stace-Smith, R., 1978b, Some hosts, properties and possible affinities of a labile virus from *Hypochoeris radicata* (Compositae), *Ann. Appl. Biol.* **90**:205.

Cai, W., Peng, X., and Mang, K., 1983, Identification of soil-borne wheat mosaic virus causing wheat mosaic in Shandong Province, China, *Acta Phytopathol. Sin.* **13**:7.

Calvert, E. L., 1968, The reaction of potato varieties to potato mop-top virus, *Rec. Agric. Res. Min. Agric. North. Ireland* **17**:31.

Calvert, E. L., and Harrison, B. D., 1966, Potato mop-top, a soil-borne virus, *Plant Pathol.* **15**:134.

Canova, A., 1959, Appunti di patologica della barbabietola, *Inf. Fitopatol.* **9**:390.

Canova, A., 1966a, Si studia la rizomania della bietola, *Inf. Fitopatol.* **16**:235.

Canova, A., 1966b, Ricerche sulle malattie da virus delle Graminacee. III. *Polymyxa graminis* Led. vettore del virus del mosaico del frumento transmissible attraverso il terrano, *Phytopathol. Mediterr.* **5**:53.

Catherall, P. L., Boulton, R. E., and Chamberlain, J. A., 1977, Oat tubular virus, *Rep. Welsh Plant Breeding Stn.* 1976, p. 186.

Cooper, J. I., and Harrison, B. D., 1972, Potato mop-top virus, *Rep. Scott. Hortic. Res. Inst.* 1971, p. 62.

Di Franco, A., and Gallitelli, D., 1980, Le virosi delle piante ortensi in Puglia. XXVI. La rizomania della bietola da coste, *Inf. Fitopatol.* **30**:9.

Dodds, J. A., and Mirkov, T. E., 1983, Association of double-stranded RNA with lettuce big vein disease, *VIth Int. Congr. Plant Pathol. Melbourne, Abstracts*, p. 113.

Dollet, M., Fauquet, C., and Thouvenel, J.-C., 1976, *Sorghum arundinacearum*, a natural host of peanut clump virus in Upper Volta, *Plant Dis. Rep.* **60**:1076.

Estes, E. P., and Brakke, M. K., 1966, Correlation of *Polymyxa graminis* with transmission of soil-borne wheat mosaic virus, *Virology* **28**:772.

Faccioli, G., and Giunchedi, L., 1974, On the viruses involved in rizomania disease of sugarbeet in Italy, *Phytopathol. Mediterr.* **13**:28.

Falk, B. W., and Duffus, J. E., 1977, The first report of *Polymyxa betae* in the Western hemisphere, *Plant Dis. Rep.* **61**:492.

Fauquet, C., and Thouvenel, J.-C., 1980, Virus diseases of crop plants in Ivory Coast, Initiations—Documentations Techniques No. 46, ORSTOM, France.

Fauquet, C., and Thouvenel, J.-C., 1983, Association d'un nouveau virus en bâtonnet avec la maladie necrose à rayures du riz en Câte d'Ivoire, *C.R. Acad Sci. Ser. D* **296**:575.

Fenner, F., 1976, Classification and nomenclature of viruses: Second Report of the International Committee on the Taxonomy of Viruses, *Intervirology* **7**:88.

Fraser, T. W., 1976, Mop-top tubules: The ultrastructure of usual tubular elements associated with different leaf symptoms of potato mop-top virus infected potato plants, *Protoplasma* **90**:15.

Fujikawa, T., 1963, On the broad bean necrosis disease and its control, *Nogyo Oyobi Engei* **38**:359.

Fujisawa, I., and Sugimoto, T., 1977, Transmission of beet necrotic yellow vein virus by *Polymyxa betae, Ann. Phytopathol. Soc. Jpn.* **43**:583.

Fukano, H., and Yokoyama, S., 1952, On a necrotic virus disease of broadbean, garden pea and spring vetch, *Kyushu Nogyo Kenkyu* **10**:133.

Gao, J. L., Deng, F., Zhai, H. Q., Liang, X. S., and Liu, Y., 1983, The occurrence of sugar beet rhizomania caused by beet necrotic yellow vein virus in China, *Acta Phytopathol. Sin.* **13**:1.

Germani, G., and Dhery, M., 1973, Observations et experimentations concernant le rôle des nématodes dans deux affections de l'arachide en Haute-Volta: la 'chlorose' et le 'clump,' *Oleagineux* **28:**235.

Germani, G., Thouvenel, J.-C., and Dhery, M., 1975, Le rabougrissement de l'arachide: une maladie à virus au Sénégal et Haute-Volta, *Oleagineux* **30:**259.

Gibbs, A. J., 1977, Tobamovirus group, *CMI/AAB Descriptions of Plant Viruses No. 184.*

Giunchedi, L., Langenberg, W. G., and Marani, F., 1981, Appearance of beet necrotic yellow vein virus (BNYVV) in host cells, *Phytopathol. Mediterr.* **20:**112.

Gold, A. H., Scott, H. A., and McKinney, H. H., 1957, Electron microscopy of several viruses occurring in wheat and other monocots, *Plant Dis. Rep.* **41:**250.

Greber, R. S., and Finlay, J. R., 1981, Hypochoeris mosaic virus in Australia, *Aust. Plant Pathol.* **10:**30.

Gumpf, D. J., 1971, Purification and properties of soil-borne wheat mosaic virus, *Virology* **43:**586.

Gibbs, A., Tien, P., Kang, L-Y., Tian, Y.-C., and Randles, J., 1982, Classification of several viruses isolated in China on the basis of the amino acid composition of their virion proteins, *Intervirology* **18:**160.

Hamdorf, G., Lesemann, D. E., and Weidemann, H. L., 1977, Untersuchungen über die Rizomania-Krankheit an Zucherrüben in der Bundesrepublik Deutschland, *Phytopathol. Z.* **90:**97.

Harrison, B. D., 1974, Potato mop-top virus, *CMI/AAB Descriptions of Plant Viruses No. 138.*

Harrison, B. D., and Jones, R. A. C., 1970, Host range and properties of potato mop-top virus, *Ann. Appl. Biol.* **65:**393.

Harrison, B. D., and Jones, R. A. C., 1971, Factors affecting the development of spraing in potato tubers infected with potato mop-top virus, *Ann. Appl. Biol.* **68:**281.

Hebert, T. T., 1963, Precipitation of plant viruses by polyethylene glycol, *Phytopathology* **53:**362.

Hebert, T. T., and Coleman, N. T., 1955, Rod-shaped particles associated with soil-borne wheat mosaic virus, *Phytopathology* **45:**348.

Hebert, T. T., and Panizo, C. H., 1975, Oat mosaic virus, *CMI/AAB Descriptions of Plant Viruses No. 145.*

Hibino, H., Tsuchizaki, T., and Saito, Y., 1974a, Comparative electron microscopy of cytoplasmic inclusions induced by nine isolates of soil-borne wheat mosaic virus, *Virology* **57:**510.

Hibino, H., Tsuchizaki, T., and Saito, Y., 1974b, Electron microscopy of inclusion development of rye leaf cells infected with soil-borne wheat mosaic virus, *Virology* **57:**522.

Hibino, H., Tsuchizaki, T., and Saito, Y., 1981, Influence of host species and infection stages on the components of soil-borne wheat mosaic virus, *Ann. Phytopathol. Soc. Jpn.* **47:**566.

Hooper, G. R., and Wiese, M. V., 1972, Cytoplasmic inclusions in wheat affected by wheat spindle streak mosaic, *Virology* **47:**664.

Hsu, Y.-H., and Brakke, M. K., 1983, Translation of soil-borne wheat mosaic virus RNAs in rabbit reticulocyte lysate, *Phytopathology* **73:**790.

Hsu, Y.-H., and Brakke, M. K., 1985, Sequence relationships among soil-borne wheat mosaic virus RNA species and terminal structures of RNA II, *J. Gen. Virol.* **66:**915.

Hull, R., 1968, A virus disease of hart's-tongue fern, *Virology* **35:**333.

Imoto, M., Tochihara, H., Iwaki, M., and Nakamura, H., 1981, Occurrence of potato mop-top disease in Japan, *Ann. Phytopathol. Soc. Jpn.* **47:**409.

Inouye, T., 1969, Filamentous particles as the causal agent of yellow mosaic disease of wheat, *Ber. Ohara Inst. Landwirtsch. Biol. Okayama Univ.* **13:**A7.

Inouye, T., 1971, Electron microscopy of sweetpea cells infected with broad bean necrosis virus, *Ber. Ohara Inst. Landwirtsch. Biol. Okayama Univ.* **15:**69.

Inouye, T., and Asatani, M., 1968, Broad-bean necrosis virus, *Ann. Phytopathol. Soc. Jpn.* **34:**317.

Inouye, T., and Fujii, S., 1977, Rice necrosis mosaic virus, *CMI/AAB Descriptions of Plant Viruses No. 172.*

Inouye, T., and Saito, Y., 1975, Barley yellow mosaic virus, *CMI/AAB Descriptions of Plant Viruses No. 143.*

Ioanides, A., 1978, Rhizomania, *To Teutho* **1.10.78:**13.

Ivanovic, M., MacFarlane, I., and Woods, R. D., 1983, Viruses transmitted by fungi: Viruses of sugarbeet associated with *Polymyxa betae, Rep. Rothamsted Exp. Stn.* 1982, p. 189.

Jones, R. A. C., and Harrison, B. D., 1969, The behavior of potato mop-top virus in soil, and evidence for its transmission by *Spongospora subterranea* (Wall.) Lagerh., *Ann. Appl. Biol.* **63:**1.

Jones, R. A. C., and Harrison, B. D., 1972, Ecological studies on potato mop-top virus in Scotland, *Ann. Appl. Biol.* **71:**47.

Kanzawa, K., 1970, On the abnormal growth of sugarbeet, *Ann. Phytopathol. Soc. Jpn.* **36:**365.

Kanzawa, K., and Ui, T., 1972, A note on rhizomania on sugarbeet in Japan, *Ann. Phytopathol. Soc. Jpn.* **38:**434.

Kassanis, B., and Woods, R. D., 1969, Properties of some defective strains of tobacco mosaic virus and their behavior as affected by inhibitors during storage in sap, *Ann. Appl. Biol.* **64:**213.

Kassanis, B., Woods, R. D., and White, R. F., 1972, Some properties of potato mop-top virus and its serological relationship to tobacco mosaic virus, *J. Gen. Virol.* **14:**123.

Koch, F., 1982, Die Rizomania der Zuckerrube, *Proc. Int. Inst. Sugar Beet Res., Brussels,* p. 211.

Koehler, B., Bever, W. M., and Bennett, O. T., 1952, Soil-borne wheat mosaic, *Univ. Ill. Exp. Stn. Bull.* **556:**567.

Koenig, R., Lesemann, D.-E., and Burgermeister, W., 1984, Beet necrotic yellow vein virus: Purification, preparation of antisera and detection by means of ELISA, immunosorbent electronmicroscopy and electro-blot immunoassay, *Phytopathol. Z.* **111:**224.

Kuwata, S., and Kubo, S., 1981, Rod-shaped particles found in tobacco plants infected with tobacco stunt agent, *Ann. Phytopathol. Soc. Jpn.* **47:**264.

Kuwata, S., and Kubo, S., 1984, Properties of two *Olpidium*-transmitted viruses: Tobacco stunt and lettuce big vein, *IVth Int. Congr. Virol., Sendai, Abstract P45-2,* p. 331.

Kuwata, S., Kubo, S., Yamashita, S., and Doi, Y., 1983, Rod-shaped particles, a probable entity of lettuce big vein virus, *Ann. Phytopathol. Soc. Jpn.* **49:**246.

MacFarlane, I., Jenkins, J. E. E., and Melville, S., 1968, Viruses infecting barley in S.W. England, *Plant Pathol.* **17:**167.

McKinney, H. H., 1923, Investigation on the rosette disease of wheat and its control, *J. Agric. Res.* **23:**771.

McKinney, H. H., Eckerson, S. H., and Webb, R. W., 1923, The intracellular bodies associated with the rosette disease and mosaic-like leaf mottling of wheat, *J. Agric. Res.* **26:**605.

McKinney, H. H., Silber, G., and Greeley, L., 1965, Longevity of some plant viruses stored in chemically dehydrated tissues, *Phytopathology* **55:**1043.

Masri, S. A., and Hiruki, C., 1983, A new group of elongated plant viruses having a capsid protein of an unusually high molecular weight, *Can. J. Plant Pathol.* **5:**208.

Matthews, R. E. F., 1982, Classification and nomenclature of viruses: Fourth Report of the International Committee on Taxonomy of Viruses, *Intervirology* **17:**158.

Mayo, M. A., and Reddy, D. V. R., 1985, Translation products of RNA from Indian peanut clump virus, *J. Gen. Virol.* **66:**1347.

Nakasone, W., and Inouye, T., 1978, On the serological relationship between broad bean necrosis virus and the tobamovirus group, *Ann. Phytopathol. Soc. Jpn.* **44:**97.

Novak, J. B., and Lanzova, J., 1984, Serological proof of beet necrotic yellow vein virus on sugarbeet in Czechoslovakia, *Proc. IXth Czechoslovak Plant Prot. Conf., Brno,* p. 253.

Plumb, R. T., and MacFarlane, I., 1977, Cereal diseases: A "new" virus of oats, *Rep. Rothamsted Exp. Stn.* 1976, p. 256.

Plumb, R. T., and MacFarlane, I., 1978, Oat golden stripe virus, *Rep. Rothamsted Exp. Stn. 1977*, p. 212.

Powell, C. A., 1976, The relationship between soil-borne wheat mosaic virus and tobacco mosaic virus, *Virology* **71**:453.

Putz, C., 1977, Composition and structure of beet necrotic yellow vein virus, *J. Gen. Virol.* **35**:397.

Putz, C., and Fritsch, C., 1978, Some properties of the RNA of beet necrotic yellow vein virus, *IIIrd Int. Congr. Plant Pathol., Abstracts*, p. 32.

Putz, C., and Kuszala, M., 1978, Le rhizomanie de la betterave sucrière en Alsace: Recherche d'une nouvelle méthode de purification du 'beet necrotic yellow vein virus,' *Ann. Phytopathol.* **10**:247.

Putz, C., and Vuittenez, A., 1974, Observation de particules virales chez des betteraves présentant en Alsace des symptômes de 'rhizomanie,' *Ann. Phytopathol.* **6**:129.

Putz, C., and Vuittenez, A., 1980, The intracellular location of beet necrotic yellow vein virus, *J. Gen. Virol.* **50**:201.

Putz, C., Collot, D., and Peter, R., 1979, Caracterisation biochimique de la sous-unite proteique des virus de la rhizomanie de la betterave, *C.R. Acad. Sci. Ser. D* **284**:1951.

Putz, C., Pinck, L., Pinck, M., and Fritsch, C., 1983, Identification of the 3′ and 5′ ends of beet necrotic yellow vein virus RNAs: Presence of 3′ poly-A sequences, *FEBS Lett.* **156**:41.

Rana, R. L., Di Franco, A., and Russo, M., 1978, La rizomania della barbabietola in Italia meridionale, *Inf. Fitopatol.* **28**:5.

Randles, J. W., 1978, Nicotiana velutina mosaic virus, *CMI/AAB Descriptions of Plant Viruses No. 189*.

Randles, J. W., Harrison, B. D., and Roberts, I. M., 1976, *Nicotiana velutina* mosaic virus: Purification, properties and affinities with other rod-shaped viruses, *Ann. Appl. Biol.* **84**:193.

Rao, A. S., and Brakke, M. K., 1969, Relation of soil-borne wheat mosaic virus and its fungal vector, *Polymyxa graminis, Phytopathology* **59**:581.

Reddy, D. V. R., Iizuka, N., Subrahmanyam, P., Rajeshwari, R., and McDonald, D., 1979, A soil-borne disease of peanuts in India, *Proc. Am. Peanut Res. Educ. Soc.* **11**:49.

Reddy, D. V. R., Rajeshwari, R., Iizuka, N., Lesemann, D. E., Nolt, B. L., and Goto, T., 1983, The occurrence of Indian peanut clump, a soil-borne virus disease of groundnuts (*Arachis hypogaea*) in India, *Ann. Appl. Biol.* **102**:305.

Richards, K., Jonard, G., Guilley, H., Ziegler, V., and Putz, C., 1985, *In vitro* translation of beet necrotic yellow vein virus RNA and studies of sequence homology among the RNA species using cloned cDNA probes, *J. Gen. Virol.* **66**:345.

Russo, M., Martelli, G. O., and Di Franco, A., 1981, The fine structure of local lesions of beet necrotic yellow vein virus in *Chenopodium amaranticolor, Physiol. Plant Pathol.* **19**:237.

Saito, Y., Takanashi, K., Iwata, Y., and Okamoto, H., 1964, Studies on the soil-borne virus disease of wheat and barley. I. Several properties of the viruses, *Bull. Natl. Inst. Agric. Sci., Tokyo, Ser. C* **1964**:1711.

Saito, Y., Tsuchizaki, T., and Hibino, H., 1968, Causal virus of wheat yellow mosaic disease, *Ann. Phytopathol. Soc. Jpn.* **34**:347.

Salazar, L. F., 1977, Studies on three viruses from South American potatoes, Ph.D. thesis, University of Dundee.

Salazar, L. F., and Jones, R. A. C., 1975, Some studies on the distribution and incidence of potato mop-top virus in Peru, *Am. Potato J.* **52**:143.

Shirako, Y., and Brakke, M. K., 1981, Genetic function of RNA components of soil-borne wheat mosaic virus, *Phytopathology* **71**:862.

Shirako, Y., and Brakke, M. K., 1984a, Two purified RNAs of soil-borne wheat mosaic virus are needed for infection, *J. Gen. Virol.* **65**:119.

Shirako, Y., and Brakke, M. K., 1984b, Spontaneous deletion mutation of soil-borne wheat mosaic virus RNA II, *J. Gen. Virol.* **65**:855.

Singh, R. P., and McDonald, J. G., 1980, A chlorotic mosaic of fall hawkbit (*Leontodon autumnalis*), *Can. Plant Dis. Surv.* **60**:47.

Steven, A. C., Trus, B. L., Putz, C., and Wurtz, M., 1981, The molecular organisation of beet necrotic yellow vein virus, *Virology* **113**:428.

Stocky, G., Vuittenez, A., and Putz, C., 1977, Mise en evidence *in situ* de particules en batonnets, correspondant probablement au virus de la rhizomanie dans les tissue de diverses chenopodiacées inoculées experimentalement, ainsi que dans le champignon *Polymyxa betae* Keskin, myxomycete parasitant les raines de plantes malades, *Ann. Phytopathol.* **9**:536.

Sundararaman, S., 1927, A clump disease of groundnuts, *Madras Agric. De. Yearb.* 1926, p. 13.

Sutic, D., and Milovanovic, M., 1978, Occurrence and significance of sugar beet root stunting, *Agrochemija* **9**:363.

Tamada, Y., 1975, Beet necrotic yellow vein virus, *CMI/AAB Descriptions of Plant Viruses No. 144.*

Tamada, Y., and Baba, T., 1973, Beet necrotic yellow vein virus from rhizomania-affected sugar beet in Japan, *Ann. Phytopathol. Soc. Jpn.* **39**:325.

Tamada, Y., Kanzawa, K., and Ui, T., 1970, A virus detected in the sugarbeet plants of abnormal growth, *Ann. Phytopathol. Soc. Jpn.* **36**:365.

Tamada, T., Abe, H., and Baba, T., 1971, Relationship between the abnormal growth of sugarbeet and a virus isolated from those sugarbeet plants, *Bull. Sugar Beet Res.* **13**:179.

Tamada, T., Abe, H., and Baba, T., 1975, Beet necrotic yellow vein virus and its relation to the fungus *Polymyxa betae, Proc. 1st Int. Congr. Int. Assoc. Microbiol. Soc.* **3**:313.

Thornberry, H. H., Vatter, A. E., and Hickman, D., 1953, Virus-like particles from some plants having symptoms of virus disease, *Phytopathology* **43**:486.

Thouvenel, J.-C., and Fauquet, C., 1980, First report of *Polymyxa graminis* in Africa, and its occurrence on new *Sorghum* Moench host species, *Plant Dis.* **64**:957.

Thouvenel, J.-C., and Fauquet, C., 1981, Further properties of peanut clump virus and studies on its natural transmission, *Ann. Appl. Biol.* **97**:99.

Thouvenel, J.-C., Germani, G., and Pfeiffer, P., 1974, Preuve de l'origine virale du rabougrissement ou "clump" de l'arachide en Haute-Volta et au Senegal, *C.R. Acad. Sci.* **278**:2847.

Thouvenel, J.-C., Dollet, M., and Fauquet, C., 1976, Some properties of peanut clump, a newly discovered virus, *Ann. Appl. Biol.* **84**:311.

Thouvenel, J.-C., Fauquet, C., and Dollet, M., 1977, Influence du pH sur le virus du robougrissement de l'arachide (peanut clump virus) et nouvelle methode rapide de purification, *Ann. Phytopathol.* **10**:93.

Thouvenel, J.-C., Fauquet, C., and Lamy, D., 1978, Transmission par la graine du virus du clump de l'arachide, *Oleagineux* **33**:503.

Tsuchizaki, T., Hibino, H., and Saito, Y., 1973, Comparisons of soil-borne wheat mosaic virus isolates from Japan and the United States, *Phytopathology* **63**:634.

Tsuchizaki, T., Hibino, H., and Saito, Y., 1975, The biological functions of short and long particles of soil-borne wheat mosaic virus, *Phytopathology* **65**:523.

Vuittenez, A., Arnold, J., Spindler, C., and de Larambergue, H., 1977, Test de transmission du virus rhizomanie de la betterave (beet necrotic yellow vein virus) par le champignon *Polymyxa betae* myxomycete associé aux racines. Application a la prospection de la maladie, *Ann. Phytopathol.* **9**:537.

White, R. F., Kassanis, B., and James, M., 1972, Potato mop-top virus in infected cells, *J. Gen. Virol.* **15**:175.

PART III

TOBRAVIRUSES

CHAPTER 17

Tobraviruses

B. D. HARRISON AND D. J. ROBINSON

I. RECOGNITION AND CHARACTERISTICS OF THE TOBRAVIRUS GROUP

A. Members and Their Geographical Distribution

The tobravirus group is now considered to have three members (Table I; Robinson & Harrison, 1985a). Tobacco rattle virus (TRV) causes a tobacco disease that was first described in Germany by Behrens (1899), who called it *Mauche*. The virus was shown to pass through a bacterium-retaining filter by Böning (1931) and was partially characterized by Quanjer (1943), who first referred to it as tobacco *ratel* (= rattle) virus. Pepper ringspot virus (PRV) was for several years known as the CAM strain of TRV (Harrison and Woods, 1966) but is now considered a separate virus (Robinson and Harrison, 1985a). Pea early-browning virus (PEBV) was first found in the Netherlands (Bos and van der Want, 1962) and now includes a range of isolates including one initially described as broad bean yellow band virus (Russo *et al.*, 1984; Robinson and Harrison, 1985b). The known geographical distributions of individual tobraviruses and the more important diseases they cause are given in Table I.

B. General Properties of Tobraviruses

TRV, PRV, and PEBV were first grouped together under the name netuvirus (*ne*matode-transmitted, *tu*bular particles) but soon afterwards this was changed to the tobravirus group (Harrison *et al.*, 1971). These viruses have wide host ranges (especially TRV) that include many wild

B. D. HARRISON AND D. J. ROBINSON • Scottish Crop Research Institute, Invergowrie, Dundee DD2 5DA, Scotland.

TABLE I. Tobraviruses and Their Geographical Distribution

Virus	Principal diseases caused	Geographical distribution
Tobacco rattle virus (type member, tobravirus group)	Aster ringspot, gladiolus notched leaf, hyacinth malaria, potato spraing (corky ringspot), potato stem mottle, streaky mottle of narcissus and tulip, sugar beet yellow blotch, tobacco rattle	Britain and continental Europe, Japan, New Zealand, North America, USSR
Pea early-browning virus	Broad bean yellow band, distorting mosaic of bean (*Phaseolus*), pea early-browning	Britain and continental Europe, north Africa (Morocco)
Pepper ringspot virus	Artichoke yellow band, pepper ringspot, tomato yellow band	Brazil

and cultivated plant species and they are transmitted by soil-inhabiting nematodes in the family Trichodoridae. They are characterized as having a genome consisting of two species of ssRNA of positive polarity, RNA 1 and RNA 2, contained respectively in straight tubular particles of two specific lengths, known as long (L) and short (S). L particles on their own are infective, whereas S particles are not (Harrison and Nixon, 1959a). All tobravirus strains can give rise to NM-type isolates (Cadman and Harrison, 1959) by loss of RNA 2, which contains the particle protein gene. NM-type isolates therefore do not produce virus particles and are maintained in plants by replication of the larger genome segment (RNA 1) alone. Such infections can spread from cell to cell but apparently not

TABLE II. Properties of M-Type and NM-Type Isolates of Tobacco Rattle Virus

Property	M-type isolates	NM-type isolates
Infective entity	Long and short nucleoprotein particles	RNA 1
Infectivity of sap		
Dilution endpoint	$< 10^6:1$	$< 10^2:1$
Longevity at 20°C	> 6 weeks	< 1 hr
Longevity at −20°C	> 10 years	Zero
Infectivity of phenol extracts of infected leaf	Less than sap	Much greater than sap
Properties in plants		
Systemic symptoms	Transient, mild–moderate severity	Persistent, more severe
Route of systemic invasion	Via phloem sieve tubes	Probably by cell-to-cell movement
Transmission by trichodorid nematodes	Transmitted in persistent manner	No evidence of transmission

via phloem sieve tubes, and they cause symptoms that typically are more severe than those induced by particle-producing (M-type) isolates of the same virus strain. M-type isolates can be regenerated by adding RNA 2 or S particles to nucleic acid extracted from plants infected with NM-type isolates, and inoculating test plants with the mixture (Lister, 1968; Sänger, 1968). The existence of NM-type isolates complicates many kinds of work with tobraviruses, including their detection and identification. The properties of M-type and NM-type isolates of TRV are compared in Table II.

Later sections of this chapter describe salient features of the biology and molecular biology of the group with some emphasis on recent work. Additional general information on the properties of tobraviruses and on methods used for their study can be found in earlier reviews (Harrison and Robinson, 1978, 1981).

II. TOBRAVIRUS PARTICLES AND THEIR COMPONENTS

A. Particle Morphology and Structure

In the electron microscope, negatively stained particles of tobraviruses appear as straight rods of two predominant lengths (Fig. 1A); the lengths are characteristic of the particular virus and strain (see Section III.A). Estimates of particle diameter vary somewhat depending on the negative stain used. In uranyl formate plus sodium hydroxide, which best preserves structural detail, TRV particles have an apparent diameter of 23 nm, and a central hole 5 nm in diameter along the axis of the rods (Cooper and Mayo, 1972). Particles of PEBV and PRV appear to be 5–10% narrower than those of TRV, and to have a narrower central hole (Harrison and Woods, 1966; Cooper and Mayo, 1972). The difference between the dimensions of TRV and PRV particles is supported by X-ray diffraction data, from which Finch (1965) calculated a diameter of 22.5 nm with a 4-nm central hole for TRV, and Tollin and Wilson (1971) obtained values of 20.5 and 3.6 nm for PRV.

The structures of L and S particles seem to be essentially the same. The diameters both of the particles and of their central holes are indistinguishable, as are their buoyant densities in CsCl solutions, and although L and S particles of PRV differ slightly in electrophoretic mobility at pH 8.6, this may reflect merely the increased contribution of the surfaces at the ends of the shorter particles (Cooper and Mayo, 1972).

Transverse striations on negatively stained particles (Fig. 1B) were first observed by Nixon and Harrison (1959), whose suggestion that the particles are of helical construction was subsequently confirmed by X-ray diffraction studies (Finch, 1965; Tollin and Wilson, 1971) and by optical diffraction from electron microscope images (Offord, 1966a). The pitch of the helix is 2.5 nm, and there is an almost, but not exactly (Finch,

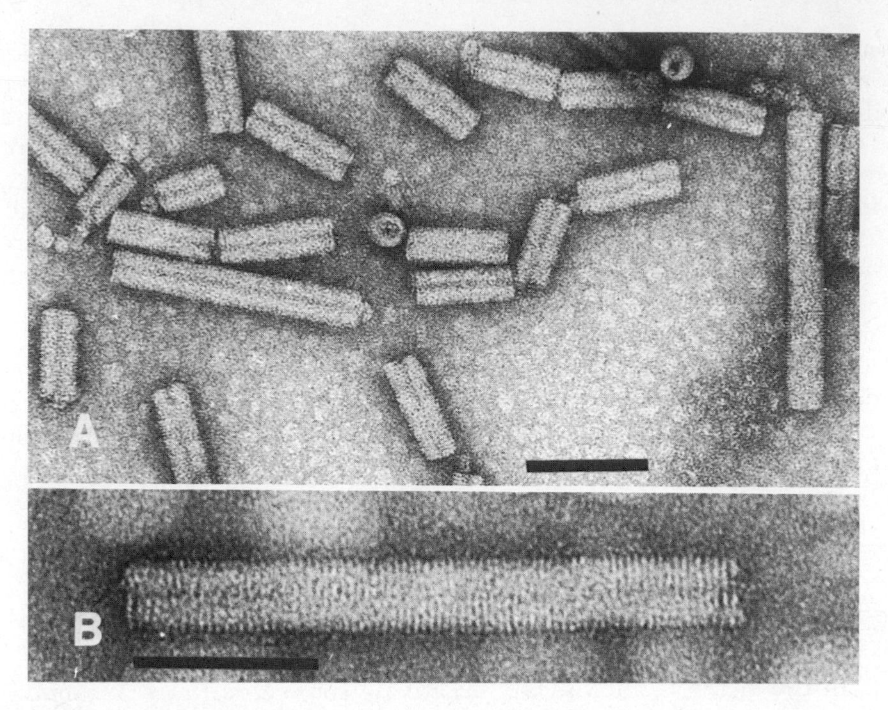

FIGURE 1. (A) Particles of TRV (strain PRN), negatively stained with 1% uranyl acetate. Bar = 100 nm. (B) Particle of PRV (strain CAM), negatively stained with uranyl formate plus formic acid. Bar = 50 nm.

1965), integral number of subunits in three turns. A value of approximately $25\frac{1}{3}$ subunits per turn was favored by Finch (1965) and has been supported by subsequent workers, but the evidence is not compelling. Sedimentation analysis and detailed electron microscopic examination of disklike protein aggregates (see Section II.D) suggested that they contain about 32 subunits per turn (Mayo and Daney de Marcillac, 1977; Roberts and Mayo, 1980), and a similar value for the nucleoprotein particles is plausible. There is no definite evidence as to the hand of the helix, although the protein aggregates are right-handed (Roberts and Mayo, 1980) as are particles of TMV (Finch, 1972), and Offord (1966a) pointed out that if the contrast in negatively stained TRV particles arises predominantly from the same side as it does in the majority of TMV particles, both have the same hand. The two ends of individual particles differ in appearance, one being slightly convex and the other slightly concave (Harrison and Woods, 1966).

There are indications in X-ray diffraction patterns (Finch, 1965; Tollin and Wilson, 1971), in transverse sections of dried, oriented particles (Tollin and Wilson, 1971), and in end views of negatively stained particles (Offord, 1966a) of a structural feature at a radius of 8–8.5 nm which may represent the position of the RNA in the helix.

There are no reports of macromolecular components other than RNA and protein in tobravirus particles, but there is evidence of a role for divalent metal ions in their stabilization. In titration experiments, Durham and Abou Haidar (1977) detected one or two binding sites for divalent ions per protein subunit in PRV particles that are not present in the isolated protein, and suggested that the hysteresis that occurred only when calcium ions were present was a result of tightening of the particle structure. Moreover, Robinson and Raschké (1977) showed that removal of divalent ions rendered PRV particles permeable to ribonuclease.

B. Particle Protein Size and Structure

In SDS–polyacrylamide gel electrophoresis, tobravirus particle proteins migrate as a single band, and proteins from L and S particles are indistinguishable (Cooper and Mayo, 1972). However, estimates of molecular weight are dependent on gel concentration (Ghabrial and Lister, 1973a) and earlier reports are therefore unreliable. Using a method which takes into account the gel concentration effect, Mayo and Robinson (1975) estimated molecular weights of 21,000–22,000 for three strains of TRV, 23,000 for PRV, and 21,000 for PEBV.

Amino acid compositions have been determined for the proteins from several strains of TRV (Offord and Harris, 1965; Semancik, 1966; Miki and Okada, 1970; Ghabrial and Lister, 1973b) and from PRV (Miki and Okada, 1970). Most are compatible with a molecular weight between 21,000 and 23,000 (Mayo and Robinson, 1975). For TRV strain PRN protein, sedimentation to equilibrium in guanidine hydrochloride solution gave a molecular weight of 21,700 (Mayo and Robinson, 1975), and a value near this seems to fit all the available data best.

No complete amino acid sequence of a tobravirus particle protein has been obtained directly, although Offord (1966b) determined much of the sequence for a Dutch strain of TRV. The C-terminal end of this protein is rich in proline, and in two other strains the C-terminal residue is proline (Semancik, 1970). The α-amino group of the N-terminal amino acid, glycine, is blocked by acetylation in the Dutch strain (Offord and Harris, 1965).

C. The RNA Species and Their Structure

Tobravirus RNA is about 5% as infective as the same amount of RNA in virus particles (Harrison and Nixon, 1959b), and contains two species (RNA 1 and RNA 2) corresponding to the L and S nucleoprotein particles, respectively. RNA 1 of each of the three viruses is of similar size (Cooper and Mayo, 1972), with a molecular weight, estimated by polyacrylamide gel electrophoresis under denaturing conditions for several strains of TRV, of 2.4×10^6 (Reijnders *et al.*, 1974; Pelham, 1979;

Robinson *et al.*, 1983). The molecular weight of RNA 2 varies between isolates, ranging from 0.6 to 1.4 $\times$ 10^6.

Abou Haidar and Hirth (1977) showed that the 5′-terminal structure of 70–75% of PRV RNA 2 molecules was $m^7G^{5'}ppp^{5'}Ap$ but were unable to find a similar cap structure in RNA 1, although their experiments may have lacked sufficient sensitivity. Translation of both genomic RNA species of TRV is inhibited by the cap analogue m^7GTP (Pelham, 1979), suggesting that both are capped.

The sequence of PRV RNA 2 is known, and comprises 1799 residues. First results from sequencing RNA 1 show that 460 residues at the 3′-end are identical to those in RNA 2; they do not include a polyadenylate tract (S. Bergh and A. Siegel, unpublished results). This accords with the results of RNA–RNA hybridization experiments (Minson and Darby, 1973; Darby and Minson, 1973), which showed that a sequence or sequences totaling about 500 nucleotide residues are substantially the same in the two RNA species of PRV. The existence of this homology was confirmed by Robinson (1983), using cDNA–RNA hybridization, but no comparable homology was detected between RNA 1 and RNA 2 of two strains of TRV. However, some cDNA clones prepared from two Dutch isolates of TRV reacted with both RNA 1 and RNA 2 in Northern blots, although no estimate of the extent of the sequence homology between the RNA species can be made from these experiments (Linthorst and Bol, 1986).

For PRV, the base composition of RNA 1 was estimated to be 25.5% G, 29.5% A, 16.5% C, and 28.5% U, and of RNA 2 to be 24.3% G, 26.1% A, 17.6% C, and 32.0% U (Minson and Darby, 1973). Values for TRV resembled those for PRV RNA 1 (Semancik and Kajiyama, 1967a).

D. Particle Protein Aggregation and Nucleoprotein Reassembly

At low pH, the particle proteins of TRV and PRV exist in solution as slowly sedimenting forms (Semancik, 1970; Fritsch *et al.*, 1973), probably containing a mixture of oligomers (Gugerli, 1976). At high pH, a faster-sedimenting form known as the disk predominates. This structure has a sedimentation coefficient ($s^0_{20,w}$) of 40 *S* and a particle weight of 2 $\times$ 10^6 (Mayo and Daney de Marcillac, 1977), and comprises three turns of a helix calculated to have about 32 subunits per turn (Roberts and Mayo, 1980). The oligomer and disk forms of the protein are antigenically distinct (Gugerli, 1976).

For PRV, the transition from oligomer to disk forms occurs between pH 4.5 and 5.5 (Fritsch *et al.*, 1973), whereas for TRV it is 0.3–0.5 pH unit higher (Morris and Semancik, 1973; Mayo and Daney de Marcillac, 1977). Within the transition range, the formation of disks is favored at lower ionic strengths (Fritsch *et al.*, 1973), and even at pH 6.7 high ionic strengths cause some dissociation of disks (Mayo and Daney de Marcillac, 1977).

Under a wide range of conditions from pH 4.5 to 8.0, ionic strength 0.01 to 1.0, and temperature 0 to 25°C, PRV protein and RNA form a complex in which a single disk of protein is bound at one end of the RNA (Abou Haidar *et al.*, 1973), and which may represent the first step of particle assembly. Over much of this range, some elongation occurs, and in these partially reconstituted particles the 5′-end but not the 3′-end of the RNA is protected (Abou Haidar, 1976; Abou Haidar and Hirth, 1977). However, reconstitution of full-length, infective PRV nucleoprotein rods requires much more closely defined conditions (0.5 M phosphate buffer, pH 4.7, at 1°C; Abou Haidar *et al.*, 1973). It is not known whether other tobraviruses will reassemble under these conditions, but conditions reported as optimal for TRV reconstitution (0.25 M glycine, pH 8.0, at 9°C; Morris and Semancik, 1973) were ineffective for PRV (Abou Haidar *et al.*, 1973).

III. VARIATION AND RELATIONSHIPS

A. Biological Variation

In this section, we are concerned mainly with the extent and nature of variation among M-type isolates of tobraviruses, but it should be borne in mind that each M-type isolate can be converted irreversibly into an NM-type isolate (Cadman and Harrison, 1959). NM-type isolates are common in nature, and their properties differ from those of M-type isolates in several ways (Table II).

TRV has a very broad host range (Schmelzer, 1957; Horváth, 1978) and isolates seem unusually variable in the type and severity of symptoms produced. Indeed, M-type single-lesion isolates from one stock culture may differ considerably with regard to the symptoms they produce in test plants (Cadman and Harrison, 1959; Lister and Bracker, 1969). A particularly noteworthy naturally occurring variant is strain SYM, which unlike most other strains of TRV infects *Chenopodium amaranticolor* systemically (Kurppa *et al.*, 1981). Variants of TRV and PRV that induce bright yellow markings in several solanaceous species are described by Lister and Bracker (1969) and Robinson (1977), respectively. PEBV has a narrower host range than TRV (Bos and van der Want, 1962) and fewer strains have been studied, but at least one, strain BBYB, is symptomatologically distinct in *Vicia faba* (Russo *et al.*, 1984).

L particles of different isolates have rather similar modal lengths: 185–197 nm for TRV (Harrison and Woods, 1966), 192–212 nm for PEBV (Harrison, 1966; van Hoof, 1969). However, the S particles of different isolates of TRV vary remarkably in length from about 50 to 115 nm, although most are between 80 and 110 nm (Harrison and Robinson, 1978).

Similarly, those of PEBV vary from about 55 to 105 nm (Bos and van der Want, 1962; van Hoof, 1969). Only one strain of PRV (strain CAM) has been studied intensively, so there is little information about strain variation in this virus, but strain CAM has L particles of 197 nm and rather short S particles of 52 nm (Harrison and Woods, 1966).

B. Antigenic Variation

TRV isolates were formerly separated into three serotypes (Harrison and Woods, 1966), but of these serotype III is now considered to be a distinct virus, PRV (Robinson and Harrison, 1985a), and further work has blurred the distinction between serotypes I and II. However, among the isolates in this group, there are many reports in the literature of strains the particles of which seem to have few if any epitopes in common (Harrison and Woods, 1966; van Hoof *et al.*, 1966; Sänger, 1968; Kurppa *et al.*, 1981). Nevertheless, clusters of serologically more closely related strains can be recognized (Robinson and Harrison, 1985a), although no attempt has been made to systematize them.

PEBV isolates from the Netherlands and Great Britain, respectively, form two clearly distinguishable serotypes (Gibbs and Harrison, 1964; van Hoof, 1969), and a third serotype has recently been recognized (Russo *et al.*, 1984; Robinson and Harrison, 1985b). Dutch PEBV is serologically distantly related to Dutch TRV (Maat, 1963), but no such relationship was detected between British isolates of the two viruses (Gibbs and Harrison, 1964). PRV is only very distantly serologically related to TRV or PEBV (Harrison and Woods, 1966; Kurppa *et al.*, 1981).

C. Cross-Protection between Isolates

Four M-type isolates of TRV each protected *Nicotiana sylvestris* plants from the effects of subsequent inoculation with any of the four (Cadman and Harrison, 1959). Also, Dutch PEBV protected plants from infection by British PEBV, whereas British PEBV and TRV isolates did not cross-protect (Gibbs and Harrison, 1964).

NM-type isolates of TRV protected systemically infected plants from the effects of subsequent inoculation with an M-type isolate, although they did not prevent the production of L and S particles of the challenge isolate (Cadman and Harrison, 1959).

D. Pseudorecombinant Formation

Pseudorecombinant isolates, in which RNA 1 and RNA 2 are derived from distinct strains, are readily produced from pairs of antigenically very

TABLE III. Apparent Nucleotide Sequence Homology between Tobravirus Genomes

cDNA to RNA of	% homology with RNA from		
	PRV (1[a])	TRV (10)	PEBV (4)
PRV	100[b]	<7	<6
TRV	<5	37–100	<12
PEBV	<1	<7	49–100

[a] Number of strains tested.
[b] Apparent % sequence homology in hybridization tests using RNA from unfractionated virus preparations.

different strains of TRV (Sänger, 1968; Lister and Bracker, 1969; Kurppa *et al.*, 1981) or of PEBV (Robinson and Harrison, 1985b). However, attempts to make pseudorecombinants between TRV and PEBV failed (Lister, 1968; Robinson and Harrison, 1985b). Similarly, Frost *et al.* (1967) failed to make pseudorecombinants between isolates of TRV and PRV, although Lister (1969) apparently succeeded in making two containing RNA 1 of a PRV strain and RNA 2 of TRV strain ORY. However, the tests used to identify these pseudorecombinants did not conclusively exclude the possibility that they were variants of strain ORY. In general, therefore, a grouping of isolates on the basis of their ability to form pseudorecombinants with one another indicates that PEBV, PRV, and TRV represent three different and distinct gene pools.

E. RNA Sequence Homologies

Hybridization experiments using cDNA copies of unfractionated virus RNA showed that 15 tobravirus strains could be divided into three groups, comprising strains of TRV, PRV, and PEBV, respectively, with little or no nucleotide sequence homology between isolates in different groups (Table III; Robinson and Harrison, 1985a). The RNA 1 species of TRV strains share extensive sequences, whereas the sequences of their RNA 2 species are diverse. For example, strains ORY and SYM have almost identical RNA 1 sequences, but no detectable sequence homology in their RNA 2 species (Robinson, 1983). However, some clusters of TRV strains were identified that share some or all of their RNA 2 sequences and these clusters correspond to those based on antigenic relatedness (Robinson and Harrison, 1985a). A similar pattern of variation was found among four strains of PEBV, with all having similar RNA 1 sequences, but strains from different serotypes differed in their RNA 2 sequences (Robinson and Harrison, 1985b). Nevertheless, there seem to be some sequences that characterize tobravirus RNA species more generally. At least 25 bases at the 3′-end are identical in PEBV RNA 1 and RNA 2, and of these the 3′-

proximal 15 are almost identical to those of PRV (G. Hughes, K. R. Wood, and J. W. Davies, unpublished results).

F. Pseudorecombination in Nature

Although viable pseudorecombinant isolates containing one RNA species from TRV and one from PEBV cannot be prepared artificially, two naturally occurring isolates have been shown to contain RNA 1 species consisting largely or completely of sequences typical of TRV, and RNA 2 species similar in sequence to those of PEBV strains (Robinson and Harrison, 1984). These atypical isolates produce symptoms in infected plants similar to those produced by TRV, but are serologically similar to PEBV strains. Viable isolates were obtained in pseudorecombination tests when one RNA species was derived from a TRV strain and the other from one of the atypical isolates, but combinations containing one RNA species from a PEBV strain and the other from an atypical isolate were not, or in one instance scarcely, viable. These results suggest that RNA 2 of the atypical isolates, although derived from the PEBV gene pool, has become adapted to enable its replication to be supported by TRV RNA 1. These isolates provide the first evidence for the occurrence of pseudorecombination in plant viruses in nature.

G. Classification of Tobraviruses

Taken together, the patterns of variation described in the foregoing sections suggest that tobraviruses can be regarded as isolates of three distinct viruses: TRV, PEBV, and PRV. Each virus has its own distinct gene pool, in which the RNA 1 genes are strongly conserved, whereas the RNA 2 genes seem very variable. Reassortment of genes within each gene pool occurs readily, but between gene pools is rarely if ever observed in experiments. Most but not all strains of each virus are serologically distinct from those of the other two viruses, but strains of each virus are also serologically diverse, and for TRV and PEBV a number of serotypes can be recognized. Only one strain of PRV, strain CAM, has been thoroughly studied, but it seems likely that other South American tobravirus isolates (e.g., Silberschmidt, 1962/63; Kitajima *et al.*, 1969; Chagas and Silberschmidt, 1972) are also PRV. The position of the two putative pseudorecombinant isolates discussed in Section III.F is less clear, but it seems that, although a portion of their genome is derived from the PEBV gene pool, they are now part of the TRV gene pool, and therefore that these isolates are best regarded as atypical TRV strains.

IV. REPLICATION AND GENOME STRATEGY

A. Multiplication in Plants

In inoculated tobacco leaves, PRV attains its maximum concentration in about 2 days at 22°C and remains at this level for the next 4 days at least (Frost and Harrison, 1967). At lower temperatures replication is slower, virus concentration still increasing after 6 days at 14°C. At 26 or 30°C the maximum concentration is lower than at 22°C and is reached in 2 days, after which the infectivity and number of particles decrease rapidly. TRV strain PRN shows similar trends, but equivalent temperatures are 4–6°C lower, and scarcely any replication is detected at 30°C.

In some conditions, nucleoprotein particles seem quite stable in inoculated leaves, numbers of infective particles decreasing only slowly over a period of several weeks (Semancik and Kajiyama, 1967b) during which the rate of ^{32}P incorporation into the particles decreases greatly (Semancik and Kajiyama, 1967a). In contrast, infectivity in leaves of *N. clevelandii* inoculated with an NM-type isolate of TRV decreased rapidly after reaching a maximum in 4 days (Semancik and Odening, 1969).

In *N. clevelandii* or *Petunia hybrida* at 20–25°C, M-type isolates of TRV rapidly spread systemically, first appearing in young tip leaves 4–5 days after inoculation (Cadman, 1962). In contrast, NM-type isolates often move initially into leaves immediately above the inoculated ones, and spread more slowly toward the shoot tip.

B. Replication in Protoplasts

Up to 98% of tobacco mesophyll protoplasts can be infected by exposure to inocula containing 1 μg/ml PRV or TRV and 1 μg/ml poly-L-ornithine (PLO) (Kubo *et al.*, 1975). Of several buffers tried, 0.025 M phosphate (pH 6.0) gives good results (Kubo *et al.*, 1974), and preincubation of virus, buffer, and PLO, probably to allow aggregates of virus particles to form, is necessary to maximize infection. At the ID_{50} in these conditions, 30 L particles and 85 S particles were taken up per protoplast (Kubo *et al.*, 1976).

In protoplasts inoculated with PRV and kept at 22°C, infective virus RNA was first detected 7 hr after inoculation and increased rapidly up to 12 hr, after which little further net synthesis took place. Coat protein antigen and infective nucleoprotein first appeared at 9 hr. At the earliest times, L particles outnumber S, but as infection progresses the proportion of S particles increases steadily (Harrison *et al.*, 1976). The final yield, which is not attained until about 48 hr after infection, is about 2×10^5 L particles and 6×10^5 S particles per infected protoplast (Kubo *et al.*, 1975).

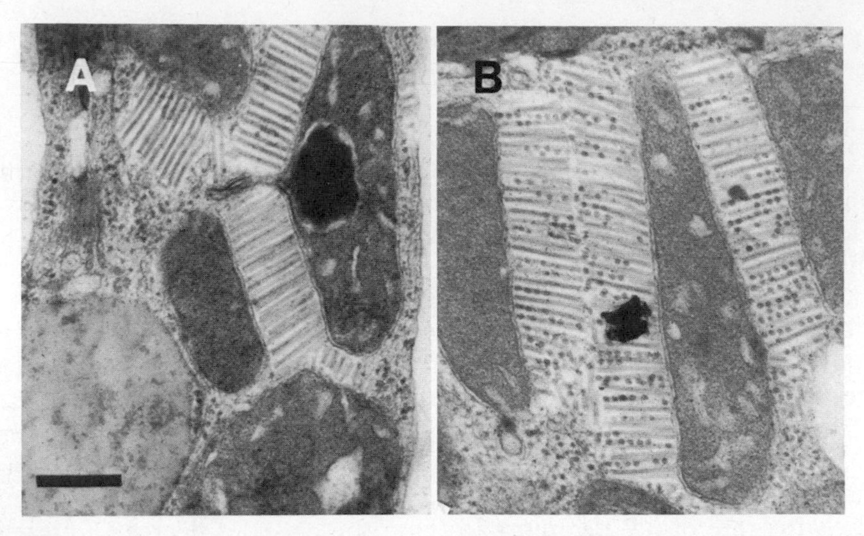

FIGURE 2. Electron micrographs showing sections of parts of protoplasts isolated from leaves of *N. benthamiana* infected with (A) PRV, (B) PRV and raspberry ringspot virus. Bar = 200 nm.

C. Interactions with Cells

In infected cells or protoplasts, L particles of PRV appear to be attached by one end to the outer membrane of mitochondria, which are often distorted (Fig. 2A; Harrison and Roberts, 1968; Kitajima and Costa, 1969; Kubo *et al.*, 1975). S particles are usually not attached to mitochondria, but are found free in the cytoplasm. In cells or protoplasts doubly infected with PRV and raspberry ringspot virus (RRV), RRV particles, which specifically form aggregates when mixed with PRV particles *in vitro* (Barker and Harrison, 1977), are found attached to PRV particles that are themselves attached to mitochondria (Fig. 2B; Harrison *et al.*, 1977).

Abnormal mitochondria, with large internal membrane-bounded vacuoles and many membranous sacs around the periphery, are also found in leaf hair cells of *N. clevelandii* infected with an NM-type isolate of PRV (Harrison *et al.*, 1970). As infection progresses, these mitochondria degenerate, aggregate, and become associated with ribosomes and other material to form increasingly large noncrystalline inclusions.

TRV particles are not associated with mitochondria in the way that PRV particles are, but sometimes form cytoplasmic aggregates. Some of these aggregates are observed in inclusions that are otherwise similar in structure to those induced by the PRV NM-type isolate but which are smaller and apparently disappear after a few days (Harrison *et al.*, 1970).

Despite these effects, there is no evidence that mitochondria are involved in virus replication. Accumulation of PRV in leaf disks of *N. clevelandii* is inhibited by cycloheximide but not chloramphenicol, suggesting that virus proteins are synthesized on cytoplasmic but not mitochondrial or chloroplast ribosomes (Harrison and Crockatt, 1971). However, the infectivity in buffered extracts of leaves infected with an NM-type isolate of TRV is more stable than would be expected for unprotected RNA, and is mostly sedimentable in 10 min at 1000*g* (Cadman, 1962). It thus seems to be associated with some subcellular structure.

Actinomycin D (25 μg/ml) or cordycepin (1 mM) added within an hour of inoculation diminished the proportion of tobacco protoplasts infected by inoculation with PRV but not with PRV RNA (Mayo and Barker, 1983). However, by 1–3 hr after inoculation, infection was resistant to inhibition, implying that new transcription of the host genome is required only for a very early step in virus replication, perhaps uncoating of the RNA in virus particles.

D. Strategy of Genome Expression

Translation of tobravirus RNA 1 in cell-free extracts from rabbit reticulocytes or wheat germ gives two products of apparent molecular weight 165,000–170,000 and 120,000–140,000 (Mayo *et al.*, 1976; Fritsch *et al.*, 1977; Pelham, 1979). Proteins of similar size were induced by PRV infection of tobacco protoplasts (Mayo, 1982). RNA 1 is not large enough for these two peptides to be coded by nonoverlapping cistrons, and the sensitivity of their relative yields to the magnesium ion concentration in translation mixtures suggested that the larger may arise by readthrough of a leaky termination codon (Pelham, 1979). Indeed, the relative yield of the 170K polypeptide from PRV RNA 1 is stimulated by addition of yeast opal, but not amber or ochre, suppressor tRNA to a reticulocyte lysate, and a similar stimulation was obtained with a tRNA fraction from tobacco (H. Beier, unpublished results).

RNA 2 of PRV yields only particle protein when translated in cell-free extracts of mouse L cells (Ball *et al.*, 1973) or of wheat germ (Mayo *et al.*, 1976). Moreover, the sequence of PRV RNA 2 (S. Bergh and A. Siegel, unpublished results) contains only one open reading frame, which codes for a protein that corresponds in size and amino acid composition to the particle protein. The sequence has long untranslatable regions of 573 residues at the 5′-end and 554 residues at the 3′-end.

RNA 2 molecules of TRV strains PRN and Lisse, and of PEBV, are also reported to be translated *in vitro* to produce particle protein (Fritsch *et al.*,1977; G. Hughes, K. R. Wood, and J. W. Davies, unpublished results), but the RNA 2 of strain SYM is not (Robinson *et al.*, 1983). Instead, a subgenomic RNA (RNA 3; M_r 600 K), which occurs in a small proportion of strain SYM particles and is derived from RNA 2, is the messenger RNA for particle protein. Strain PRN particles also contain an RNA component slightly smaller than RNA 2 (Ramirez-Baudrit, 1981;

Robinson *et al.*, 1983), and it is possible that this, rather than RNA 2 itself, is the particle protein messenger.

Another subgenomic RNA species (RNA-4; M_r 550K), which occurs in some particles of TRV strains SYM, PRN, and Lisse (Pelham, 1979; Ramirez-Baudrit, 1981; Robinson *et al.*, 1983), codes for a protein of molecular weight about 30K. For strain SYM, the 5′-end of RNA 4 has been mapped on the partial sequence of RNA 1, and RNA 4 has been shown to contain an open reading frame for a protein of 252 amino acid residues (M. Boccara and D. Baulcombe, unpublished results). For strain PRN, a similar sized polypeptide was detected as a minor product of RNA 1 translation in reticulocyte lysates, but unlike the 170K and 120K products its synthesis was resistant to inhibition by a cap analogue (Pelham, 1979). RNA 4 therefore seems to represent the 3′-portion of RNA 1. Contamination of RNA 2 preparations with this RNA species probably accounts for the 31K protein reported by Fritsch *et al.* (1977) to be produced in cell-free translation of strain PRN RNA 2.

A subgenomic RNA species of molecular weight 500K extracted from PRV-infected leaves can be translated *in vitro* to give a 30K protein (Bisaro and Siegel, 1980). Although this RNA was originally thought to be derived from RNA 2 on the basis of Northern blotting experiments, the sequence of PRV RNA 2 (S. Bergh and A. Siegel, unpublished results) does not contain an open reading frame that could code for such a protein. Thus, it seems likely that the 30K protein of PRV is a product of RNA 1, and may be analogous to the similar sized protein of TRV. *In vitro* synthesis of a PEBV-specified 30K protein has also been detected, but it is not clear what mRNA species is involved (G. Hughes, K. R. Wood, and J. W. Davies, unpublished results).

Current knowledge of the genome strategy of tobraviruses is summarized in Fig. 3.

E. Distribution of Genetic Determinants

RNA 1 can replicate in plants that do not contain RNA 2, and therefore presumably contains a cistron directing synthesis of an RNA replicase or replicase component. The fact that two temperature-sensitive mutants of PRV that have mutations in RNA 1 failed to synthesize virus RNA at the nonpermissive temperature (Robinson, 1974) supports this idea. In infections by RNA 1 alone, no particle protein is produced, suggesting that the particle protein gene is on RNA 2. This was confirmed by Sänger (1968, 1969) who showed that in pseudorecombinant isolates between serologically distinctive strains, the particles were serologically indistinguishable from those of the strain donating RNA 2.

Both RNA species contain genes affecting the types of symptom produced in infected plants. The morphology of lesions induced by Sänger's (1969) pseudorecombinants in *N. tabacum* cv. Xanthi-nc was always that

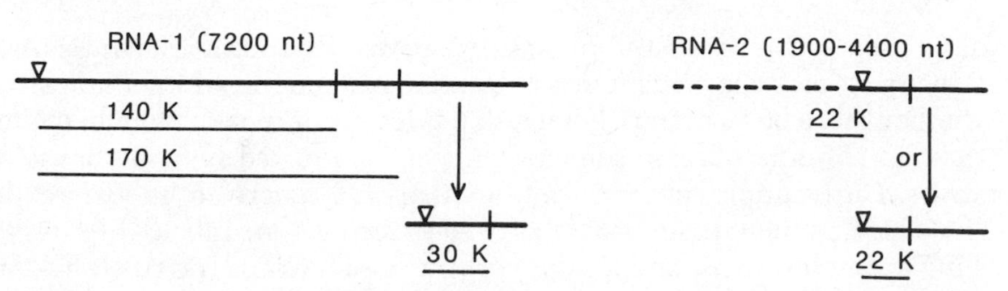

FIGURE 3. Current ideas of the strategy of genome expression of TRV. Thick lines represent RNA species with: ∇ = translation initiation site; vertical line = translation termination site. Thin lines represent polypeptide translation products. Vertical arrows indicate the derivation of subgenomic RNA species. RNA sizes are given in numbers of nucleotide residues (nt), and protein molecular weights in kilodaltons (K). The positions relative to one another of the genes on RNA 1 are known (see text), but the positions indicated for translation initiation sites are otherwise arbitrary. For RNA 2, the variation in size between strains is indicated by the broken thick line and is arbitrarily positioned; the alternative strategies indicated are apparently used by different strains. For PEBV and PRV, the available data are consistent with the strategies shown.

of the isolate that donated RNA 1. Moreover, NM-type isolates of TRV retain some of the characteristics of their M-type parents, such as the ability to invade *N. glutinosa* systemically (Cadman and Harrison, 1959), indicating that this character is controlled by RNA 1. However, in both TRV (Lister and Bracker, 1969) and PRV (Robinson, 1977), the types of systemic symptom in *N. clevelandii* infected with pseudorecombinant isolates were those characteristic of the isolate from which RNA 2 was derived. Furthermore, the ability of TRV strain SYM to invade *C. amaranticolor* systemically is controlled by a determinant in RNA 2 (Kurppa *et al.*, 1981).

In some instances, symptom type is affected by determinants in both genome parts. M- and NM-type inocula both induce lesions in *C. amaranticolor*, but with PRV (Frost *et al.*, 1967) and with some isolates of PEBV (Robinson and Harrison, 1985b), lesions that contain virus particles appear less necrotic than those that do not. Thus, RNA 1 carries a determinant for lesion production in this host, whereas a determinant in RNA 2 modifies the lesion type. Some pseudorecombinants between isolates of PEBV induced symptoms unlike those of either parent, pinpoint necrotic lesions in *Phaseolus vulgaris* and systemic mosaic in *N. clevelandii* (Robinson and Harrison, 1985b). Subtle interactions between determinants in the two parts of the genome are therefore involved in the production of some kinds of symptom.

V. NATURAL HISTORY

A. Diseases Caused

TRV has the widest known host range of any plant virus (Uschdraweit and Valentin, 1956; Noordam, 1956; Schmelzer, 1957), and infects many crop species (Table I) in addition to weeds and other wild plants. In many species the virus remains localized at the initial infection sites, whether these be in manually inoculated leaves or in roots inoculated by trichodorid nematodes. In some other species, limited systemic invasion occurs. For example, in many potato cultivars, tubers are inoculated by TRV-carrying nematodes (van Hoof, 1964), some of the inoculation points being at the lenticels, and arcs of corky tissue form in the tuber flesh at some distance from the initial infection sites (Fig. 4B). The resulting symptoms, known as spraing (= corky ringspot, kringerigheid, Pfropfenbildung), disfigure the tubers and thereby substantially decrease the value of potato crops in several parts of western Europe and North America. When planted, some (usually only a few) of the infected tubers produce foliage showing symptoms (Fig. 4A) that typically are confined to one or a few of the shoots, which are the only one(s) to be infected. In Britain, nearly all plants with these symptoms, known as stem-mottle, contain NM-type isolates, although in other countries M-type isolates may more often be found in potato foliage. Yet other species, such as the weed *Stellaria media* and bulbous ornamentals such as narcissus (Fig. 4C), hyacinth, and tulip, are readily invaded systemically. Moreover, in contrast to potato, tests on narcissus detected only M-type isolates, and infection of these ornamental species is passed efficiently through daughter bulbs and corms to progeny plants. TRV also has a few woody hosts. These include *Hydrangea arborescens* and *Ribes sanguineum*, in which yellow spots and line patterns develop (Schmelzer, 1970).

TRV is therefore associated with several types of disease. Its symptoms in naturally infected leaves range from necrosis to yellowish blotches, ringspots, and other patterns, and among underground plant parts it causes corky arcs in potato tubers and necrotic spots (malaria) in hyacinth bulbs. In several species, infection tends not to be fully systemic, and may not be passed on to all the vegetative progeny of infected mother plants.

PEBV has a narrower natural host range and is particularly adapted to leguminous species. Like TRV, it can cause sporadic systemic lesions, especially in pea (Fig. 4D); other leguminous crop plants in which natural infection is systemic include broad bean (Fig. 4E), *Phaseolus* bean, and lucerne (alfalfa) (Harrison and Robinson, 1981). The resulting diseases can be locally important but only pea early-browning is known to be at all widespread (Bos and van der Want, 1962). Like TRV, PEBV probably has weed hosts but these have not been well studied.

The crop diseases caused by PRV in artichoke and tomato are characterized by yellow leaf markings (Silberschmidt, 1962/63; Chagas and Silberschmidt, 1972). The effects on tomato are of local concern in Brazil but are not reported to be widespread. In addition, wild hosts, such as *Bidens pilosa*, have been reported.

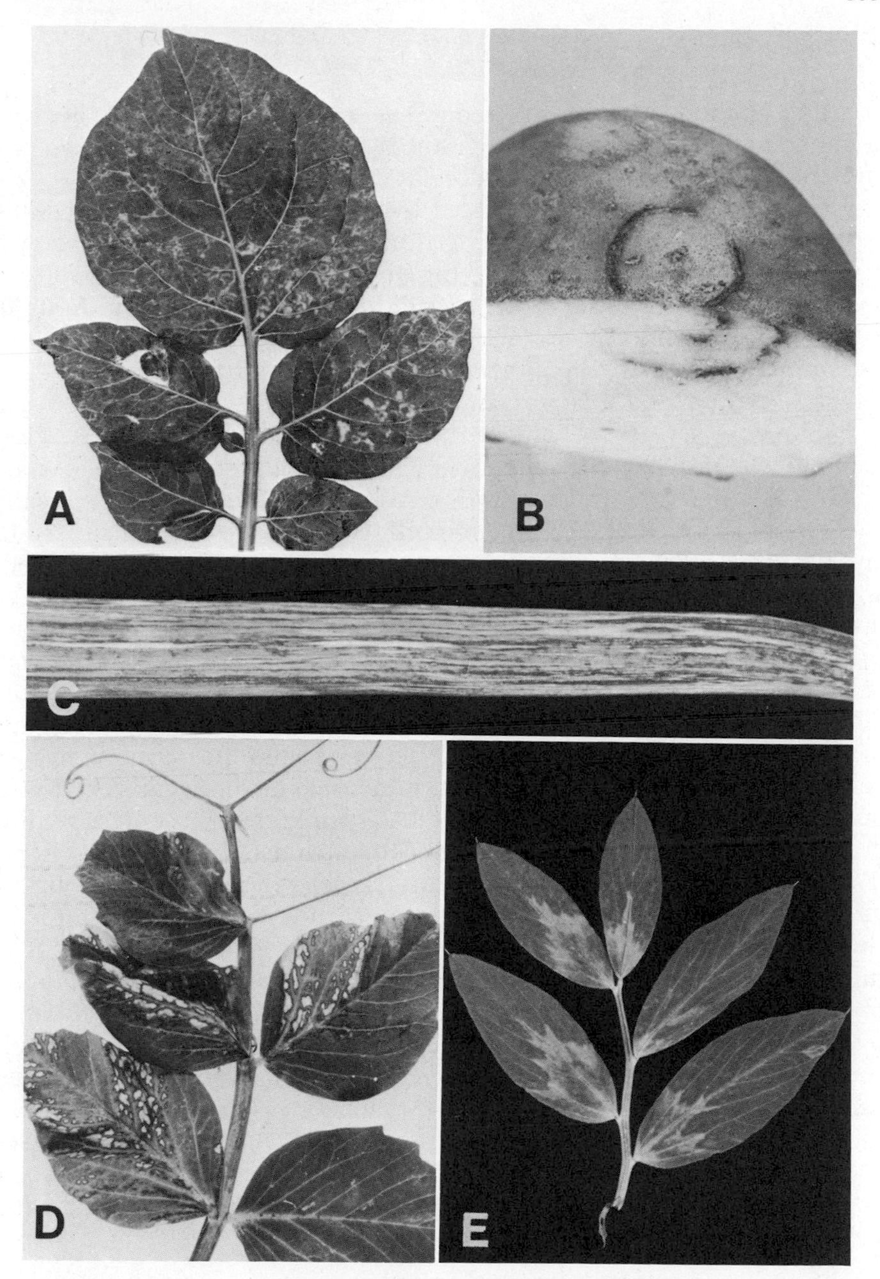

FIGURE 4. Symptoms of diseases induced by TRV (A–C) and PEBV (D, E). (A) Leaf of potato (*Solanum tuberosum* cv. Majestic) with stem-mottle; (B) tuber of potato cv. Pentland Dell showing internal arcs of corky tissue (spraing) and surface ring marking; (C) leaf of *Narcissus pseudonarcissus* with streaky mottle; (D) pea (*Pisum sativum*) leaf with necrotic sectors typical of early-browning disease; and (E) broad bean (*Vicia faba*) leaf showing yellow bands. A, from Harrison (1971); C, courtesy of W. P. Mowat; D, from Gibbs and Harrison (1976).

B. Nematode Transmission

TRV and PEBV are transmitted by several soil-inhabiting trichodorid nematodes but no vector is known for PRV despite attempts to find one (Silberschmidt, 1962/63). Vectors of TRV include several species of *Paratrichodorus* and *Trichodorus* listed by Harrison and Robinson (1978), especially *P. pachydermus* and *T. primitivus* in Europe, and *P. christiei* and *P. porosus* in North America; those of PEBV include *P. pachydermus* and *P. teres* in the Netherlands, and *P. anemones* and *T. viruliferus* in Britain. There are indications of specificity between virus strain and vector species (van Hoof, 1968), but this topic has not been studied thoroughly. The nematodes acquire and inoculate M-type virus isolates when they puncture superficial cells of the underground parts of plants. They feed especially at the root tips, where individuals of some species are known to congregate (Pitcher, 1967). Although the nematodes typically feed for only about 3 min on any one cell (Wyss, 1971), few transmissions of TRV occur when the virus acquisition and inoculation access periods are less than 1 hr, and the probability of transmission increases with increase in access period to 2 days (Ayala and Allen, 1968). TRV particles can be observed by electron microscopy to become attached to the wall of the nematode's pharynx and esophagus (Fig. 5; Taylor and Robertson, 1970). They are thought to be released in small numbers into the nematode's saliva as it passes forward and is ejected into punctured root cells. Whether the root cells that are punctured or those adjacent to the puncture are the first to become infected is not known but unless feeding is interrupted, the nematodes usually ingest most of the cell contents (Wyss, 1971), which probably renders such cells incapable of supporting virus replication. The virus can be retained in a transmissible form by the nematodes for months or years (van Hoof, 1970) but seems not to infect them. There is no evidence that the virus is passed from mother to progeny nematode (Ayala and Allen, 1968) or that it can be retained by nematodes through the molt; indeed, the lining of the esophagus is shed during molting, presumably together with any attached virus particles. There is no evidence that NM-type isolates can be acquired and inoculated by nematodes, and it seems unlikely that this happens. However, many plants infected by means of nematodes contain NM-type isolates, presumably as a result of the chance separation of long and short virus particles during transmission.

C. Seed Transmission

TRV is transmitted through seed to a small proportion of progeny plants of several weed species, and to a somewhat larger and more ecologically significant proportion of seedlings of *Viola arvensis* (Cooper and

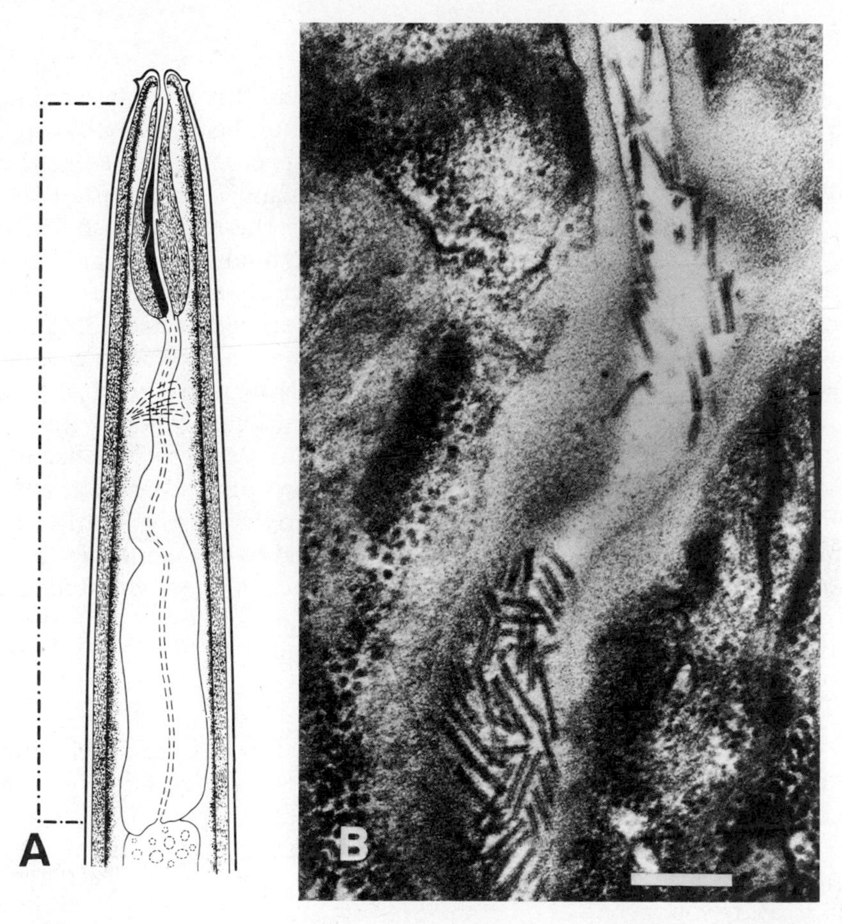

FIGURE 5. Sites of retention of TRV particles in trichodorid nematodes. (A) Diagram of head end of *Trichodorus* showing the region of the pharynx and esophagus where the virus particles are found (broken line); (B) nearly longitudinal section through the esophagus showing many TRV particles associated with the esophageal wall. Bar = 200 nm. Courtesy of W. M. Robertson.

Harrison, 1973). The infected seedlings may or may not develop symptoms. Although seed transmission in crop plants is not known to occur with TRV, it certainly occurs with PEBV in pea and probably plays a role in disseminating the virus to new sites. In pea cv. Rondo, about 10% seed transmission of PEBV is reported in the Netherlands and infected seed may be abnormally wrinkled (Bos and van der Want, 1962). However, only 1–2% transmission was found with British isolates of PEBV in several other cultivars (Harrison and Robinson, 1981). PRV too is seedborne; 15–30% seed transmission was reported for three isolates of the virus in tomato (Costa and Kitajima, 1968).

D. Ecology

The distribution of tobraviruses depends on that of their vectors. For example, in Scotland, TRV was not detected on heavy soils, whereas it was common on sandy soils or light loams of specific soil series, reflecting the prevalence of the vectors *P. pachydermus* and *T. primitivus* (Cooper, 1971). The incidence of infection depends on the numbers and activity of vector species and on the proportion of individuals that are carrying virus. On agricultural land, vector numbers are affected by previous cropping history and by cultural practices. For example, in the United States, numbers of *P. christiei* increased rapidly on land cropped with cotton and decreased when *Crotalaria spectabilis* was grown (Brodie *et al.*, 1970). Moreover, in winter, trichodorids tend to occur mainly below the top 15 cm of soil, at the soil pan where there is one, and seem to be redistributed through the top soil more effectively by deep ploughing than shallow ploughing (Cooper and Harrison, 1973). Activity of the nematodes is greatly affected by the soil water content, and to a lesser extent by soil temperature. To be able to move through the soil, the nematodes need a water film on the soil particles, and where this is lacking they are virtually immobilized. In tests in the glasshouse, TRV was rarely transmitted when the soil water content was below 15% (Cooper and Harrison, 1973; van Hoof, 1976), and in the field potato spraing disease is most severe in irrigated crops (Weingartner *et al.*, 1983). In rain-fed crops its prevalence can be related to the number of weeks in the crop's life in which substantial rainfall occurs (Cooper and Harrison,1973; Engsbro, 1973). Effects of soil temperature can also be substantial. In two British soils, TRV and PEBV were transmitted less frequently at 24°C than at 20°C (Harrison, 1966; Cooper and Harrison, 1973).

The proportion of virus-carrying individuals in populations of vector trichodorids varies greatly. It is affected, for example, by the abundance of virus sources. TRV is essentially a virus of weeds and other wild plants, that infects crop plants when they are grown on land infested with virus-carrying trichodorids. Weeds such as *Stellaria media* may have a high incidence of systemic infection and so act as important virus sources. Indeed, tests on the naturally occurring *S. media* plants at a site provide a good indication of whether or not TRV occurs there (Eibner, 1959). *S. media* and other weed plants also harbor TRV through the winter in northern Britain although presumably not in areas with more extreme climates. Finally, dormant infected seed of weeds such as *Viola arvensis* not only maintain inoculum when conditions are unfavorable for plant growth but also are a probable means of virus dissemination to other sites (Cooper and Harrison, 1973). Infected vegetative propagules of ornamentals, such as narcissus, are probably another inoculum source for previously TRV-free populations of trichodorids, and thereby disseminate the virus over large distances. Infected pea seed is thought to play a similar

TABLE IV. Effects of Nematicides on Incidence of Tobacco Rattle Virus Symptoms in Potato Tubers in Florida[a]

Oxamyl sprays (3.4 + 2.2 kg a.i./ha)	Soil treatment		
	1,3-Dichloropropene (56 l/ha)	Aldicarb (3.4 kg a.i./ha)	None
41 + 50 days postplanting	5[b]	0	7
None	43	5	53

[a] Data from Weingartner *et al.* (1983).
[b] % tubers with virus symptoms.

role in disseminating PEBV. In contrast, infected seed-potato tubers seem to play at most a minor role in distributing TRV in Europe (Engsbro, 1976) because the majority contain NM-type isolates (Harrison *et al.*, 1983).

E. Control

The main methods of controlling the spread of tobraviruses involve decreasing (1) the numbers or activity of vector nematodes, (2) the susceptibility of host plants, or (3) the availability of virus sources.

Populations of trichodorids can be decreased to low levels by treating field soils with fumigant nematicides such as 1,3-dichloropropene, and this can give good control of TRV (Cooper and Thomas, 1971). A drawback to this approach is that the treatment is never totally effective, and the few surviving trichodorids, which are mostly near the soil surface or at depths exceeding 25 cm, may build up to reach the initial population size within a few years. An alternative approach is to make soil applications of granular formulations of chemicals such as aldicarb, which become systemic in plants and prevent the nematodes from feeding but may not kill them (Steudel, 1974; Alphey *et al.*, 1975). This has the disadvantage that the nematode population can regain its original size within a year. A further variation is to spray systemic nematostatic chemicals, such as oxamyl, on the growing crop. Infection of potatoes with TRV can be controlled in this way (Alphey, 1978), although timing the spray to precede immediately a rainfall-dependent period of maximum vector activity may be difficult. In irrigated potato crops in Florida, where the infection pressure is especially great, fumigant nematicides have proved inadequate, and the best control of TRV has been achieved by combining an application of aldicarb to the soil at planting time with repeated foliar sprays of oxamyl to the growing crop (Table IV).

TRV can also be controlled by using resistant cultivars. Potato cultivars show a wide range of sensitivity and susceptibility to TRV (Harrison, 1968); there are several, such as Arran Pilot, Bintje, and Stormont Enterprise, whose tubers very rarely develop spraing although the symptoms are obvious in the few instances where infection occurs. Similarly, several pea cultivars are reported to be resistant to PEBV, both in the Netherlands (Hubbeling and Kooistra, 1963) and in England. A complication, however, is that all the cultivars that were resistant at one site in England became infected at another, presumably because of the occurrence there of a different virus strain (Harrison, 1966).

The availability of virus sources can be decreased in several ways, although such precautions can be slow to take effect. Indeed, in the short term, rigorous control of weeds by herbicides can actually increase TRV infection of potato, probably because virus-carrying trichodorids prefer weeds as food sources but on herbicide-treated land are obliged to feed on potato (Cooper and Harrison, 1973). Finally, virus dissemination can be minimized by using virus-free planting material, whether this be pea seed free from PEBV or flower bulbs free from TRV.

VI. DETECTION AND IDENTIFICATION

M-type isolates of tobraviruses differ from other sap-transmissible plant viruses in having rod-shaped particles 20–23 nm wide and of two predominant lengths, the larger measuring 180–215 nm. Electron microscopy of crude plant extracts and measurements of the sizes of particles observed are therefore an aid to tobravirus detection and identification. Identification of individual tobraviruses is not straightforward, although symptoms in manually inoculated test plants are usually a guide. PEBV is characterized by its ability to infect *Pisum sativum* and *Vicia faba* systemically, and by the larger type of local lesion induced in *Phaseolus vulgaris* (Fig. 6B). However, some strains of PEBV induce pinpoint lesions in inoculated *P. vulgaris* leaves, resembling those of PRV and TRV (Fig. 6A; Robinson and Harrison, 1985b). Although TRV and PRV also produce similar symptoms to one another in several other species, they can usually be distinguished in White Burley tobacco (Kurppa *et al.*, 1981). PRV is also serologically distinct from TRV and PEBV, and is reported only from Brazil.

Serological tests are of limited value for differentiating and identifying M-type isolates of PEBV and TRV because (1) there is great antigenic variation among isolates of each virus (Harrison and Woods, 1966; Robinson and Harrison, 1985b), and (2) some virus isolates that produce pseudorecombinants with TRV, and therefore share its gene pool, have antigenic determinants typical of PEBV (Robinson and Harrison, 1984). In contrast, nucleic acid hybridization tests with probes specific for RNA 1 have provided accurate and unequivocal identifications of all three tobraviruses (see Sections III.E and III.G).

Detection and identification of NM-type isolates present special problems because infected plants contain no virus particles, and the isolates are not readily transmitted by inoculation with sap unless nucleic

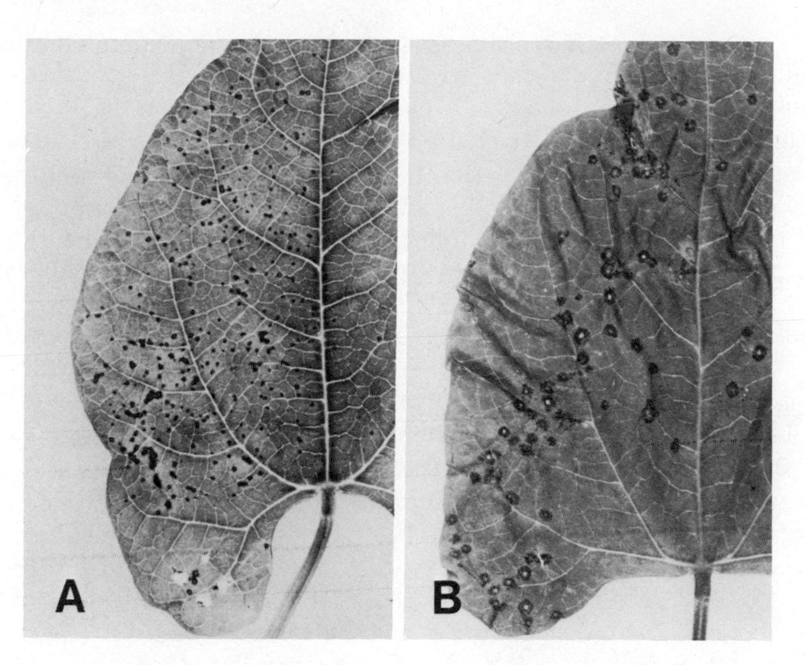

FIGURE 6. Symptoms in inoculated leaves of French bean (*Phaseolus vulgaris* cv. The Prince). (A) TRV; (B) PEBV.

acid inocula are prepared with the aid of phenol. Even then, the resulting symptoms produced in test plants, although typically more severe than those caused by M-type isolates, are not sufficiently diagnostic to be attributed safely to a specific virus; they provide circumstantial evidence only. Two approaches to identifying NM-type isolates have been adopted. One, called the genome reconstitution test (Harrison and Robinson, 1982), consists of adding S particles or RNA 2 of known identity to nucleic acid extracts from plants containing the putative NM-type isolate and inoculating test plants with the mixture. When the RNA 2 is compatible with an RNA 1 in the extract, tobravirus particles are produced. In practice, however, it is difficult to obtain S particles or RNA 2 absolutely free from L particles or RNA 1, and carefully designed control tests are always needed to assess the extent of the contamination. The second approach, which is the method of choice and has given reliable results although it would benefit from some extra sensitivity, is to do nucleic acid hybridization tests in liquid with probes for tobravirus RNA 1 (Harrison and Robinson, 1982; Harrison *et al.*, 1983). Spot hybridization tests have proved less reliable for NM-type isolates although they can be useful for M-type isolates (Robinson, 1985).

VII. AFFINITIES WITH OTHER VIRUS GROUPS

Most other plant viruses with bipartite ssRNA genomes, such as comoviruses and nepoviruses, differ from tobraviruses in having isometric particles, and RNA that has 3′-terminal polyadenylate and a 5′-terminal covalently linked protein. Among other viruses with rod-shaped particles, only soilborne wheat mosaic virus (Shirako and Brakke, 1984) and peanut clump virus (Reddy *et al.*, 1985) are known to have bipartite genomes. These viruses also resemble tobraviruses in several other ways: their genomic RNA lacks 3′-polyadenylate and apparently has no genome-linked protein needed for infectivity, their RNA 2 contains the particle protein gene, and overlapping sequences of their RNA 1 are translated *in vitro* to give two large polypeptides (Hsu and Brakke, 1983; Shirako and Brakke, 1984; Mayo and Reddy, 1985). However, they differ from tobraviruses in needing both genome segments to induce symptoms in plants (Shirako and Brakke, 1984; Reddy *et al.*, 1985) and they have fungus, not nematode, vectors (Rao and Brakke, 1969; Thouvenel and Fauquet, 1981). Soilborne wheat mosaic and peanut clump viruses were previously classified as tentative members of the tobamovirus group, but we agree with Shirako and Brakke (1984) that they now seem sufficiently distinct to warrant the establishment of the proposed furovirus group.

Definitive tobamoviruses resemble tobraviruses in particle shape but have an ssRNA genome in one piece. However, isolates of TMV that produce defective particle protein, such as strain PM2 (Siegel *et al.*, 1962), resemble NM-type isolates of tobraviruses in their instability in leaf extracts and their pattern of invasion of plants. Indeed, tobraviruses could have arisen from ancestral tobamoviruses by steps involving the cutting of the genome at a point to the 5′-side of the particle protein gene, leaving the genes for the overlapping 183K and 126K proteins, together with that for the 30K protein, in the larger genome segment. If it is true that the 183K and 126K proteins enable virus RNA replication to occur, and the 30K protein makes cell-to-cell movement possible (Leonard and Zaitlin, 1982), this RNA species would have properties very similar to those of TRV RNA 1, with the smaller segment resembling TRV RNA 2. However, if the cut was made to the 5′-side of the gene for the tobamovirus 30K intercellular movement protein, the result could be a bipartite genome organized as in the furoviruses. Such a hypothesis predicts that furovirus RNA 1 can replicate in protoplasts independently of RNA 2 but, as observed (Reddy *et al.*, 1985), cannot spread from cell to cell and induce symptoms in intact tissue. It therefore seems possible that tobraviruses and furoviruses are derived from ancestral tobamoviruses by different pathways.

An alternative scheme, in which tobraviruses would be more closely allied to bipartite ssRNA genome viruses with isometric particles than to tobamoviruses, should not be discarded prematurely. However, recent nucleotide sequence data indicate that TRV RNA 1 contains open reading frames for products which have substantial amino acid sequence homology with the 30K protein and the C-terminal region of the 183K protein of TMV and that this homology is greater than that with the corresponding proteins of alfalfa mosaic and brome mosaic viruses (M. Boccara and D. Baulcombe, unpublished results).

A feature of the tobraviruses and of peanut clump furovirus is that the particle protein in different isolates differs greatly in antigenic specificity (Robinson and Harrison, 1985a; Reddy *et al.*, 1985). Indeed, much of the nucleotide sequence of RNA 2 seems so different in some isolates of TRV as to suggest that these RNA 2 species may not be derived from the same ancestral molecule (Robinson and Harrison, 1985a). However, a possibility more in line with the hypothesis outlined above is that they are so derived but that there has been little selection pressure acting to conserve the amino acid sequence on the outer surface of the virus particle protein. In general, conservation of the particle proteins of plant viruses is particularly strong where they play key roles in determining vector transmissibility and specificity (Harrison and Murant, 1984). The variation in the tobravirus particle protein therefore suggests that it does not have these functions.

VIII. SPECIAL FEATURES

The preceding pages of this chapter have described the main characteristics of tobraviruses as a group and individually. The group displays several notable features not found in other plant virus groups. Foremost among these is the ability of RNA 1 to infect plants and induce symptoms on its own. In addition, the variation in size and nucleotide sequence of the RNA 2 of different strains of the same tobravirus is characteristic and results in a pattern of genome variation unlike any pattern yet described for other plant viruses with bipartite genomes. Whereas the RNA 1 of tobraviruses is strongly conserved in different strains and the RNA 2 is very variable, in nepoviruses both parts of the genome diverge more or less in parallel (Dodd and Robinson, 1984) and in geminiviruses the part of the genome that contains the particle protein gene is the more strongly conserved (Harrison, 1985). A consequence of this pattern of variation in tobraviruses, and of the existence of NM-type isolates, is that special methods are needed to detect and identify tobraviruses. Nucleic acid hybridization using probes for RNA 1 is the most reliable test currently available.

Tobraviruses also have wide host ranges and are the only plant viruses known to have trichodorid nematodes as vectors. Although they do not replicate in these nematodes, they are retained by them for weeks or months. They are therefore well adapted for survival in communities of wild plants (WILPAD viruses; Harrison, 1981), which are probably their main hosts. Their mode of transmission is one that allows several virus particles to be delivered to the same host cell, important for a virus whose genome parts are in different particles. In the field they persist efficiently at the sites where they and their vectors occur, and they are spread rather inefficiently to other sites, probably mainly in infected seed and vegetative planting material. They cause a few diseases of local, and in some instances very considerable, economic importance in crops. Their control depends largely on the use of resistant cultivars, where these are known and agriculturally acceptable, and on the application of expensive nematicidal and nematostatic chemicals to vector-infested land.

Among viruses with rod-shaped particles, tobraviruses, tobamoviruses, and furoviruses have related but different genome strategies and spread in different ways. However, the three groups have enough in common to suggest that they could be derived from a common ancestor.

ACKNOWLEDGMENTS. We are most grateful to the following for providing unpublished results or illustrations: D. Baulcombe and M. Boccara, H. Beier, J. W. Davies, H. Linthorst, W. P. Mowat, I. M. Roberts, W. M. Robertson, and A. Siegel.

REFERENCES

Abou Haidar, M., 1976, Tobacco rattle virus RNA–protein interactions, *Philos. Trans. R. Soc. London Ser. B* **276:**165.

Abou Haidar, M., and Hirth, L., 1977, 5′-terminal structure of tobacco rattle virus RNA: Evidence for polarity of reconstitution, *Virology* **76:**173.

Abou Haidar, M., Pfeiffer, P., Fritsch, C., and Hirth, L., 1973, Sequential reconstitution of tobacco rattle virus, *J. Gen. Virol.* **21:**83.

Alphey, T. J. W., 1978, Oxamyl sprays for the control of potato spraing disease caused by nematode-transmitted tobacco rattle virus, *Ann. Appl. Biol.* **88:**75.

Alphey, T. J. W., Cooper, J. I., and Harrison, B. D., 1975, Systemic nematicides for the control of trichodorid nematodes and of potato spraing disease caused by tobacco rattle virus, *Plant Pathol.* **24:**117.

Ayala, A., and Allen, M. W., 1968, Transmission of the California tobacco rattle virus (CTRV) by three species of the nematode genus Trichodorus, *J. Agric. Univ. P.R.* **52:**101.

Ball, L. A., Minson, A. C., and Shih, D. S., 1973, Synthesis of plant virus coat proteins in an animal cell-free system, *Nature New Biol.* **246:**206.

Barker, H., and Harrison, B. D., 1977, The interaction between raspberry ringspot and tobacco rattle viruses in doubly infected protoplasts, *J. Gen. Virol.* **35:**135.

Behrens, J., 1899, Die Mauche (Mauke) des Tabaks, *Landwirtsch. Vers. Stn.* **52:**442.

Bisaro, D. M., and Siegel, A., 1980, A new viral RNA species in tobacco rattle virus-infected tissue, *Virology* **107:**194.

Böning, K., 1931, Zur Ätiologie der Streifen- und Kräuselkrankheit des Tabaks, *Z. Parasitenkd.* **3:**103.

Bos, L., and van der Want, J. P. H., 1962, Early browning of pea, a disease caused by a soil- and seed-borne virus, *Tijdschr. Plantenziekten* **68:**368.

Brodie, B. B., Good, J. M., and Jaworski, C. A., 1970, Population dynamics of plant nematodes in cultivated soil: Effect of summer cover crops on newly cleared land, *J. Nematol.* **2:**217.

Cadman, C. H., 1962, Evidence for association of tobacco rattle virus nucleic acid with a cell component, *Nature (London)* **193:**49.

Cadman, C. H., and Harrison, B. D., 1959, Studies on the properties of soil-borne viruses of the tobacco-rattle type occurring in Scotland, *Ann. Appl. Biol.* **47:**542.

Chagas, C. M., and Silberschmidt, K. M., 1972, "Virus da faixa amarela da alcachôfra": Ocorrência, toansmissão mecânica e propriedades físicas, *Biologico* **38:**35.

Cooper, J. I., 1971, The distribution in Scotland of tobacco rattle virus and its nematode vectors in relation to soil type, *Plant Pathol.* **20:**51.
Cooper, J. I., and Harrison, B. D., 1973, The role of weed hosts and the distribution and activity of vector nematodes in the ecology of tobacco rattle virus, *Ann. Appl. Biol.* **73:**53.
Cooper, J. I., and Mayo, M. A., 1972, Some properties of the particles of three tobravirus isolates, *J. Gen. Virol.* **16:**285.
Cooper, J. I., and Thomas, P. R., 1971, Chemical treatment of soil to prevent transmission of tobacco rattle virus to potatoes by *Trichodorus* spp., *Ann. Appl. Biol.* **69:**23.
Costa, A. S., and Kitajima, E. W., 1968, Transmissão do vírus do anel do pimentão através da semente do tomate, *Rev. Soc. Bras. Fitopatol.* **2:**25.
Darby, G., and Minson, A. C., 1973, The structure of tobacco rattle virus ribonucleic acids: Common nucleotide sequences in the RNA species, *J. Gen. Virol.* **21:**285.
Dodd, S. M., and Robinson, D. J., 1984, Nucleotide sequence homologies among RNA species of strains of tomato black ring virus and other nepoviruses, *J. Gen. Virol.* **65:**1731.
Durham, A. C. H., and Abou Haidar, M., 1977, Cation binding by tobacco rattle virus, *Virology* **77:**520.
Eibner, R., 1959, Untersuchungen über die "Eisenfleckigkeit" der Kartoffel, Doctoral dissertation, Justus-Liebig-Universität, Giessen, West Germany.
Engsbro, B., 1973, Undersogelser og forsog vedrorende jordbarne vira. I. Rattlevirus fortsatte undersogelser i kartofler, *Tidsskr. Planteavl* **77:**103.
Engsbro, B., 1976, Soil-borne viruses. I. Rattle virus (continued investigations 2), *Tidsskr. Planteavl* **80:**405.
Finch, J. T., 1965, Preliminary X-ray diffraction studies on tobacco rattle and barley stripe mosaic viruses, *J. Mol. Biol.* **12:**612.
Finch, J. T., 1972, The hand of the helix of tobacco mosaic virus, *J. Mol. Biol.* **66:**291.
Fritsch, C., Witz, J., Abou Haidar, M., and Hirth, L., 1973, Polymerization of tobacco rattle virus protein, *FEBS Lett.* **29:**211.
Fritsch, C., Mayo, M. A., and Hirth, L., 1977, Further studies on the translation products of tobacco rattle virus RNA *in vitro*, *Virology* **77:**722.
Frost, R. R., and Harrison, B. D., 1967, Comparative effects of temperature on the multiplication in tobacco leaves of two tobacco rattle viruses, *J. Gen. Virol.* **1:**455.
Frost, R. R., Harrison, B. D., and Woods, R. D., 1967, Apparent symbiotic interaction between particles of tobacco rattle virus, *J. Gen. Virol.* **1:**57.
Ghabrial, S. A., and Lister, R. M., 1973a, Anomalies in molecular weight determinations of tobacco rattle virus protein by SDS-polyacrylamide gel electrophoresis, *Virology* **51:**485.
Ghabrial, S. A., and Lister, R. M., 1973b, Coat protein and symptom specification in tobacco rattle virus, *Virology* **52:**1.
Gibbs, A. J., and Harrison, B. D., 1964, A form of pea early-browning virus found in Britain, *Ann. Appl. Biol.* **54:**1.
Gibbs, A., and Harrison, B., 1976, *Plant Virology: The principles*, Arnold, London.
Gugerli, P., 1976, Different states of aggregation of tobacco rattle virus coat protein, *J. Gen. Virol.* **33:**297.
Harrison, B. D., 1966, Further studies on a British form of pea early-browning virus, *Ann. Appl. Biol.* **57:**121.
Harrison, B. D., 1968, Reactions of some old and new British potato cultivars to tobacco rattle virus, *Eur. Potato J.* **11:**165.
Harrison, B. D., 1971, Potato viruses in Britain, in: *Diseases of Crop Plants* (J. H. Western, ed.), pp. 123–159, Macmillan & Co., London.
Harrison, B. D., 1981, Plant virus ecology: Ingredients, interactions and environmental influences, *Ann. Appl. Biol.* **99:**195.
Harrison, B. D., 1985, Advances in geminivirus research, *Annu. Rev. Phytopathol.* **23:**55.
Harrison, B. D., and Crockatt, A. A., 1971, Effects of cycloheximide on the accumulation of tobacco rattle virus in leaf discs of *Nicotiana clevelandii*, *J. Gen. Virol.* **12:**183.

Harrison, B. D., and Murant, A. F., 1984, Involvement of virus-coded proteins in transmission of plant viruses by vectors, in: *Vectors in Virus Biology* (M. A. Mayo and K. A. Harrap, eds.), pp. 1–36, Academic Press, New York.
Harrison, B. D., and Nixon, H. L., 1959a, Separation and properties of particles of tobacco rattle virus with different lengths, *J. Gen. Microbiol.* **21**:569.
Harrison, B. D., and Nixon, H. L., 1959b, Some properties of infective preparations made by disrupting tobacco rattle virus with phenol, *J. Gen. Microbiol.* **21**:591.
Harrison, B. D., and Roberts, I. M., 1968, Association of tobacco rattle virus with mitochondria, *J. Gen. Virol.* **3**:121.
Harrison, B. D., and Robinson, D. J., 1978, The tobraviruses, *Adv. Virus Res.* **23**:25.
Harrison, B. D., and Robinson, D. J., 1981, Tobraviruses, in: *Handbook of Plant Virus Infections and Comparative Diagnosis* (E. Kurstak, ed.), pp. 515–540, Elsevier/North-Holland, Amsterdam.
Harrison, B. D., and Robinson, D. J., 1982, Genome reconstitution and nucleic acid hybridization as methods of identifying particle-deficient isolates of tobacco rattle virus in potato plants with stem-mottle disease, *J. Virol. Methods* **5**:255.
Harrison, B. D., and Woods, R. D., 1966, Serotypes and particle dimensions of tobacco rattle viruses from Europe and America, *Virology* **28**:610.
Harrison, B. D., Stefanac, Z., and Roberts, I. M., 1970, Role of mitochondria in the formation of X-bodies in cells of *Nicotiana clevelandii* infected by tobacco rattle virus, *J. Gen. Virol.* **6**:127.
Harrison, B. D., Finch, J. T., Gibbs, A. J., Hollings, M., Shepherd, R. J., Valenta, V., and Wetter, C., 1971, Sixteen groups of plant viruses, *Virology* **45**:356.
Harrison, B. D., Kubo, S., Robinson, D. J., and Hutcheson, A. M., 1976, The multiplication cycle of tobacco rattle virus in tobacco mesophyll protoplasts, *J. Gen. Virol.* **33**:237.
Harrison, B. D., Hutcheson, A. M., and Barker, H., 1977, Association between the particles of raspberry ringspot and tobacco rattle viruses in doubly infected *Nicotiana benthamiana* cells and protoplasts, *J. Gen. Virol.* **36**:535.
Harrison, B. D., Robinson, D. J., Mowat, W. P., and Duncan, G. H., 1983, Comparison of nucleic acid hybridisation and other tests for detecting tobacco rattle virus in narcissus plants and potato tubers, *Ann. Appl. Biol.* **102**:331.
Horváth, J., 1978, New artifical hosts and non-hosts of plant viruses and their role in the identification and separation of viruses. III. Tobravirus group: Tobacco rattle virus, *Acta Phytopathol. Acad. Sci. Hung.* **13**:51.
Hsu, Y. H., and Brakke, M. K., 1983, Translation of soil-borne wheat mosaic virus RNAs in rabbit reticulocyte lysate, *Phytopathology* **73**:790.
Hubbeling, N., and Kooistra, E., 1963, Resistentie tegen vroege verbruining bij erwten, *Zaadbelangen* **17**:256.
Kitajima, E. W., and Costa, A. S., 1969, Association of pepper ringspot virus (Brazilian tobacco rattle virus) and host cell mitochondria, *J. Gen. Virol.* **4**:177.
Kitajima, E. W., Oliveira, A. R., and Costa, A. S., 1969, Morfologia das partículas do vírus do anel do pimentão, *Bragantia* **28**:1.
Kubo, S., Harrison, B. D., and Robinson, D. J., 1974, Effect of phosphate on the infection of tobacco protoplasts by tobacco rattle virus, *Intervirology* **3**:382.
Kubo, S., Harrison, B. D., Robinson, D. J., and Mayo, M. A., 1975, Tobacco rattle virus in tobacco mesophyll protoplasts: Infection and virus multiplication, *J. Gen. Virol.* **27**:293.
Kubo, S., Robinson, D. J., Harrison, B. D., and Hutcheson, A. M., 1976, Uptake of tobacco rattle virus by tobacco protoplasts, and the effect of phosphate on infection, *J. Gen. Virol.* **30**:287.
Kurppa, A., Jones, A. T., Harrison, B. D., and Bailiss, K. W., 1981, Properties of spinach yellow mottle, a distinctive strain of tobacco rattle virus, *Ann. Appl. Biol.* **98**:243.
Leonard, D. A., and Zaitlin, M., 1982, A temperature-sensitive strain of tobacco mosaic virus defective in cell-to-cell movement generates an altered viral-coded protein, *Virology* **117**:416.
Linthorst, H. J. M., and Bol, J. F., 1986, cDNA hybridisation as a means of detection of

tobacco rattle virus in potato and tulip, in: *Developments and Applications in Virus Testing* (R. A. C. Jones and L. Torrance, eds.), Association of Applied Biologists, Wellesbourne (in press).
Lister, R. M., 1968, Functional relationships between virus-specific products of infection by viruses of the tobacco rattle type, *J. Gen. Virol.* **2**:43.
Lister, R. M., 1969, Tobacco rattle, NETU, viruses in relation to functional heterogeneity in plant viruses, *Fed. Proc.* **28**:1875.
Lister, R. M., and Bracker, C. E., 1969, Defectiveness and dependence in three related strains of tobacco rattle virus, *Virology* **37**:262.
Maat, D. Z., 1963, Pea early-browning virus and tobacco rattle virus—Two different, but serologically related viruses, *Neth. J. Plant Pathol.* **69**:287.
Mayo, M. A., 1982, Polypeptides induced by tobacco rattle virus during multiplication in tobacco protoplasts, *Intervirology* **17**:240.
Mayo, M. A., and Barker, H., 1983, Effects of actinomycin D on the infection of tobacco protoplasts by four viruses, *J. Gen. Virol.* **64**:1775.
Mayo, M. A., and Daney de Marcillac, G., 1977, Analytical centrifugation of the disk aggregates of tobacco rattle virus protein, *Virology* **76**:560.
Mayo, M. A., and Reddy, D. V. R., 1985, Translation products of RNA from Indian peanut clump virus, *J. Gen. Virol.* **66**:1347.
Mayo, M. A., and Robinson, D. J., 1975, Revision of estimates of the molecular weights of tobravirus coat proteins, *Intervirology* **5**:313.
Mayo, M. A., Fritsch, C., and Hirth, L., 1976, Translation of tobacco rattle virus RNA *in vitro* using wheat germ extracts, *Virology* **69**:408.
Miki, T., and Okada, Y., 1970, Comparative studies on some strains of tobacco rattle virus, *Virology* **42**:993.
Minson, A. C., and Darby, G., 1973, A study of sequence homology between tobacco rattle virus ribonucleic acids, *J. Gen. Virol.* **19**:253.
Morris, T. J., and Semancik, J. S., 1973, *In vitro* protein polymerization and nucleoprotein reconstitution of tobacco rattle virus, *Virology* **53**:215.
Nixon, H. L., and Harrison, B. D., 1959, Electron microscopic evidence on the structure of the particles of tobacco rattle virus, *J. Gen. Microbiol.* **21**:582.
Noordam, D., 1956, Waardplanten en toetsplanten van het ratelvirus van de tabak, *Tijdschr. Plantenziekten* **62**:219.
Offord, R. E., 1966a, Electron microscopic observations on the substructure of tobacco rattle virus, *J. Mol. Biol.* **17**:370.
Offord, R. E., 1966b, Studies on plant viruses, Ph.D. thesis, University of Cambridge.
Offord, R. E., and Harris, J. I., 1965, The protein sub-unit of tobacco rattle virus, *Proceedings of the 2nd FEBS Meeting*, pp. 216–217.
Pelham, H. R. B., 1979, Translation of tobacco rattle virus RNAs *in vitro*: Four proteins from three RNAs, *Virology* **97**:256.
Pitcher, R. S., 1967, The host–parasite relations and ecology of *Trichodorus viruliferus* on apple roots, as observed from an underground laboratory, *Nematologica* **13**:547.
Quanjer, H. M., 1943, Bijdrage tot de kennis van de in Nederland voorkomende ziekten van tabak en van de tabaksteelt op kleigrond, *Tijdschr. Plantenziekten* **49**:37.
Ramirez-Baudrit, M. P., 1981, Contribution à l'étude du génome du "tobacco rattle virus," Thèse de 3ème cycle, Université Louis Pasteur de Strasbourg.
Rao, A. S., and Brakke, M. K., 1969, Relation of soil-borne wheat mosaic virus and its fungal vector, *Polymyxa graminis, Phytopathology* **59**:581.
Reddy, D. V. R., Robinson, D. J., Roberts, I. M., and Harrison, B. D., 1985, Genome properties and relationships of Indian peanut clump virus, *J. Gen. Virol.* **66**:2011.
Reijnders, L., Aalbers, A. M. J., van Kammen, A., and Thuring, R. W. J., 1974, Molecular weights of plant viral RNAs determined by gel electrophoresis under denaturing conditions, *Virology* **60**:515.
Roberts, I. M., and Mayo, M. A., 1980, Electron microscope studies of the structure of the disk aggregate of tobacco rattle virus protein, *J. Ultrastruct. Res.* **71**:49.

Robinson, D. J., 1974, Early events in local infection of *Chenopodium amaranticolor* leaves by mutant and wild-type strains of tobacco rattle virus, *J. Gen. Virol.* **24**:391.
Robinson, D. J., 1977, A variant of tobacco rattle virus: Evidence for a second gene in RNA-2, *J. Gen. Virol.* **35**:37.
Robinson, D. J., 1983, RNA species of tobacco rattle virus strains and their nucleotide sequence relationships, *J. Gen. Virol.* **64**:657.
Robinson, D. J., 1985, Detection of tobacco rattle virus (TRV) infections by spot hybridization, Annual Report, Scottish Crop Research Institute, 1984, p. 187.
Robinson, D., and Harrison, B., 1984, The biochemical and genetic bases of variation between tobravirus isolates, *Abstracts, 6th International Congress of Virology, Sendai,* p. 207.
Robinson, D. J., and Harrison, B. D., 1985a, Unequal variation in the two genome parts of tobraviruses and evidence for the existence of three separate viruses, *J. Gen. Virol.* **66**:171.
Robinson, D. J., and Harrison, B. D., 1985b, Evidence that broad bean yellow band virus is a new serotype of pea early-browning virus, *J. Gen. Virol.* **66**:2003.
Robinson, D. J., and Raschké, J. H., 1977, Inactivation of tobacco rattle virus by EDTA, and the role of divalent metal ions in the stability of the virus, *J. Gen. Virol.* **34**:547.
Robinson, D. J., Mayo, M. A., Fritsch, C., Jones, A. T., and Raschké, J. H., 1983, Origin and messenger activity of two small RNA species found in particles of tobacco rattle virus strain SYM, *J. Gen. Virol.* **64**:1591.
Russo, M., Gallitelli, D., Vovlas, C., and Savino, V., 1984, Properties of broad bean yellow band virus, a possible new tobravirus, *Ann. Appl. Biol.* **105**:223.
Sänger, H. L., 1968, Characteristics of tobacco rattle virus. I. Evidence that its two particles are functionally defective and mutually complementing, *Mol. Gen. Genet.* **101**:346.
Sänger, H. L., 1969, Functions of the two particles of tobacco rattle virus, *J. Virol.* **3**:304.
Schmelzer, K., 1957, Untersuchungen über den Wirtspflanzenkreis des Tabakmauche-Virus, *Phytopathol. Z.* **30**:281.
Schmelzer, K., 1970, Untersuchungen an Viren der Zier- und Wildgehölze. 7. Mitteilung:Weitere Befunde an *Buddleja, Viburnum, Caryopteris* und *Philadelphus* sowie Viren an *Leycesteria, Chionanthus, Ribes, Hydrangea, Syringa, Spiraea* und *Catalpa, Phytopathol. Z.* **67**:285.
Semancik, J. S., 1966, Purification and properties of two isolates of tobacco rattle virus from pepper in California, *Phytopathology* **56**:1190.
Semancik, J. S., 1970, Identity of structural protein from two isolates of TRV with different length of associated short particles, *Virology* **40**:618.
Semancik, J. S., and Kajiyama, M. R., 1967a, Properties and relationships among RNA species from tobacco rattle virus, *Virology* **33**:523.
Semancik, J. S., and Kajiyama, M. R., 1967b, Comparative studies on two strains of tobacco rattle virus, *J. Gen. Virol.* **1**:153.
Semancik, J. S., and Odening, L. A., 1969, Growth curve of unstable form of TRV infection and conversion of unstable to stable form infection, *Virology* **39**:613.
Shirako, Y., and Brakke, M. K., 1984, Two purified RNAs of soil-borne wheat mosaic virus are needed for infection, *J. Gen. Virol.* **65**:119.
Siegel, A., Zaitlin, M., and Sehgal, O. P., 1962, The isolation of defective tobacco mosaic virus strains, *Proc. Natl. Acad. Sci. USA* **48**:1845.
Silberschmidt, K., 1962/63, Studies on a Brazilian tomato ringspot disease, *Phytopathol. Z.* **46**:209.
Steudel, W., 1974, Versuchsergebnisse zur Wirkung einiger systemischer Nematizide auf das Auftreten der durch das Tabak-Rattle-Virus verursachten Stippigkeit der Kartoffel, *Nachrichtenbl. Dtsch. Pflanzenschutzdienstes (Braunschweig)* **26**:165.
Taylor, C. E., and Robertson, W. M., 1970, Location of tobacco rattle virus in the nematode vector, *Trichodorus pachydermus* Seinhorst, *J. Gen. Virol.* **6**:179.
Thouvenel, J. C., and Fauquet, C., 1981, Further properties of peanut clump virus and studies on its natural transmission, *Ann. Appl. Biol.* **97**:99.

Tollin, P., and Wilson, H. R., 1971, Some observations on the structure of the Campinas strain of tobacco rattle virus, *J. Gen. Virol.* **13:**433.

Uschdraweit, H. A., and Valentin, H., 1956, Das Tabakmauchevirus an Zierpflanzen, *Nachrichtenbl. Dtsch. Pflanzenschutzdienstes (Braunschweig)* **8:**132.

van Hoof, H. A., 1964, Het tijdstip van infectie en veranderingen in de concentratie van ratelvirus (kringerigheid) in de aardappelknol, *Meded. Landbouwhogesch. Opzoekingsstn. Staat Gent* **29:**944.

van Hoof, H. A., 1968, Transmission of tobacco rattle by *Trichodorus* species, *Nematologica* **14:**20.

van Hoof, H. A., 1969, Enige eigenschappen van Nederlandse isolaten van het vroege-verbruiningsvirus van de erwt, *Meded. Rijksfac. Landbouwwet. Gent* **34:**888.

van Hoof, H. A., 1970, Some observations on retention of tobacco rattle virus in nematodes, *Neth. J. Plant Pathol.* **76:**329.

van Hoof, H. A., 1976, The effect of soil moisture content on the activity of trichodorid nematodes, *Nematologica* **22:**260.

van Hoof, H. A., Maat, D. Z., and Seinhorst, J. W., 1966, Viruses of the tobacco rattle virus group in northern Italy: Their vectors and serological relationships, *Neth. J. Plant Pathol.* **72:**253.

Weingartner, D. P., Shumaker, J. R., and Smart, G. C., 1983, Why soil fumigation fails to control potato corky ringspot disease in Florida, *Plant Dis.* **67:**130.

Wyss, U., 1971, Der Mechanismus der Nahrungsaufnahme bei *Trichodorus similis*, *Nematologica* **17:**508.

PART IV

HORDEIVIRUSES

CHAPTER 18

HORDEIVIRUSES

Biology and Pathology

T. W. CARROLL

I. INTRODUCTION

Three viruses are presently classified as members of the hordeivirus group (Matthews, 1982): barley stripe mosaic virus (BSMV), Lychnis ringspot virus (LRSV), and poa semilatent virus (PSLV). Their particles are rod- or tubule-shaped and lack envelopes (Francki, *et al.*, 1985). Particle diameters range from about 19 to 25 nm and lengths extend from about 100 to over 165 nm. The genome of each virus consists of a single-stranded (ss) RNA and is believed to be tripartite. All three viruses are transmissible by mechanical inoculation. BSMV and LRSV are also transmitted through seed and pollen. LRSV and PSLV are distantly related to BSMV serologically. The type member of the hordeivirus group is BSMV which was first described in 1951 as barley false stripe mosaic virus (McKinney). Anthoxanthum latent bleaching virus (ALBV) is considered to be a possible member of this group (Catherall and Chamberlain, 1980). The name *hordei* was derived from the Latin *hordeum*, meaning barley. A thorough review of the hordeiviruses is given by Jackson and Lane (1981).

II. BARLEY STRIPE MOSAIC VIRUS

A. Main Characteristics

BSMV consists of different variants or strains (McKinney and Greeley, 1965). Differentiation of these strains is based upon host range and

T. W. CARROLL • Department of Plant Pathology, Montana State University, Bozeman, Montana 59717.

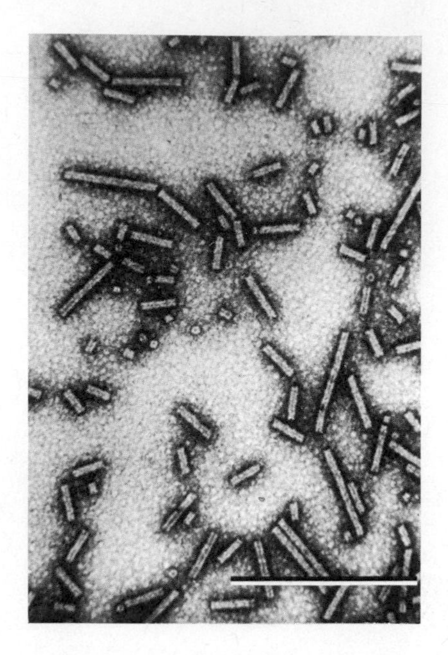

FIGURE 1. Electron micrograph of an isolate or strain of barley stripe mosaic virus showing rod-shaped particles of variable lengths. From Carroll (1980).

symptomatology. The rod- or tubule-shaped virus particles of each strain are about 20 nm in diameter and 100–150 nm long (Fig. 1). They have a helical symmetry with a pitch of about 2.5 nm (Matthews, 1982). The weight of the long particles is about 26×10^6 dalton (Atabekov and Novikov, 1971). Viral coat protein, which is glycosylated in some strains, is a single polypeptide of molecular weight of about 21×10^3. For several strains, three genomic (g) RNAs are necessary for maximum infection. Each of these g RNAs is a distinct molecule of linear positive-sense RNA. According to Gustafson *et al.* (1982), the α, β, and γ g RNA of the type, ND18, and Norwich strains have molecular weights of 1.43, 1.24, and 1.1×10^6, respectively. A subgenomic (sg) RNA, molecular weight 2.8×10^5, is also associated with several strains of BSMV (Jackson *et al.*, 1983). These sg RNAs have sequence homologies with γ and β g RNA depending on the virus strain. Whereas the 3′-end of the α, β, and γ g RNA of two virus strains consists of a conserved icosanucleotide sequence terminating with CCA_{OH}, the 3′-end of the sg RNAs is polyadenylated (Stanley *et al.*, 1984).

B. Geographical Distribution

Initially, BSMV was reported to occur in North America (Afanasiev, 1956; Timian and Sisler, 1955; Hagborg, 1954; McKinney, 1953) and Japan (Inouye, 1962). Subsequently, it was discovered in the British Isles, several European countries, and Israel. In 1965, McKinney and Greeley wrote that BSMV was worldwide in distribution. Since then, the occurrence of the virus has been further noted in Denmark (Sandfaer and Haahr, 1975), Korea (La and Park, 1979), and China (Yu *et al.*, 1980). In addition, the prevalence of BSMV in California (Slack and Shepherd, 1975), and the distribution of the virus in North Dakota (Timian, 1971) and in Canada (Chiko, 1971a,b, 1973, 1974, 1976, 1978, 1980) have been studied in some detail.

C. Host Range and Symptomatology

In nature, only three hosts have been shown to be important for BSMV. All are members of the Gramineae. Barley (*Hordeum vulgare*) is the principal host (Slack and Shepherd, 1975; Timian, 1971; McKinney and Greeley, 1965; Afanasiev, 1956; Hagborg, 1954) of the virus, but wheat (*Triticum aestivum*) is occasionally infected (Yu *et al.*, 1980; La and Park, 1979; McKinney and Greeley, 1965; Nitzany and Kenneth, 1960; Walters, 1954). Certain strains of BSMV have also been detected by Chiko (1975) in wild oats (*Avena fatua*).

Symptom expression in BSMV-infected plants depends on the virus strain, host cultivar or species, and environmental interaction (McKinney and Greeley, 1965). Warm temperatures (25–30°C) and daylight were shown to be optimal for development of symptoms in barley grown from infected seed (McKinney, 1954). Color deviations and anatomical abnormalities, including necrosis, are the main visual symptoms of BSMV-infected plants.

About 237 members of the Gramineae have been shown to be experimental hosts of various strains of BSMV (Jackson and Lane, 1981). After mechanical inoculation with the virus, most hosts developed systemic stripe mosaic symptoms. A few infected host members, however, did not express macroscopic symptoms. Many of the hosts are members of four tribes of Gramineae. The number of hosts reported for each tribe is as follows: Hordeae (89), Festuceae (65), Avenae (25), and Agrostideae (22). Interestingly enough, BSMV, unlike PSLV, does not infect *Agropyron trachycaulum* or *Poa palustris* (Polak and Slykhuis, 1972).

Some 13 dicotyledonous plants have also been shown to be experimental hosts of certain strains of BSMV (Jackson and Lane, 1981). Nine of them are members of the Chenopodiaceae: Eight are *Chenopodium* spp. that form yellow or necrotic local lesions in response to mechanical inoculation with BSMV; the ninth is *Spinacia oleracea* which, when mechanically inoculated, may or may not manifest systemic mosaic symptoms. In addition, the Solanaceae, Amaranthaceae, and Primulaceae are reported to contain one host each of BSMV (Jackson and Lane, 1981).

About 10 diagnostic hosts are commonly used for some strains of BSMV according to Atabekov and Novikov (1971). Barley (*Hordeum vulgare*), wheat (*Triticum aestivum*), and oat (*Avena sativa*) usually develop systemic stripe mosaics following inoculation by mechanical means.

Chenopodium amaranticolor, C. quinoa, C. album, and beet (*Beta vulgaris*) form yellow local lesions when inoculated mechanically. Some virus strains cause stripe mosaic symptoms in maize (*Zea mays*), systemic mosaic in spinach (*Spinacia oleracea*), and yellow local lesions in tobacco (*Nicotiana tabacum* cv. Samsun). For propagation purposes, barley (*H. vulgare*) and wheat (*T. aestivum*) are the best hosts for producing virus for purification. *C. amaranticolor* and *C. quinoa* have been used as assay hosts for certain strains of BSMV.

McKinney and Greeley (1965) noted acute and chronic phases of symptom development in barley mechanically inoculated with BSMV. Chlorotic and necrotic stripe, streak, and spot symptoms of the primary acute phase, and necrotic gray lesion, scald, and stripe symptoms of the secondary acute phase appeared on inoculated leaves. Chronic symptoms of blade chlorosis and chlorotic and necrotic stripes and chevrons were observed on noninoculated, systemically infected leaves. Nonfoliar symptoms also evident during the chronic disease phase include plant dwarfing (Inouye, 1962; Hagborg, 1954), anther and pollen sterility (Carroll and Mayhew, 1976a; Slack *et al.*, 1975; Inouye, 1962), ovule sterility (Slack *et al.*, 1975), seed-set reduction (Slack *et al.*, 1975), seed shriveling (Inouye, 1962; Hagborg, 1954), increased frequency of triploid and aneuploid seeds (Sandfaer, 1973), reduced seed germinability (Slack *et al.*, 1975), and general unthriftiness of plants (Hagborg, 1954).

D. Strains

Biological variants of BSMV have been designated as different strains of the virus (McKinney and Greeley, 1965), depending on their host range and symptomatology in barley, wheat, and oats. They were named the latent, mild, moderate, severe, yellow-leaf, white-leaf, dwarf, and necrotic strains. The mild and moderate strains were the most prevalent. The kind and extent of chlorosis that appeared in systemically infected leaf blades during the chronic phase of disease in barley (*H. vulgare* cv. Atsel) or oats (*A. sativa* cv. Statesville) were used routinely to differentiate the mild, moderate, severe, yellow-leaf, and white-leaf strains.

Different combinations of indigenous virus strains, barley cultivars, and environmental factors in Montana (Carroll, 1980) produced symptom variations such as striping, streaking, chlorosis, and necrosis in systemically infected leaf blades during the chronic phase of disease (Fig. 2).

While most strains of BSMV maintain their infectivity in culture for many transfers via mechanical inoculation, some become completely noninfectious after five or more transfers (McKinney and Greeley, 1965).

The phenomenon of cross-protection has been demonstrated for some strains of BSMV (McKinney and Greeley, 1965). Infection by a second or challenging strain was not shown when plants infected by a first or established strain were mechanically inoculated.

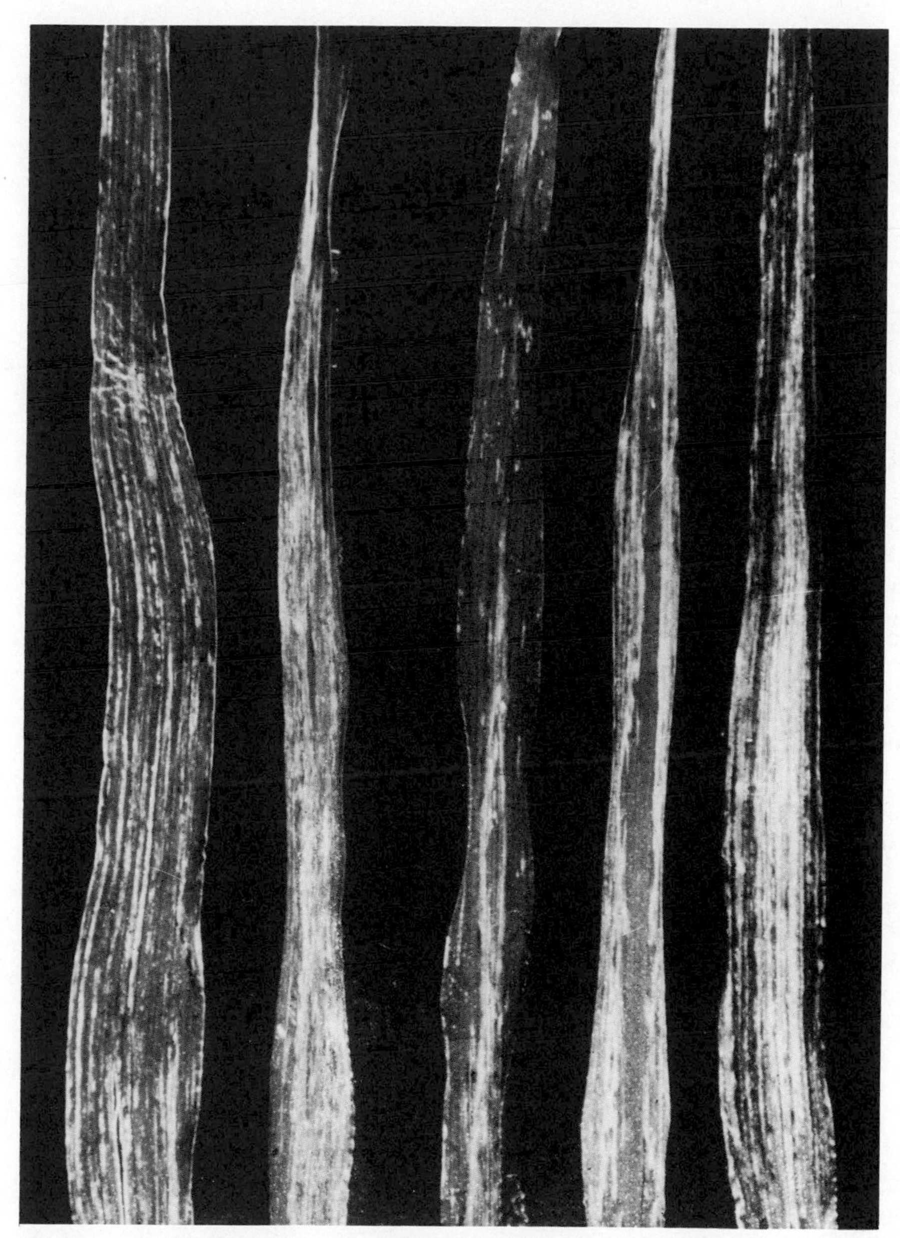

FIGURE 2. Symptom variations in barley leaves due to barley stripe mosaic virus. Different combinations of virus strain, barley cultivar, and environmental factors produce symptom variations such as striping, streaking, chlorosis, and necrosis. From Carroll (1980).

E. Cytological Effects on Host

Light and electron microscopic studies have shown that pathological alterations of host cells resulted from BSMV infection. Many cells within infected tissues contained rod-shaped virus particles and structurally abnormal cell components.

BSMV particles were first discovered by Shalla (1959) in the cytoplasm of mesophyll and epidermal cells of infected barley leaves. Subsequently, it was determined that most virus particles in the cytoplasm occurred as aggregates or as individuals scattered among the ribosomes (Gardner, 1967; Shalla, 1966). Some particles, however, were apparently attached to the surface of chloroplasts (Carroll, 1970; Shalla, 1966). Aggregates of BSMV particles were also seen in nuclei (Gardner, 1967). Those nuclear particles and ones in cytoplasmic areas of the same cells had a diameter of about 12 nm. In 1983, it was shown by Lin and Langenberg that intracellular particles of similar size and morphology in BSMV-infected wheat or barley leaves stained specifically with either antiserum or IgG to BSMV using the immunocolloidal gold method. In floral cells, BSMV particles have been seen not only in close association with plastids, but also attached to microtubules (Mayhew and Carroll, 1974a; Carroll and Mayhew, 1976a,b). BSMV particle relations with plastids and microtubules are illustrated in Fig. 3.

Alterations in normal cell structure caused by BSMV infection were found most frequently in the cytoplasm and cell wall. Ribosome numbers were severely reduced in some mesophyll cells (Shalla, 1966). Disorganized membranes, large globular inclusions, and deformed chloroplasts were observed in the cytoplasm (Gardner, 1967). The chloroplasts were swollen and contained unusually developed granal systems. Further examination of abnormal chloroplasts (Carroll, 1970) revealed that single membrane-bound vesicles were present in the periplastidial gap of the organelles. Recently, peripheral vesicles in proplastids of wheat cells infected with BSMV have been found to contain double-stranded RNA (Lin and Langenberg, 1985). Structurally abnormal chloroplasts in infected barley leaves have been classified into three morphologically aberrant types (McMullen *et al.*, 1978a).

Thickened cell walls often containing modified plasmodesmata and frequently associated with membranous or vesicular paramural bodies have been seen in infected tissues (McMullen *et al.*, 1977a,b).

BSMV particles have been detected in mesophyll, subsidiary, guard, epidermal, vascular parenchyma, sieve, and tracheary cells within infected leaf tissue of barley (McMullen *et al.*, 1978b).

F. Physiological Effects on Host

Several physiological abnormalities, apparently not directly involved with virus multiplication, occur in BSMV-infected tissue. Gordon (1966)

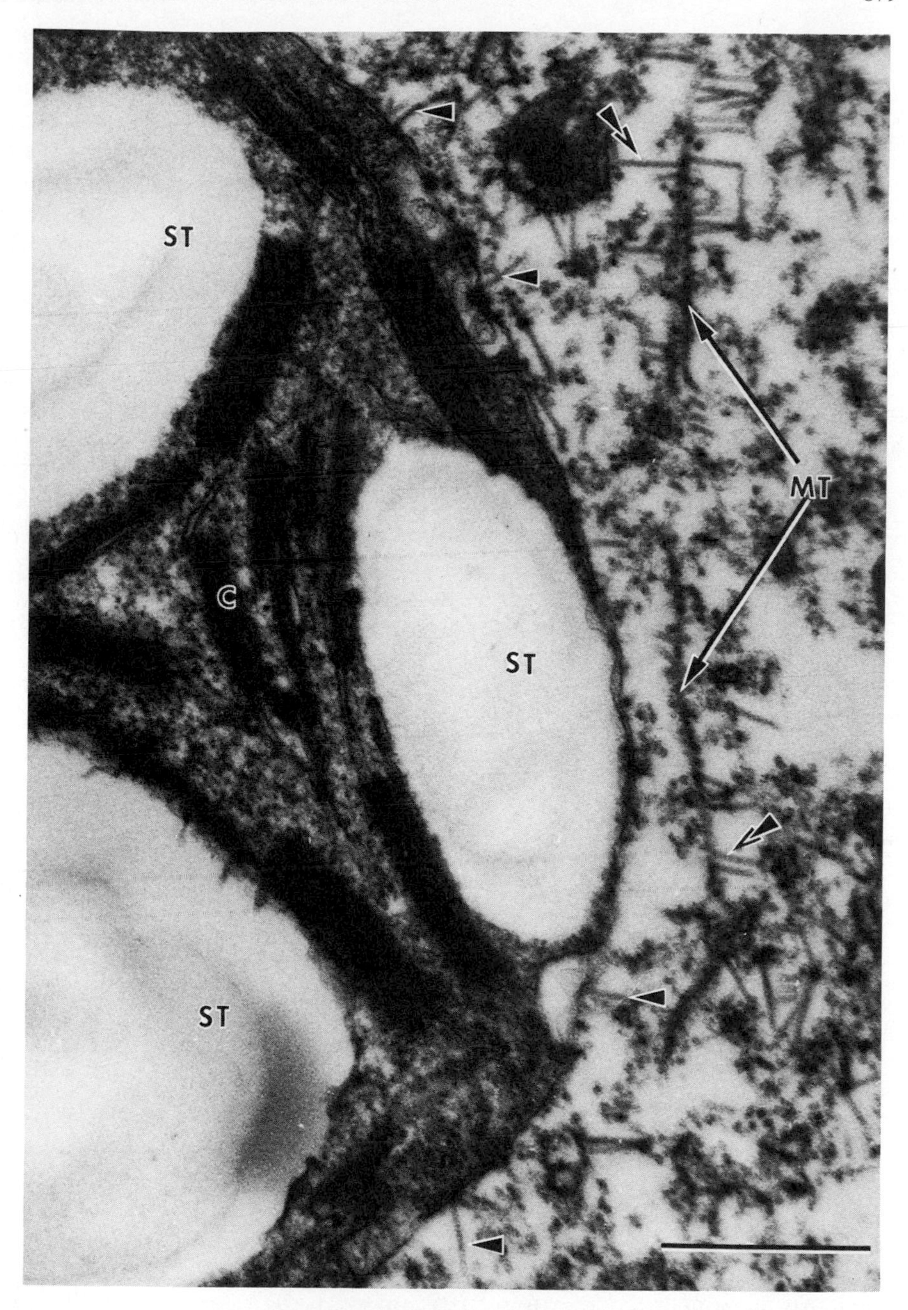

FIGURE 3. Association of particles of barley stripe mosaic virus to cytological components in an ovary cell during reproduction in barley. Some particles (single arrowheads) are seen in close association with the outer membrane of the chloroplast (C), while others (double arrowheads) are attached to microtubules (MT). Also shown are starch granules (ST) in the chloroplast.

reported that there was a twofold increase in respiration in infected tissue over that in healthy tissue. A loss of fresh weight per unit area and a reduction in chlorophyll content (Pring and Timian, 1969), as well as a reduction in chloroplast Fraction I protein (Kato and Yamaguchi, 1971) were shown in fully expanded leaf 1 after it was mechanically inoculated with BSMV. Chloroplast rRNA became reduced in leaves mechanically inoculated with BSMV (White and Brakke, 1982; Suzuki and Taniguchi, 1973). When the expanding third leaf of barley systemically infected with BSMV was analyzed by SDS gel electrophoresis, a decrease in chloroplast polypeptides was found (White and Brakke, 1982). The decrease occurred when proplastids were developing into chloroplasts. The large-subunit RuBPCase had definitely been identified as one of these polypeptides. The small subunit of the same enzyme and the chlorophyll a/b binding protein have also been tentatively identified. The mechanism for these protein changes associated with plastids is unknown. In addition to the reduction of plastid rRNA, cytoplasmic rRNA degradation has been detected (Suzuki and Taniguchi, 1973). The reported physiological abnormalities of plastids and cytoplasmic ribosomes correlate nicely with the ultrastructural alterations of those same cell components observed with the electron microscope in BSMV-infected tissues.

G. Transmission through Pollen and Seed

The seed transmissibility of many strains of BSMV in barley has been demonstrated (McKinney and Greeley, 1965; McKinney, 1951, 1953). The type strain was the first reported to be seedborne. Eslick (1953), suspecting since 1947 that some seedborne virus was present in poorly yielding selections of Glacier barley grown in Montana, sent seed to McKinney (1953) for confirmation. McKinney determined that the seed was infected with BSMV. In 1954, Hagborg observed 71% transmission in wheat grown from seed originating from mechanically inoculated plants.

The level of seed transmission of BSMV varies widely with the barley cultivar or genotype. In general, barleys expressing mild or no symptoms after mechanical inoculation have shown low levels (up to 10%) of seed transmission (Inouye, 1962). By contrast, many barleys that display moderate to severe symptoms after mechanical inoculation, have had percentages of seed transmission ranging from 11 to 90% (Inouye, 1962; Singh *et al.*, 1960; Eslick and Afanasiev, 1955). In 1974, Timian studied the interaction of four barley cultivars with 20 strains of BSMV, and virtually no seed transmission resulted because the plants were highly sensitive to infection and thus produced few seeds.

A single recessive gene was found to condition the resistance to the seed transmission of BSMV in barley (Carroll *et al.*, 1979b). Seed transmissibility is also dependent on the virus strain involved (Timian, 1974;

Carroll, 1972, 1981; Shivanathan, 1970; Hamilton, 1965; McKinney and Greeley, 1965). While most strains were found to be seed transmitted in barley and wheat, several strains were discovered that were not seed transmitted (non-seed-passage strains).

Temperature influences seed transmission also. Singh *et al.* (1960) observed that levels of seed transmission for four tolerant barley cultivars were highest at temperatures ranging from 20 to 24°C. The stage of development of the parent plant at the time of infection also appears to affect seed transmission. Some workers (Singh *et al.*, 1960; Crowley, 1959; Eslick and Afanasiev, 1955) found that when the virus infected mother plants after flowering, some seed transmission resulted. Others, however (Carroll, 1972; Timian, 1967; Inouye, 1962), failed to detect seed transmission in barley plants inoculated after flowering.

Transmission of BSMV occurs not only in seed, but also in pollen (Carroll and Mayhew, 1976a; Slack and Shepherd, 1975; Timian, 1967; Inouye, 1962; Gold *et al.*, 1954) and ovules (Carroll and Mayhew, 1976a; Slack and Shepherd, 1975; Timian, 1967; Inouye, 1962; Gold *et al.*, 1954). BSMV particles have been observed in extracts of pollen and unfertilized pistils (Gold *et al.*, 1954), and in thin sections of pollen (Carroll and Mayhew, 1976a; Carroll, 1972, 1974; Gardner, 1967) and ovules (Mayhew and Carroll, 1974b). In pollen, BSMV particles were located in the cytoplasm of vegetative cells (Carroll, 1972; Gardner, 1967) and in the cytoplasm and nucleus of sperms (Carroll, 1974). Figure 4 shows the presence of BSMV particles in a sperm cell. In ovules, virus particles were found in the egg cell and embryo sac (Mayhew and Carroll, 1974b). The percentage of pollen (Carroll and Mayhew, 1976a; Inouye, 1962; Gold *et al.*, 1954) and ovule (Carroll and Mayhew, 1976b; Slack *et al.*, 1975; Timian, 1967) transmission has varied from 10 to 35 and 17 to 66, respectively.

To date, all evidence indicates that seed transmission of BSMV in barley is due to virus carried in the embryo. Inouye (1966) determined that infected embryos, not infected endosperms, resulted in seed transmission. In electron microscopic studies, virus particles were seen in extracts (Gold *et al.*, 1954) and thin sections (Carroll, 1969) of embryos. Aggregates of BSMV particles in an embryo are shown in Fig. 5. Infectious virus was detected in embryos by Inouye (1962) and Crowley (1959). A non-seed-passage strain (NSP) of BSMV (McKinney and Greeley, 1965) was found not to infect the embryos (Carroll, 1972; Hamilton, 1965), pollen (Carroll, 1972; Carroll and Mayhew, 1976a; Shivanathan, 1970), or ovules (Carroll, 1972; Carroll and Mayhew, 1976b; Shivanathan, 1970) of Atlas barley. In all probability, every seed-transmitted strain of BSMV invades the juvenile primary meristems of its host and this results in the infection of both sperms and eggs. Infected gametes then give rise to infected zygotes and embryos during fertilization and embryogenesis (Carroll, 1981). Glycoprotein in the viral capsid (Partridge *et al.*, 1974) may also influence the seed transmissibility of BSMV.

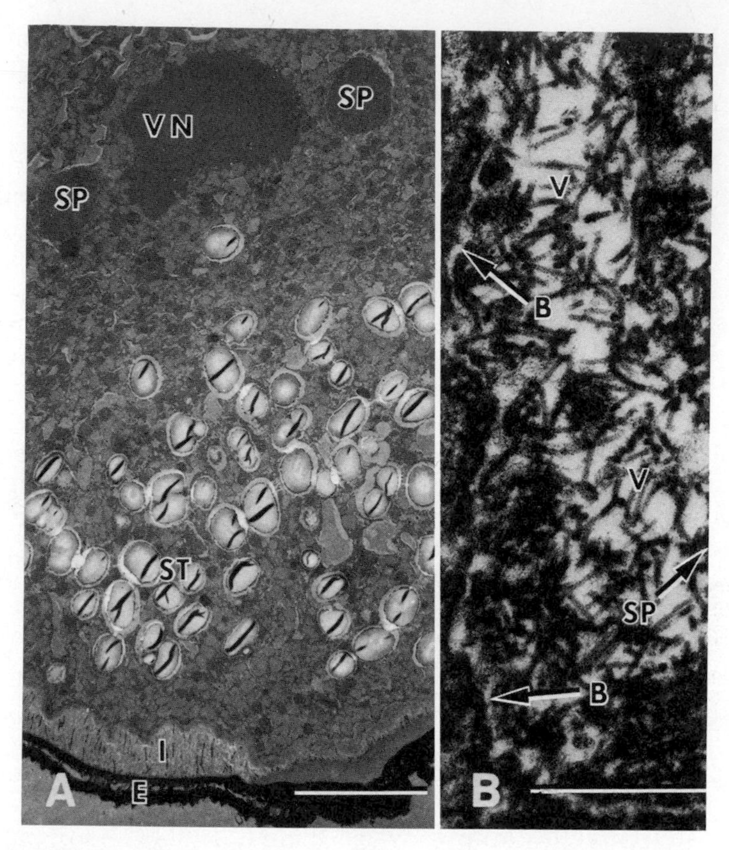

FIGURE 4. Barley stripe mosaic virus in relation to barley pollen. (A) Cross section of a mature pollen grain in which two sperms (SP) and a vegetative nucleus (VN) are apparent. Also visible are starch grains (ST), intine (I), and exine (E). From Carroll (1981). (B) Scattered virus particles (V) in the cytoplasm of a sperm (SP) cell. A special boundary (B) separates the sperm cell from the vegetative cytoplasm. From Carroll (1981).

H. Methods of Disease Diagnosis and Virus Detection

Field and greenhouse diagnosis of BSM has been made simply by observing plants with characteristic stripe mosaic symptoms (McKinney and Greeley, 1965; Inouye, 1962; Hagborg, 1954). In the field, symptoms are generally noted in plants up until the time they turn brown and senesce. Many barleys become dwarfed in the field if infected prior to heading. In the greenhouse, seedlings in 3–6 leaves are usually read for symptoms. Diagnosis of BSM in individual plants or in groups (lines, cultivars, or populations) of plants has also been accomplished by observing stripe mosaic symptoms in the seedlings that develop from seed believed to be infected with BSMV (McKinney, 1951, 1954; Inouye, 1962; Hagborg, 1954).

The principal methods of detecting and identifying BSMV have relied

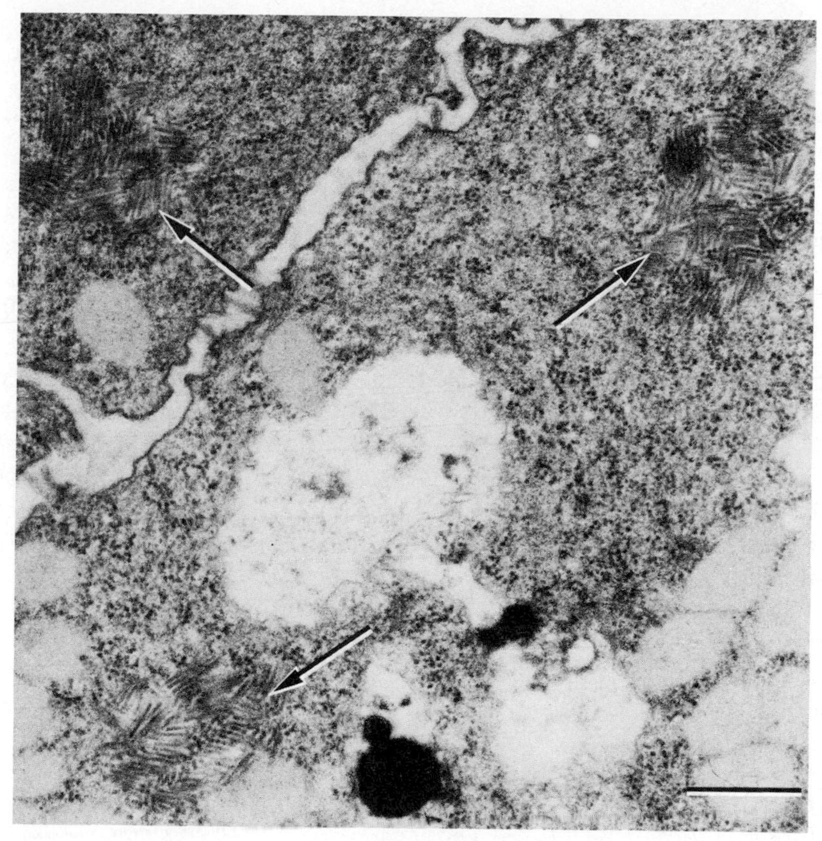

FIGURE 5. Barley stripe mosaic virus in relation to the barley embryo. Aggregates of virus particles (arrows) in an immature embryo. From Carroll (1981).

on key properties of the virus, such as its sap transmissibility, antigenicity, and its distinctive particle morphology. Infectivity assays have been made by inoculating sap from suspect plants mechanically onto host range or indicator test plants (McKinney, 1951; Inouye, 1962; Hagborg, 1954). Frequently, barley seedlings highly susceptible to infection by BSMV are inoculated when they are in 3–6 leaves.

Most serological tests have been designed to detect BSMV in seed lots. In comparison to the biological methods, these serological tests are more rapid and sensitive and require less space. The whole seed, embryo, or seedling is usually assayed. These tests have been effective because antibodies elicited by one strain recognize all strains of BSMV in high concentration. To date, all strains encountered are related serologically (Carroll, 1980, 1983).

For over 20 years, double-immunodiffusion methods have been used in Montana for the detection of BSMV in embryos (Carroll *et al.*, 1979a; Hamilton, 1965). A single-diffusion method, about ten times more sensitive than the double-diffusion methods, was developed by Slack and Shepherd (1975) for detecting BSMV in individual germinated barley seed. It could not, however, be used for direct assay of seed because of associated nonspecific precipitations that occurred in the agar. The nonspecific reaction was also a problem when the whole seed or the excised barley embryo was tested by the latex agglutination (Lundsgaard, 1976); hence, only the barley seedling was recommended for assay. Although ELISA was shown to have considerable potential for the routine testing of seed (Lister *et al.*, 1981), it is more reliable when applied to the detection of BSMV in the seedling. A procedure currently under evaluation for possible use in the Montana Seed Certification Program (Zaske *et al.*, 1985) can detect about 0.5 mg/ml BSMV or 1 infected seedling within 250 barley seedlings. This latter test not only is highly sensitive, but is also antiserum-conserving and utilizes intact virus as an immunogen. Serologically specific electron microscopy (SSEM) has also been used to detect and identify BSMV in extracts of barley seed (Brlansky and Derrick, 1979). Virus particles were detected in 1 part of infected seed extract per 1000 parts of healthy seed extract. Disadvantages of SSEM for large-scale testing of BSMV include the need for an electron microscope, cost of sample processing, and time required for sample preparation and observation.

More recently, BSMV has been detected in ultrathin sections of tissue (Lin and Langenberg, 1983) and leaf-dip preparations (Lin, 1984) with the electron microscope using immunochemical methods. In addition, Lin and Langenberg (1984) have also detected BSMV protein in embedded plant tissue at the light microscopic level using a gold–silver staining procedure specific for BSMV antigen.

In the future, detection and identification of BSMV may perhaps be accomplished by suitable forms of monoclonal antibody tests (Van Regenmortel, 1984), radioimmunoassays (Ball, 1973), and molecular biological approaches such as hybridization analysis of labeled cDNAs to ssRNAs of BSMV (Gould and Symons, 1983), and dsRNA analysis of BSMV-infected plant tissues (Dodds *et al.*, 1984).

I. Economic Significance

Economically significant yield losses have been attributed to BSMV in commercial barley. In Montana, experimental field trials (Eslick, 1953; Eslick and Afanasiev, 1955) and surveys of commercial barley seed lots and fields indicated that the total loss in barley due to BSMV exceeded $30 million during 1953–1970. The peak year for crop loss was 1964, when the virus caused over $3.1 million in damage (Carroll, 1980). Losses of 31% were reported by Eslick (1953) for Glacier barley in a yield nursery infected naturally by seed and grown under rainfed conditions. In irrigated plots, the average yield reduction in the cultivars Betzes, Compana, and

Vantage ranged from 24 to 35% when plants were infected by mechanical inoculation (Carroll, 1980).

Timian (1971) estimated that losses in barley in North Dakota were at least $3.0 million annually prior to initiation of control measures. On the basis of replicated field trials and an extensive survey of commercial fields, he concluded that losses to growers declined from a high of at least 3.5% to 0% in 1966. In field nurseries planted with different percentages of BSMV-infected seed, the greatest yield reduction (25–30%) occurred in those nurseries planted with seed having 25–30% infection. A loss of 24% had previously been reported (Timian and Sisler, 1955) in field plots naturally infected via seed.

Chiko and Baker (1978) described a method for estimating yield losses due to BSMV in commercially grown two-row barley in Canada. Estimates were based on data obtained from surveys of commercial fields and experimental field trials of two-row barley inoculated with BSMV at the tillering stage. Field trial losses to Betzes, Herta, and Fergus barley averaged over 26, 29, and 41% in 1975 and 1976.

Other yield loss experiments have been conducted in greenhouses or experimental field nurseries during the past 34 years, ever since McKinney (1951) discovered BSMV. Mechanical BSMV inoculation of barley caused up to 90% reduction in yield in pot tests (McKinney, 1953) and over 60% crop loss in field plots (Hagborg, 1954). In California (Slack *et al.*, 1975), a field plot planted with 36% BSMV-infected seed suffered a yield loss of 46%. Nutter *et al.* (1984) investigated the effect of BSMV seed infection levels on grain yield and other barley traits in North Dakota. They found that the reduction in yield per plot in response to increasing levels of seed infection by BSMV was best described by a quadratic model in 1981, and a linear model in 1982. Decreased plant vigor, sterility, and small and shriveled seed account for the losses in barley (Chiko and Baker, 1978; Inouye, 1962; Hagborg, 1954).

Apparently, no serious economic losses due to BSMV have been documented in commercial wheat, although the virus has been shown to be responsible for significant yield reductions in experimental field plots. Hagborg observed that wheat plots mechanically inoculated early with BSMV had a reduction of 75% in yield. Using BSMV-infected seed, McNeal *et al.* (1958) noted an average yield field reduction of 32.8% in eight cultivars of spring wheat having an average of 28.5% plants with symptoms of BSM.

Obviously, losses in commercial crops due to BSMV are influenced by the many and varied factors associated with the virus–host plant–environment interaction.

J. Epidemiology

Most data support the hypothesis that spread of BSMV into commercial fields of uninfected barley is due solely to the movement of infected barley seed. Attempts to transmit BSMV by mites, insects, or soil have been unsuccessful (Paliwal, 1980; Chiko, 1973; Inouye, 1962). Infected barley seed not only is a vehicle that ensures transportation of the virus to new locations, but it also enables BSMV to survive from year to year and crop to crop (Bennett, 1969).

The proportion of infected seed and the subsequent spread of the virus from infected seedlings account for the incidence of BSMV within a given commercial barley field (Timian, 1971). Results from field nursery experiments (Hockett and Davis, 1970; Inouye, 1962) imply that secondary spread of BSMV from infected to healthy plants in a commercial field occurs as a result of mechanical transmission by leaf contact from neighboring plants. The rate of virus spread (Timian, 1974) is dependent on the virus strain–host genotype combination. Light (Hampton *et al.*, 1957), temperature (Slack *et al.*, 1975; Singh *et al.*, 1960), plant density or spacing (Hockett and Davis, 1970; Inouye, 1962) are environmental factors believed to influence the secondary spread of BSMV to healthy plants. Wind and rain may also be factors affecting secondary spread, because they would increase contact between infected and healthy plants.

Although wild oats naturally infected with BSMV have been reported in Canada (Chiko, 1975, 1983), they probably play an insignificant role in the epidemiology of the virus in commercial barley (Chiko, 1984). Most strains of BSMV from barley fail to infect wild oats.

Pollen transmission is unimportant in the epidemiology of BSMV in normal self-pollinated barley (Slack *et al.*, 1975). For pollen transmission to occur, infected pollen must be directly applied to the stigmata of a healthy maternal plant (Inouye, 1962; Gold *et al.*, 1954). Spread of BSMV in pollen, however, is a significant factor in the epidemiology of the virus when male-sterile plants are used to develop barley germ plasm or hybrids. Infected pollen from normal self-fertile plants can cross-pollinate or outcross onto male-sterile plants and thereby infect the resulting seed. In Montana, considerable time and effort have been devoted in an attempt to eradicate BSMV in a composite cross population of barley having male sterility. After 7 years of extensive plant rogueing in field nurseries and serological screening of the seed with the SDS disk test (Carroll *et al.*, 1979a), the level of virus infection was only reduced from about 5% to approximately 1%. In 1985, ELISA (Zaske *et al.*, 1985), which is far more sensitive than the disk test, will be used to detect BSMV in groups of plants within the field population, so any infected group can be destroyed during the growing season. Other male-sterile facilitated recurrent selection populations of barley being developed in Montana are isolated from known BSMV-infected barley.

K. Control

The elimination of infected seed has been the only approach taken to control BSMV in commercial barley. In the United States, two states have perceived an economic need for some kind of certification scheme against the virus in barley (Carroll, 1983). Currently, Montana operates a complete generation certification scheme covering the production of foundation, registered, and certified classes of seed. It entails two procedures, the first of which is the inspection of fields for plants exhibiting BSM symptoms. If any symptomatic plants are found in a given field, that field is no longer eligible for certification. The second procedure is the inspection of seed from each potentially certifiable field and this is made by assaying barley embryos serologically, by the SDS disk test (Carroll *et al.*, 1979a). Two hundred embryos are tested per seed lot from each certifiable field. No infected embryos are permitted in any seed lot (zero tolerance for seedborne BSMV). In 1966 when the Seed Laboratory at Montana State University began testing, over 40% of the more than 400 seed lots inspected were infected (Carroll, 1980). By contrast, in 1985, only 0.8% of the 147 seed lots eligible for certification were infected. The certification program has markedly reduced the incidence and severity of BSMV in Montana. The virus, however, has by no means been eliminated from the state, because the planting of virus-free seed is not legally required. Nevertheless, BSMV has been brought to a reasonable economical level of control in Montana. Factors contributing to this control include the release of new BSMV-free cultivars and the wide planting of certified seed lots devoid of the virus.

North Dakota has a limited generation certification program wherein only foundation seed fields and foundation seed lots of barley are inspected (Carroll, 1983). Seed from fields revealing no BSM symptomatic plants is tested serologically by the latex flocculation method. Only foundation seed determined by test to be BSMV-free can be used to produce certified seed, which is then made available to growers. In 1954, surveys in North Dakota showed that 93% of 214 barley fields had BSM (Timian and Sisler, 1955), but by 1971 it was reported that for all practical purposes, BSMV had been eliminated from the state (Timian, 1971). Other states in the United States and various countries around the world have also controlled BSMV by field and/or seed inspections.

The use of barley cultivars resistant to BSMV for commercial production has also been considered as a possible approach by which to control the virus. Four barleys, Moreval (CI 5724), Modjo (CI 3212), Modjo 1 (CI 14080), and CI 4219, from the USDA world collection found to be resistant to several strains of BSMV after mechanical inoculation, have been used experimentally in plant breeding programs in North Dakota, Montana, and Japan. In one study, Modjo and Modjo 1 seemed to contain a single recessive gene (Timian and Sisler, 1955), and in another study they appeared to have at least two recessive genes (Sisler and Timian, 1956). Inouye (1962) determined that resistance in the barley cultivar Wien was conditioned by a single dominant gene, but resistance in the cultivar Imperial was recessive.

Vasquez *et al.* (1974) investigated the inheritance of resistance to three strains of BSMV (type, ND1, and CV 52) in two resistant cultivars (Modjo 1 and Moreval) crossed with a susceptible cultivar (Traill), after mechanical inoculation of plants. Based on the leaf symptoms observed, the cultivar genotypes proposed for reaction to the type strain of BSMV were (Sm = susceptible dominant gene, sm = resistant recessive gene) Sm3bSm3b for Traill, Sm3aSm3a for Modjo 1, and Sm3cSm3c for Moreval (genes in a multiple allelic series). Genotypes Sm5Sm5 for Traill and Modjo 1, and sm5sm5 for Moreval were proposed for reaction to BSMV strain ND1. All three cultivars were susceptible to strain CV 52.

In Montana, Mobet barley germ plasm (PI 467884) was developed for its resistance to the seed transmission of three isolates of BSMV indigenous to Montana (MI-1, MI-2, and MI-3). Mobet was derived from a cross between Modjo (CI 3212) and Betzes (CI 6398) followed by six backcrosses to Betzes and subsequent self-pollination 11 times (Carroll *et al.*, 1983). A single recessive gene in Modjo had been shown earlier (Carroll *et al.*, 1979b) to condition resistance to seed transmission of MI-3 isolate of BSMV in Modjo crossed with susceptible Vantage (CI 7324) barley. After the F_{11} plants were mechanically inoculated in the field, a low number (6.7%) of the plants expressed BSM symptoms. Most of these symptoms did not appear until the plants were flowering. When F_{12} seed was sown in the greenhouse, only 3 out of 2610 (0.11%) of the F_{12} seedlings expressed BSM symptoms and tested positive serologically for BSMV.

In spite of the different gene sources known to control resistance to BSMV, no resistant commercial barley cultivar has yet been developed and released. Inasmuch as the gene sources are in barleys of poor agronomic performance, and a resistant gene effective against all recognized strains of BSMV has not yet been found, the feasibility of breeding for resistance to BSMV in barley is dubious.

III. POA SEMILATENT VIRUS

PSLV was discovered in 1966 by Slykhuis (1972) in *Poa palustris* and *Agropyron trachycaulum* plants growing in Alberta, Canada. The original isolate of the virus segregated into severe and mild isolates in mechanically inoculated wheat (*Triticum aestivum* cv. Kent) and oats (*Avena sativa* cv. Clintland 60). All isolates of PSLV, however, caused only a mild transitory mottling on *P. palustris* after mechanical inoculation of the virus. Kent wheat and Clintland 60 oats developed mosaic, chlorotic, and blighting symptoms when inoculated with a severe isolate of PSLV (Fig. 6). Less extensive areas of chlorosis and less necrosis were observed on Kent wheat inoculated with milder isolates. PSLV also caused some plants of *A. trachycaulum*, barley (*Hordeum vulgare* L., 11 cultivars), durum wheat (*Triticum durum* Desf. cv. Ramsey), wheat (*T. aestivum* L., spring types—2 cultivars; winter types—in addition to Kent, 6 cultivars), and corn (*Zea mays* L. cv. Golden Cross Bantam), to develop

FIGURE 6. Symptom variations in oat leaves due to Poa semilatent virus. The seven infected leaves to the right of the healthy leaf reveal mosaic, chlorotic, and necrotic symptoms. From Slykhuis (1972).

chlorotic mottling and sometimes necrosis after mechanical inoculation. Other hosts included *A. cristatum, Alopecurus aequalis, Avena* (15 spp. in addition to *A. sativa*), *Elymus canadensis, Hordeum marinum, Lolium multiflorum, Phleum pratense, Poa compressa*, and *Secale cereale*. Transmission by plant-to-plant contact was shown by sowing Kent wheat and Clintland 60 oats in pots of soil with diseased *P. palustris* and *A. trachycaulum* transplanted from the field. No symptoms developed on Kent wheat, Clintland 60 oats, *P. palustris*, and *A. trachycaulum* seeded and grown in soil obtained from around virus-infected field plants. No PSLV was detected in about 200 *P. palustris* plants grown from seed produced by naturally infected plants.

The rod-shaped particles of PSLV are about 25 nm in diameter and from 50 to 600 nm long. They have a normal length of about 161 nm (Polak and Slykhuis, 1972; Slykhuis, 1972). Microprecipitation and ring-interface tests showed that serologically PSLV was distantly related to BSMV (Slykhuis, 1972). By comparison, no serological relationship was observed by double-immunodiffusion or leaf-dip serology (Polak and Slykhuis, 1972). The molecular weights reported for the four RNA components of PSLV are 1.5, 1.35, 1.20, and 1.15 × 10^6 (Lane, 1974).

Wheat and oats infected with a severe isolate of PSLV developed a mosaic 6–10 days after inoculation followed by rapid chlorosis and necrosis (Slykhuis, 1972). Infected wheat plants usually died within 2 weeks of inoculation, but oats frequently recovered and developed new tillers which were chlorotic, mottled, or symptomless.

Little is known about the cytopathological or physiological effects of PSLV on hosts, or the economic importance, epidemiology, and control of the virus. The distance separating the virus collection sites suggests a widespread distribution in Alberta, Canada. Results of the pot test, wherein indicator plants became infected by PSLV through contact with diseased plants, strongly suggest that plant contact transmission is responsible, to some degree, for field spread.

IV. LYCHNIS RINGSPOT VIRUS

In 1955, LRSV was discovered by Bennett (1959) in a single plant of *Lychnis divaricata* growing in a greenhouse at Salinas, California. Rings and zonate spots were observed on the leaves of this plant. When LRSV was sap inoculated to determine its host range, the virus was found to be confined to several dicotyledonous families. LRSV produced systemic infections consisting of ringspots, spots, rings, and lines in *L. divaricata, Silene noctiflora, S. gallica, Beta vulgaris* (sugar beet), *Spinacia oleracea* (spinach), *Callistephus chinensis*, and certain other plants. It caused local lesions in several species in which it never became systemic. No seeds were produced by infected plants of *Tetragonia expansa* or *Stellaria media*. Seed production was significantly reduced in sugar beet, *C. chinensis*, and *S. noctiflora*. Insect transmission tests with the beet leafhopper, the green peach, bean, and corn leaf aphids were negative. LRSV was transmitted through 58.4 and 41.7% of *L. divaricata* and *S. noctiflora* seed, respectively, and lower percentages in seed of sugar beet, spinach, and several other hosts. It was also transmitted through pollen of *L. divaricata* and *S. noctiflora*. The rod-shaped particles of LRSV have a modal length of 125 nm and a diameter of 18–19 nm (Gibbs *et al.*, 1963). LRSV is distantly related serologically to BSMV as shown by precipitation and gel diffusion tests (Gibbs *et al.*, 1963). According to Lane (1974), the estimated molecular weights of the four RNAs of LRSV are 1.50, 1.30, 1.15, and 0.90 × 10^6.

The cytopathological and physiological effects of LRSV, and the distribution, economic importance, epidemiology, and control of the virus do not appear to have been investigated. Nevertheless, one can speculate that in nature, LRSV probably survives from year to year in seed, and spreads from plant to plant by contact transmission.

V. CLOSING REMARKS

To characterize the known members of the hordeiviruses more fully will require not only the excellent research on the many different properties and behavior of BSMV to continue, but also new research into virtually all aspects of PSLV and LRSV. For example, more knowledge is needed on genome strategy, RNA and protein synthesis during infection, particle assembly, and pollen and seed transmission of BSMV. For PSLV and LRSV, it is essential that new information be obtained on their biology, pathology, particle structure, and genome structure and expression. Finally, we need data to enhance our understanding of the chemical and serological relatedness of PSLV and LRSV to BSMV and to each other.

Recently, a new virus with the proposed name of Anthoxanthum latent blanching virus (ALBV) was isolated from symptomless and indistinctly streaked plants of *Anthoxanthum odoratum* (sweet vernal grass) growing in a pasture in Wales (Catherall and Chamberlain, 1980). ALBV was transmitted to 27 out of 156 *A. odoratum* plants after mechanical inoculation. Fifteen of the infected plants developed transitory indistinct, pale yellow or gray streaks, while the rest remained symptomless. When *Lagurus ovatus* was mechanically inoculated, all of the infected tillers became blanched and died. ALBV particles had a variable length with a main model peak at 135 nm and a diameter of 22 nm. On the basis of sap transmissibility, narrow host range, lack of known vector, and particle morphology, ALBV appears to be a hordeivirus.

ACKNOWLEDGMENTS. Many thanks to Drs. A. O. Jackson, W. S. Gardner, R. G. Timian, and W. G. Langenberg for their pertinent reprints and to J. T. Slykhuis for his photograph pertaining to PSLV. I am grateful to Drs. D. E. Mathre and A. L. Scharen and to S. K. Zaske for critical review of the manuscript. The excellent technical assistance of C. Bradbury with the photographic work and figure labeling is acknowledged. The help of T. Dysart was invaluable in the preparation of the manuscript. Reports on BSMV by the author and his colleagues resulted from studies of the Agricultural Experiment Station at Montana State University. Support for these studies came, in part, from NSF Grants GB-8082 and GB-35323, and Contract AID/DSAN-C0024, Agency for International Development, U.S. Department of State, Washington, D.C. Published with the approval of the director as Journal Series Paper J-1730, Montana Agricultural Experiment Station. Accepted for publication 8-28-85.

REFERENCES

Afanasiev, M. M., 1956, Occurrence of barley stripe mosaic in Montana, *Plant Dis. Rep.* **40:**142.

Atabekov, J. G., and Novikov, V. K., 1971, Barley stripe mosaic virus, *CMI/AAB Descriptions of Plant Viruses No. 68.*

Ball, E. M., 1973, Solid phase radioimmunoassay for plant viruses, *Virology* **55:**516–520.

Bennett, C. W., 1959, Lychnis ringspot, *Phytopathology* **49:**706–713.

Bennett, C. W., 1969, Seed transmission of plant viruses, *Adv. Virus Res.* **14:**221–261.

Brlansky, R. H., and Derrick, K. S., 1979, Detection of seedborne plant viruses using serologically specific electron microscopy, *Phytopathology* **69:**96–100.

Carroll, T. W., 1969, Electron microscopic evidence for the presence of barley stripe mosaic virus in cells of barley embryos, *Virology* **37:**649–657.

Carroll, T. W., 1970, Relation of barley stripe mosaic virus to plastids, *Virology* **42:**1015–1022.

Carroll, T. W., 1972, Seed transmissibility of two strains of barley stripe mosaic virus, *Virology* **48:**323–336.

Carroll, T. W., 1974, Barley stripe mosaic virus in sperm and vegetative cells of barley pollen, *Virology* **60:**21–28.

Carroll, T. W., 1980, Barley stripe mosaic virus: Its economic importance and control in Montana, *Plant Dis* **64:**136–140.

Carroll, T. W., 1981, Seedborne viruses: Virus–host interactions, in: *Plant Diseases and Vectors: Ecology and Epidemiology* (K. Maramorosch and K. F. Harris, eds.), pp. 293–317, Academic Press, New York.

Carroll, T. W., 1983, Certification schemes against barley stripe mosaic, *Seed Sci. Technol.* **11:**1033–1042.

Carroll, T. W., and Mayhew, D. E., 1976a, Anther and pollen infection in relation to the pollen and seed transmissibility of two strains of barley stripe mosaic virus in barley, *Can. J. Bot.* **54:**1604–1621.

Carroll, T. W., and Mayhew, D. E., 1976b, Occurrence of virions in developing ovules and embryo sacs of barley in relation to the seed transmissibility of barley stripe mosaic virus, *Can. J. Bot.* **54:**2497–2512.

Carroll, T. W., Gossel, P. L., and Batchelor, D. L., 1979a, Use of sodium dodecyl sulfate in serodiagnosis of barley stripe mosaic virus in embryos and leaves, *Phytopathology* **69:**12–14.

Carroll, T. W., Gossel, P. L., and Hockett, E. A., 1979b, Inheritance of resistance to seed transmission of barley stripe mosaic virus in barley, *Phytopathology* **69:**431–433.

Carroll, T. W., Hockett, E. A., and Zaske, S. K., 1983, Registration of Mobet barley germplasm (Reg. No. GP61), **Crop Sci. 23**(3):599.

Catherall, P. L., and Chamberlain, J. A., 1980, A new virus disease of sweet vernal grass (*Anthoxanthum odoratum* L.,), in: *Proc. 3rd Conf. Virus Diseases of Gramineae in Europe* (R. T. Plumb, ed.), pp. 83–85, Rothamsted, England.

Chiko, A. W., 1971a, Distribution of barley stripe mosaic virus in Manitoba in 1970, *Can. Plant Dis. Surv.* **51:**111–115.

Chiko, A. W., 1971b, Barley stripe Mosaic virus in Manitoba in 1971, *Can. Plant Dis. Surv.* **51:**159–160.

Chiko, A. W., 1973, Barley stripe mosaic in the Canadian Prairies in 1972, *Can. Plant Dis. Surv.* **53:**107–111.

Chiko, A. W., 1974, Barley stripe mosaic in Manitoba in 1973, *Can. Plant Dis. Surv.* **54:**21.

Chiko, A. W., 1975, Natural occurrence of barley stripe mosaic virus in wild oats (*Avena fatua*), *Can. J. Bot.* **53:**417–420.

Chiko, A. W., 1976, Barley stripe mosaic in the Canadian Prairies, 1974–75, *Can. Plant Dis. Surv.* **56:**53–55.

Chiko, A. W., 1978, Barley stripe mosaic in Saskatchewan in 1977, *Can. Plant Dis. Surv.* **58:**29–30.

Chiko, A. W., 1980, Barley stripe mosaic in Manitoba in 1978, *Can. Plant Dis. Surv.* **60:**11–12.

Chiko, A. W., 1983, Reciprocal contact transmission of barley stripe mosaic virus between wild oats and barley, *Plant Dis.* **67:**207–208.

Chiko, A. W., 1984, Increased virulence of barley stripe mosaic virus for wild oats: Evidence of strain selection by host passage, *Phytopathology* **74:**595–599.

Chiko, A. W., and Baker, R. J., 1978, Economic significance of barley stripe mosaic virus in the Canadian Prairies, *Can. J. Plant Sci.* **58:**331–340.

Crowley, N. C., 1959, Studies on the time of embryo infection by seed-transmitted viruses, *Virology* **8:**116–123.

Dodds, J. A., Morris, T. J., and Jordan, R. L., 1984, Plant viral double-stranded RNA, *Annu. Rev. Phytopathol.* **22:**151–168.

Eslick, R. F., 1953, Yield reductions in Glacier barley associated with a virus infection, *Plant Dis. Rep.* **37:**290–291.

Eslick, R. F., and Afanasiev, M. M., 1955, Influence of time of infection with barley stripe mosaic on symptoms, plant yield, and seed infection of barley, *Plant Dis. Rep.* **39:**722–724.

Francki, R. I. B., Milne, R. G., and Hatta, T., 1985, Hordeivirus group, in: *Atlas of Plant Viruses*, Volume II, pp. 133–146, CRC Press, Boca Raton, Fla.

Gardner, W. S., 1967, Electron microscopy of barley stripe mosaic virus: Comparative cytology of tissues infected during different stages of maturity, *Phytopathology* **57:**1315–1326.

Gibbs, A. J., Kassanis, B., Nixon, H. L., and Woods, R. D., 1963, The relationship between barley stripe mosaic and Lychnis ringspot viruses, *Virology* **20:**194–198.

Gold, A. H., Suneson, C. A., Houston, B. R., and Oswald, J. W., 1954, Electron microscopy and seed and pollen transmission of rod-shaped particles associated with the false stripe virus disease of barley, *Phytopathology* **44:**115–117.

Gordon, D. T., 1966, Effects of barley stripe mosaic virus infection on the respiratory metabolism of barley leaves, Ph.D. thesis, University of Wisconsin, Madison.

Gould, A. R., and Symons, R. H., 1983, A molecular biological approach to relationships among viruses, *Annu. Rev. Phytopathol.* **21:**179–199.

Gustafson, G. D., Milner, J. J., McFarland, J. E., Pedersen, K., Larkins, B. A., and Jackson, A. O., 1982, Investigation of the complexity of barley stripe mosaic virus RNAs with recombinant DNA clones, *Virology* **120:**182–193.

Hagborg, W. A. F., 1954, Dwarfing of wheat and barley by the barley stripe mosaic (false stripe) virus, *Can. J. Bot.* **32:**24–37.

Hamilton, R. I., 1965, An embryo test for detecting seed-borne barley stripe mosaic virus in barley, *Phytopathology* **55:**798–799.

Hampton, R. E., Sill, W. H., and Hansing, E. D., 1957, Barley stripe mosaic virus in Kansas and its control by a greenhouse seed-lot testing technic, *Plant Dis. Rep.* **41:**735–738.

Hockett, E. A., and Davis, D. J., 1970, Infection levels and field spread of barley stripe mosaic virus in Compana barley, *Plant Dis. Rep.* **54:**749–751.

Inouye, T., 1962, Studies on barley stripe mosaic in Japan, *Ohara (Japan) Inst. Landw. Biol. Ber. Univ.* **11:**413–496.

Inouye, T., 1966, Some experiments on the seed transmission of barley stripe mosaic virus in barley with electron microscopy, *Ohara (Japan) Inst. Landw. Biol. Ber. Univ.* **13:**111–121.

Jackson, A. O., and Lane, L. C., 1981, Hordeiviruses, in: *Handbook of Plant Virus Infections and Comparative Diagnosis* (E. Kwistakk, ed.), pp. 565–625, Elsevier/North-Holland, Amsterdam.

Jackson, A. O., Dawson, J. R. O., Covey, S. N., Hull, R., Davies, J. W., McFarland, J. E., and Gustafson, G. D., 1983, Sequence relations and coding properties of a subgenomic RNA isolated from barley stripe mosaic virus, *Virology* **127:**37–44.

Kato, T., and Yamaguchi, A., 1971, Soluble antigens from barley leaves infected with barley stripe mosaic virus, *Ann. Phytopathol. Soc. Jpn.* **37:**117–123.

La, Y. J., and Park, Y. K., 1979, Serological detection of barley stripe mosaic virus infection in the seeds of barley and wheat cultivars grown in Korea, *Korean J. Plant Prot.* **18:**29–33.

Lane, L. C., 1974, The components of barley stripe mosaic and related viruses, *Virology* **58:**323–333.

Lin, N.-S., 1984, Gold–IgG complexes improve the detection and identification of viruses in leaf dip preparations, *J. Virol. Methods* **8:**181–190.

Lin, N.-S., and Langenberg, W. G., 1983, Immunohistochemical localization of barley stripe mosaic virions in infected wheat cells, *J. Ultrastruct. Res.* **84:**16–23.

Lin, N.-S., and Langenberg, W. G., 1984, Distribution of barley stripe mosaic virus protein in infected wheat root and shoot tips, *J. Gen. Virol.* **65:**2217–2224.

Lin, N.-S., and Langenberg, W. G., 1985, Peripheral vesicles in proplastids of barley stripe mosaic virus-infected wheat cells contain double-stranded RNA, *Virology* **142:**291–298.

Lister, R. M., Carroll, T. W., and Zaske, S. K., 1981, Sensitive serologic detection of barley stripe mosaic virus in barley seed. *Plant Dis.* **65:**809–814.

Lundsgaard, T., 1976, Routine seed health testing for barley stripe mosaic virus in barley seeds using the latex-test, *Z. Pflanzenkr. Pflanzenschutz* **83:**278–283.

McKinney, H. H., 1951, A seed-borne virus causing false-stripe symptoms in barley, *Plant Dis. Rep.* **35:**48.

McKinney, H. H., 1953, New evidence on virus diseases in barley, *Plant Dis. Rep.* **37:**292–295.

McKinney, H. H., 1954, Culture methods of detecting seed-borne virus in Glacier barley seedlings, *Plant Dis. Rep.* **38:**152–162.

McKinney, H. H., and Greeley, L. W., 1965, Biological Characteristics of Barley Stripe-Mosaic Virus Strains and their Evolution, U.S. Dep. Agric. Tech. Bull. No. 1324.

McMullen, C. R., Gardner, W. S., and Myers, G. A., 1977a, Ultrastructure of cell-wall thickenings and paramural bodies induced by barley stripe mosaic virus, *Phytopathology* **67:**462–467.

McMullen, C. R., Gardner, W. S., and Myers, G. A., 1977b, Ultrastructure of corn leaf tissue infected with the ND18 strain of barley stripe mosaic virus, *Proc. S.D. Acad. Sci.* **56:**100–104.

McMullen, C. R., Gardner, W. S., and Myers, G. A., 1978a, Aberrant plastids in barley leaf tissues infected with barley stripe mosaic virus, *Phytopathology* **68:**317–325.

McMullen, C. R., Gardner, W. S., and Myers, G. A., 1978b, Distribution of the ND18 strain of barley stripe mosaic virus in barley leaf tissue, *Proc. S.D. Acad. Sci.* **57:**92–107.

McNeal, F. H., Berg, M. A., Afanasiev, M. M., and Army, T. J., 1958, The influence of barley stripe mosaic on yield and other plant characters of 8 spring wheat varieties grown at 4 nitrogen levels, *Agron. J.* **50:**103–105.

Matthews, R. E. F., 1982, *Classification and Nomenclature of Viruses,* Fourth Report of the International Committee on Taxonomy of Viruses, Karger, Basel.

Mayhew, D. E., and Carroll, T. W., 1974a, Barley stripe mosaic virions associated with spindle microtubules, *Science* **185:**597–598.

Mayhew, D. E., and Carroll, T. W., 1974b, Barley stripe mosaic virus in the egg cell and egg sac of infected barley, *Virology* **58:**561–567.

Nitzany, F. E., and Kenneth, R., 1960, The identification of barley stripe mosaic virus in Israel, *FAO Plant Prot. Bull.* **8:**31–32.

Nutter, F. W., Jr., Pederson, V. D., and Timian, R. G. 1984, Relationship between seed infection by barley stripe mosaic virus and yield loss, *Phytopathology* **74:**363–366.

Paliwal, Y. C., 1980, Relationship of wheat streak mosaic and barley stripe mosaic viruses to vector and nonvector eriophyid mites, *Arch. Virol.* 63:123–132.

Partridge, J. E., Shannon, L. M., Gumpf, D. J., and Colbaugh, P., 1974, Glycoprotein in the capsid of plant viruses as a possible determinant of seed transmissibility, *Nature (London)* **247:**391–392.

Polak, Z., and Slykhuis, J. T., 1972, Comparisons of Poa semilatent and barley stripe mosaic viruses, *Can. J. Bot.* **50:**263–267.

Pring, D. R., and Timian, R. G., 1969, Physiological effects of barley stripe mosaic virus infection, *Phytopathology* **59:**1381–1386.

Sandfaer, J., 1973, Barley stripe mosaic virus and the frequency of triploids and aneuploids in barley, *Genetics* **73:**597–603.

Sandfaer, J., and Haahr, V., 1975, Barley stripe mosaic virus and the yield of old and new barley varieties, *Z. Pflanzenzuecht.* **74:**211–222.

Shalla, T. A., 1959, Relations of tobacco mosaic virus and barley stripe mosaic virus to their host cells as revealed by ultrathin sectioning for the electron microscope, *Virology* **7:**193–219.

Shalla, T. A., 1966, Electron microscopy of cells infected with barley stripe mosaic virus as a result of mechanical and seed transmission, in: *Viruses of Plants* (A.B.R. Beemster and J. Dijkstra, eds.), pp. 94–97, North-Holland, Amsterdam.

Shivanathan, P., 1970, Studies on the non-seed transmissibility of the NSP strain of barley stripe mosaic virus, Ph.D. thesis, University of McGill, Montreal, Canada.

Singh, G. P., Arny, D. C., and Pound, G. S., 1960, Studies on the stripe mosaic of barley, including effects of temperature and age of host on disease development and seed infection, *Phytopathology* **50:**290–296.

Sisler, W. W., and Timian, R. G., 1956, Inheritance of the barley stripe mosaic resistance of Modjo (CI 3212) and CI 3212-1, *Plant Dis. Rep.* **40:**1106–1108.

Slack, S. A., and Shepherd, R. J., 1975, Serological detection of seed-borne barley stripe mosaic virus by a simplified radial-diffusion technique, *Phytopathology* **65:**948–955.

Slack, S. A., Shepherd, R. J., and Hall, D. H., 1975, Spread of seed-borne barley stripe mosaic virus and effects of the virus on barley in California, *Phytopathology* **65:**1218–1223.

Slykhuis, J. T., 1972, Poa semilatent virus from native grasses, *Phytopathology* **62:**508–513.

Stanley, J., Hanau, R., and Jackson, A. O., 1984, Sequence comparison of the 3′ ends of a subgenomic RNA and the genomic RNAs of barley stripe mosaic virus, *Virology* **139:**375–383.

Suzuki, T., and Taniguchi, T., 1973, Degradation of host high molecular weight RNA in barley leaves infected with barley stripe mosaic virus, *Phytopathol. Z.* **77:**55–64.

Timian, R. G., 1967, Barley stripe mosaic virus seed transmission and barley yield as influenced by time of infection, *Phytopathology* **57:**1375–1377.

Timian, R. G., 1971, Barley stripe mosaic virus in North Dakota, *N.D. Agric. Exp. Stn. Farm Res.* **28:**3–6.

Timian, R. G., 1974, The range of symbiosis of barley and barley stripe mosaic virus, *Phytopathology* **64:**342–345.

Timian, R. G., and Sisler, W. W., 1955, Prevalence, sources of resistance, and inheritance of resistance to barley stripe mosaic (false stripe), *Plant Dis. Rep.* **39:**550–552.

Van Regenmortel, M. H. V., 1984, Monoclonal antibodies in plant virology, *Microbiol. Sci.* **1:**73–78.

Vasquez, G., Peterson, G. A., and Timian, R. G., 1974, Inheritance of barley stripe mosaic reaction in crosses among three barley varieties, *Crop Sci.* **14:**429–432.

Walters, H. J., 1954, Virus diseases in small grains in Wyoming, *Plant Dis. Rep.* **38:**836–837.

White, J., and Brakke, M., 1982, Chloroplast RNA and proteins decrease as wheat streak and barley stripe mosaic viruses multiply in expanding, systemically infected leaves, *Phytopathology* **72:**939.

Yu, S. G., Wang, M. G., Chen, Z. Y., and Zhang, R. P., 1980, Studies on seed-borne virus diseases of barley and wheat (I)—Occurrence of barley stripe mosaic virus in China, *J. (Acta) Fudan Univ. (Nat. Sci. Rep.)* **19:**373–379.

Zaske, S. K., Carroll, T. W., and Sipes, J. K., 1985, An enzyme-linked immunosorbent assay to detect barley stripe mosaic virus in barley for use in the Montana Seed Certification Program, *Phytopathology* **75:**1359 (Abst.).

CHAPTER 19

HORDEIVIRUSES

Structure and Replication

J. G. ATABEKOV AND V. V. DOLJA

I. INTRODUCTION

The hordeivirus group includes three rod-shaped viruses with multipartite genomes: barley stripe mosaic virus (BSMV), Poa semilatent virus, and Lychnis ringspot virus, which are distantly related serologically and partly differ in their host ranges.

The more important and typical features of the hordeiviruses (or at least of their most studied representative, BSMV) are as follows:

1. Multipartite nature of the genome—hordeiviruses are the only known group of rod-shaped viruses with a tripartite genome (Lane, 1974; Gustafson *et al.*, 1982).

2. Different BSMV strains have different sets of virion RNA species (Jackson and Brakke, 1973; Lane, 1974). The natural tripartite (orthotripartite) strains contain RNA 1 (4000 nucleotides), RNA 2a (3300 nucleotides), and RNA 3 (3000 nucleotides). In pseudobipartite strains the second and third components of the genome are of the same size and are designated RNA 2a and RNA 2b (3300 nucleotides). RNA 2b is homologous to RNA 3 and appears to be its functional analogue (Gustafson *et al.*, 1982; Dolja *et al.*, 1983b).

There also are pseudoquadripartite BSMV strains containing an additional virion RNA, RNA 4 (2600 nucleotides) homologous to RNA 3 (and RNA 2b). RNA 4 is functionally nonessential and can be eliminated from the BSMV genome (McFarland *et al.*, 1983) or artificially added to the natural tripartite strain (Dolja *et al.*, unpublished).

3. The BSMV coat protein gene is located in the 5′-terminal region

J. G. ATABEKOV AND V. V. DOLJA • Department of Virology, Moscow State University, Moscow 119899, USSR.

of the second genomic component—RNA 2a—and is actively translated therefrom (Dolja *et al.*, 1979, 1982, 1983b).

4. Unique arrangement of the 3′-terminal region of BSMV RNA—this region contains a tandem of an internal poly(A) tract of varying length and a 3′-terminal tRNA-like structure accepting tyrosine (Agranovsky *et al.*, 1981, 1982; Kozlov *et al.*, 1984).

5. BSMV coat protein is a glycoprotein (Partridge *et al.*, 1974); it can be repolymerized *in vitro* to yield a number of stable intermediate aggregates (Atabekov *et al.*, 1968a,b,c). The BSMV protein interacts nonselectively with RNAs of different foreign viruses *in vitro* (Atabekov *et al.*, 1970a,b) and *in vivo* (Dodds and Hamilton, 1971, 1974).

6. Hordeiviruses can be vertically transmitted (McKinney and Greeley, 1965; Carroll, 1974).

7. BSMV infection induces mutations in host plants (Sprague *et al.*, 1963; Sprague and McKinney, 1966, 1971). This effect was designated "aberrant ratio" because such mutations distort Mendelian ratios for several phenotypic characteristics.

Comprehensive reviews on hordeiviruses have been published by McKinney and Greeley (1965) and Jackson and Lane (1981). We have little to add to their discussions; furthermore, the biology of the viruses is treated in detail by Carroll in this volume. Thus, this chapter deals with some aspects of BSMV structure, and first to be considered are the structure and expression of its genome, to illustrate current progress in this field.

II. GENERAL BIOLOGICAL CHARACTERISTICS

BSMV, the most studied hordeivirus, was recognized by McKinney (1953) as the causative agent of the disease that for many years had been known as "false stripe" in barley.

BSMV has a worldwide distribution. Natural hosts are members of the Gramineae, but species of Chenopodiaceae and Solanaceae can be experimentally infected (for review see Jackson and Lane, 1981; Atabekov and Novikov, 1971).

Numerous BSMV strains have been isolated which differ in symptoms induced in barley, wheat, and oat: stripe mosaic, from very mild to yellow-leaf and white-leaf types, extreme dwarf, and lethal necrosis (McKinney and Greeley, 1965). Mild and moderate strains are the most common.

Most BSMV strains have been successfully maintained in culture for many transfers, but some variants become difficult or impossible to transfer by sap inoculation (McKinney and Greeley, 1965). Why this happens is not known.

No vectors have been reported for BSMV transmission. The virus is pollen- and seed-transmitted with up to 90% of seedlings from the seeds of diseased plants being infected (McKinney, 1953; Hagborg, 1954; Singh *et al.*, 1960; McKinney and Greeley, 1965; Carroll and Chapman, 1970).

Seed transmissibility of a number of BSMV strains has been studied (Singh *et al.*, 1960; Hamilton, 1965; McKinney and Greeley, 1965; Slack *et al.*, 1975); it is of interest that some strains are inherently weak in infecting the embryo or are inactivated upon seed maturation (Hamilton, 1965; Carroll, 1972). Factors determining seed transmissibility or lack of it are obscure.

Unfortunately, the other representatives of the hordeiviruses have not been studied extensively.

The second hordeivirus, Poa semilatent virus (PSLV), was isolated by Slykhuis and Atkinson (1966). Its host range includes some Gramineae (*Triticum aestivum*, *T. durum*, *Secale cereale* L., *Zea mays*, *Avena sativa* L., *Hordeum vulgare*, *Poa palustris* L., among others) (Slykhuis, 1972). PSLV causes mosaic, chlorosis, and blight on wheat and oat but only mild transitory mottling on *P. palustris*. In contrast to BSMV, PSLV is not seed-transmitted (at least in *P. palustris*). PSLV is serologically distantly related to BSMV and resembles it in host range, symptomatology, and morphology of virus particles (Slykhuis, 1972).

The third and last hordeivirus, Lychnis ringspot virus (LRSV), was isolated by Bennett (1959) from *Lychnis divaricata* Reichenb. grown in a greenhouse. Later, Gibbs *et al.* (1963) found that LRSV and BSMV are similar in virion structure and are distantly related serologically. LRSV differs considerably from BSMV and PSLV in host range; the virus infects various dicotyledonous plants including some crop plants (*Beta vulgaris* L., *Spinacia oleracea* L., *Tetragonia expansa* L.) (Bennett, 1959).

III. THE VIRIONS AND THEIR SUBSTRUCTURAL COMPONENTS

Among hordeiviruses, only BSMV has been subjected to physicochemical and structural studies. However, even for BSMV the information concerning the structural organization of virions is scanty.

A. Purified BSMV

Preparations of BSMV were purified and partly analyzed in the late 1950s (Moorhead, 1956; Brakke, 1959; Kassanis and Slykhuis, 1959). BSMV accumulates in the infected cells in rather large amounts, average yields being 100–200 mg/kg fresh leaf material (Atabekov and Novikov, 1971).

The infective virus stock contains particles of different lengths corresponding to different virion RNA species of the multipartite BSMV genome separately encapsidated with the coat protein. Strain Norwich comprises particles of 143, 128, and 112 nm (Brakke and Palomar, 1976).

It has been repeatedly reported that infectivity of BSMV (or at least of some BSMV strains) decreases upon purification (Kassanis and Slykhuis, 1959; Brakke, 1962). Purified preparations frequently show little infectivity or even none, yet when the RNA of totally noninfective preparations is extracted it is often found to be infective (Atabekov and Novikov, 1966; Novikov and Atabekov, 1970). Several authors have found that linear aggregation of BSMV virions occurs upon virus purification (Kassanis and Slykhuis, 1959; Brakke, 1962; Harrison *et al.*, 1965).

Purified BSMV is most stable at pH 6.0–7.6; its stability drops sharply at pH below 5.5 or above 7.5; the thermal inactivation point for BSMV is about 70°C (Kassanis and Slykhuis, 1959; Brakke, 1962; Kiselev *et al.*, 1966).

BSMV particles are tubular rods 20 nm in diameter, helically constructed with the pitch of the basic helix 2.5–2.6 nm; the diameter of the inner canal is 3.4 nm (Gibbs *et al.*, 1963; Harrison *et al.*, 1965; Finch, 1965).

The RNA chain seems to be located at a radius of 5.5 nm from the long axis of the virion (Finch, 1965). The structural details of the packing of BSMV RNA within the virion are obscure. It can be suggested that virion RNA molecules lie in the groove formed by the helically arrayed protein subunits and make no secondary foldings. Brakke (1979) reported that RNA in the BSMV virion is hyperchromic; the estimated hyperchromicity was 116%, compared to 130% in TMV, which suggests some structural differences between these viruses. The RNA content in BSMV particles was estimated at 3.8–4.0% (Atabekov and Novikov, 1971) or 3.7% (Brakke, 1979).

B. Coat Protein

The BSMV protein dissociates completely to monomeric units ($s^0_{20,w}$ = 1.8 S) under the influence of various agents including high concentrations of divalent cations or KCl, 1% sodium dodecyl sulfate (SDS), acetic acid, and alkaline pH (Atabekov and Novikov, 1966, 1971; Kiselev *et al.*, 1966; Atabekov *et al.*, 1968a; Gumpf and Hamilton, 1968).

The molecular weight of the monomeric BSMV protein determined by different methods is about 22–23 × 10^3 (Kiselev *et al.*, 1966; Atabekov *et al.*, 1968a; Dolja *et al.*, 1979), which corresponds to 187–190 amino acid residues. The α-helical content of the BSMV protein does not exceed 7% (Atabekov *et al.*, 1968a). The protein contains relatively large amounts of polar amino acids and proline but no cysteine, cystine, or methionine. Unlike the TMV protein, the BSMV protein contains histidine (Atabekov *et al.*, 1968a).

The N-terminal amino acid sequence of two BSMV strains has recently been determined (Baratova *et al.*, 1983; Dolja *et al.*, 1982):

10
Pro-Gln-Val-Ser-Leu-Thr-Ala-Lys-Gly-Gly-Gly-His-Tyr-Asn-Glu

20 30
Asp-Gln-Trp-Asp-Thr-Gln-Val-Val-Glu-Ala-Gly-Val-Phe-Asp-Asp

Partridge *et al.* (1974) found the BSMV protein to be a glycoprotein, the carbohydrate consisting of glucose, mannose, xylose, galactosamine, and glucosamine and linked through an amide bond involving asparagine (Gumpf *et al.*, 1977).

The capsid protein isolated from a fresh highly purified BSMV preparation yields a single band in gel electrophoresis under denaturing conditions; however, pronounced degradation takes place in the course of its storage at 0°C if SDS is not added. The breakdown of the 22K BSMV protein probably results from contamination of the preparations by host-cell proteases, and proceeds by discrete steps (Negruk *et al.*, 1974).

C. Polymerization of the Capsid Protein and Virion Reassembly

The viral protein exists as a monomer only in the presence of disaggregating agents; after their removal it can be assembled *in vitro* into structures resembling those of the intact virus (Kiselev *et al.*, 1966, 1969; Atabekov *et al.*, 1968a,b,c; Gumpf and Hamilton, 1968). Polymerization of the BSMV protein proceeds in a stepwise fashion, and a number of stable intermediates of increasing size may be formed. The following BSMV protein aggregates have been defined (Atabekov *et al.*, 1968a): (i) 9–12 S (designated 10 S), (ii) 20–23 S (designated 20 S), (iii) 27–31 S (designated 30 S), (iv) 34–36 S, and (v) 40–42 S (designated 40 S). It is probable that all protein intermediates always coexist in solutions of the BSMV protein in equilibrium with each other, but that different aggregates prevail under different conditions (Atabekov *et al.*, 1968a).

The actual structure of each aggregate is not clear, nor is it possible to say in which way subunits actually polymerize. However, it is certain that the 30 S aggregate is a double disk, each disk appearing to be composed of 21 protein subunits (Atabekov *et al.*, 1968a). It should be noted that monodisks and triple disks, as well as disks consisting of 4–6 layers of subunits, were observed together with double-disk particles. Various types of crystalline aggregates can be formed by aggregation of disks (see, e.g., Fig. 1a, b).

Cations may play a specific role in the organization of the macromolecular structure of BSMV protein aggregates. At a comparatively low ionic strength, divalent cations are extremely effective in polymerizing the BSMV protein: calcium ions at an ionic strength of about 0.01 or less

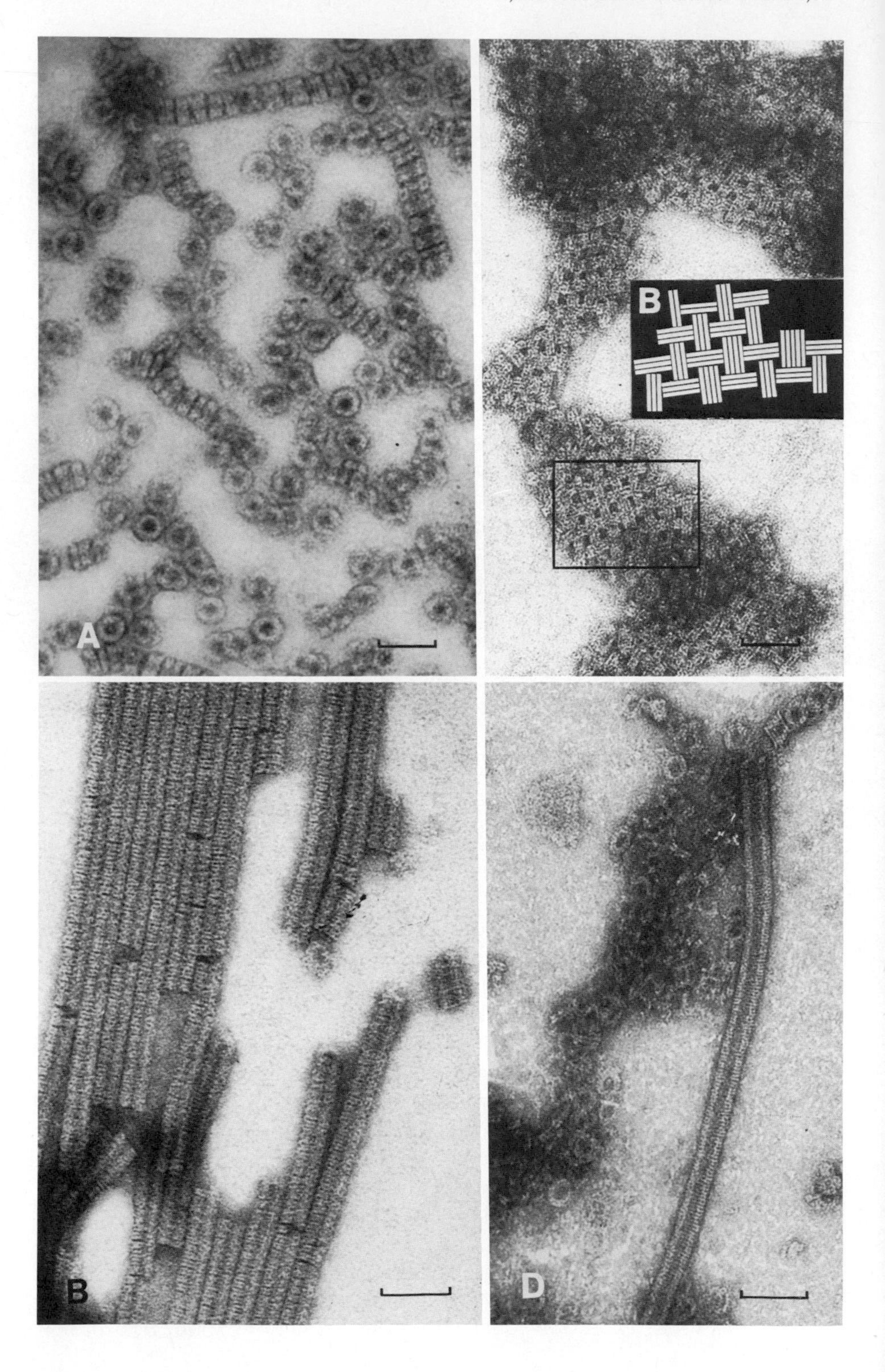
A
B
B
D

can induce the formation of rodlike particles of helically polymerized protein 1 μm or more in length. (see Fig. 1d). Under the same conditions (pH, ionic strength, time of incubation) monovalent cations induce only the first polymerization steps, yielding the 10 S aggregate.

By contrast, high ionic strength brings about dissociation of the BSMV protein aggregates as well as of intact BSMV; it is of interest that Ca^{2+} is much more potent than monovalent cations both in its polymerizing and depolymerizing capacity (Atabekov *et al.*, 1968a).

Macromolecular aggregates of another type can occasionally be observed in repolymerized protein preparations (Fig. 1c); Kiselev *et al.* (1969) have found these particles to have an extraordinary two-start helical structure.

The rodlike particles of polymerized BSMV protein can interact (in 0.01–0.05 M $CaCl_2$) by their sides, giving rise to paracrystalline aggregates. Such paracrystals are needlelike and have large enough dimensions to be seen in the light microscope.

It seems probable that the viral protein undergoes conformational changes during polymerization; the immunological aspects of the BSMV protein polymerization have been studied by Atabekov *et al.* (1968b,c) who found that all protein aggregates from 10 S up and including 20 S, 30 S, 40 S, and helically polymerized particles are antigenically identical with each other and with intact BSMV. This means that no neotopes are formed in the course of polymerization, starting from the 10 S-aggregate level. The monomeric BSMV protein subunit is serologically related to, but not identical with, intact BSMV (Atabekov *et al.*, 1968b,c).

In contrast to BSMV, polymerization of the TMV protein is accompanied by multistep antigenic shifts. The degree of antigenic similarity between intact TMV and various protein intermediates appears to increase as the aggregate grows in size. Only macromolecular rodlike aggregates of helically polymerized TMV protein were found to be serologically identical to intact TMV. Even double-disk and stacked-disk structures do not show identity to the intact virus (Rappaport, 1965).

The 20 S double disk is transformed into a helical structure on interaction with RNA molecules. This transformation could be mimicked with short RNA fragments, giving rise to nucleoprotein "helical disks" antigenically identical to intact TMV (Tyulkina *et al.*, 1975; Mizenina *et al.*, 1984).

Atabekov *et al.* (1970b) demonstrated the possibility of *in vitro* reassembly between the BSMV protein and viral RNA giving rise to rodlike

←

FIGURE 1. Electron micrographs of BSMV protein aggregates. (a) Disklike structures; disks connected in pairs and "stringing" of such pairs of disks (0.05 M $CaCl_2$, pH 8.0). (b) Protein aggregates built of disklike particles of different thickness joined edge-to-plane, and the scheme corresponding to the outlined section (0.05 M $CaCl_2$, pH 9.0). (c) Rod-shaped aggregates of polymerized protein with double-helical structure (0.01 M $CaCl_2$, pH 9.0). (d) Putative helical rod-shaped particles of polymerized protein (0.01 M $CaCl_2$, pH 7.5).

nucleoprotein particles; the efficiency of reassembly did not depend on the state of protein aggregation (10 S, 20 S, 30 S) or the RNA source (BSMV, TMV, brome mosaic virus, cucumber virus 4, potato virus X). Unfortunately, no infective virus could be reassembled *in vitro* from BSMV protein and viral RNA, including its own RNA (Atabekov *et al.*, 1970a).

It is of interest that *in vivo*, in doubly infected barley plants, the BSMV protein can interact not only with its homologous RNA but also with TMV RNA, thereby producing viral particles with masked genome (TMV RNA encapsidated in BSMV protein) (Dodds and Hamilton, 1971, 1974). The reverse situation, i.e., masking of BSMV genome by TMV protein, does not occur in the mixedly infected cells. This observation may reflect the fact that the TMV protein recognizes specifically its homologous RNA, whereas the BSMV protein interacts nonselectively with viral RNAs of different origin.

IV. GENOME STRUCTURE AND EXPRESSION

A. Multipartite Nature of the Genome

Pring (1972) was the first to suggest the multipartite nature of the BSMV genome. Working with the strain North Dakota 18 (ND 18), he discovered the heterogeneity of the BSMV RNA preparation: sucrose gradient centrifugation resolved it into two components with sedimentation coefficients of 21.3 S and 19.5 S.

Electrophoretic analysis made it possible to reveal two to four components in the preparations of virion RNA from different hordeiviruses (Jackson and Brakke, 1973; Lane, 1974). Virion RNA of PSLV and LRSV were shown to contain three major components and a minor one. For BSMV, "bipartite" (type, Compana), tripartite (Norwich, ND 18), and even "quadripartite" strains (Argentina Mild or AM) have been described (see Fig. 2) The BSMV virion RNA species are designated in the order of decreasing length as RNA 1, 2, 3, and 4. RNAs 1 and 2 are present in all BSMV strains without exception; tripartite strains also contain RNA 3, and "quadripartite," RNAs 3 and 4. Strains having RNAs 1, 2, and 4 but devoid of RNA 3 have not been found. The molecular weights of BSMV RNAs determined by gel electrophoresis in denaturing conditions are: RNA 1, 1.35×10^6 (ca. 4000 nucleotides); RNA 2, 1.1×10^6 (3300 nucleotides); RNA 3, 1.0×10^6 (3000 nucleotides); RNA 4, 0.85×10^6 (2600 nucleotides) (Dolja *et al.*, 1983a).

Studies of the functional role of individual BSMV RNAs showed that replication of the "bipartite" strain (i.e., the one displaying two RNA bands in gel electrophoresis) requires the presence of both RNA fractions in the inoculum (Jackson and Brakke, 1973). The tripartite strain efficiently propagates only in the presence of all three virion RNAs (Lane, 1974). The question of the actual number of components of the multi-

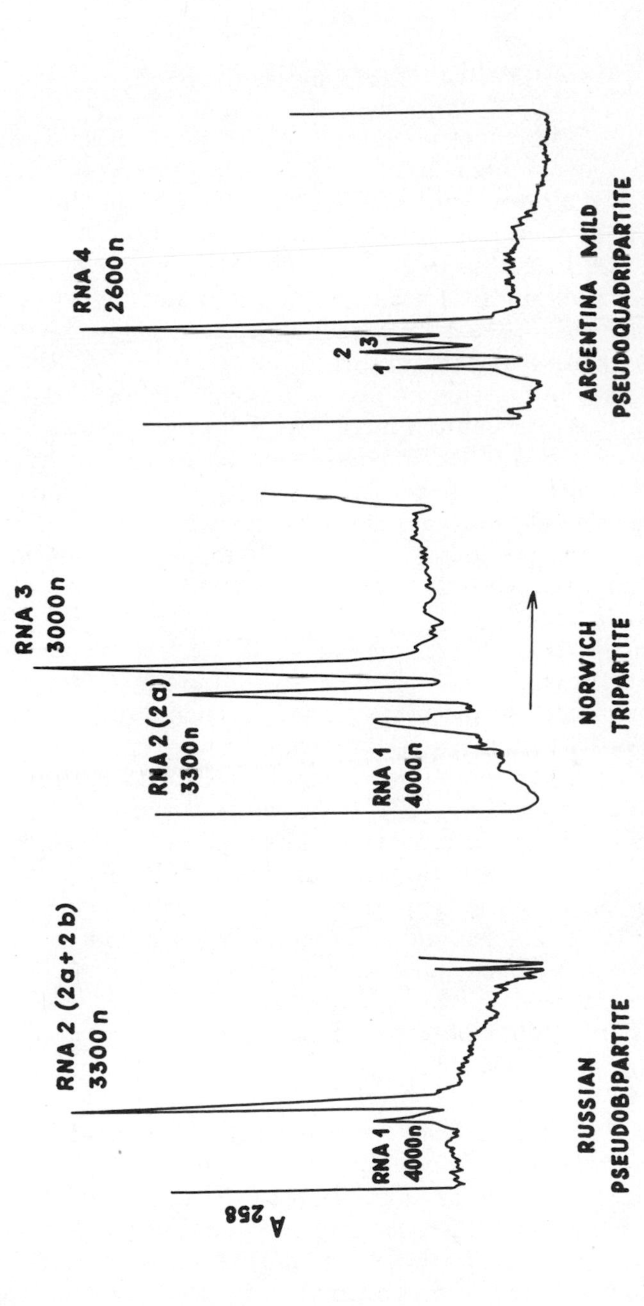

FIGURE 2. Densitometer tracing of 3% polyacrylamide gels after the separation of RNA from different BSMV strains. The arrow indicates the direction of electrophoresis. The vertical line formed by recorder deflection at the left edge of each pattern denotes the origin.

partite BSMV genome was solved on the basis of the studies of nucleotide sequence homology between particular RNAs.

B. Sequence Relationship among BSMV RNAs

Despite the differences in the number of virion RNA components, BSMV strains have a high degree of genomic homology, i.e., they are closely related (Palomar *et al.*, 1977; McFarland *et al.*, 1983); they are also antigenically identical (Novikov, unpublished).

BSMV RNA 1 appears to have a unique nucleotide sequence, or at least does not contain significant regions homologous to other virion RNAs (Palomar *et al.*, 1977; Gustafson *et al.*, 1982). On the other hand, the sequence of RNA 3 is highly homologous to that of RNA 2 of the "bipartite" strain but not homologous to the sequence of RNA 2 of the tripartite strain. At the same time, RNAs 2 of the two strains are highly homologous to one another (Gustafson *et al.*, 1982; McFarland *et al.*, 1983). These results prompted a suggestion that RNA 2 of the "bipartite" strain is a mixture of two types of molecules which have similar length but different primary structure: RNA 2a homologous to RNA 2 of the tripartite strain, and RNA 2b homologous to RNA 3 (Gustafson *et al.*, 1982; Dolja *et al.*, 1983b).

Thus, the functionally segmented BSMV genome is tripartite, but at the strain level it can be composed of four different RNA segments: (1) natural tripartite (orthotripartite) strains (Norwich, ND 18) contain RNA 1, RNA 2a, and RNA 3; (2) pseudobipartite strains (type, Russian) contain RNA 1, RNA 2a, and RNA 2b which appears to be an analogue of, though 300 nucleotides longer than, RNA 3 of natural tripartite strains; (3) pseudoquadripartite strains (AM) contain RNA 1, RNA 2a, RNA 3, and RNA 4 (see Fig. 5). RNA 4 is functionally nonessential, highly homologous to RNA 3 (and to RNA 2b), and appears to be its naturally occurring fragment (Dolja *et al.*, 1983b; McFarland *et al.*, 1983) (see Section IV c and f).

C. Changes in the Set of Virion RNAs upon Laboratory Passaging of BSMV

Infection of plants with a dilute inoculum of a "quadripartite" BSMV strain leads to total loss of RNA 4 from the viral progeny, while the virus yield is unaffected (Palomar *et al.*, 1977). If RNA 4 of the AM strain is artificially added to the three-component mixture of RNA from strain Norwich, it can replicate in the infected plant, apparently owing to complementation (Dolja *et al.*, unpublished). It is interesting to note that the pseudoquadripartite strain AM passaged under laboratory conditions contains a mixture of RNA 2b and RNA 3. Its subcloning yields pseudobipartite and tripartite isolates having no RNA 4 (McFarland *et al.*, 1983).

In our hands the original AM isolate obtained from Dr. M. K. Brakke contained no RNA 2b (Dolja *et al.*, 1983b); however, RNA 2b appeared after 2 years of passaging.

A peculiarity of BSMV is the variable relation between individual virion RNAs in the viral progeny. This relation can be affected by a number of factors: the temperature of infection and its duration, the degree of dilution of the inoculum (Palomar and Brakke, 1976), and probably the influence of the host, its species, and cultivar. Thus, prolonged passaging of the natural tripartite strain Norwich (Norwich III) on wheat led to a marked decrease in the content of RNA 3. The ratio RNA 1:RNA 2:RNA 3 in the initial isolate was 2:3:5 and in 2 years became 2:5:3. Passaging of the same initial isolate on barley cultivar Xenia already after ten passages resulted in total loss of RNA 3 (Dolja *et al.*, 1983b).

During laboratory passaging of the pseudobipartite strain, appearance of RNA 3 was occasionally observed in the progeny (Jackson and Brakke, 1973). All these variants of interconversions between pseudobi-, pseudoquadri-, and tripartite isolates are poorly documented since they occur rarely, under insufficiently controlled conditions, and for obscure reasons. Nevertheless, they are hardly likely to always result merely from selection of admixtures. This point of view is also supported by the finding, in some BSMV isolates, of RNAs that are intermediate between RNA 2b and RNA 3 in size (McFarland *et al.*, 1983) or coding properties (Dolja *et al.*, 1983b). It seems that, unlike the genetically stable BSMV genomic components (RNAs 1 and 2a), RNA 2b is labile.

D. Structure of the 3′-Terminal Region of BSMV RNAs

1. Tandem Coupling of the Poly(A) Tract and the tRNA-Like Structure

The 3′-terminal region of BSMV virion RNAs contains both a poly(A) tract (Agranovsky *et al.*, 1978) and a tRNA-like structure capable of specifically binding tyrosine (Agranovsky *et al.*, 1981). Interestingly, the poly(A) tract is heterogeneous in length (8–40 nucleotides) and is located inside the RNA molecules, more than 200 nucleotides from the 3′-end represented by the tRNA-like structure (Agranovsky *et al.*, 1982, 1983). Sequencing of a recombinant clone containing a copy of the 3′-terminal region of BSMV RNA revealed a poly(A) tract of 21 nucleotides at a distance of 236 nucleotides from the 3′-terminal adenosine (Kozlov *et al.*, 1984). BSMV RNA is the only RNA as yet found to show such tandem arrangement of two functionally important structures.

2. Primary and Secondary Structure of the Tyrosine-Accepting Region

All four individual RNAs of various BSMV strains are capable of accepting tyrosine (Agranovsky *et al.*, 1981). The sequence of at least 100

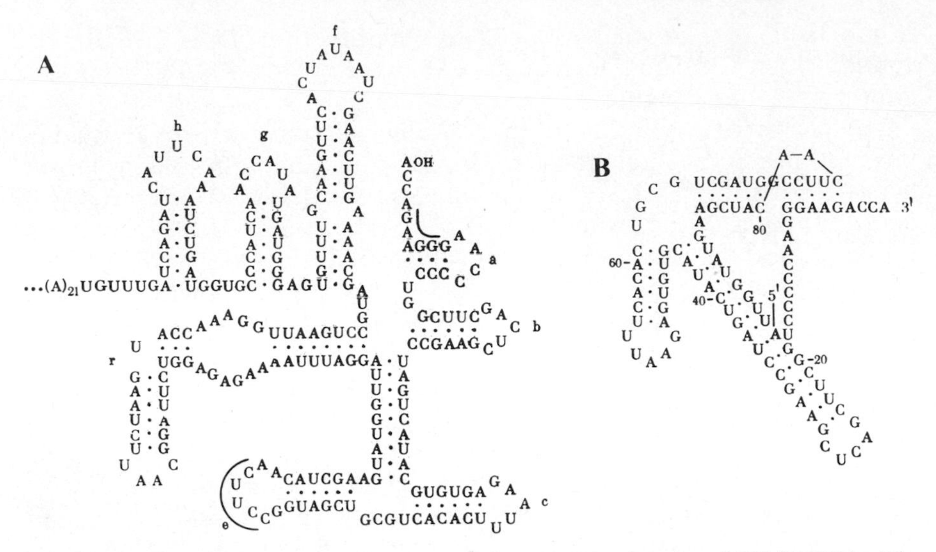

FIGURE 3. Possible secondary structures of the 3′-terminal region of BSMV RNAs. The stems and loops (A) are designated with lowercase letters in accordance with Ahlquist *et al.* (1981a). The sequences of the stem "a" and the loop "e" involved in the secondary structure of the other type (B) are indicated by lines.

nucleotides from the 3′-end is the same in all genomic RNAs of strain Norwich and differs by only several nucleotides from that of another BSMV strain, AM (Kozlov *et al.*, 1984).

The primary structure of this region of BSMV RNA shares some features with $tRNA^{Tyr}$ from yeast (the position of the tyrosine anticodon, tetranucleotide UUCG—an analogue of TΨ CG), but cannot be folded into the classical tRNA cloverleaf model. The secondary structure of the tRNA-like region of BSMV RNA (see Fig. 3) is very similar to that of the RNA of bromo- and cucumoviruses also accepting tyrosine (cf. Rietveld *et al.*, 1983; Joshi *et al.*, 1983; Kozlov *et al.*, 1984). By the absence of the "d" loop, it most of all resembles the RNA of broad bean mosaic virus (Ahlquist *et al.*, 1981a). However, the minimal length predicted from the secondary structure for a sequence packed in accordance with the criteria of Joshi *et al.* (1983) and still capable of being aminoacylated is, for BSMV, only 95 nucleotides—the smallest tyrosine-accepting tRNA-like structure known heretofore (Kozlov *et al.*, 1984).

Removal of the terminal tRNA-like structure by specific cleavage of the poly(A) tract with RNase H in the presence of oligo(dT) leads to total loss of infectivity of BSMV RNA. At the same time, its coding properties in the cell-free system are not changed (Agranovsky *et al.*, 1982). Genomic and at least some subgenomic viral RNAs are known to be tyrosylated in BSMV-infected barley protoplasts, whereas in the virion they are deacylated (Loesh-Fries and Hall, 1982). The tRNA-like structure appears to play a prominent role in virus reproduction, most probably in the replication of its RNA (Haenni *et al.*, 1982).

3. Internal Heterogeneous Poly(A) Tract

When chromatographed on poly(A)-specific sorbents, BSMV RNA resolves into poly(A)$^+$ (adsorbed) and poly (A)$^-$ (unadsorbed) fractions (Agranovsky *et al.*, 1978). The difference between poly(A)$^+$ and poly(A)$^-$ RNA was shown to be rather quantitative than qualitative, both containing poly(A) but of different length: poly (A)$^+$ of 20–40 nucleotides and poly(A)$^-$ of 8–25 nucleotides (Agranovsky *et al.*, 1983; Qi *et al.*, 1984).

RNAs of the poly(A)$^+$ and poly(A)$^-$ fractions are equally infective (Agranovsky *et al.*, 1978). Surprisingly, the progeny from either fraction again shows the initial distribution into poly(A)$^+$ and poly(A)$^-$ fractions when chromatographed on oligo(dT)-cellulose. In other words, replication of RNA molecules with short polyadenylates may give rise to molecules with long polyadenylates and vice versa (Agranovsky, Dolja, and Atabekov, unpublished).

This unexpected observation was confirmed in model experiments. As already mentioned, upon cleavage at the poly(A) tract with RNase H and dT_{10}, BSMV RNA loses infectivity. If the two parts are again joined with RNA ligase, infectivity is restored. However, such RNA is completely unable to bind to oligo(dT)-cellulose, since the poly(A) tract after cleavage and ligation, if at all present, does not exceed four residues (Rodionova *et al.*, unpublished). The progeny of this RNA with the artificially shortened poly(A) again displays a standard resolution into poly(A)$^+$ and poly(A)$^-$ fractions. Thus, replication of the internal variable-length poly(A) sequence appears to involve a mechanism other than the orthodox template copying.

It must be noted that the individual genomic components are represented in the poly(A)$^+$ and poly(A)$^-$ fractions at markedly altered ratios as compared to the total preparation of virion RNA. The bulk of RNA 1 is found in the poly(A)$^+$ fraction whereas that of RNA 2a is found in the poly(A)$^-$ fraction of total RNA. RNA 3 is distributed roughly equally between these fractions and RNA 4 is mostly found in the poly(A)$^-$ fraction (Agranovsky *et al.*, 1978, 1983). In other words, RNA 1 of all BSMV strains studied on average contains longer poly(A) sequences than RNAs 2 and 4. In the progeny of the artificially prepared poly(A)$^-$ RNA (see above), all individual RNAs restore their natural average lengths of poly(A). Thus, the bulk of RNA 1 again goes into the poly(A)$^+$ fraction (Dolja *et al.*, 1986b). The functional significance of such genetically conserved differences in the structure of the 3′-terminal regions of individual BSMV RNAs is unknown.

E. Translation *in Vitro* of Individual BSMV RNAs

Individual BSMV RNAs were translated in cell-free systems from wheat embryos and rabbit reticulocytes (Dolja *et al.*, 1979, 1983b). Since

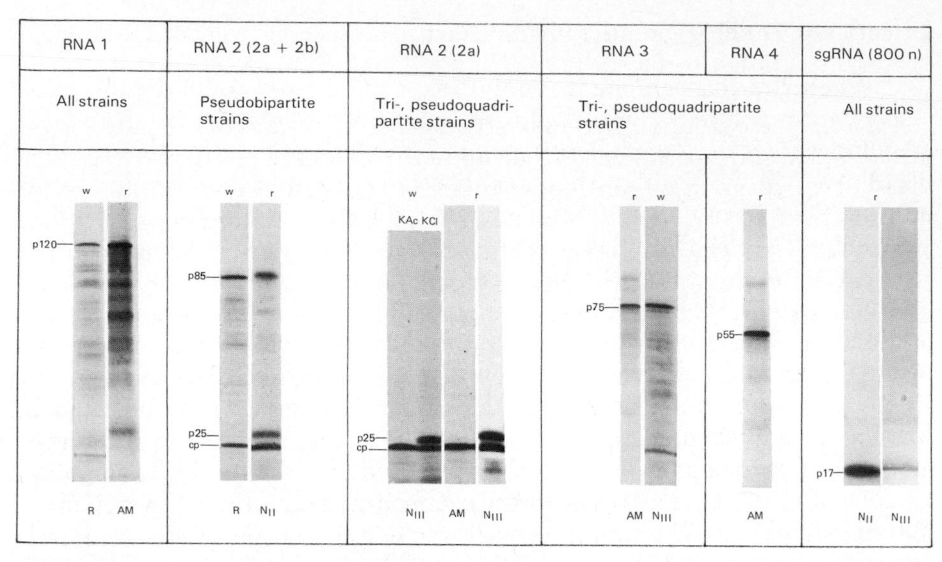

FIGURE 4. Electrophoretic analysis of the products translated from individual BSMV RNAs in the wheat embryo cell-free systems (w) and in the reticulocyte lysate (r). R, Russian; AM, Argentina mild; N_{II}, pseudobipartite isolate; N_{III}, original tripartite isolate of Norwich strain. For details see text (Sections IV.E, IV.F).

BSMV infects wheat, the first system can be considered homologous for this virus. No significant differences could be found in the sets of polypeptides produced upon BSMV RNA translation in the two systems (Dolja *et al.*, 1983b).

1. Translation Products of RNA 1

RNA 1 has the same size in all BSMV strains and codes *in vitro* for a polypeptide of molecular weight 120,000 (p120) (see Fig. 4) (Dolja *et al.*, 1979, 1983b). The sequence of 300 nucleotides from the 3′-end of the p120 gene was determined (Kozlov *et al.*, personal communication). The amino acid sequence of this region is homologous to 110 K and 130 K polypeptides of brome mosaic and tobacco mosaic viruses; these polypeptides have been shown to be the components of RNA replicases (Bujarsky *et al.*, 1982; Cornelissen and Bol, 1984). By analogy with these plant viruses, p120 coded for by BSMV RNA 1 can be thought to be a virus-specific replicase. The gene coding for p120 exhausts about 80% of the coding capacity of RNA 1. It is not yet clear whether the RNA is monocistronic or the remaining part codes for some additional viral proteins.

2. Translation Products of RNA 2a and Exact Location of the Coat Protein Gene

The RNAs of all BSMV strains carry and actively express *in vitro* the coat protein gene (Negruk *et al.*, 1974; Dolja *et al.*, 1979, 1983b). The coat protein, or p23, is the main translation product of RNA 2 of tripartite and pseudoquadripartite BSMV strains, that is, of RNA 2a (Fig. 4). However, certain conditions induce the synthesis of another polypeptide (p25) coded for by RNA 2a. This polypeptide is not precipitated with antiserum to BSMV and, unlike the coat protein, contains methionine (Dolja *et al.*, 1983b). The intensity of p25 synthesis varies with the ionic conditions of *in vitro* translation and with the stock of the cell-free system. Comparison of tryptic fingerprints of the coat protein and p25 revealed a high degree of similarity (Lunina and Dolja, unpublished). p25 probably is the precursor of the mature coat protein or a product of the ribosomes reading through a weak terminator.

To exactly locate the coat protein gene on RNA 2a, the sequence of 30 N-terminal amino acids in the native coat protein has been determined (Baratova *et al.*, 1983). From this sequence, three amino acids were selected (residues 16 to 18, Asp-Gln-Trp) for which the code is the least degenerated: GA$^{A}_{C}$CA$^{A}_{C}$UGG. Then a nonadeoxyribonucleotide was synthesized which is complementary to the corresponding site in the coat protein gene: dCTG GTT ACC. This oligonucleotide was used as a guide for specific RNA cleavage with RNase H from *E. coli* according to Stepanova *et al.* (1979) and as a primer for reverse transcription. Both approaches gave the same result: the codon for Trp 18 of the coat protein is at a distance of about 145 nucleotides from the 5′-end of RNA 2a both of tri- and "bipartite" BSMV strains (Dolja *et al.*, 1982). It must be noted that the efficiency of cleavage and reverse transcription was only about half with RNA 2 of pseudobipartite strains. This is apparently due to the fact that RNA 2a carrying the coat protein gene constitutes only a portion of the RNA population of this strain.

The cleaved-off 5′-terminal fragment of RNA, as well as cDNA to this region, were sequenced. The noncoding leader sequence of RNA 2a is 90 nucleotides long and is enriched in uracil and adenine (about 65%) (Dolja *et al.*, 1982). The codon for proline, the first amino acid of the native coat protein (Baratova *et al.*, 1983), is immediately preceded by the initiatory methionine codon. Since the BSMV coat protein does not contain methionine, this amino acid is apparently removed after protein synthesis both *in vitro* and *in vivo*.

The AUG codon of the coat protein (p23) is the first from the 5′-end of RNA 2a, and since fingerprints of p23 and p25 differ only in 2 spots out of 27, it is highly probable that these polypeptides have the same N-terminus. P25 appears to contain an extension of some 20 amino acids at the C-terminus. It is not yet clear whether synthesis of p25 has any physiological significance.

3. Translation Products of RNAs 2b, 3, and 4; Comparative Peptide Analysis of Polypeptides Encoded in BSMV Genomes

The RNAs 2 of all pseudobipartite BSMV strains we have studied (Russian, type, Norwich II), unlike the RNAs 2 of the tripartite strains, code *in vitro* besides for the coat protein and p25, for another polypeptide (p85) (Fig. 4) (Dolja *et al.*, 1979, 1983b). As mentioned above, RNA 2 of the "bipartite" strains is a population of molecules of the same size but different primary structure, RNAs 2a and 2b (Gustafson *et al.*, 1982). Hence, a reasonable suggestion is that it is RNA 2b that carries the gene coding for p85.

Recently we were succeeded in separating of RNAs 2a and 2b of pseudo-bipartite Norwich II strain. Due to the differences in poly(A) length, these RNAs are found in different fractions under the chromatography on oligo(dT)-cellulose: RNA 2a mostly in poly(A)$^-$ fraction but RNA 2b in poly(A)$^+$-fraction. Translation in the cell-free system demonstrated that it is RNA 2b that codes for p85 and RNA 2a for coat protein (Dolja *et al.*, 1986b).

Translation of RNA 3 of both tripartite and pseudoquadripartite BSMV strains gives rise to a single major product—p75 (Fig. 4) (Dolja *et al.*, 1979, 1983b).

Sequencing of the 3′-half of RNA 3 molecule made it possible to determine the carboxy-terminal amino acids of p75 (Kozlov *et al.*, personal communication). The sequence contained a pair of asparagines surrounded by nonpolar amino acids and was highly homologous to the conservative domain found by Kamer and Argos (1984) in quite a number of RNA replicases of plant, animal and bacterial viruses. P75 as well as its functional analog p85 seems to participate in the replication of BSMV RNA. If this is true, BSMV like bromo-, cucumo-, and tobamoviruses, replicates with the help of two virus-specific proteins; here they are p120 and p75 (or p85).

Finally, RNA 4 of strain AM is translated in the cell-free system as a monocistronic template to yield p55 (Fig. 4) (Dolja *et al.*, 1983b).

Since RNA 2b, RNA 3, and RNA 4 are highly homologous (Palomar *et al.*, 1977; McFarland *et al.*, 1983), a question arises whether the genes coding for p85, p75, and p55 overlap. Comparison of their fingerprints showed that at least 26 out of 34–37 tryptic peptides labeled with [^{35}S]methionine are common for all three proteins. On the other hand, fingerprinting of p120 revealed almost no peptides common with these three proteins (Dolja *et al.*, 1983b). The position of the overlapping region in the amino acid sequences of p85, p75, and p55 is not clear; nevertheless, it can be concluded that homologous regions are open for translation on RNAs 2b, 3, and 4, and in the same reading frame.

Thus, all BSMV virion RNAs are translated *in vitro* as monocistronic molecules. RNA 1 codes for p120, RNA 2a for the coat protein (and for its extended variety p25), and RNA 2b and homologous RNAs 3 and 4 for p85, p75, and p55, respectively. Amino acid sequences of the last three proteins overlap.

It must be noted that, besides the coat protein, three BSMV-coded (or BSMV-induced) proteins have been found *in vivo* in BSMV-infected leaves, having molecular weights of 19,000, 67,000, and 120,000 (Gustafson *et al.*, 1981). The last polypeptide probably corresponds to p120 translated from RNA 1 *in vitro* (Dolja *et al.*, 1979) and the 19K product to p17 coded for by the subgenomic RNA derived from RNA 3 and RNA 2b (see Section IV.F). The origin of the 67K protein is obscure.

BSMV infection induces considerable changes in the set of proteins synthesized in the host cell (White and Brakke, 1983).

F. Subgenomic RNAs of BSMV

If from the coding capacity of BSMV genomic RNAs we subtract the 6–8% occupied by the no coding tRNA-like structure and the length of the genes open for *in vitro* translation, there will still be left about 15% of RNA 1, 75% of RNA 2a, 25% of RNA 2b, and 30% of RNA 3. Thus, in principle, subgenomic RNAs may exist through which the information closed for translation on the genomic RNAs can be realized in the infected cell.

One such RNA some 800 residues long was found in preparations of virion RNA of orthotripartite and pseudobipartite BSMV strains (Dolja *et al.*, 1983a; Jackson *et al.*, 1983) as well as in total RNA preparations from plants infected with several virus strains (Dolja *et al.*, 1986a). This RNA is homologous to RNA 3 of the orthotripartite strain and to RNA 2b of the pseudobipartite one (Jackson *et al.*, 1983). In a cell-free system, this RNA codes for a 17.5K protein p17 (Fig. 4) (Dolja *et al.*, 1983a).

Very recently the p17 gene was sequenced (Kozlov *et al.*, personal communication). This gene represents the 3′-terminal part of the coding body of RNA 3. It is interesting that the terminator UAA-codon of the p17 gene is the beginning of the internal poly(A) sequence. The same situation is found also in RNA 1: the terminator UAA-codon of the p120 gene comes into the poly(A) tract.

Stanley *et al.*, (1984), in their work with a partially purified preparation of subgenomic RNA, came to the conclusion that its 3′-end is polyadenylated and carries no tRNA-like structure. However, with pure subgenomic RNA we have determined that its 3′-terminal region is arranged just as in genomic RNAs, i.e., it comprises an internal poly(A) sequence and the terminal tRNA-like structure (Dolja *et al.*, 1986a). This is corroborated by the finding in BSMV-infected barley protoplasts of an RNA of the same size which proved capable of being aminoacylated with tyrosine *in vivo* (Loesh-Fries and Hall, 1982). Thus, RNAs 2b and 3 carry in their 3′-terminal region the closed gene for p17, which is autonomized and expressed in the subgenomic RNA. In BSMV-infected plants, besides the 800-nucleotide-long RNA, at least three minor RNAs are found composed of 1200, 1700, and 2600 nucleotides and also homologous to RNAs 3 and 2b. All these RNAs are present in tripartite (ND 18) and pseudobipartite (Norwich II) BSMV strains and are found in virion RNA preparations, i.e., they are coatable (Dolja *et al.*, 1986a). Probably analogous

RNA were observed by McFarland *et al.* (1983) in virion RNA of strain AM subclones.

It should be noted that upon oligo(dT)-cellulose chromatography, all subgenomic RNAs go, as a rule, into the poly(A)$^+$ fraction, i.e., they retain the 3′-terminal RNA region. Hybridization analysis with the use of recombinant clones also confirmed that these RNAs all have overlapping sequences (Dolja *et al.*, unpublished). Thus, it can be suggested that the RNAs homologous to RNAs 2b and 3 are 3′-terminal fragments of these genomic RNAs differing in length—a situation common enough for plant viruses.

RNA 4 can hardly be considered subgenomic since pseudobipartite and tripartite strains do not contain it either in the virion or *in vivo*, and pseudoquadripartite ones lose it readily and irreversibly.

It remains to be determined whether there are any subgenomic RNAs homologous to RNA 1 and RNA 2a.

G. Replication of BSMV

Upon plant infection by BSMV, as during replication of other RNA-containing viruses, one can isolate from the infected cells, besides the single-stranded RNA, the double-stranded replicative form (RF). The molecular weight of the RF is twice that of the virion RNA (Pring, 1971, 1972). The RF efficiently hybridizes with the virion RNA, which confirms its virus-specific nature (Palomar *et al.*, 1977). The number of RF species for each BSMV strain corresponds to the number of virion RNA components; each individual RNA appears to be able to replicate autonomously (Pring, 1971; Jackson and Brakke, 1973; Palomar *et al.*, 1977). However, relative quantities of virion RNAs and their RFs in the preparations may vary markedly.

BSMV-infected plant cells are known to develop vesiculated chloroplasts and proplastids (Carroll, 1970; Matthews, 1981). Quite recently, Lin and Langenberg (1984) have found that the first cytological alteration evolving in infected root tip cells is the appearance of peripheral vesicles in proplastids. By immunoelectron microscopy, these vesicles are known to contain double-stranded RNA, presumably virus-specific (Lin and Langenberg, 1985). The BSMV protein and then the viral particles also first appear in association with proplastid membranes and later on the endoplasmic reticulum. These findings may mean that BSMV replication is associated with plastids.

It is interesting to note that chloroplast RNA, but not cytoplasmic, was reduced in BSMV-infected leaves (White and Brakke, 1982).

H. BSMV-Induced Mutagenesis in Host Plants

Systemic infection by BSMV can evoke unusual genetic alterations in host plants. Thus, the frequency of triploidy and aneuploidy is enhanced in barley (Sandfaer, 1973). For several phenotypic characteristics of maize corn, deviations from the classical (Mendelian) ratios of marker inheritance were observed by Sprague *et al.* (1963). Crosses between homozygous dominant plants systemically infected with BSMV and homozygous recessive plants gave F_2 and further generations with a higher-than-expected frequency of recessive phenotypes (Sprague and McKinney, 1966, 1971). This phenomenon was called "aberrant ratio."

The mechanism of the mutagenic action of BSMV infection is still obscure. Attempts to find direct integration of virus-specific sequences into the plant genome were unsuccessful (Mottinger *et al.*, 1984a,b; Saedler, personal communication). Direct studies of the mutations in the alcohol dehydrogenase gene in the progeny of systemically BSMV-infected maize revealed inserts not homologous to the viral genome (Mottinger *et al.*, 1984b).

A similar situation was found in a study of virus-induced mutations in the Shrunken and Bronze loci of the maize chromosome 9 (Mottinger *et al.*, 1984a); it was shown that genes are inactivated in an insertional mode, and the insert appears to be part of the normal maize genome. The authors believe that the viral infection mobilizes endogenous "dormant" transposons in the maize genome. Mutagenesis appears to evolve from the general "genomic stress" which, immediately or in further generations, wakes up elements capable of transposition.

V. CONCLUDING REMARKS

Rod-shaped plant viruses with rigid helically built virions (tobamoviruses, tobraviruses, soilborne wheat mosaic virus, and hordeiviruses) differ fundamentally in the structure of their genome, which can be composed of one or several components.

Formally, hordeiviruses belong to tripartite viruses (tricornaviruses); however, the BSMV genome is arranged in a substantially different way. In tripartite viruses (bromoviruses, cucumoviruses, and ilarviruses), each of the two larger RNAs contains only information for a single large protein produced by direct translation. The third genomic component carries two cistrons, one of which (the coat protein gene) is located near the 3′-end of RNA 3 and is closed for direct translation; in infected cells a subgenomic mRNA is made to direct the synthesis of the coat protein [for review, see van Kammen (1984) and *The Plant Viruses*, Volume 1 of this series].

The data on the structure of the BSMV genome are summarized in Fig. 5. Only RNA 1 in BSMV resembles RNA 1 of the above-mentioned viruses: it codes for a single heavy protein, p120. In BSMV the coat protein gene is contained in the 5′-terminal part of the second genomic component.

As in all other known plant tricornaviruses (van Kammen, 1984), the 5′-end of BSMV RNA is capped (Agranovsky *et al.*, 1979).

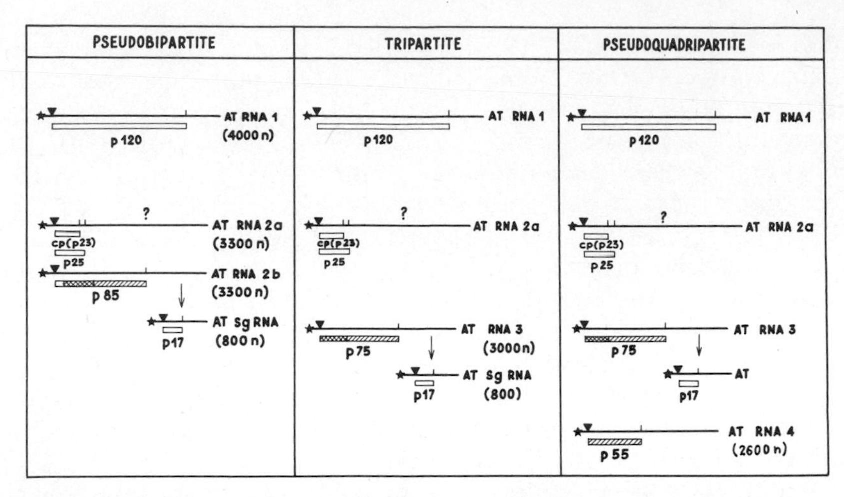

FIGURE 5. Scheme of the genomes of different BSMV strains. Solid lines represent viral RNAs; stars denote the "cap" structures at the 5'-termini (for RNA 2a and sg RNA these are presumed); triangles denote the initiator codons; short vertical lines denote the terminators. Uppercase A's are for the internal poly(A) tracts, and T's are for tRNA-like structure. Blocks show the translation products; cp, coat protein. The sequence supposedly common for p85, p75, and p55 is hatched, the one shared by p75 and p85 is cross-hatched, and the sequence specific for p85 is represented by the empty part of the block. Modified from Dolja *et al.* (1983b).

A unique feature of all BSMV RNAs is a poly(A) tract of variable length intercalated between the 3'-terminal tRNA-like tyrosine-accepting structure and the coding body of RNA molecules. Besides BSMV, an internal poly(A) sequence has been found only in brome mosaic virus between the genes of the bicistronic RNA 3 (Ahlquist *et al.*, 1981b).

An intriguing fact calling for explanation is that the number and length of BSMV RNA species may change upon subculturing. Firstly, RNA 4 of a pseudoquadripartite strain can be irreversibly eliminated without loss of infectivity. RNA 4 appears to be an RNA 3 fragment, a defective RNA probably analogous to defective interfering RNAs of some animal viruses (for review, see Huang and Baltimore, 1977). Secondly, a normal tripartite strain can be transformed into a pseudobipartite one and vice versa. As such conversions take place, the length of the third genomic component (RNA 3 and 2b) changes by some 300 nucleotides. Variations in size have also been described for RNA 2 of bipartite rod-shaped viruses, tobacco rattle virus (Robinson, 1983), and soilborne wheat mosaic virus (Shirako and Brakke, 1984). In the first case RNA 2 is shortened by approximately 150 nucleotides, in the second by 1000 nucleotides whereby the viruses retain their replicating ability.

On the whole, although we are far from understanding all the details of structure of the hordeivirus genome, it is clear that we are dealing with rather unusual genetically unstable viruses.

REFERENCES

Agranovsky, A. A., Dolja, V. V., Kavsan, V. M., and Atabekov, J. G., 1978, Detection of polyadenylate sequences in RNA components of barley stripe mosaic virus, *Virology* **91**:95.

Agranovsky, A. A., Dolja, V. V., Kagramanova, V. K., and Atabekov, J. G., 1979, The presence of a cap structure at the 5′-end of barley stripe mosaic virus RNA, *Virology* **95**:208.

Agranovsky, A. A., Dolja, V. V., Gorbulev, V. G., Kozlov, Y. V., and Atabekov, J. G., 1981, Aminoacylation of barley stripe mosaic virus RNA: Polyadenylate-containing RNA has a 3′-terminal tyrosine-accepting structure, *Virology* **113**:174.

Agranovsky, A. A., Dolja, V. V., and Atabekov, J. G., 1982, Structure of the 3′ extremity of barley stripe mosaic virus RNA: Evidence for internal poly(A) and a 3′-terminal tRNA-like structure, *Virology* **119**:51.

Agranovsky, A. A., Dolja, V. V., and Atabekov, J. G., 1983, Differences in polyadenylate length between individual barley stripe mosaic virus RNA species, *Virology* **129**:344.

Ahlquist, P., Dasgupta, R., and Kaesberg, P., 1981a, Near identity of 3′ RNA secondary structure in bromoviruses and cucumber mosaic virus, *Cell* **23**:183.

Ahlquist, P., Luckow, V., and Kaesberg, P., 1981b, Complete nucleotide sequence of brome mosaic virus RNA3, *J. Mol. Biol.* **153**:23.

Atabekov, J. G., and Novikov, V. K., 1966, Some properties of the nucleoprotein of barley stripe mosaic virus, and its structural components, *Biokhimiya* **31**:157.

Atabekov, J. G., and Novikov, V. K., 1971, Barley stripe mosaic virus, *CMI/AAB Descriptions of Plant Viruses No. 68.*

Atabekov, J. G., Novikov, V. K., Kiselev, N. A., Kaftanova, A. S., and Egorov, A. M., 1968a, Stable intermediate aggregates formed by the polymerization of barley stripe mosaic virus protein, *Virology* **36**:620.

Atabekov, J. G., Schaskolskaya, N. D., Dementyeva, S. P., Sacharovskaya, G. N., and Senchenkov, E. P., 1968b, Serological study on barley stripe mosaic virus protein polymerization. I. Immunodiffusion, immunoelectrophoretic characteristics and absorption experiments, *Virology* **36**:587.

Atabekov, J. G., Dementyeva, S. P., Schaskolskaya, N. D., and Sacharovskaya, G. N., 1968c, Serological study on barley stripe mosaic virus protein polymerization. II. Comparative antigenic analysis of intact virus and some stable protein intermediates, *Virology* **36**:601.

Atabekov, J. G., Novikov, V. K., Vishnichenko, V. K., and Javakhia, V. G., 1970a, A study of the mechanisms controlling the host range of plant viruses, *Virology* **41**:108.

Atabekov, J. G., Novikov, V. K., Vishnichenko, V. K., and Kaftanova, A. S., 1970b, Some properties of hybrid viruses reassembled in vitro, *Virology* **41**:519.

Baratova, L. A., Belanova, L. P., Chodova, O. H., Lunina, N. A., Dolja, V. V., and Atabekov, J. G., 1983, The N-terminal amino acid sequences of the capsid proteins of two strains of barley stripe mosaic virus, *Biologicheskije Nauki* **4**:23.

Bennett, C. W., 1959, Lychnis ringspot, *Phytopathology* **49**:706.

Brakke, M. K., 1959, Dispersion of aggregated barley stripe mosaic virus by detergents, *Virology* **9**:506.

Brakke, M. K., 1962, Stability of purified barley stripe mosaic virus, *Virology* **17**:131.

Brakke, M. K., 1979, Ultraviolet absorption spectra and difference spectra of barley stripe mosaic and tobacco mosaic viruses in buffer and sodium dodecyl sulfate, *Virology* **98**:76.

Brakke, M. K., and Palomar, M. K., 1976, Separation of components of barley stripe mosaic virions by density-gradient centrifugation, *Virology* **71**:255.

Bujarsky, J. J., Hardy, S. F., Miller, W. A., and Hall, T. C., 1982, Use of dodecyl-β-d-maltoside in the purification and stabilization of RNA polymerase from brome mosaic virus-infected barley, *Virology* **119**:465.

Carroll, T. W., 1970, Relation of barley stripe mosaic virus to plastids, *Virology* **42**:1015.

Carroll, T. W., 1972, Seed transmissibility of two strains of barley stripe mosaic virus, *Virology* **48**:323.

Carroll, T. W., 1974, Barley stripe mosaic virus in sperm and vegetative cells of barley pollen, *Virology* **60:**21.

Carroll, T. W., and Chapman, S. R., 1970, Variation in embryo infection and seed transmission of barley stripe mosaic virus within and between two cultivars of barley, *Phytopathology* **60:**1079.

Cornelissen, B. J. C., and Bol, J. F., 1984, Homology between the proteins encoded by tobacco mosaic virus and two tricornaviruses, *Plant Mol. Biol.* **3:**379.

Dodds, J. A., and Hamilton, R. I., 1971, Evidence for possible genomic masking between two unrelated plant viruses, *Phytopathology* **61:**889.

Dodds, J. A., and Hamilton, R. I., 1974, Masking of the RNA genome of tobacco mosaic virus by the protein of barley stripe mosaic virus in doubly infected barley, *Virology* **59:**418.

Dolja, V. V., Sokolova, N. A., Tjulkina, L. G., and Atabekov, J. G. 1979, A study of barley stripe mosaic virus (BSMV) genome. II. Translation of individual RNA species of two BSMV strains in a homologous cell-free system, *Mol. Gen. Genet.* **175:**93.

Dolja, V. V., Lunina, N. A., Smirnov, V. D., Karpov, V. A., Chudiakov, Y. E., Kozlov, Y. V., Bayev, A. A., and Atabekov, J. G., 1982, Location of genes in polycistronic mRNA by oligonucleotide probing as exemplified with barley stripe mosaic virus coat protein gene, *Dokl. Akad. Nauk SSSR* **265:**474.

Dolja, V. V., Agranovsky, A. A., Lunina, N. A., and Atabekov, J. G., 1983a, Short virion RNA in barley stripe mosaic virus, *FEBS Lett.* **151:** 215.

Dolja, V. V., Lunina, N. A., Leiser, R.-M., Stanarius, T., Belzhelarskaya, S. N., Kozlov, Y. V., and Atabekov, J. G., 1983b, A comparative study on the *in vitro* translation products of individual RNAs from two-, three- and four-component strains of barley stripe mosaic virus, *Virology* **127:**1.

Dolja, V. V., Lunina, N. A., Chumakov, K. M., Ziegler, A., and Atabekov, J. G., 1986a, Subgenomic RNAs of barley stripe mosaic virus, *Dokl. Akad. Nauk SSSR,* in press.

Dolja, V. V., Lunina, N. A., Karpova, O. V., Rodionova, N. P., Tjulkina, L. G., Kulayeva, O. I., and Atabekov, J. G., 1986b, Length of internal poly(A)-tract in individual RNAs of barley stripe mosaic virus is genetically conserved, *Investiya Akad. Nauk SSSR,* in press.

Finch, J. T., 1965, Preliminary X-ray diffraction studies on tobacco rattle and barley stripe mosaic viruses, *J. Mol. Biol.* **12:**612.

Gibbs, A. J., Kassanis, B., Nixon, H. L., and Woods, R. D., 1963, The relationship between barley stripe mosaic virus and Lychnis ringspot virus, *Virology* **20:**194.

Gumpf, D. J., and Hamilton, R. I., 1968, Isolation and characterization of barley stripe mosaic virus protein, *Virology* **35:** 87.

Gumpf, D. J., Cunningham, D. S., Heick, J. A., and Shannon, L. M., 1977, Amino acid sequence in the proteolytic glycopeptide of barley stripe mosaic virus, *Virology* **78:**328.

Gustafson, G. D., Larkius, B. A., and Jackson, A. O., 1981, Comparative analysis of polypeptides synthesized *in vivo* and *in vitro* by two strains of barley stripe mosaic virus, *Virology* **111:**579.

Gustafson, G. D., Milner, J. J., McFarland, J. E., Pedersen, K., Larkins, B. A., and Jackson, A. O., 1982, Investigation of the complexity of barley stripe mosaic virus RNAs with recombinant DNA clones, *Virology* **120:**182.

Haenni, A. L., Joshi, S., and Chapeville, F., 1982, tRNA-like structures in the genomes of RNA viruses, *Prog. Nucleic Acid Res. Mol. Biol.* **27:**85.

Hagborg, W. A. F., 1954, Dwarfing of wheat and barley by the barley stripe mosaic (false stripe) virus, *Can. J. Bot.* **32:**24.

Hamilton, R. I., 1965, An embryo test for detecting seed borne barley stripe mosaic virus in barley, *Phytopathology* **55:**798.

Harrison, B. D., Nixon, H. L., and Woods, R. D., 1965, Lengths and structure of particles of barley stripe mosaic virus, *Virology* **26:**284.

Huang, A. S., and Baltimore, D., 1977, Defective interfering animal viruses, in: *Compre-*

hensive Virology (H. Fraenkel-Conrat and R. R. Wagner, eds.), Volume 10, pp. 73–116, Plenum Press, New York.

Jackson, A. O., and Brakke, M. K., 1973, Multicomponent properties of barley stripe mosaic virus ribonucleic acid, *Virology* **55**:483.

Jackson, A. O., and Lane, L. C., 1981, Hordeiviruses, in: *Handbook of Plant Virus Infections* (E. Kurstak, ed.), pp. 565–625, Elsevier/North-Holland, Amsterdam.

Jackson, A. O., Dawson, J. R. O., Covey, S. N., Hull, R., Davies, J. W., McFarland, J. E., and Gustafson, G. D., 1983, Sequence relations and coding properties of a subgenomic RNA isolated from barley stripe mosaic virus, *Virology* **127**:37.

Joshi, R. L., Joshi, S., Chapeville, F., and Haenni, A. L., 1983, tRNA-like structures of plant viral RNAs: Conformational requirements for adenylation and aminoacylation, *EMBO J.* **2**:1123.

Kamer, G., and Argos, P., 1984, Primary structural comparison of RNA-dependent polymerases from plant, animal and bacterial viruses, *Nucleic Acids Res.* **12**:7269.

Kassanis, B., and Slykhuis, J. T., 1959, Some properties of barley stripe mosaic virus, *Ann. Appl. Biol.* **47**:254.

Kiselev, N. A., Atabekov, J. G., Kaftanova, A. S., and Novikov, V. K., 1966, Study of virus protein repolymerization and reconstruction of some rod-like viruses, *Biokhimiya* **31**:670.

Kiselev, N. A., De Rosier, D. J., and Atabekov, J. G., 1969, A double-helical structure found on the re-aggregation of the protein of barley stripe mosaic virus, *J. Mol. Biol.* **39**:673.

Kozlov, Y. V., Rupasov, V. V., Adyshev, D. M., Belgelarskaya, S. N., Agranovsky, A. A., Mankin, A. S., Morozov, S. Y., Dolja, V. V., and Atabekov, J. G., 1984, Nucleotide sequence of the 3′-terminal tRNA-like structure in barley stripe mosaic virus genome, *Nucleic Acids Res.* **12**:4001.

Lane, L. C., 1974, The components of barley stripe mosaic and related viruses, *Virology* **58**:323.

Lin, N.-S., and Langenberg, W. G., 1984, Chronology of appearance of barley stripe mosaic virus protein in infected wheat cells, *J. Ultrastruct. Res.* **89**:309.

Lin, N.-S., and Langenberg, W. G., 1985, Peripheral vesicles in proplastids of barley stripe mosaic virus-infected wheat cells contain double-stranded RNA, *Virology* **142**:291.

Loesh-Fries, L. S., and Hall, T. C., 1982, *In vivo* amino-acylation of brome mosaic and barley stripe mosaic virus RNAs, *Nature (London)* **298**:771.

McFarland, J. E., Brakke, M. K., and Jackson, A. O., 1983, Complexity of the Argentina Mild strain of barley stripe mosaic virus, *Virology* **130**:397.

McKinney, H. H., 1953, New evidence on virus diseases in barley, *Plant Dis. Rep.* **37**:292.

McKinney, H. H., and Greeley, L. W., 1965, Biological characteristics of barley stripe mosaic virus strains and their evolution, *Tech. Bull. U.S. Dep. Agr.* 1324.

Matthews, R. E. F., 1981, *Plant Virology*, 2nd ed., Academic Press, New York.

Mizenina, O. A., Kiseleva, N. P., Kaftanova, A. S., and Dobrov, E. N., 1984, Formaldehyde-induced RNA–protein cross-linking in short ribonucleoprotein particles reconstituted from tobacco mosaic virus protein and short fragments of TMV RNA, *Biokhimiya* **49**:787.

Moorhead, E., 1956, Serological studies of viruses infecting the cereal crops, *Phytopathology* **46**:498.

Mottinger, J. P., Dellaporta, S. L., and Keller, P. B., 1984a, Stable and unstable mutations in aberrant ratio stocks of maize, *Genetics* **106**:751.

Mottinger, J. P., Johns, M. A., and Freeling, M., 1984b, Mutations of the Adh1 gene in maize following infection with barley stripe mosaic virus, *Mol. Gen. Genet.* **195**:367.

Negruk, V. I., Novikov, V. K., and Atabekov, J. G., 1974, Translation of barley stripe mosaic virus RNA in cell-free system from wheat germs, *Dokl. Acad. Nauk SSSR* **218**:489.

Novikov, V. K., and Atabekov, J. G., 1970, A study of the mechanisms controlling the host range of plant viruses. I. Virus-specific receptors of *Chenopodium amaranticolor*, *Virology* **41**:101.

Palomar, M. K., and Brakke, M. K., 1976, Concentration and infectivity of barley stripe mosaic virus in barley, *Phytopathology* **66:**1422.

Palomar, M. K., Brakke, M. K., and Jackson, A. O., 1977, Base sequence homology in the RNAs of barley stripe mosaic virus, *Virology* **77:**471.

Partridge, J. E., Shannon, L. M., Gumpf, D. J., and Colbaugh, P., 1974, Glycoprotein in the capsid of plant viruses as a possible determinant of seed transmissibility, *Nature (London)* **247:**391.

Pring, D. R., 1971, Viral and host RNA synthesis in BSMV-infected barley, *Virology* **44:**54.

Pring, D. R., 1972, Barley stripe mosaic virus replicative form RNA: Preparation and characterization, *Virology* **48:**22.

Qi, G., Xie, D., and Pei, M., 1984, The content, infectivity and distribution of poly(A) at the 3′-terminus of polyadenylated RNA from barley stripe mosaic virus (Xinjiang strain), *Acta Biochim. Biophys. Sin.* **16:**216.

Rappaport, I., 1965, The antigenic structure of tobacco mosaic virus, *Adv. Virus Res.* **11:**223.

Rietveld, K., Pleij, C. W. A., and Bosch, L., 1983, Three-dimensional models of the tRNA-like 3′-termini of some plant viral RNAs, *EMBO J.* **2:**1079.

Robinson, D. J., 1983, RNA species of tobacco rattle virus strains and their nucleotide sequence relationships, *J. Gen. Virol.* **64:**657.

Sandfaer, J., 1973, Barley stripe mosaic virus and the frequency of triploids and aneuploids in barley, *Genetics* **73:**597.

Shirako, Y., and Brakke, M. K., 1984, Spontaneous deletion mutation of soil-borne wheat mosaic virus RNA II, *J. Gen. Virol.* **65:**855.

Singh, G. P., Arny, D. C., and Pound, G. S., 1960, Studies on the stripe mosaic of barley, including effects of temperature and age of host on disease development and seed infection, *Phytopathology* **50:**290.

Slack, S. A., Shepherd, R. J., and Hall, D. H., 1975, Spread of seed-borne barley stripe mosaic virus and effects of the virus on barley in California, *Phytopathology* **65:**1218.

Slykhuis, J. T., 1972, Poa semilatent virus from native grasses, *Phytopathology* **62:**508.

Slykhuis, J. T., and Atkinson, T. K., 1966, A mosaic disease of *Poa palustris* in Alberta, *Can. Plant Dis. Surv.* **46:**147.

Sprague, G. F., and McKinney, H. H., 1966, Aberrant ratio: An anomaly in maize associated with virus infection, *Genetics* **54:**1287.

Sprague, G. F., and McKinney, H. H., 1971, Further evidence on the genetic behaviour of AR in maize, *Genetics* **67:**533.

Sprague, G. F., McKinney, H. H., and Greeley, L. W., 1963, Virus as a mutagenic agent in maize, *Science* **141:**1052.

Stanley, J., Hanau, R., and Jackson, A. O., 1984, Sequence comparison of the 3′ ends of a subgenomic RNA and the genomic RNAs of barley stripe mosaic virus, *Virology* **139:**375.

Stepanova, O. B., Metelev, V. G., Chichkova, N. V., Smirnov, V. D., Rodionova, N. P., Atabekov, J. G., Bogdanov, A. A., and Shabarova, Z. A., 1979, Addressed fragmentation of RNA molecules, *FEBS Lett.* **103:**197.

Tyulkina, L. G., Nazarova, G. N., Kaftanova, A. S., Ledneva, R. K., Bogdanov, A. A., and Atabekov, J. G., 1975, Reassembly of TMV 20S protein discs with 3S RNA fragments, *Virology* **63:**15.

van Kammen, A., 1984, Expression of functions encoded on genomic RNAs of multiparticulate plant viruses, in: *Control of Virus Diseases* (E. Kurstak and R. G. Marusyk, eds.), pp. 301–316, Marcel Dekker, New York.

White, J. L., and Brakke, M. K., 1982, Chloroplast RNA and proteins decrease as wheat streak and barley stripe mosaic viruses multiply in expanding, systemically infected leaves, *Phytopathology* **72:**939.

White, J. L., and Brakke, M. K., 1983, Protein changes in wheat infected with wheat streak mosaic virus and in barley infected with barley stripe mosaic virus, *Physiol. Plant Pathol.* **22:**87.

Index